Freshwater Biodiversity

Growing human populations and higher demands for water impose increasing impacts and stresses upon freshwater biodiversity. Their combined effects have made freshwater animals more endangered than their terrestrial or marine counterparts. Overuse and contamination of water, overexploitation and over-fishing, introduction of alien species, and alteration of natural flow regimes have led to a 'great thinning' and declines in abundance of freshwater animals, a 'great shrinking' in body size with reductions in large species, and a 'great mixing' whereby the spread of introduced species has tended to homogenize previously dissimilar communities in different parts of the world. Climate change and warming temperatures will alter global water availability and exacerbate the other threat factors. What conservation action is needed to halt or reverse these trends, and preserve freshwater biodiversity in a rapidly changing world? This book offers the tools and approaches that can be deployed to help conserve freshwater biodiversity.

DAVID DUDGEON is Chair Professor in Ecology and Biodiversity at the University of Hong Kong. He has spent almost 40 years researching and writing about the streams and rivers of monsoonal East Asia, and the animals that live in and around them. Dudgeon is well known and respected internationally in the field of freshwater ecology, on which he has published extensively. Dudgeon received the Biwako Prize in Ecology in 2000 and was Editor-in-Chief of *Freshwater Biology* between 2015 and 2017.

ECOLOGY, BIODIVERSITY AND CONSERVATION

The world's biological diversity faces unprecedented threats. The urgent challenge facing the concerned biologist is to understand ecological processes well enough to maintain their functioning in the face of the pressures resulting from human population growth. Those concerned with the conservation of biodiversity and with restoration also need to be acquainted with the political, social, historical, economic and legal frameworks within which ecological and conservation practice must be developed. The new Ecology, Biodiversity, and Conservation series will present balanced, comprehensive, up-to-date, and critical reviews of selected topics within the sciences of ecology and conservation biology, both botanical and zoological, and both 'pure' and 'applied'. It is aimed at advanced final-year undergraduates, graduate students, researchers, and university teachers, as well as ecologists and conservationists in industry, government and the voluntary sectors. The series encompasses a wide range of approaches and scales (spatial, temporal, and taxonomic), including quantitative, theoretical, population, community, ecosystem, landscape, historical, experimental, behavioural and evolutionary studies. The emphasis is on science related to the real world of plants and animals rather than on purely theoretical abstractions and mathematical models. Books in this series will, wherever possible , consider issues from a broad perspective. Some books will challenge existing paradigms and present new ecological concepts, empirical or theoretical models, and testable hypotheses. Other books will explore new approaches and present syntheses on topics of ecological importance.

Ecology and Control of Introduced Plants
Judith H. Myers and Dawn Bazely

Invertebrate Conservation and Agricultural Ecosystems
T. R. New

Freshwater Biodiversity

Status, Threats and Conservation

DAVID DUDGEON
The University of Hong Kong

CAMBRIDGE
UNIVERSITY PRESS

CAMBRIDGE
UNIVERSITY PRESS

University Printing House, Cambridge CB2 8BS, United Kingdom

One Liberty Plaza, 20th Floor, New York, NY 10006, USA

477 Williamstown Road, Port Melbourne, VIC 3207, Australia

314–321, 3rd Floor, Plot 3, Splendor Forum, Jasola District Centre, New Delhi – 110025, India

79 Anson Road, #06-04/06, Singapore 079906

Cambridge University Press is part of the University of Cambridge.

It furthers the University's mission by disseminating knowledge in the pursuit of
education, learning, and research at the highest international levels of excellence.

www.cambridge.org
Information on this title: www.cambridge.org/9780521768030
DOI: 10.1017/9781139032759

First published 2020

Printed in the United Kingdom by TJ International Ltd. Padstow Cornwall

A catalogue record for this publication is available from the British Library.

Library of Congress Cataloging-in-Publication Data
Names: Dudgeon, David, author.
Title: Freshwater biodiversity : status, threats and conservation / David Dudgeon.
Description: Cambridge ; New York, NY : Cambridge University Press, 2020. | Series: Ecology,
  biodiversity and conservation | Includes bibliographical references and index.
Identifiers: LCCN 2019049505 (print) | LCCN 2019049506 (ebook) | ISBN 9780521768030
  (hardback) | ISBN 9780521745192 (paperback) | ISBN 9781139032759 (epub)
Subjects: LCSH: Freshwater biodiversity. | Freshwater biodiversity conservation.
Classification: LCC QH96.8.B53 D83 2020 (print) | LCC QH96.8.B53 (ebook) | DDC 333.95/
  28–dc23
LC record available at https://lccn.loc.gov/2019049505
LC ebook record available at https://lccn.loc.gov/2019049506

ISBN 978-0-521-76803-0 Hardback
ISBN 978-0-521-74519-2 Paperback

I cannot but remember such things were, That were most precious to me. Did heaven look on, And would not take their part?

(William Shakespeare (1606), *Macbeth*

... it is essential to recognize the probable result of what we have done and are doing, but when we have seen that ... the menaced world may seem to be more treasured than ever. Certainly, the anguish we feel at the threat to it and the sleepless despoiling of it can lose their tragic complexity and become mere bitterness when we forget that their origin is a passion for the momentary countenance of the unrepeatable world.

W.S. Merwin (1993; p. 5), *The Second Four Books of Poems*

Contents

Colour plates can be found between pages 272 and 273

Foreword

Why is a book on freshwater biodiversity conservation necessary? Less than 3% of the Earth's water is fresh, and most of it is frozen as ice or is inaccessible underground. A mere 0.3% of global fresh water is available as habitat for plants and animals, and much of it constitutes the only water that is readily accessible to meet human needs. Since human use reduces the amount of water in rivers, lakes and wetlands, or degrades its quality, or both, growing human populations and higher demands for water impose increasing impacts and stresses upon freshwater biodiversity. One reason this matters is that a significant proportion of global biodiversity is confined to fresh water. As a rough approximation, in terms of round numbers, these environments cover no more than 1% of the Earth's surface, but host almost 10% of all non-microbial species described by scientists. The fact that fresh waters are disproportionately rich in biodiversity, relative to their small global area, is inescapable. When it is viewed in the context of the many threats human activities pose to the integrity of freshwater bodies, and the knowledge that humans already appropriate more than half of global surface runoff, the prognosis for species loss is grim. In essence, freshwater biodiversity is imperilled due to its dependence on a resource subject to unprecedented and ever-increasing human demands. Neither the terrestrial nor the marine realm combines such richness of biodiversity with the magnitude and intensity of anthropogenic threat, and, as this book will show, the rates of population decline and endangerment of freshwater species are both high and far greater than their counterparts on land or in the oceans. The endemic and monotypic Yangtze River dolphin (*Lipotes vexillifer*: Lipotidae), which became the first cetacean to be driven to extinction by humans (Turvey *et al.*, 2007), was not only emblematic of a failure to conserve charismatic freshwater megafauna, but also indicative of the conservation challenges posed by the transformation and degradation of inland waters. If we accept that species persistence is a reasonable criterion for measuring sustainability (defined, broadly, as meeting the needs of the present without compromising future needs), then the loss of this dolphin not only curtails 20 million years of evolution, but also is clear evidence that human activities are unsustainable from the perspective of preserving freshwater biodiversity.

Why are such species losses taking place? Conservation of freshwater biodiversity is difficult because anthropogenic pressures on fresh waters are already intense and, as human populations continue to grow, clean fresh water will become an ever-more scarce resource. Pollution and habitat modification will also progressively reduce the suitability of fresh waters for aquatic plants and animals. Moreover, since freshwater bodies are downhill from – and embedded within – terrestrial landscapes, conservation efforts require large-scale management of entire drainage basins rather than just the particular localities where imperilled species occur. This is nothing if not challenging: for instance, the basin of the Yangtze and its now-extinct river dolphin is inhabited by more than 400 million people.

To make matters worse, ongoing reductions in the quality and quantity of water available to sustain freshwater ecosystems will interact with global climate change, which, itself, will have consequences for the patterns of evaporation, transpiration and rainfall, and hence the flow and inundation cycles in rivers, lakes and wetlands. Many places in a warmer world will experience a greater frequency of floods and droughts, as well as hastened glacial melting. Subsequent shifts in hydrological regimes, and human responses to them (such as dams built for water storage and flood control), will greatly alter conditions in freshwater ecosystems. Furthermore, there is a very real possibility that the ranges of cold-blooded (ectothermic) animals will have to shift upward in latitude or altitude to compensate for warming. That will be especially problematic for freshwater species. Many cannot disperse overland or along the coast to potentially habitable, cooler water bodies and, hence, will be 'stranded' in conditions of ever-decreasing suitability.

The synergistic interactions of global climate change with other human-induced threats or stressors will be profound, with the potential to change the rules of existence for freshwater biota. If current trajectories of threat and endangerment continue, they will engender losses of many freshwater species, representing a significant component of what is being recognized as an ongoing, sixth mass extinction event in the Earth's geological history. To put this another way, human transformation of the Earth system and the global water system may well give rise to the first mass extinction event of the Anthropocene (*sensu* Zalasiewicz *et al.*, 2011). This term was first coined by Nobel Prize winner Paul Crutzen to mark the current epoch and thereby draw attention to the pervasive role of humans in the geology and ecology of the Earth system (Crutzen, 2002).

Freshwater Biodiversity: Status, Threats and Conservation will highlight, explain and account for the present situation with regard to freshwater ecosystems and their biota, and will indicate how the worst effects of anthropogenic impacts might be mitigated or reduced. The message is urgent: there are limited opportunities to protect much of what remains of global freshwater biodiversity. Education and enhanced awareness of the threats and possible solutions to them will be an essential prerequisite for conservation action. However, the general nature of these threats is known, and they are manifest in all non-polar regions of the Earth, although their relative magnitude varies significantly from place to place. But identifying threats has resulted in little action to mitigate or alleviate them. The transfer of knowledge to conservation action has been largely unsuccessful. This failure is related to the special features of freshwater ecosystems – and the biodiversity they support – that makes them especially vulnerable to human activities. There are many instances where humans have caused rapid and significant declines in freshwater species and habitats. These represent opportunity costs that are magnified by losses in the option values of species or reductions in the provision of ecological services that may well be irreversible. The link between freshwater biodiversity and the provision of ecosystem services that underpins human livelihoods will be discussed in some detail herein, as it provides an important justification for conservation. Nonetheless, if conservation action is to be effective, it will require implementation of different components of a variety of management options, as described in Chapter 9. There, I will also highlight the need for a major change in attitude towards freshwater biodiversity and ecosystem management. Bringing about that shift would do much to remove the greatest obstacle to effective conservation.

There have been many books written about the ever-expanding list of dead and dying species. They may have the effect of lessening the abhorrence with which we view extinction, if repetition begins to trivialize these events. But there is a justification for such tomes. If human carelessness or stupidity results in the loss of a species, its evolution will stop, its unique genotype is obliterated, and its name will have no more significance to the next generation than any of the other plants and animals that became extinct long ago – through natural processes.

Forgetting is a luxury we cannot afford: telling and retelling can be seen as a moral imperative. And we must hope that our writings will 'stimulate a unity of compassion' (Schaller, 1993: p. xvi) and promote action to preserve species. This, it seems to me, is a responsibility we must fulfil, because the opportunity to do so exists now; it will diminish the longer that we delay.

This book is an entirely original text representing many years of work, but some ideas and examples have been reworked from some of my earlier publications (e.g. Dudgeon, 2011, 2013, 2014). Threat status, as used herein, follows the IUCN Red List (www.iucnredlist.org). Critically endangered, endangered, vulnerable and near-threatened species are denoted CR, EN, VU or NT – as appropriate – upon first mention, whereupon the scientific binomial name, family designation and, in the case of vertebrates and some plants, the common English name are given also. After stating the scientific name, I have preferentially used the common name where one exists – hence North American beaver rather than *Castor canadensis*. Inevitably, a book on freshwater biodiversity makes frequent mention of fishes, and I have derived biometric data (size and weight) for particular species from FishBase (Froese & Pauly, 2018: www.fishbase.org), which also served as a guide for nomenclature.

The literature on freshwater ecology and conservation is vast, and I have focused mainly on sources post-dating 2005, encompassing the period after publication of the review by Dudgeon *et al.* (2006), in which this book has its (distant) origins, until the end of August 2018 (including some papers in press at that time). Nonetheless, I have referred to earlier key sources where they provide information that has yet to be superseded, or present particularly informative case studies. This timeframe has meant that certain very recent publications have been omitted, including the report setting out the Intergovernmental Panel on Climate Change (IPCC)'s advice that global carbon-neutrality is needed by 2050 to keep the average temperature increase of the Earth to 1.5°C (IPCC, 2018). This document has, understandably, received wide publicity in the public domain and is readily available from online sources (e.g. www .ipcc.ch/sr15/chapter/summary-for-policy-makers/), because it makes a compelling case that both the speed of global warming and the seriousness of its consequences have been underestimated by scientists. Readers are warned that the changes projected in Chapter 7 will inevitably require updating in the light of new information and research findings. The August 2018 cutoff date has meant that some recent work, such as the review by Reid *et al.* (2019) intended as an update to Dudgeon *et al.*

(2006), has been omitted, although contributions I made to that article (e.g. Fig. 2.1 and some topics – e.g. dam removal and environmental DNA – covered in Chapter 9) have been included. It has not been possible to explore the implications of a planned clean up of the Yangtze announced by the Chinese Government in January 2019, nor the intention to introduce a 10-year ban on fishing in the river, although both initiatives are laudable.

Wherever possible, I have drawn examples from places outside the relatively well researched north-temperate zone, although that region has certainly not been neglected. I cannot claim to offer a comprehensive survey of all of the relevant literature – that would necessitate a much longer and rather different book. Chapter 4, which deals with non-native or alien species, is something of an exception in that regard, because I wanted to illustrate the variety of freshwater species that have become established beyond their native ranges, and so draw attention to the 'great mixing' of biotas that has become such a distinctive signature of the Anthropocene.

The gestation of this book has been lengthy. I am grateful to Cambridge University Press for their patience during this period, and to Professor Michael Usher (Stirling University) for judicious application of pointed sticks that ensured the work progressed (albeit slowly). Dominic Lewis, Aleksandra Serocka, Samuel Fearnly, Orvil Matthews and Penny Lyons were collectively responsible for transforming my original manuscript into a published volume. I am also grateful to my colleagues in the Research Division of Ecology and Biodiversity at the University of Hong Kong, especially Billy Hau, Kenny Leung, Gray Williams and Yvonne Sadovy, for their collegiality and encouragement, and to Lily Ng for day-to-day logistical assistance and research support. Zeb Hogan and Yik-Hei Sung were kind enough to allow me to use their photographs, while Steve Ormerod provided helpful email insights into environmental optimism. I also thank my students for their patience and willingness to listen while I tried out some of the ideas and examples included in this volume. Finally, the book would not have been possible without the inspiration, care and kindness of Amanda Whitfort; it is dedicated to her and our daughter Lucinda.

Permissions for Epigraphs

Doubleday Publishing Group, a division of Penguin Random House LLC. All rights reserved.

From *The Abstract Wild* by Jack Turner. © 1996 Arizona Board of Regents. Reprinted by permission of the University of Arizona Press.

From *Resources of Hope: Culture, Democracy, Socialism* by Raymond Williams. Copyright © 1989. Verso Books, London. Used with permission.

Excerpt from *Natural History*, September 1991. Used with permission.

Credit notices (plates)

Plate 4. Westend61/Getty Images.

Plate 5. Ivan/Moment/Getty Images.

Plate 6. South China Morning Post/Getty Images.

Plates 7 and 8. Huët (1868) *Nouvelles Archives du Muséum d'histoire Naturelle*.

Plates 9, 10 and 11. Courtesy of Zeb Hogan.

Plate 12. DigiPub/Moment Open/Getty Images.

Plate 13. Jean-Claude Soboul/Moment Open/Getty Images.

Plate 14. Grant Ordelheide/Aurora Open/Getty Images.

Plate 15. Sian Seabrook/500px/Getty Images.

Plate 16. Gabrielle Therin-Weisse/Photographer's Choice RF/Getty Images.

Plate 17. Courtesy of Yik-Hei Sung.

Plate 18. Arjun Kumar/500px/Getty Images.

Plate 19. Gail Shotlander/Moment Open/Getty Images.

Plate 20. Enrique Diaz/7cero/Moment/Getty Images.

Plate 21. Cuvier & Valenciennes (1835) *Histoire Naturelle des Poissons, Volume 18*.

Plate 22. Mohd Hazrizal Mohd Hanapiah/EyeEM/Getty Images.

Plate 23. Sherif A. Wagih/Moment/Getty Images.

Plate 24. Javier Fernandez Sanchez/Moment/Getty Images.

Plate 25. Jennifer Idol/Stocktrek Images/Getty Images.

Plate 27. Danita Delimont/Gallo Images ROOTS RF Collection/Getty Images.

Plate 28. Huët (1866) *Nouvelles Archives du Muséum d'histoire Naturelle*.

1 · *The Freshwater Commons*

That which is common to the greatest number has the least care bestowed upon it. Every one thinks chiefly of his own, hardly at all of the common interest . . .

Aristotle (350 BCE), Politics Book II, Part 3

These considerations should lead us to look upon all the works of nature, animate or inanimate, as invested with a certain sanctity, to be *used* by us but not *abused*, and never to be recklessly destroyed or defaced. To pollute a spring or a river, to exterminate a bird or beast, should be treated as moral offences and as social crimes . . .

Alfred Russel Wallace (1914: p. 278), The World of Life

. . . this is a problem of ecology, of interrelationships, of interdependence. We poison the caddis flies in a stream and the salmon runs dwindle and die. We poison the gnats in a lake and the poison travels from link to link of the food chain and soon the birds of the lake margins become its victims. . . . They reflect the web of life — or death — that scientists know as ecology.

Rachel Carson (1962: p. 189), Silent Spring

The Medium Is the Message

Water is essential for life in general, and for humans in particular. Food production in the form of rain-fed and irrigated agriculture, livestock production, fisheries and aquaculture depend upon the availability of fresh water. This scarce resource also sustains a significant amount of animal biodiversity, much of which is now threatened. Ichthyologist Melanie Stiassny (1999) co-opted Marshall McLuhan's 1964 catchphrase 'the medium is the message' to encapsulate the notion that freshwater biodiversity is imperilled due to dependence on a resource subject to unprecedented and ever-increasing human demands. And the trade-offs between human use of water and the water needed for nature have increasing been skewed in favour of the former. 'The medium is the message' also because fresh water is a more limiting resource than the supply of land for humans, while sea water is not limiting at all; because the land is not subject to comparable patterns of consumption, use or contamination; and because freshwater animals have far more restricted distributions than their terrestrial (or marine) counterparts. The inherent connectivity between fresh waters and their surrounding catchments, with receiving waters almost always located at the lowest point in basins (volcanic crater lakes are an exception), bestows vulnerability since threats to biodiversity can originate uphill well beyond lake or river banks. Within-river hydrological connectivity allows insults to be transmitted both down- and upstream. This is markedly different from the relatively localized effects of most human impacts in terrestrial landscapes. In short, 'the medium is the message' serves as an uncomplicated summary of the existential threat to freshwater biodiversity.

It is but a short step from Marshall McLuhan to ecologist Garrett Hardin, whose popularization of the notion of 'the tragedy of the commons' dates from a 1968 article, although the origins of this idea can be traced at least as far back as Aristotle (see chapter epigraph). The now-familiar story goes something like this. A villager puts a goat out to graze on common land around the settlement so that his family can have a regular supply of milk. Seeing their neighbour enjoying this benefit, each of the other villagers sets their own goat to graze. All goes well until one villager realizes that he can gain more milk by putting out two goats. He does so, and soon his observant neighbours do the same. As the number of goats increases, there is less grass for each to eat and their individual milk yield declines. Nonetheless, the combined yield of two goats is greater than that from a single goat, so the villagers are better off if each grazes two goats. One of the villagers is then tempted to put a third goat on the commons; his neighbours follow suit. A fourth goat is added . . . and so on. The additional increment of milk from each additional goat declines as their number increases, but so long as the villagers obtain some benefit from adding another animal, the goat population of the commons will continue to rise. A critical point is reached where the grass can no longer withstand the intensity of grazing: it dies back, the goats starve and the supply of milk dries up. The lesson here is that protection of the environmental commons requires individuals to forego some gain. Instead of maximizing the amount of milk obtained over the short term, it is wiser to limit the number of goats and optimize the long-term gain of milk by ensuring the commons is not overgrazed (Hardin, 1968).

Why is the tragedy of the commons relevant to fresh water? Water is an irreplaceable resource for both humans and biodiversity, and consumption or contamination of water by one group of human users renders it unavailable or unfit for other users – including ecosystems that sustain biodiversity. For instance, the extraction of river water for irrigation is incompatible with its role in preserving fish stocks, and therefore agriculture has impacts on those who make a living from fishing. Other uses of the same water, if it had remained in the river channel, might include generating hydropower, flushing wastes downstream, allowing navigation or sustaining biodiversity. Our warming atmosphere is another manifestation of this underlying tragedy, but the scope for conflict over freshwater use makes it the common resource *ne plus ultra*. Equitable sharing of water requires human users to forego gains: the farmer must limit the water he extracts for irrigation so that users downstream can enjoy some benefit; likewise, the industrialist must treat effluent – thereby limiting profits – rather than simply discharging untreated waste water. It is in the interest of individual water users to overextract or to contaminate because they profit more from doing so than from refraining; polluters also benefit from the convenient fact that river water flows downhill, so their impacts are felt elsewhere. Overextraction affects the commons whether water is taken from rivers and lakes or from underground aquifers, since depletion of the latter (see Dalin *et al.*, 2017) affects the former. Overexploitation of fishes and other economically valuable animals represents yet another expression of the tragedy of the freshwater commons since it is the short-term interests of the individual to capture yet one more fish now rather than leaving it in the river where it could contribute to maintaining the population. The tragedy of the freshwater commons is that individual users rarely forego gains voluntarily, while the majority of the community of users, and the ecosystems upon which they depend, share the negative consequences of those gains.

The view of fresh waters as a commons is, of course, something of a caricature. Fresh water is a complicated natural resource (Lodge, 2010): some sources are renewable (rivers and streams, for example) while others are not (fossil ground water); some uses are substitutable (flushing toilets) while others are not (drinking); and some benefits accrue as public goods (aesthetics, recreation, fisheries) while others reflect private goods (a drinking-water supply). Nevertheless, consideration of trade-offs among these different aspects of water reveals the potential for conflicts among users, as manifest in, for example, construction of a hydropower dam.

People dwelling downstream of the dam, or in cities some distance away, receive the benefits of flood control and electricity. But farmland and forest may be inundated by the impoundment formed behind the dam, and the livelihoods of fishers are compromised by changes in river ecology. Importantly, the impacts of the dam are felt locally, typically by the rural poor, whereas the benefits accrue some distance from the dam site. Decisions about dam building tend to be made by city-dwellers who have more political and economic influence than people who are directly impacted by the dam and, typically, receive no benefit from it. Thus, the interests of parties who stand to gain economically from generating electricity override concerns of others who derive livelihoods from the intact river. In any case, scant consideration is given to the need to conserve aquatic biodiversity or preserve ecosystems when conflicting human interests are at stake. Only the water which remains after human needs have been satisfied will be available to sustain ecosystems, and this may be a vanishingly small amount with compromised quality. Unless some external control is imposed, water is monopolized by the most powerful human users, leaving little or nothing for weaker parties, or for nature. Arguably, this is the real tragedy of the freshwater commons.

How much fresh water is available on Earth? And, importantly, what proportion of that is already appropriated by humans? In addition, how does human use of water compromise its quality? The answers to these questions will determine the amounts and likely condition of water available to ecosystems after humans have extracted and consumed their share. The matter of water quality will be addressed later in this chapter. Its quantity, however, sets the context in which biodiversity conservation must take place and focuses attention on the most salient characteristic of fresh water: that of absolute scarcity.

The Scarcity Issue

Almost all (97%) of the Earth's water is in the oceans and, of the ~3% that is fresh (i.e. its salt content is less than 1 part per thousand), around two thirds (69%) is frozen solid at the poles (mostly in Antarctica) and the remaining third (31%) is deep underground. Estimates of the quantities in liquid form on the surface vary slightly among authorities (e.g. Shiklomanov, 1993; Gleick, 1996) and are, for instance, sensitive to scaling of lake volume–depth relationships (Cael et al., 2017). Irrespective, this volume is but a tiny fraction of global water: around 0.01% − or 0.3% of all fresh water − covering 0.8% of the Earth's surface. It amounts to

~105 000 km^3, much of it in lakes and fully 22% in Lake Baikal, Siberia. This volumetric estimate excludes water in the endorheic Caspian Sea, the world's largest inland-water body in terms of area (371 000 km^2), where many centuries of evaporation has rendered its contents saline – around one-third that of sea water – and only in the shallow northern basin, where it receives inflow from the Volga, could the Caspian be considered 'fresh'.

In terms of areal extent, lakes cover approximately 3.7% of the Earth's continental land surface (Verpoorter et al., 2014), whereas rivers encompass a very minor (almost negligible) proportion. They contain a mere 2% of surface fresh water (i.e. 0.006% of total fresh water, or 0.0002% of global water) equivalent, at any moment in time, to a standing volume of around 2120 km^3. A further 11% is in swamps of various types, including floodplain water bodies. The minute fraction in rivers is the source of most water used by humans and serves as habitat for many organisms found nowhere else. The total volume of water for sustaining humans and ecosystems, which consists of surface fresh water plus an estimate of the accessible ground water, amounts to a standing volume of around 200 000 km^3 – less than 1% of all freshwater resources.

Key Point

Freshwater ecosystems constitute no more than 0.01% of total global water volume and occupy less than 1% of the Earth's surface, equivalent to ~3% of the land area. The tiny amount of fresh water that is actually available as habitat, in combination with the number (and proportion) of species living in this water, makes them hotspots of global biodiversity. This association also goes some way towards explaining why fresh waters are hotspots of threatened species.

The Hydrological Cycle

Surface fresh water is not static, but a component of the global hydrological cycle which describes the movement of water above, on and below the Earth's surface driven by heat from the sun. However, the amount of water participating in the global hydrological cycle at a temporal scale relevant to nature and humans is a small proportion of the Earth's total – a mere 0.1%. In absolute terms, it represents a huge volume of around $520\,000\,km^3$ annually. One of the major constituents of the hydrological cycle is atmospheric water vapour representing, in large part (87%), evaporation from the ocean plus evapotranspiration from land (13%). Most of it (79%) returns as precipitation to the oceans, but a slightly larger share (21%) than originated from terrestrial evapotranspiration falls as precipitation over land (~$110\,000\,km^3$ each year). This represents an annual net transport of ~$40\,700\,km^3$ of water from the oceans to the land. The hydrological cycle is closed when the water that does not return to the atmosphere as evapotranspiration flows to the oceans as river runoff or as renewable ground water; the former is the larger fraction (70% of the total). In contrast to those short-term dynamics, most water in the oceans, icecaps or under ground cycles at timescales in the order of thousands of years. The underground reserves mostly constitute non-renewable 'fossil' water (perhaps $15\,000\,000\,km^3$) accumulated in aquifers during the past when conditions were wetter: for example, after the melting of Pleistocene ice sheets.

Two components of the global hydrological cycle are particularly relevant to humans and nature (Falkenmark & Rockström, 2006). The water that flows downhill to the sea, whether by surface or underground routes, is known as 'blue water'. It is this that sustains freshwater biodiversity, and it amounts to approximately $28\,000\,km^3$ annually. Soil water that passes from the land to the atmosphere by evapotranspiration is more than twice that volume (around $70\,000\,km^3$ annually). Termed 'green water', it plays an essential role in supporting terrestrial biodiversity, the transpiration component contributing to the production of plant biomass. The extent to which precipitation is transformed into either green or blue water – or, more typically, a mixture of both – depends on local circumstances of climate and vegetation type, and the extent to which land has been converted to agriculture. In arid areas such as savannahs or places with highly seasonal rainfall like the Australian outback, blue-water flow may cease entirely for part of the year. In humid tropical areas, by contrast, the volumes of blue water are very substantial, as evinced by mighty rivers such as the Amazon and Congo. Human modification of vegetation cover and land use profoundly affect the trade-off between percolation into the soil and runoff, and the proportion of soil water that is transpired by plants, and hence the relative proportions of blue and green water. Blue water appropriated by humans for irrigation is the major component (over 60%) of global water withdrawals, and this diversion into the green-water pathway reduces the quantities available to sustain freshwater ecosystems. The relative proportions of blue and green water are particularly affected by the amount of land used to grow food, which now covers some 25% of continental land areas, although by no means all of that is irrigated. Much of the withdrawn irrigation water is converted by crops to water vapour, but some proportion is returned via percolation or runoff from agricultural land. The quality of this so-called grey water is often compromised by pesticides or fertilizers, and can degrade the receiving blue-water ecosystems. Blue water always flows downhill, but green-water movements lack this directionality, and repeated withdrawal of blue water along its passage to the sea results in progressive contamination and deterioration in quality.

In the Anthropocene world, we face a 'pandemic array' of human transformations of fresh waters globally (Vörösmarty *et al.*, 2004; Alcamo *et al.*, 2008), including changes in their physical characteristics, and their biogeochemical and biological processes. The future health and sustainability of freshwater ecosystems and the biodiversity they support will depend upon how humans use water and manage drainage basins, and any changes in the global water supply as a whole (see Box 1.1).

Quantifying Human Water Use

How much of the accessible blue-water supply or available runoff is appropriated by humans? Estimates of the proportion withdrawn – 54% is widely quoted (Jackson *et al.*, 2001; see Table 1.1) – are sensitive to assumptions about how much of a river or its flow can be regarded as accessible or, conversely, too remote from major population centres. Rivers in far northern latitudes are mostly untapped, representing slightly more than 20% of the inaccessible supply, as are large rivers such as the Congo and Amazon that drain landscapes where relatively few people live. Floodwaters are also typically unavailable for capture, and they represent about half of the estimated total global annual runoff of $\sim$40 700 km^3 (see above). A consensus figure is that the 'available' remainder constitutes approximately 12 500–15 000 km^3 each year (Jackson *et al.*, 2001). Around two-thirds of the total runoff appropriated by humans is withdrawn for irrigation, which is by far the largest user, as well as for industries and municipalities, with another 6% evaporating from the surface of reservoirs (but see Table 1.1). The remaining 35% or thereabouts supports in-stream uses of rivers by humans, mainly through dilution of pollutants but also navigation, recreation, fisheries and so on.

Box 1.1 *The Global Water System*

The concept of a global water system (GWS) as a suite of water-related human, biological, biogeochemical and physical components – together with their interactions – provides a useful organizing framework that places emphasis on the primary objective of meeting the need for development within the bounds of planetary sustainability of water (Vörösmarty *et al.*, 2004; Alcamo *et al.*, 2008). The GWS connects several socio-ecological, economic and geophysical components at multiple scales: firstly, water in all its forms (liquid, vapour and ice) as part of the global hydrological cycle, including transport, precipitation, flow and storage; secondly, biological systems as integral transformers of water and the constituent fluxes that determine biogeochemical cycling and water quality; and thirdly, human beings and their water-related institutions, which are agents of environmental change through engineering works, as well as entities that experience and respond to shifts in pathways and thresholds within the GWS. A corollary of the GWS concept is that it is no longer sufficient (if it ever was) for water scientists and managers to focus solely on local processes, as there is a serious risk of overlooking important global dynamics with large and possibly irreversible impacts on society and nature (Alcamo *et al.*, 2008). Water security for humans and nature in the twenty-first century will require better linkage of science and policy, as well as innovative and cross-sectoral management initiatives and polycentric governance models (for elaboration, see Bogardi *et al.*, 2012).

The amount of water that is available for human use in the form of annual precipitation on a global scale is currently far in excess of demand, but the spatial and temporal patterns of accessibility of water derived from that precipitation are not well-matched to the distribution of people. Blue water is overextracted in some areas, often accompanied by pumping and consequent depletion of non-renewable underground water but is relatively underused in places where conditions do not favour settlement or agriculture. The demand for water has increased four-fold during the last half century, and the global population, which is now 7.5 billion, is projected to reach 9 billion by 2050 or thereabouts at which time up to 70% of the available supply may be appropriated. Human populations are growing faster than potential increases in

Table 1.1 *Blue-water runoff, withdrawals and appropriation by humans. Data from Jackson et al. (2001) and sources therein. Note that these figures are broadly indicative, as estimates the volume of runoff, appropriation, consumption, etc. vary among sources, and have been subject to updating (see text and, for example, Cael et al., 2017).*

	Volume (km^3/y)
Total runoff (rivers + renewable ground water)	40 700
Remote (geographically inaccessible) flow	7800
Flood water	20 400
Accessible runoff	12 500
Human appropriation	6780
	(54% of accessible runoff)
Withdrawals	4430
Irrigation (agriculture)	2880
Industry	975
Urban (domestic)	300
Losses from reservoirs	275
In-stream uses (e.g. dilution of pollution, navigation, etc.)	2350
Consumption (after withdrawal, converted to green water)	2285
	(18% of accessible runoff; 33% of appropriation)

blue-water supply, which can only be brought about by building dams and reservoirs to trap and store flood flows, and rapid shifts in anthropogenic water use are causing dramatic changes in patterns of water stress (Alcamo *et al.*, 2008).

One-fifth of cultivated land is irrigated, and it is highly productive yielding around 40% of the world's food. An increase in its extent will likely be needed to feed the 1.5 billion additional people expected by 2050 and improve the nutritional status of many others currently undernourished. This will increase withdrawal and consumption of blue water. Water-saving and irrigation technologies could slow the rate at which demand for water grows, and more efficient application of fertilizers might reduce waste and the consequent pollution of fresh water (Foley *et al.*, 2011). Because many rivers and lakes are situated in the far north where the inhospitable climate limits agricultural potential, the demands for blue water in latitudes suitable for agriculture will undoubtedly increase. Shifts towards diets incorporating greater amounts of animal protein will exacerbate this situation. As a rough indication, the annual water requirements of a meat-rich diet are $12\,000\,m^3$ per person, whereas an adequate mixed diet can be produced using only $1300\,m^3$; a 3000 kcal-per-day purely vegetarian diet needs only $500\,m^3$. Supplying the food needed to eradicate hunger, feed the Earth's growing population and provide for dietary shifts could increase both green-water use and blue-water consumption by 50% during the next two decades This scenario could result in overstepping of the 'planetary boundary' for sustainable water use by humans estimated at $4000\,km^3/y$ globally, with an upper uncertainty bound of $6000\,km^3/y$ (Rockström *et al.*, 2009).

The planetary boundary for blue water is of profound relevance here because (and, again, the medium is the message) as more water is consumed by humans, there is less available to sustain freshwater bio-diversity. A conservative estimate of this planetary boundary should take account of both human needs and of environmental-flow requirements; i.e. it would include some allocation of water for nature to protect freshwater ecosystems. Depending on the method used to assess ecosys-tem needs (see Gerten *et al.*, 2013), that global boundary is ~$2800\,km^3/y$ (the average of an uncertainty range between 1100 and $4500\,km^3/y$), considerably less than the earlier threshold of $4000\,km^3/y$ estimated by Rockström *et al.* (2009). Global blue-water consumption is more than $1700\,km^3/y$ at present, and exceeds the lower end of the estimated planetary boundary, amounting to 61% of the $2800\,km^3/y$ threshold (Gerten *et al.*, 2013; this proportion is higher than the estimate of 54% in Table 1.1).

Another estimate of planetary boundaries for blue water confirms the $4000\,km^3/year$ global value and gives a value of ~$2600\,km^3/y$ for current consumptive use (Steffen *et al.*, 2015); this is 65% of the $4000\,km^3/y$ threshold, close to the estimate of 61% by Gerten *et al.* (2013). This planetary boundary for water indicates there is still some scope for expansion of water use, but that impression may be misleading since water is extracted from rivers or lakes at a local rather than a global scale. At the river-basin scale, and especially in arid regions, there are likely to be many places where blue-water consumption is in excess of what would be envisaged as ecologically sustainable. The effects of decisions about how much water can be extracted locally depend on assumptions – assuming these are made at all – about the environmental water require-ments of individual rivers, which will differ between those with stable flow regimes and ample year-round rainfall, monsoonal rivers with most runoff concentrated during a wet season, and ephemeral or intermittent rivers. This matter is discussed more fully in Chapter 5.

To take account of environmental needs, the planetary-scale boundary for water has been complemented by formulation of basin-scale boundaries for water withdrawal, again invoking the concept of environmental water flows. These are the magnitude of flows required to maintain rivers '. . . in a fair-to-good ecosystem state . . .' and '. . . to avoid regime shifts in the functioning of flow-dependent ecosystems' (Steffen *et al.*, 2015: p. 1259855–7). Although it is theoretically possible to calculate a one-size-fits-all water allocation for river basins globally (an average 37% of mean annual flow has been mooted: Pastor *et al.*, 2014), the actual amount required to maintain ecosystem health will depend, as mentioned above, on the hydrological characteristics of individual rivers as reflected in their flow seasonality. Some of this variability can be taken into account by unpacking the annual flow regime into high-, intermediate- and low-flow months, with the environmental water allocation determined as a shifting percentage of the mean monthly flow. To remain within drainage-basin scale boundaries, the blue–water withdrawals should not exceed 25% (range of uncertainty 25–55%) of mean flow during low-flow periods, 30% (30–60%) during intermediate-flow periods, and 55% (55–85%) during high-flow periods (Steffen *et al.*, 2015). As a precautionary measure, the basin-scale water boundary is placed at the lower end of the uncertainty range. Applying this procedure to river basins globally, it is evident that withdrawals already exceed the allowed volumes for parts of the Murray–Darling and Colorado River basins, although not for the entire basins, as well as in extensive areas of northeastern China, much of India, the Iberian Peninsula, Mexico and elsewhere (Plate 1). These locations also transgress the freshwater boundary during more than half of the year when high withdrawals coincide with the low-flow period (Steffen *et al.*, 2015).

The 2600 km^3/y figure for current consumptive use of fresh water has been criticized because it underestimates the amount lost through evapotranspiration and evaporation associated with reservoirs. Taking such losses more fully into account brings the total consumptive use of fresh water to as much as 3569 km^3/y (Jaramillo & Destouni, 2015). The uncertainty range around this estimate is large – at least 1300 km^3/y. If added to the consumption estimate, it would breach the planetary boundary for water. Further uncertainty about proximity to the planetary boundary is contributed by the effects of land-use change, particularly forest clearance, which results in reduced evapotranspiration of green water, but greater blue-water runoff. The effects of deforestation are, however, more than offset by increased evapotranspiration associated with intensification of irrigated and non-irrigated agriculture, and, after this has been factored into calculations, the estimate for global consumptive use of fresh water could reach 4664 km^3/y (Jaramillo & Destouni, 2015). Close examination of an estimate of an upper limit for water withdrawals made in the context of the United Nations Sustainable Development Goals, adopted in 2015, likewise suggests rates of consumptive use could be high (Box 1.2). Irrespective of the precise figure, humans seem to be far closer to the planetary boundary than envisaged

---

Box. 1.2 *Global Water Withdrawal and Sustainable Development Goals*

The United Nations Rio+20 summit in Brazil in 2012 committed governments to draw up a set of Sustainable Development Goals that would replace the eight Millennium Development Goals (themselves derived from targets in the 2000 United Nations Millennium Declaration) upon their expiry in 2015. One of the sustainability targets adopted by the UN for 2030 is to '. . . limit volumes withdrawn from river basins to no more than 50–80% of mean annual flow . . .' (a component of Goal 3 of Griggs *et al.*, 2013; p. 307). This target seems inconsistent with the need to mimic natural flow regimes envisaged under current best practice for environmental water allocations, and such mimicry seems hardly achievable in instances where only 20% of mean annual flow remains in rivers. The same goal of sustainable water security proposes restricting consumptive use of runoff resources to <4000 km^3 annually (Griggs *et al.*, 2013; importantly, this is not to '. . . restrict global water runoff to less than 4000 km^3 a year. . .' as originally written in the paper: D. Griggs, pers. comm.). Thus, the goal envisages that humans would consume no more than one third of the accessible global runoff of 12 500 km^3/y (see Table 1.1), assuming we have not transgressed this limit already. However, some rivers will retain only 20% of their mean annual flow under the proposed target, so 80% of water could be extracted at the scale of individual rivers. Even though not all extracted water is consumed, the proportion of annual flow that it is envisaged could be withdrawn could well result in global consumption rates in excess of one third of accessible runoff.

by those who promulgated the original $4000\,km^3/y$ threshold. This is unwelcome news for humans but bodes even worse for freshwater biodiversity.

As humans approach the planetary boundary, increased appropriation of blue water will intensify competition for water among groups of humans, and between humans and nature, with serious implications for freshwater ecosystems and their biodiversity. Such competition is always highly asymmetric: as human requirements for water go up, that which remains for nature declines; the converse is *never* true. It may, in fact, be inaccurate to use the term 'competition' when referring to water that sustains ecosystems. While there is a genuine struggle on the part of humans to obtain a larger share of the limited blue-water resource at the expense of that available for nature, there is no real sense in which nature is attempting to wrest a larger share of that water to the detriment of human needs. In addition, reductions in the absolute availability of blue water — and increases in the ratio of green to blue water — can be anticipated as the Earth's climate warms, due to the influence of temperature on the hydrological cycle, even if all other things (e.g. land use) remain unchanged. Higher temperatures equate to greater transpiration rates by plants, and hence less blue water will be available to fill lakes and rivers. The scope for further appropriation of blue water by humans may be very limited if water scarcity and further degradation of freshwater ecosystems and declines in their biota are to be avoided.

Conflicts Over Water: Transboundary Rivers, Connectivity and Hydropolitics

The absolute scarcity of fresh water combined with the hydrological connectivity of rivers that transmits changes in water quantity and quality downstream has significant implications for stakeholders of the freshwater commons, especially in basins where national interests intensify conflicts among water users. There are 286 large transboundary rivers that flow through two or more countries, and around 60% of all water is drawn from them. They span 151 countries, including more than 40% of the Earth's land area, almost 3 billion people, and a high proportion of the global biodiversity (UNEP, 2016). Pre-eminent among these is the Danube with a mainstream that flows through or along the borders of 10 European countries; its drainage basin extends into a further nine. The Congo, Niger, Nile, Rhine and Zambezi are all shared by at least nine nations.

International disputes or disagreements over the freshwater commons can arise, even in cases where there are only two or three countries involved, in part because of differences in national priorities, development needs and water-management practices. But, almost inevitably, water-related schemes intended to benefit an upstream country have implications for supplies downstream, representing the main source of 'hydropolitical' tension and potential conflict. For example, the construction of large dams affects the quantity, quality or flow variability of water in downstream nations, as well as the transport of sediment and the movement of animals, with consequences for ecology and livelihoods. The ongoing controversy among riparian nations over mainstream dams along the Mekong River and the implications for fish and fisheries (described in Chapter 5) is one such example, while disputes about water also feature in relations between Afghanistan and (downstream) Iran (see Chapter 6).

Another instance of transboundary conflict is the long-simmering dispute between India and Bangladesh over the construction of the Farakka Barrage (a 2.2-km-long dam completed in 1975) on the Indian section of the Ganges less than 20 km from the border with Bangladesh. The barrage diverts more than half of the dry-season flow of the Ganges, leading to complaints from Bangladesh about the declining supply of irrigation water, increased in-stream sedimentation and saline intrusion into the Sundarbans mangrove forests and freshwater channels upstream. The barrage also had effects upstream as it formed a barrier to migrations of anadromous hilsa shad (*Tenualosa ilisha*: Clupeidae), almost wiping out the commercial fishery for this species in India (see also and Box 3.3); catches have likewise declined in Bangladesh where hilsa formerly constituted about half of landings. The sharing of Ganges water was partially resolved in 1996 when a treaty between the two nations was approved although, as would be expected where the freshwater commons is concerned, both nations continue to claim rights to unreasonably large quantities of water. India also has an international water-sharing agreement with Pakistan – The Indus Waters Treaty 1960 – that has generally been successful, although Pakistan took India to the Court of Arbitration in The Hague in 2010 in a successful effort to secure their water rights in the face of hydropower development upstream. Further information on hydropolitics and the rivers of South Asia is given in Box 5.2. An even more complicated set of transboundary issues involves the Nile, as explained in Box 1.3.

Box 1.3 *The Nile: Hydropolitics of a Great Transboundary River*

The Nile is the longest (6800 km) river on Earth. It flows through 11 countries and drains around 10% of the African continent. The basin contains a number of the world's least developed countries, and poverty is widespread among the 280 million inhabitants. Ethiopia, which is the source of much of the Nile's water, has ambitious plans for dam construction to promote irrigated agriculture and hydropower generation. Unlike some other transboundary rivers that traverse several countries, the Nile has no institutional mechanism to ensure equitable sharing or sustainable management of water. In contrast, management of the Danube basin is overseen by the International Commission for the Protection of the Danube that was established in 1998, with origins in a nineteenth-century alliance, and equivalent organizations (albeit of varying effectiveness) have been established for other transboundary rivers such as the Mekong. The Nile Waters Agreement, signed in 1959, involved only the two most downstream nations – Sudan and Egypt – and essentially gave them rights to the entirety of the Nile flow. This was always untenable, but became more so as Ethiopian plans for the upper (Blue) Nile developed. Egypt, Sudan and Ethiopia have been in acrimonious discussions over water management and cooperation since 2013, and an initial agreement was signed in 2015, but other national stakeholders have yet to become involved. Furthermore, both Sudan and Egypt want to increase their allocation from the river, and they and the other riparian nations have rapidly growing human populations. A welcome initiative in the region is the developing Nile Basin Initiative, involving 10 riparian nations, plus Eritrea that has observer status. The Initiative is currently in a '... primarily direction-setting stage' (see www.nilebasin.org/) but has the eventual goal of facilitating basin-scale cooperative planning.

A major concern of countries downstream of Ethiopia is the construction of four large hydropower dams on the Blue Nile, including the 155-m tall Grand Ethiopian Renaissance Dam, originally scheduled for completion in 2017 when it would have become the largest dam in Africa. It could take more than five years to fill, during which time amounts of water flowing to the Mediterranean would be substantially reduced. Evapotranspiration from a greatly increased expanse of irrigated land, combined with evaporation from the surface of reservoirs, would also diminish downstream flows after completion of the four dams. Evaporative losses and seepage from Lake Nasser behind the Aswan High Dam already represent significant losses of water for Egypt.

An earlier plan to increase the downstream supply of Nile water, conceived over a century ago, envisioned construction of a 300-km-long canal through the Sudd, a vast seasonally fluctuating grassy swamp and floodplain wetland in what is now South Sudan. It is around twice the area of Spain, includes a $57\,000$-km^2 Ramsar site (https://rsis.ramsar.org/ris/1622), and is the second largest wetland in the world (after the Pantanal). The Sudd is drained by the Bahr el Jebel, the local name of the White Nile, and the project aim was to limit the extent of inundation of the swamp by diverting water through a canal so that it would flow rapidly northward. Water loss attributable to transpiration by wetland plants, which amounts to around half of the volume of water in the White Nile tributaries flowing into the Sudd, would thereby be minimized.

Following agreement between Sudan and Egypt, construction of the Jonglei Canal began in the 1970s, but work was abandoned in 1984 following the outbreak of the Sudanese civil war when around two thirds of its length had been completed. In 2011, the new South Sudanese government reached an agreement, in principle, with Egypt to resume work on the canal, but the project remains stalled. There is concern in the south that the project would degrade the capacity of the Sudd to providing ecosystem services for humans (fisheries, grazing land, thatch) and reduce habitat for Nile crocodile (*Crocodylus niloticus*: Crocodylidae), hippo (*Hippopotamus amphibius*: Hippopotamidae; VU) and other large mammals, as well as for migrating and resident birds such as the great white pelican (*Pelecanus onocrotalus*: Pelecanidae), black tern (*Chlidonias nigra*: Sternidae), white-backed duck (*Thalassornis leuconotus*: Anatidae), African skimmer (*Rynchops flavirostris*: Rhyncopidae), African darter (*Anhinga rufa*: Anhingidae), black-crowned crane (*Balearica pavonina*: Gruidae; VU), white, saddle-billed and open-billed storks (*Ciconia ciconia*, *Ephippiorhynchus senegalensis* and *Anastomus lamelligerus*: Ciconiidae), shoebill stork (*Balaeniceps rex*: Balaenicipitidae; VU) and huge numbers of glossy ibis (*Plegadis falcinellus*: Threskiornithidae). The swamp is also home to an endemic genus of wetland plant (*Suddia*: Poaceae) and more than 100 fish species.

Although Sudan and Egypt would certainly stand to benefit from completion of the Jonglei Canal, all impacts of the project would be borne by South Sudan which, even leaving aside current political instability in the region, probably renders the further work on the project untenable. The consequences for the ecology and biodiversity of the Sudd – if the Jonglei Canal were ever to come into operation – would be devastating. Furthermore, the Sudd is a major economic asset for South Sudan and, if properly managed, could make potential economic contributions that could amount to almost US$1 billion annually (Gowdy & Lang, 2016). This sum represents only a fraction of the total value of the Sudd, which would include its potential as a symbol of national identity, its role in climate change mitigation, regulation of the flow of the White Nile, and support of biodiversity and human cultures in South Sudan.

The Nile offers a very clear example of the potential for international hydropolitics to cause basin-scale changes to rivers, in which the beneficial interests of certain human stakeholders are traded off against biodiversity losses that would be detrimental, both in and of themselves and in terms of their impact on ecosystem services upon which other human stakeholders depend.

A Right to Water? A Matter of Quantity and Quality

The scarcity of fresh water described above has obvious and important implications for humans, and concerns about water quantity are paralleled by those pertaining to quality. Although 71% of the global population used a safely managed drinking-water service – contamination-free and located on the premises – in 2015, and 89% had access to a basic service within a 30-minute round trip, 844 million people did not have, including 159 million dependent on surface water (WHO, 2018a). Since 1990, the proportion of people benefitting from improved sanitation rose from 54% to 68%, but, by 2015, 2.3 billion people lacked toilets or latrines; inadequate wastewater management polluted drinking water, causing 361 000 child deaths annually (WHO, 2018b). This is unacceptable but, little more than a decade ago, child deaths attributable to contaminated water and resultant diarrheal diseases were as high 1.5 million annually – 5000 *each day* (WHO/UNICEF, 2008). In symbolic response to this parlous situation, in 2010 the UN General Assembly passed a resolution declaring that access to clean water and sanitation is a fundamental human right. Some countries abstained, but there were no votes against the resolution. Self-evidently, the matter was (and remains) urgent, not least because halving the number of people without access to clean water and sanitation was one of the Millennium Development Goals intended to stand as a major achievement of the UN-designated 'Water for Life' International Decade for Action (2005–2015). Considerable progress was, in fact made, so that by 2015 the percent of the global population using 'improved' drinking water sources exceeded the 2015 Millennium Development Goal target of 88%. This is a praiseworthy feat, but there is still a considerable way to go. 'Improved' drinking water sources may not always be free of con- taminants and/or provide a reliable supply of water throughout the year. The extent of coverage by improved drinking water sources varies widely in developing regions, and is lowest in sub-Saharan nations, especially in rural areas. Thus, although the estimated number of diar- rheal deaths in children younger than five has fallen considerably, it was nonetheless 0.578 million (uncertainty range 0.45–0·75 million) in 2013, amounting to approximately 10% of all deaths in this age group (Liu *et al.*, 2015).

While the legal implications of any UN human water rights are uncertain, implementation of this resolution will be problematic given the possibility that, within current human lifetimes, demand for blue-water resources may exceed the supply from runoff, and hence transgress planetary boundaries for water consumption. Thus, it is far from clear whether and how the water needs of burgeoning human populations can be satisfied. Meeting these needs will have major impli- cations for the supply of water required by ecosystems, especially what is sometimes thought of as the 'wasted' water remaining in rivers when they flow into the sea. Moreover, despite impressive progress on access to improved water for drinking, the Millennium Development Goal for sanitation fell short of its target by nearly 700 million people. While 68% of the global population have access to toilets or latrines (WHO, 2018b), coverage in sub-Saharan Africa and southern Asia is lower. There is significant general uncertainty over the completeness of the sanitation chain (especially the extents of treatment), and whether excreta are safely reused or returned to the environment where they compromise water quality and affect human health. This is but one of many sources of contamination that degrade water quality and are generally detrimental to freshwater biodiversity. Other sources of such vitiation are introduced below.

The Quality Issue: Pollution and Contamination

Pollution can be defined broadly as something occurring in the wrong place, or at the wrong time, in the wrong (usually excessive) amount. This definition has the advantage that it avoids equating a pollutant with an un-natural or man-made contaminant, and can include non-chemical alteration of environments in which pollution is not caused by a substance added to the water. That is helpful because pollution can be caused by something naturally occurring, such as sediments from soil or riverbank erosion, or may involve the receipt of heated effluent (warm water with no substance added) from a power-station cooling plant that results in thermal pollution. Pollutants may originate as 'end-of-the-pipe' point-source discharges from industrial or mining operations and sewage treatment works, or they may be relatively diffuse in the case of runoff or percolation of contaminated water from agricultural land. Pollution seldom presents as a single compound or contaminant, and more often represents a mixture of organic and inorganic substances, with concentrations of both the combined and individual constituents subject to substantial spatiotemporal variability.

The list of potential ingredients in the cocktail of pollutants that can occur in a given lake or river is almost infinitely long. It can include some or all of the following: human and livestock wastes as well as partially treated sewage that may contain pharmaceuticals (and their degradation products) and endocrine disrupters; discharges from chemical factories or food-processing industries; seepage from landfills; oily runoff from roads and impermeable surfaces; agrochemicals such as fertilizers, herbicides or pesticides; microplastics (Box 1.4) and nanoparticles; heavy metals from mining and industrial processes; acid-rain deposition, and many others. For example, chloride levels have been increasing in North American lakes due to urbanization of shorelines and concomitantly greater use of salt for road deicing in winter: 44% of 371 lakes studied (mostly in the vicinity of the Great Lakes) have undergone long-term (>10 years) salinization. If this trend continues, within the next 50 years lake biota will be exposed to chloride concentrations in excess of the critical threshold (230 mg/L) set by the US Environmental Protection Agency (Dugan *et al.*, 2017). In short, many lakes – especially those associated with urbanization – will no longer be 'fresh'. Salinization of other fresh waters is also occurring widely as a result of poor irrigation practices and other human activities, with detrimental effects on biodiversity (e.g. Karraker *et al.*, 2008), and its extent can be expected to increase owing

Box 1.4 *The Emerging Threat of Microplastics Pollution*

We live in an age of 'peak plastic': around 8300 million t of virgin plastics has been produced globally, almost all of it since the end of World War II (Geyer *et al.*, 2017). Around 80% has amassed in landfills or the natural environment; more than half of the rest has been incinerated, while relatively little has been recycled. If current production and waste management trends continue, roughly 12 000 million t of plastic waste will be in landfills or the environment by 2050. This material has become so ubiquitous, it has even been proposed as a geological indicator of the Anthropocene era (Zalasiewicz *et al.*, 2016). Accumulation of plastic waste has given rise to concern over microplastic (particles less than 5 mm diameter) pollution of aquatic environments. They are most frequently in the form of pellets and beads, or fibres, with the former of a size and shape that can be readily ingested by consumers and thus become incorporated into food webs. The high sorption capacity of microplastics enables the accumulation of persistent organic pollutants on their surfaces, which may also attract pathogens and metals (Baldwin *et al.*, 2016). Thus, both physical damage of the animals that ingest them and chemical transfer of toxicants are potential sources of impacts from microplastics (Eerkes-Medrano *et al.*, 2015).

Microplastics enter freshwater environments in a number of ways. Photodegradation and/or mechanical breakdown of larger items is a major source; mismanagement of waste containing plastics (especially those present in personal care products), washing of synthetic textiles, and abrasion of particles from road markings and tyres are important also. Removal at wastewater treatment works can be effective for larger microplastic particles, especially fibres, but they may be transferred back into the environment via the spreading of sewage sludge on land for agriculture, which allows remobilization by runoff into receiving waters (Baldwin *et al.*, 2016). Although most research undertaken thus far has focused on marine environments, inland waters may be at greater risk from microplastic contamination due to their closer proximity to point sources (e.g. wastewater treatment plants and plastic processing factories), the typically smaller size of freshwater systems, and the ready transport of particles through stream drainages (Eerkes-Medrano *et al.*, 2015). Recent reviews (Wagner *et al.*, 2014; Eerkes-Medrano *et al.*, 2015; Horton *et al.*, 2017) stress our lack of knowledge of the impacts of microplastic pollution on freshwater ecosystems, and on the uptake of microplastics by invertebrates and fishes. They also highlight concerns over the prevalence of microplastics, which are seemingly both abundant and ubiquitous (see also Baldwin *et al.*, 2016), as well as if and how they might affect human health. A notable recent study (Windsor *et al.*, 2019) demonstrates the ubiquity of microplastics in a range of macroinvertebrates from different feeding guilds in three Welsh river catchments, by highlighting the existence of multiple points of entry and subsequent transfer of microplastics within freshwater food webs.

to climate change and increased water demand. Salinized rivers represent novel ecosystems as, unlike lakes which can be naturally salty, they have no ecological counterpart and, hence, no preadapted fauna (Kefford *et al.*, 2016).

The particular mixture of pollutants received by a lake, river or wetland is often unique to that particular water body, as even adjacent farms of the same type will tend to release slightly different mixtures of nutrients, livestock waste, agrochemicals and sediments according to individual management practices, stocking levels and so on. The effects of pollution are likewise multifarious: they can be lethal or sublethal, and acute or chronic, reducing fitness and population performance via reduced growth or reproduction. The so-called jellification of fresh waters (see Box 1.5) is an indirect outcome of atmospheric pollution and acid rain that greatly reduces the availability of calcium ions to organisms living in affected lakes. While pollutants frequently affect organisms directly, through toxicity or other physiological effects such as sexual disruption and intersex development, they also act indirectly by compromising environmental quality: for example, where fine sediments clog stream beds, or eutrophication reduces dissolved oxygen concentrations, or contaminants affect the quality and quality of food (or prey) required by consumers. In certain cases, as with mercury, organochlorines and dioxin, concentrations may be biomagnified in food webs so exerting greater influence on upper trophic levels, especially top predators (fish-eating eagles, for instance). Furthermore, because of synergistic interactions, the combined effects of mixtures of pollutants can be more damaging than the sum of the individual effects and give rise to unexpected consequences. Chemical pollutants and toxicants frequently interact synergistically with natural and anthropogenic stressors (Meis *et al.*,

Box 1.5 *'Jellification' of Northern Lakes*

A rapid and severe decline in calcium concentrations, attributable to long-range atmospheric transport of sulphuric and nitric acid precursors, has taken place in the waters of lakes with low buffering capacity (reviewed by Jeziorski *et al.*, 2008); boreal soft-water lakes have been particularly affected, as the acids in rain combine with calcium and remove it from the water column. Examination of zooplankton remains preserved in cores of lake-bed sediment have revealed near-extirpations of various species of *Daphnia* (Daphniidae). These cladocerans have much higher calcium concentrations than other crustacean zooplankton, with declines taking place concurrent with reductions in calcium and increased acidification of lake water. Since *Daphnia* are keystone grazers of phytoplankton, the sensitivity of their populations to declining calcium concentrations could have implications for energy flow though pelagic food webs.

Many soft-water lakes, including more than half of those on the Canadian Shield, have experienced calcium reductions to levels that would compromise survival and fecundity of *Daphnia* under laboratory conditions, and are likely to have the same effect in nature (Jeziorski *et al.*, 2008; Cairns & Yan 2009). In some Canadian lakes, *Daphnia* spp., are being replaced by another cladoceran, *Holopedium glacialis* (Holopediidae). It lacks a calcified carapace and, instead, is enclosed within a mucopolysaccharide jelly capsule (providing one reason for the use of the term 'jellification'). The reductions in calcium that drive this transition in the zooplankton assemblage are facilitated by the protection that the jelly coat provides *H. glacialis* from predation, so that less energy flows to predatory macroinvertebrates and fishes that would readily eat daphniids (Jeziorski *et al.*, 2015). Since *H. glacialis* also contains less phosphorus than *Daphnia* spp., both nutrient and energy availability to upper trophic levels is reduced where *H. glacialis* dominates lake zooplankton.

Increased acidification is likely to impact other species with high calcium demands, especially molluscs and larger crustaceans in potentially susceptible water bodies. For instance, some crayfishes are highly sensitive to reduced calcium (Cairns & Yan 2009), and some northern populations of *Cambarus bartonii* (Cambaridae) that were already experiencing calcium-poor conditions have almost disappeared (Hadley *et al.*, 2015).

2009; Rohr & Palmer, 2013), and their effects may be worsened by climate warming (Moe *et al.*, 2013; Wang *et al.*, 2019). One meta-analysis of the combined effects of stressors on freshwater ecosystems suggests they were more often antagonistic than synergistic (Jackson *et al.*, 2016), but mixtures of chemicals can have greater impacts than would be expected from their simple additive effects (Holmstrup *et al.*, 2010; but see Reid *et al.*, 2019), and better understanding of the ecotoxicological effects of these cocktails is needed (Klecka *et al.*, 2010).

Consideration of the toxicological and ecological effects of the myriad pollutants and contaminants in fresh waters lies far beyond the scope of this book, and there is a colossal specialist literature on the biology of pollution. Reviews of the pollution of fresh waters by nutrients (Stevenson & Esselman, 2013), pharmaceuticals, recently developed pesticides such as neonicotinoids (Goulsen, 2013) and other emerging contaminants (e.g. Petrovic *et al.*, 2013; Stamm *et al.*, 2016) offer a gateway to the literature and highlight the ubiquity and geographic extent of the threats they pose and our limited ability to predict their potential ecological consequences. For instance, a recent global scan of the incidence of antihistamines indicates these compounds can exceed therapeutic hazard thresholds in surface waters (Kristofco & Brooks, 2017), but their ecotoxicological effects have received scant attention. And, our ability to detect a wide variety of chemicals of emerging concern in surface waters far surpasses understanding of the implications of such detection for ecosystem health (Klecka *et al.*, 2010). Without going into further detail, it is sufficient to say that there are few – if any – inland waters that are entirely unaffected by pollution, and a vast majority where the extent and variety of contaminants have yet to be documented fully.

Pollution is probably the best-known and widely appreciated threat to freshwater ecosystems, and readily comprehended and generally understood by most citizens. There is also clear and direct relevance to public health: pollution of fresh water, often by sewage, has been linked to 1.8 million deaths annually caused by gastrointestinal diseases and parasitic infections (Landigran *et al.*, 2017); this is equivalent to approximately 3% of global deaths (or one in every 30), but is likely conservative given lack of knowledge about the toxic effects of many newer chemicals. Many of the sources of pollution (especially where there is an end-of-the-pipe source) are relatively easy to abate and tractable to control by legislation, particularly where there is public support for measures to improve water quality. A striking example of such occurred in April 2017, when El Salvador became the first country in the world to ban metal mining, with the aim of reducing pollution of rivers by heavy metals and toxic chemicals. Partial bans had already been introduced in Argentina, Colombia and Costa Rica, and a similar prohibition has been mooted for implementation in the Philippines.

Despite such measures and successes in reducing pollution loads in European and North American fresh waters, some of the largest rivers in the world (including the Ganges, Yangtze and their tributaries) are grossly polluted, and experience periodic fish kills as a result of deoxygenation or discharge of toxic effluents. Historically, the Thames and the Rhine were in a comparably poor condition in the nineteenth and twentieth centuries respectively. Even in Europe, where legislation to limit water pollution is now well entrenched, around half of rivers and lakes are still polluted (Gilbert, 2015), especially by nitrate. This is despite a target promulgated as part of the 2000 EU Water Framework Directive (WFD) to restore the continent's waters to good ecological health within 15 years. Legislation that carries the threat of penalties for non-compliance with environmental targets is incorporated within the WFD and represents an essential part of it. Despite this, the WFD has not delivered its main objectives of non-deterioration of water quality and the achievement of good ecological status for EU waters (for details, see Vouvoulis *et al.*, 2017). Many countries outside Europe also have legislation intended to limit pollution, but enforcement may be lax or lacking almost entirely in some countries with rapidly developing economies (where the mantra seems to be 'pollute now, clean up later'). Inevitably, therefore, the threats to freshwater biodiversity from a range of disparate sources often play out against a backdrop of pollution-imposed stress.

Other Threats to Fresh Waters

Irrespective of minor differences in the precise nature or relative intensity of threats to individual water bodies, the six general categories of threats are fairly uniform the world over. In addition to pollution of all types, they comprise:

- Flow alteration and regulation, water extraction and transfers, channelization and dam building;
- Degradation or alteration of drainage basins (essentially, land-use change);
- Overexploitation of fishes and other animals;

- Introduced, alien or non-native organisms, especially those that become invasive;
- Global warming and climate change.

Some threats do not fit neatly into one of these categories. For instance, mining of sand for use as a raw material in concrete can have devastating effects on river and lakes, especially in countries with fast-growing economies and a booming construction industry such as China and India. It could fall into an additional category − degradation of aquatic habitat − which might also accommodate some of the items included within flow alteration (e.g. channelization, levees and replacement of 'soft' banks or shores with hard man-made surfaces) or pollution (e.g. gold mining). Nonetheless, the six categories listed offer a reasonably parsimonious and comprehensive grouping of the main threats to fresh-water biodiversity, and this overall classification is used throughout this book.

The effects of drainage-basin degradation and land-use change depend on the extent and type of the alteration, and where it occurs within the catchment. Urbanization is associated with the development of expanses of impermeable surfaces that dramatically increase the magnitude and rates of runoff, which carries pollutants of many sorts. Runoff from agricultural land is also higher and faster than from natural vegetated land, and transports soil eroded from farmland during high-rainfall events. That runoff and the accompanying below-ground seepage also contains significant concentrations of agrichemicals and nutrients from fertilizers or animal wastes. Changes in land cover that involve total or partial removal of natural vegetation − for instance, by deforestation or timber extraction − alter runoff patterns to a degree that is largely dependent on the extent of change, and may lead to soil erosion and in-stream sedimentation. Replacement of natural vegetation with plants that have different water requirements alters the amounts of water supplied from the soil and runoff. It can also affects the types and amounts of organic matter (e.g. leaf litter and wood debris) entering water bodies, as well as extent of shading and, hence, water temperatures. Clearance or thinning of riparian vegetation can be associated with livestock grazing along banks that can lead to erosion, sedimentation and, sometimes, eutrophication of receiving waters. Unsurprisingly, then, land-use change degrades biological communities in fresh waters leading to species loss: a large-scale survey of Appalachian streams showed disturbance of land cover reduced richness of insects (Plecoptera and Trichoptera) by an average of 70%; most species extirpations occurred adjacent to residential areas and coal mining (Pond, 2012).

Deterioration in the condition of fresh waters associated with changes in land use within a catchment reflects the fundamental fact that streams, rivers, lakes and wetlands are landscape 'receivers' within drainage basins and exhibit lateral connectivity with their surroundings. Under the influence of gravity, any increases in soil erosion, nutrient loads and contaminants that accompany land-use change are transported downhill into valley bottoms and hence into rivers. Their landscape position not only makes fresh waters vulnerable to whatever changes occur within the drainage basin. Rivers and streams are also downstream transmitters of the material they receive so that human impacts do not remain locally within a particular section. The hierarchical architecture of rivers and their tributaries which ensures that this transmission takes place has the effect of heightening the vulnerability of biodiversity throughout the drainage network. However, land-use change is probably the primary threat to low-order streams and headwaters globally (Freeman et al., 2007), in part because these waters are less liable to suffer from contamination, impoundment or the other threat factors that are ubiquitous further downstream. This is of importance given that headwater streams may contribute more than three quarters of stream channel length in drainage basins, and support macroinvertebrate communities with high β-diversity at the landscape scale and, especially, across river networks (Clarke et al., 2008). Strong linkages between land and water in low-order streams make them particularly vulnerable to transformation or degradation of the surrounding catchment, and more susceptible to biodiversity loss from this disturbance than larger streams. Appropriate land management, incorporating, at the very least, undisturbed riparian buffers, will be an essential component of measures to conserve biodiversity in headwaters and other small streams.

As mentioned in the context of pollution, stressors and threats to freshwater biodiversity do not act in isolation, and their multiple combined effects may be hard to predict and greater than the simple sum of their individual impacts. To give one example, increased water abstraction from rivers will reduce their capacity to dilute pollutants, effectively amplifying the concentrations of any contaminants entering downstream of the off-take point. This interaction will be exacerbated by climate change: warmer temperatures, greater evapotranspiration and reduced river flows will likely increase the physiological burden of pollution on the aquatic biota, and biological feedback between stressors (e.g. climate change and nutrient pollution) may produce unexpected outcomes (Moss, 2010). To provide a second example, pollution and habitat degradation (including flow regulation) may limit ability of populations to recover from or compensate for human exploitation. Furthermore, the reduced abundance of native species may lower biotic resistance to invasion, offering non-native species the opportunity to establish themselves. Given this array of threats, it is hardly surprising that Convention on Biological Diversity (2016) concluded that inland-water ecosystems are frequently more extensively modified by humans than marine or terrestrial systems, and are amongst the most threatened of all ecosystems.

While citizens and policy-makers are generally well aware of the risks to environmental and human health posed by water pollution, general knowledge of the impacts on freshwater ecosystems associated with the other threat categories seems relatively limited. For instance, many people know that the oceans are overfished, but fewer understand that populations of numerous freshwater species are greatly depleted by historic overexploitation, and that some of them have disappeared from parts of their former ranges. That topic will be elaborated in Chapter 3. With a few notable exceptions, invasive alien freshwater species have not impinged upon the public consciousness, and a lot of people are unaware that some widespread and popular sport fishes (particularly trout) have established self-sustaining populations far beyond their native ranges. That topic, and related matters, will be considered at length in Chapter 4. Dams and pharaonic water-engineering schemes do receive periodic media coverage, especially where they involve relocation of the affected people within project footprints, but despite the profound changes frequently wrought by dams on river ecosystems, biodiversity considerations receive scant mention. The general impacts of flow regulation, and measures that can help to mitigate them, are described in Chapter 5, while some of the effects of dwindling supplies of flows to lakes (largely due to human overabstraction of water) are examined in Chapter 6. And, while few people can fail to be aware of the looming threat of climate change (irrespective of whether they expect that it will affect them directly, or they have anything to do with causing it), discussion of the various consequences that pertain to freshwater ecosystems has been largely confined to the specialist technical literature. That topic is addressed in Chapter 7.

The Freshwater Commons *Redux*

To return to the issue of the freshwater commons, four of the six threat categories listed above arise directly from the abuse of this commons since both overextraction and contamination of water are in the interests of the individual but disadvantage the wider community of users. Drainage-basin degradation and habitat destruction are another aspect of the same phenomenon whereby individuals maximize the use of land for cultivation, grazing, timber harvest and so on. Overfishing can, as mentioned above, be viewed in the same way, and climate change is likewise a consequence of human misuse of the global atmospheric commons, and the inability or unwillingness of individual states (or their citizens) to limit carbon emissions.

Of the six general categories of threat, only those arising from introduced or non-native species does not directly involve (mis)treatment of fresh waters as though they were a commons. Even in this instance, however, it is possible to argue that the individual introducing a bucket of live non-native sport fish to a river has an expectation of gaining future catches of that fish at the expense of other users' enjoyment of resident native species that might be affected by the aliens. Moreover, the chances of successful establishment of these aliens can be enhanced by interaction with the other five threat factors. Thus disturbed or degraded water bodies are more susceptible to invasion by non-native or alien species than intact systems, while reservoirs and man-made lakes created behind dams can serve as stepping stones for the spread of invaders to other water bodies (for details, see Chapter 4). The ongoing global epidemic of dam construction and fragmentation of rivers by impoundments (Zarfl *et al.*, 2014) not only has direct effects on biodiversity through changed flow and habitat conditions, but also facilitates invaders and their various impacts on native species. Box 1.6 summarizes the major categories of threats to fresh waters in general and indicates some of the interactions that take place between them. The remainder of this chapter describes the main findings of a global analysis of some of the many stressors and drivers that threaten rivers in particular, taking account of the potential consequences for biodiversity and for humans.

A Global Geography of River Threat

A global analysis of threats to river health (Vörösmarty *et al.*, 2010: Box 1.7) underscores the consequences of conflicts over the freshwater commons, and the consequences of the scant consideration given to biodiversity in explicit or implicit decisions about water-resource management or water engineering. The analysis addressed threats to human water security (i.e. a reliable supply of clean water plus protection against floods) and threats to riverine biodiversity separately, since the impacts of

Box 1.6. *In a Nutshell: The Combination of Threats to Freshwater Ecosystems*

Rivers and most lakes are components of an interconnected drainage system. Disruption of longitudinal connectivity in that system, typically due to dams, leads to changes in flow downstream usually involving reductions in quantity and, sometimes, quality with, typically, a reduction in the extent of variability. In turn, this means that the magnitude of lateral connectivity between water and land (e.g. through floodplain inundation) is reduced. Channelization together with related hard engineering exacerbates this disconnection, and simplifies in-stream habitat as rivers are modified to facilitate navigation and enhance rates of flow and runoff. Conversion of the disconnected floodplains and upstream areas of the catchment for agriculture, which may be associated with irrigation, and settlement or urbanization, contributes a host of contaminants, as well as nutrients and eroded sediments, to receiving waters thereby reducing their quality. The interactions between reductions in water quantity and quality attributable to humans, and our efforts to increase the predictability of spatial and temporal dynamics of fresh waters, thereby reducing their natural variability, will be compounded by shifts in climate. They will likely result in adaptive responses through water-engineering works that are directed towards maintaining stationarity in conditions of water flow and availability (Milly *et al.*, 2008). The outcome will be further modification of the physical conditions of surface waters, and greater simplification of habits that facilitate the establishment of invasive species and biotic homogenization. The potential ramifications of these changes cannot be predicted with accuracy, but are unlikely to result in anything other than declines in indigenous freshwater biodiversity and concomitant species losses.

a particular stressor will differ greatly depending on whether it affects biodiversity or humans. For instance, as mentioned above, the construction of a dam will benefit some human stakeholders and disadvantage others, whereas the effects on (say) fishes – due to altered river flow and habitat conditions, blocked migration routes and so on – are generally detrimental (see also Box 1.7). To give other examples, mercury deposition poses a greater threat to humans who are at the apex of the food chain than it does to most freshwater plants and animals, whereas acid

Box 1.7 *Analysis of Aggregate Threats to Rivers Water at a Global Scale*

Vörösmarty *et al.* (2010) set out to map the aggregate effects of a range of threat factors and stressors (termed drivers) on human water security and freshwater biodiversity. The two analyses combined 23 weighted drivers within four categories, as set out below, to provide a global geography of threats to rivers. Note that Category 3 is essentially equivalent to the extent of water-resource development.

Category 1: drainage-basin
    disturbance
    Cropland area
    Impervious surfaces
    Livestock density
    Wetland discontinuity
Category 2: pollutants
    Soil salinization
    Nitrogen loading
    Phosphorus loading
    Mercury deposition
    Pesticide loading
    Sediment loading
    Organic loading
    Potential acidification
    Thermal alteration

Category 3: dams and flow regulation
    Dam density
    River fragmentation
    Consumptive water loss
    Human water stress
    Agricultural water stress
    Flow disruption
Category 4: biotic threats
    Number of non-native fish richness
    % of non-native fish species
    Fishing pressure
    Aquaculture pressure

This list of drivers does not encompass all potential threats or stressors, in part because of the shortage of global datasets at a pixel-scale resolution of 0.5° (i.e. grids of 55.5 × 55.5 km), especially those relating to biotic threats; physicochemical threats are much better represented. Nonetheless, the range of drivers is wide and, incidentally, indicates the variety of threats to rivers and their biodiversity. Some drivers were routed downstream (if their effects were not inherently local) or divided by annual discharge (if their effects were subject to dilution by river flow volume), and all were weighted according to their relative impacts.

The weightings that were assigned to each driver within each theme, and then assigned to each theme overall, depended on whether their impacts were on biodiversity or on human water security. For instance, the weightings assigned to the number of dams and the extent of river network fragmentation in the context of human water security were quite different from their weightings in calculations of impacts on biodiversity, because dams can benefit humans but are detrimental to riverine biodiversity. Weightings assigned to other drivers that were detrimental for both humans and biodiversity, such as pollutants, also differed between the two analyses since, for example, high loadings of phosphorus and, especially, suspended solids, are relatively more detrimental to biodiversity. In addition, the beneficial impacts of technological advances in engineering and regulatory approaches that enhance human water security were accounted for in order to map 'adjusted' human water security; no such adjustment was possible for aggregate threats to biodiversity (for details, see Vörösmarty *et al.*, 2010). Note that these two analyses summarize levels of relative threat to biodiversity and human water security; they do not show the status of human or animal populations as a result of these threats.

rain or thermal pollution (arising from water used to cool industrial processes) can have profound impacts on freshwater biodiversity, but much smaller effects on humans. This means that the various threat factors must be weighted differently according to their relative impacts on human water security or biodiversity.

A surprising outcome of the global analysis of river-related threats for humans and biodiversity is the similarity of the patterns seen: in many parts of the world, low levels of human water security and high endangerment of biodiversity (i.e. the areas coloured red in Plate 2) are correlated. And, rivers draining large areas of the Earth's surface (especially in India, China and the Middle East) experience comparable and acute levels of threat to both humans and biodiversity. While sources of degradation in most rivers are alike, their engineered amelioration (included in the 'adjusted' upper map in Plate 2), which emphasize treatment of the symptoms rather than protection of resources, reduces the imposed threat in Europe, Australia and North America. However, such technological fixes are either too costly for many other nations or have yet to be adopted. In addition, a lack of comparable investments to conserve biodiversity account for the observed declines in freshwater species globally (see Chapter 2). Accordingly, while areas of low human water security and high threat to biodiversity tend to coincide, the match between the two is far from complete (Plate 3). In places where threats to human water security have been ameliorated by considerable investment in hard engineering solutions and water treatment, biodiversity remains imperilled. Thus conditions are 'good' for humans and 'bad' for biodiversity. But over much of the rest of the globe, and especially in densely populated parts of the developing world, the spatial pattern of threats to human water security and biodiversity are remarkably congruent: conditions are 'bad' for both humans and biodiversity (Plate 3). In places where there are relatively few humans, such as the Amazon, and the far northern portions of Asia, North America and Australia, rivers experience generally low levels of threat (things are 'good' for humans and biodiversity). However, this state of affairs is increasingly the exception rather than the rule. And it is notable that there are no places on Earth (except those lacking significant surface flows) where human water security is at risk but threats to freshwater biodiversity are absent (Plate 3). In short, this global analysis reinforces the conclusion that human demands upon the freshwater commons invariably trump those of nature. Furthermore, the reliance of some nations on costly technological remedies to safeguard human water security fails to address the underlying threat factors or stressors that put freshwater biodiversity at risk.

Although the global analysis incorporated a range of threats under four different general categories, they represent only those variables for which data were available at a global scale. Other human activities which degrade freshwater ecosystems (extraction of river sand for use by the construction industry, the pollution caused by mining activities) could not be included as data on their spatial extent was incomplete. Nor did the work take any account of the likely consequences of climate change (see Chapter 7). Suffice to say here that climate change projections do not augur well for riverine biota in regions where the human footprint is pervasive, since this is where conflicts over water are likely to be most intense and, thus, the scope for protection of biodiversity will be relatively limited.

Since there is potential for misunderstanding, it must be emphasized that the lower map in Plate 2 shows only aggregate threats to biodiversity, and not the consequences for populations and species (i.e. it does not show the responses of species or populations to these threats). The best current source of such data are species-level assessments in the IUCN Red List (www.redlist.org), and these and other indicators of actual threat status are considered in Chapter 2. A comparable analysis showing the aggregate impacts of these 23 drivers on biodiversity would be desirable, and may soon be possible, at least for vertebrates for which the IUCN assessment data are far more complete and comprehensive than for invertebrates. A related issue is the need to translate the results of such analyses into action intended to transform current practices of water management. That feat remains a major challenge.

2 · *Global Endangerment of Freshwater Biodiversity*

How will we teach the children to speak when all the animals are gone? Because animals are what they want to talk about first . . . animals are what they break their silence for.

Martin Amis (1991: p. 64), *London Fields*

Once there were brook trout in the streams in the mountains. You could see them standing in the amber current where the white edges of their fins wimpled softly in the flow . . .. On their backs were vermiculate patterns that were maps of the world in its becoming. Maps and mazes. Of a thing which could not be put back. Not be made right again. In the deep glens where they lived all things were older than man and they hummed of mystery.

Cormac McCarthy (2006: pp. 306–307), *The Road*

A Global Mass Extinction?

The biosphere is undergoing an epidemic of human-caused extinctions (e.g. Ehrlich & Pringle, 2008; Butchart *et al.*, 2010; Mace *et al.*, 2010). They exceed background (non-anthropogenic) extinction rates to such an extent that humans have overstepped planetary boundaries for marine and terrestrial biodiversity loss (Rockström *et al.*, 2009), with likely consequences for ecosystem functioning and societal prosperity (Cardinale *et al.*, 2012). It therefore seems reasonable to ask: are we now in the midst of a mass extinction event? Five such extinction events have taken place during the Earth's history, each one representing the loss of some 75% of known species from the fossil record. The latest of them, the Cretaceous mass extinction, ended around 65 million years ago but, given known species losses during relatively recent time, there are signs that the Anthropocene epoch represents the sixth such extinction event. Reviewing the evidence for this, Barnosky *et al.* (2011) concluded that current extinction rates are far higher than would be expected from the fossil record, with dramatic recent losses. Although such changes may not yet be sufficiently widespread or rapid to represent a sixth mass extinction, the loss of species classified as

CR by the IUCN would precipitate just such an event; it would be brought about sooner if threatened species in general (those that are EN and VU, as well as CR) were to disappear due to human activities.

Inevitably, forecasts of this type are sensitive to comprehensiveness of databases, taxonomic biases, completeness of the fossil record, and coverage of IUCN conservation assessments, but – irrespective of assumptions and the precision of estimates about the magnitude of losses – additional extinctions in the near- to mid-term seem unavoidable. The combined effects of multiple stressors, including climate change and expanding human populations, which have become more intense than any that living species have experienced previously, will cause an acceleration in extinction rates (Barnosky *et al.*, 2011). The full extent of any current mass extinction event may well have been underestimated because the understandable focus on species losses has meant that the accelerating extinction of populations – i.e. reductions in density – has received less attention. This destruction of wildlife abundance has been termed 'the great thinning' (McCarthy, 2015). It is manifest in widespread declines of insect populations in Europe and elsewhere (Fox *et al.*, 2006; Vogel, 2017), with biomass reductions of flying insects exceeding 75% within three decades (1989 to 2016: Hallmann *et al.*, 2017). Turning to larger animals, dwindling abundance and range shrinkage among terrestrial vertebrates (especially mammals) at the global scale has been described as '... a biological annihilation ... "representing" ... a massive anthropogenic erosion of biodiversity and of the ecosystem services essential to civilization ...' (Ceballos *et al.*, 2017: E6089). Similar warnings have been sounded for migratory shorebirds along the East Asian–Australasian flyway, where declines in species abundance estimated at 5–9% per year, and even greater for the spoon-billed sandpiper (*Eurynorhynchus pygmeus*: Scolopacidae; CR), are projected to result in '... extinctions and associated collapses of essential and valuable ecosystem services in the near future ...' (MacKinnon *et al.*, 2012: p. ii). Whether such trends represent an ongoing mass extinction event, or a bellwether of its initial phases, the question for the future is not whether more species losses will occur but, rather, how rapidly will they take place? And, which taxa and ecosystems will be most affected?

What Is Happening in Fresh Waters?

The Freshwater Animal Biodiversity Assessment (FADA; Balian *et al.*, 2008a), constituting the first global inventory of these organisms (although it actually incorporated aquatic macrophytes also), revealed that inland waters host almost 9.5% of the Earth's species, including one third of the vertebrates. These proportions are particularly striking given that lakes and rivers contain no more than 0.01% of the water on Earth and encompass only around 0.8% of the planet's surface area. Moreover, despite the much greater extent and total production of marine environments, the species richness of marine and freshwater fishes (Actinopterygii) is similar (14 736 and 15 149, respectively), with all of the saltwater species derived from a freshwater ancestor (Carrete & Wiens, 2012). Although currently far from comprehensive, knowledge of freshwater biodiversity has been improving (e.g. Darwall *et al.*, 2009, 2011a; Allen *et al.*, 2012). Global databases on fishes and herpetofauna have become available (e.g. Brosse *et al.*, 2013; Tedesco *et al.*, 2017; www.amphibiaweb.org/), and there is a growing number of checklists for invertebrate taxa (e.g. Trichoptera http://entweb .sites.clemson.edu/database/trichopt/) many of which have been consolidated by FADA (http://fada.biodiversity.be/). Sizeable gaps nonetheless remain among the invertebrates and, for freshwater animals more generally, in tropical latitudes (Balian *et al.*, 2008b; Böhm *et al.*, 2013).

Even relatively well-studied groups such as fishes are incompletely known: between 1976 and 2000, more than 300 new fish species were formally described or resurrected from synonymy each year (Lundberg *et al.*, 2000). More strikingly, almost 40% of the 6695 amphibian species recognized in 2010 had been described during the preceding two decades. Estimates of amphibian richness have continued to rise, reaching 7681 as of July 1, 2017 (www.amphibiaweb.org/), and the 2017 global total had increased by 91% since 1990 – almost doubling in less than 30 years! The as-yet-undocumented richness of freshwater invertebrates is even more substantial, with tens of thousands of species awaiting description (Balian *et al.*, 2008b), and some large and conspicuous invertebrates have yet to be inventoried adequately. To give one example, the colourful and polymorphic crayfish *Cherax pulcher* (Parastacidae) was described (from Irian Jaya) as new to science (Lukhaup, 2015) only after having being internationally traded as an aquarium ornamental for at least a decade; it is threatened by overcollection for export, and local subsistence consumption for food.

While we lack complete inventories of any freshwater taxon, it is obvious that much less is known about invertebrates than vertebrates. In practical terms, however, the types of organisms that people tend to value most highly – and hence want to conserve – are vertebrates. Accordingly, improving the comprehensiveness of vertebrate inventories is a more pressing matter than filling the gaps in knowledge of invertebrates, plants and other freshwater organisms. Nonetheless, inclusion of a variety of taxa in biogeographical inventories of freshwater biodiversity is necessary, as spatial analyses show that no individual taxon is an adequate surrogate for the others, either in terms of total richness or threatened species (Darwall *et al.*, 2011b).

Just as with taxon-specific differences, regional discovery rates of new freshwater species, and geographical coverage of conservation assessments, also vary widely (e.g. Hermoso *et al.*, 2017), but there have been recent improvements in coverage of Africa (Darwall *et al.*, 2011a) and parts of Southeast Asia (Allen *et al.*, 2012). Global biodiversity assessments have largely ignored freshwater species (e.g. Myers *et al.*, 2000; Brooks *et al.*, 2006), and the first attempt to identify areas that support particularly high freshwater vertebrate richness – the Freshwater Ecoregions of the World database (www.feow.org/) – was made public fairly recently (Abell *et al.*, 2008). It was a significant development, particularly in view of the absence of confirmation that terrestrial and freshwater biodiversity hotspots overlap (Strayer & Dudgeon, 2010; Darwall *et al.*, 2011b), and relatively low representation of fresh waters in protected-areas systems that are designed primarily for terrestrial habitats (Roux *et al.*, 2008; Darwall *et al.*, 2011b). The FEOW database serves as a useful tool to identify biodiverse and imperilled freshwater ecosystems and has the potential to inform global and regional conservation planning. While it is still far from complete, with taxonomic and geographical biases, the database can be updated as more information becomes available. Importantly, the FEOW initiative marks significant progress towards a global-scale tool for improving biogeographical literacy and knowledge of freshwater biodiversity in general.

Despite the many concerns about incompleteness of taxonomic and geographical coverage, to which can be added shortcomings in – or complete absence of – population trend data, the general picture of the threat status of freshwater biodiversity is clear. There is general consensus that population declines and species losses from inland waters are greater than in terrestrial or marine realms, with both vertebrates and invertebrates faring worse than their land-based or oceanic counterparts (see, for example, Lydeard *et al.*, 2004; Dudgeon *et al.*, 2006; Darwall *et al.*, 2009, Strayer & Dudgeon, 2010; Collen *et al.*, 2014). Investigation of a subset of 20 well-sampled central and North American river basins has shown that fish extinctions caused by humans are some 150 times greater than natural extinction rates (Tedesco *et al.*, 2013). And even one of the most iconic and charismatic groups – the freshwater cetaceans – have been subject to species loss (Box 2.1), with most Asian river dolphins declining in abundance (e.g. Braulik *et al.*, 2014).

Box 2.1 *Extinction of the 'Yangtze Goddess'*

The deteriorating condition of the Yangtze was responsible for the loss of one of the rivers' most iconic species: the Yangtze River dolphin or baiji, known colloquially as the 'Yangtze goddess'. The last documented sighting was in 2002, but the baiji was presumed to have become extinct (or 'functionally extinct') in 2006 when comprehensive surveys in known habitat failed to locate a single individual. This loss represented both the first human-caused extinction of any cetacean – freshwater or marine. Furthermore, because the Lipotidae contained only one extant species, the extinction of the baiji results in the disappearance of an entire evolutionary lineage (for details, see Turvey *et al.*, 2007). Pollution, dams and flow regulation, overfishing, sand mining and other drivers of river degradation contributed indirectly to the decline of the baiji, but many deaths were directly attributable to human activities, including illegal – but widely used – fishing methods (rolling hooks that ensnare dolphins, explosives and electrofishing), entanglement in nets, collisions with boats, injuries from propellers, and underwater blasting for channel and harbour construction. Baiji populations declined sharply during the 'Great Leap Forward' (1958–1961) when they were hunted for food and hide. Despite full legal protection after 1975, numbers declined from around 6000 in the 1950s to fewer than 100 in 1994, with subsequent surveys yielding population estimates of only a handful of animals.

Since the late 1980s, the primary strategy adopted by the Ministry of Agriculture for conserving the baiji was to capture as many individuals as possible in order to establish *ex situ* breeding populations. These would eventually provide surplus animals for replenishment of wild populations or reintroduction of baiji to rehabilitated sections of the Yangtze. To this end, a dolphinarium for research on captive animals was built at the Institute of Hydrobiology of the Chinese Academy of Sciences at Wuhan in 1992. Prior to this, captive baiji had been held in outdoor pools lacking filtration. The most famous of these was a young male named Qi-Qi caught accidentally by fishermen in 1980. Initial hopes for captive breeding of Qi-Qi were dashed by the difficulty of obtaining potential mates and poor survival of the few animals that were brought into captivity. Also, in 1992, a national reserve for baiji was established at Tian'e-Zhou in Shishou County,

Hubei Province, based around an oxbow lake cut off from the Yangtze. This, it was hoped, could support a breeding population of baiji. A single female captured from the Yangtze was introduced to the reserve in 1995, and it was intended that Qi-Qi be relocated from Wuhan to the oxbow also. Before this could occur, however, the female's emaciated carcass was found tangled in a net, and the plan to move Qi-Qi was abandoned. Some time later, Qi-Qi died in the Wuhan dolphinarium – after 23 years of solitary captivity.

Other indicators point to a general decline in freshwater populations. For example, the WWF Living Planet Index (LPI; Collen *et al.*, 2009), which is roughly comparable to the Dow Jones Industrial Average Index but representing animals instead of equities, shows that population trends for freshwater species between 1970 and 2012 have fallen far more steeply than those of their marine or terrestrial equivalents (Fig. 2.1). Freshwater animals have declined 81% (confidence range −68 to −89%) relative to −38% and −36% for those on land and in the sea, and −58% for the global index as a whole. This represents an average 3.9% annual decline in abundance of monitored freshwater populations since 1970, nearly four times the 1.1% reduction in terrestrial vertebrates (WWF/ZSL, 2016). Furthermore, rates of decline have steepened recently (Fig. 2.1), particularly between 2012 and 2014, reflecting reductions in populations censused between 2008 and 2010. The latest LPI report (WWF, 2018) states that between 1970 and 2014 the freshwater index values across the globe declined by 83% (range −73% to −90%) – equivalent to a fall to 17 in Fig. 2.1 (compared to 19 in the 2016 report). The values of the global indices for terrestrial and marine realms are not given in the 2018 report.

All 880 freshwater species (and 3358 populations) in the LPI are vertebrates, with waterbirds especially well represented (WWF/ZSL, 2016; WWF, 2018). Few tropical representatives are included, but the index is weighted to take account of taxonomic and biogeographical biases (Collen *et al.*, 2009) and the temporal trends are likely to be broadly indicative of freshwater biodiversity in general. Studies using different data sources (such as conservation status assessed by the IUCN: see below) have likewise found a high proportion of threatened species among freshwater-associated vertebrates (Collen *et al.*, 2014). For instance, around one quarter of the world's freshwater fishes – almost

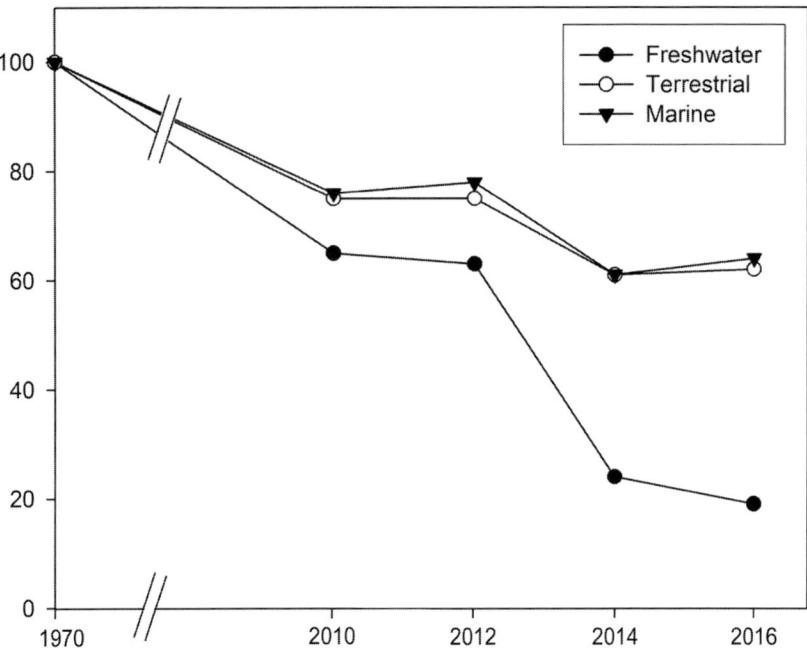

Fig. 2.1 The WWF Living Planet Index (LPI) shows population trend data for a collective 'basket' of vertebrates in the freshwater, marine and terrestrial realms, revealing remarkable decreases among freshwater species. Declines are relative to a benchmark value of 100 in 1970. Dates of the LPI reports typically refer to data from four years earlier (thus the 2016 LPI is based on estimates of abundance from 2012). The 2016 value of 19 for freshwater populations has confidence limits ranging from 11 to 32; limits for the value of 62 for terrestrial populations range from 49 to 79 (WWF, 2016). From Reid *et al.* (2019) with permission.

40% in Europe and the United States – are threatened, as are perhaps 20 000 freshwater macroinvertebrate species, especially crayfishes and pearly (unionid) mussels (Box 2.2; see also Strayer, 2006; Taylor *et al.*, 2007). There is some evidence that habitat occupancy may increase the susceptibility of some freshwater animals to anthropogenic impacts, as there are more threatened species associated with streams and rivers than there are in marshland or lacustrine habitats (Collen *et al.*, 2014).

Shifting from the global to a national focus, the same general trends of taxon-specific endangerment are evident. For instance, Elston *et al.* (2015) estimate that 74% of native freshwater fishes, pearly mussels and crayfish species in New Zealand are threatened with extinction; 25 years

Box 2.2 *The Particular Vulnerability of Unionids*

The bivalve order Unionida (sometimes Unionoida) consists of around 850 species in six families, the majority in the Unionidae (Fig. 2.2). Pearly mussels are especially diverse in large rivers in China and in those parts of the United States that escaped glaciation during the Ice Age, but they occur also in Southeast Asia and elsewhere; the tropical species are especially poorly known. All threatened freshwater bivalves are pearly mussels, and of the 533 species that have been assessed, 6% of them (32 species) have disappeared within the last century as a result of human activities, while another 32% are threatened (CR, EN or VU). Most of them are in the United States, where around 70% of the 281 species of Unionida (plus 16 subspecies) are classified as federally endangered; almost 10% (27 species) are already extinct, rising to as many as 37 species if those 'presumed extinct' are included (Lydeard *et al.*, 2004). IUCN assessments of 233 species of North American union-ids suggest 40% are threatened, of which half are CR. For this reason, they have been characterized as North America's most imperilled animals (Strayer *et al.*, 2004). To put this into a global context, the unionid mussels in North America may represent around one third of the entire group. Comprehensive assessments of pearly mussels in China have yet to be undertaken, but anecdotal reports suggest widespread declines (see also Shu *et al.*, 2009), and a detailed account of the group in Europe, where there have been large reductions in range and abundance, is given by Lopes-Lima *et al.* (2017). A recent report of declines of unionids in rivers of Southeast Asia (Zieritz *et al.*, 2016), where data on these animals is especially poor, includes evidence of the replacement of native species by invasive *Sinanodonta woodiana* from China. The eco-logical consequences of reductions in pearly-mussel populations are discussed in Box 3.10.

Unionids are vulnerable due to a combination of their own inherent attributes and their interactions with other species (Strayer *et al.*, 2004). Many have restricted distributions and a high degree of endemism compared to (say) aquatic insects that are relatively vagile and have higher dispersal ability. Moreover, because pearly mussels are filter feeders and burrow in sand and gravel of river beds, they are acutely sensitive to water quality and sedimentation resulting from

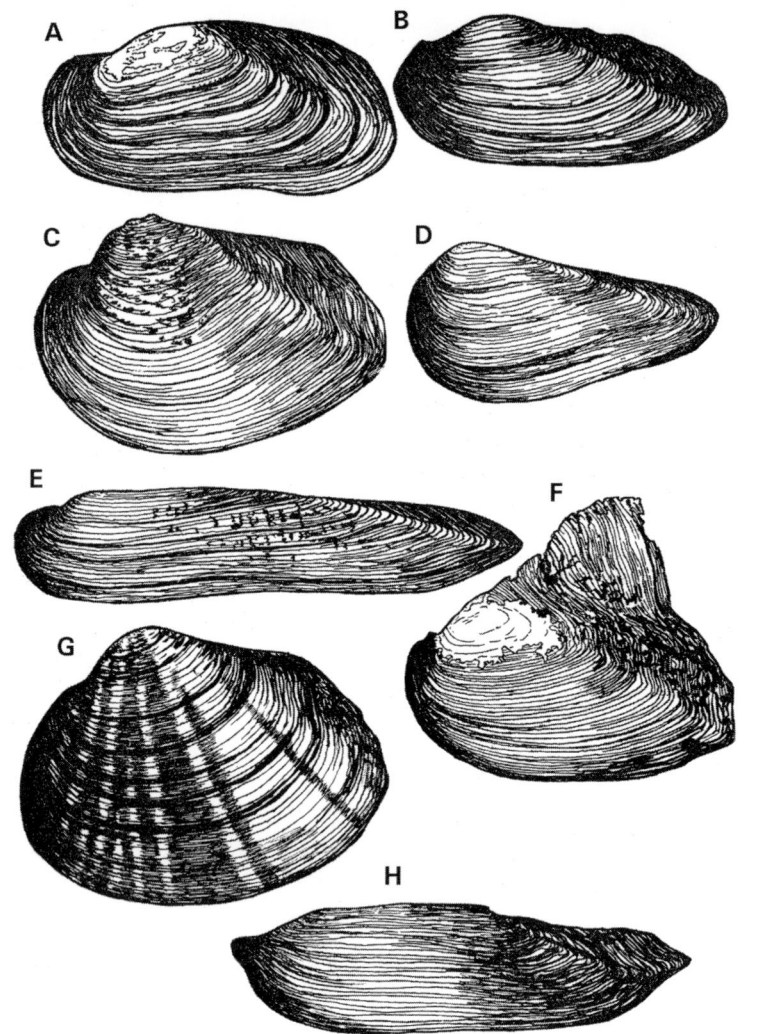

Fig. 2.2 Examples of the shell forms of various genera of pearly mussels – mainly Unionidae – from China. (A) *Margaritifera* (Margaritiferidae; shell length 115 mm); (B) *Unio* (55 mm); (C) *Acuticosta* (285 mm); (D) *Cuneopsis* (70 mm), (E) *Lanceolaria* (100 mm); (F) *Hyriopsis* (140 mm); (G) *Schistodesmus* (35 mm); (H) *Arconaia* (90 mm). Drawings by the author.

the wash-off of soil and silt from agricultural land. Flow modification or channelization that affect patterns of riverbed erosion and deposition also reduce habitat suitability. To make matters worse, since 1985, unionids in the eastern United States have suffered from competition with non-native and highly invasive dreissenid mussels that foul their shells, compete for food and have been implicated in extirpations of pearly-mussel populations (see Chapter 4). Furthermore, and unusually for a freshwater invertebrate, many unionids have been historically depleted by human exploitation, driven by demand for their nacreous shells, the quest for pearls and, in some countries, consumption of their flesh. The monetary value of this 'fishery' was once substantial: between 1897 and 1963, when unionids in the United States were heavily exploited for nacre, a total of US$6 billion (in 2017 dollars) of buttons was produced; the pearl fishery began around four decades earlier, and was probably worth another $4 billion (Strayer, 2017). The current wholesale value of unionid shells is relatively small, in the order of millions of dollars annually.

The life cycle of unionids makes them susceptible to further risk since they rely on the presence of a suitable vertebrate host to complete larval development. Female mussels incubate fertilized eggs in modified gills (marsupia) until they are released as glochidia larvae. The glochidia are expelled into the surrounding water whereupon they attach to the fins, gills or skin of a fish host; a few species also attach to amphibians or turtles. There, the glochidia live as parasites for several days or weeks (sometimes longer), subsequently metamorphosing into a tiny mussel that abandons the host to live on the river bed. Some unionids rely on little more than chance to locate a host, and therefore produce prodigious numbers of glochidia. In others, the swollen tissues of the mantle margins protrude from the shell of gravid females, and may display contrasting colours or form more elaborate lures that mimic fish shapes and eyespots. Some unionids eject edible clumps of sterile eggs to serve as a bribe for potential hosts, or enclose masses of glochidia within an ovisac that resembles prey items such as aquatic insects (for examples, see Barnhart, 2008). These adaptations serve to attract the attention of potential hosts in search of a mate or a meal, with glochidia being expelled as the host approaches, or released when a fish strikes at the lure. Without the appropriate host, pearly mussels cannot breed successfully. While there is rarely an exclusive one-to-one relationship between fish host and unionid, the majority of pearly mussels depend on a few hosts only and, in extreme cases, the match between fish and mussel can be specific to a particular drainage basin. Reproductive failure need not involve complete disappearance of the preferred hosts: once encounters between glochidia and hosts fail to exceed some critical threshold, the probability of successful recruitment falls to zero. For these mussels, population sustainability – and susceptibility to extirpation – is not merely a matter of suitable habitat conditions for themselves but, also, for their hosts.

earlier only a quarter of them were considered threatened. Excessive nutrient runoff from intensive agriculture is cited as a primary cause, but river engineering and commercial exploitation of declining fishes have also been influential. Among them, the New Zealand longfin eel (*Anguilla dieffenbachia*: Anguillidae) has been strongly affected by dams that block movements of these catadromous fish, because seaward migration of females occurs only after several decades of growth to sexual maturity. The prolonged pre-reproductive period makes this eel highly susceptible to mortality from fishing. The association between age (or size) at maturity and vulnerability to population decline as a result of overexploitation will be described in detail elsewhere, but Box 2.3 sets out how a clade of large-bodied Palaearctic salmonids have been significantly depleted by human activities. This susceptibility of the bigger freshwater animals – especially fishes – to population reduction has been pointed out previously (e.g. Dudgeon *et al.*, 2006; Dudgeon, 2011), and represents a 'great shrinking' in size, analogous to the 'great thinning' in abundance seen in terrestrial and aquatic species (McCarthy, 2015). This shrinking reflects both a reduction in the average size of individual species and a reduction in the mean body size of species making up freshwater assemblages. Further examples of declines in the abundance of large animals and 'fishing down' food chains are given in Chapter 3.

What Does the IUCN Red List Reveal about Threatened Freshwater Animals?

One way to obtain an overview of the threats to freshwater biodiversity, and compare them with the threats facing biodiversity in general as well as species in the marine and terrestrial realms, is to use data included in

Box 2.3 *Size, Age and Multiple Threats Contribute to Extinction Risk of* Hucho *spp.*

The vulnerability that large body size combined with slow growth or late maturity bestows upon freshwater animals, as well as the variety of threats that affect their populations, is shown clearly by the salmonid genus *Hucho*, comprising five species distributed across Eurasia. They are piscivorous and iteroparous, generally remaining in fresh water throughout their lives. Among them, the Danube salmon (*Hucho hucho*; EN) is, as its name suggests, native to the Danube drainage, although some stocking has occurred in headwaters of the Rhône and Vistula. It reaches up to 1.5 m in length, and adults migrate each year into upstream tributaries for spawning. Declines in populations attributable to overfishing began a century ago, and were exacerbated by pollution, but the primary threat to Danube salmon has been and is population fragmentation and flow alteration resulting from hydropower dams and channelization. Natural reproduction is so limited that the fish now occupies no more than 10% of its former range. Few populations are self-sustaining, and most depend on stocking of aquacultured individuals intended to facilitate sport fishing and conservation (Freyhof & Kottelat, 2008). Restoration of natural reproduction by Danube salmon will involve removing barriers that fragment populations and limit access to upstream spawning sites, combined with maintenance of sufficient in-stream flow to allow these large fishes and their prey to coexist within regulated reaches.

Further east, the Siberian salmon or taimen (*Hucho taimen*; VU) is much more widely distributed than its Danube counterpart, and there remains some dispute among anglers over which of the two is, in fact, the bigger fish. The taimen formerly occurred in the Caspian and Arctic drainages of Eurasia (i.e. in the Volga and beyond to the Lena), and in portions of the Pacific drainage in Mongolia, Russia and China. It has been largely eliminated from the Volga and westerns portions of its range by pollution and overfishing; further east, dams and mining are locally significant threats but, again, overfishing is the main driver of declines of this long-lived salmonid (Hogan & Jensen, 2013). Introduction and enforcement of fishery regulations combined with establishment of protected areas and riverine reserves that protect migration routes and spawning areas could facilitate the recovery of taimen populations.

Still further east, the Sakhalin taimen (*Hucho* [= *Parahucho*] *perryi*; CR), which spawns in rivers in Japan (Hokkaido) and the Russian Far East, is severely threatened by overfishing as well as transformation of riverine spawning reaches, and most Japanese populations are now extinct (Fukushima *et al.*, 2011). Like other *Hucho* spp., the Sakhalin taimen is iteroparous and takes a considerable time to reach sexual maturity. However, some populations are at least partially anadromous, and it is thus at risk from coastal as well as inland fisheries. *Hucho bleekeri* – the Sichuan taimen (CR) – endemic to the upper Yangtze in China, is threatened by overfishing (involving electricity, piscicides and explosives) and dam construction, but apart from a trend of continually waning abundance, there are scant data on its conservation status. Even less is known about the fifth species in the genus, the Korean taimen (*H. shikawae*; DD), which has a historical range encompassing rivers that traverse the border between China and North Korea, and may well be at risk.

the IUCN Red List to determine the percentage of extinct or threatened species at risk from a particular threat factor. The results (**Table 2.1**) show that while there are more threatened terrestrial species than freshwater species, the numbers of the latter is high relative to the area their habitats occupy: there are two thirds as many threatened animals in fresh waters than there are on land. This finding is likely to be relatively robust as there is no reason to suppose that conservation assessments for terrestrial animals are any less complete than those of freshwater species; the contrary is more likely to be true. It also accords with the pattern of high endangerment of freshwater animals evident from the LPI (and Fig. 2.1).

All threatened species tend to be affected by two or more factors acting in combination (2.3 on average: Table 2.1), biological resource use (i.e. overexploitation) and agriculture being the most important drivers across all realms. In fresh waters, pollution and overexploitation are the most frequent threats; pollution is far less important on land. Agriculture and natural-system modification (essentially, land-use change) are also significant threats to freshwater species, but not for those living in the sea. Urban development and invasive species are ranked slightly lower in terms of the proportion of threatened freshwater species they affect, with climate change seemingly constituting a less important threat to

Table 2.1 *Relative intensity of the main threats to animals in fresh waters, on land and in the sea as indicated by data extracted from the IUCN Red List (2017). Threats to species across all realms are shown also. Data are mainly percentage of threatened (CR, EN and VU) species identified as being at risk from drivers or stressors associated with a particular category. Threat categories that affect fewer than 5% of species in any realm (i.e. human intrusion and mining, both 4% across all realms, and other factors, all <1%) are not shown.*

Threat categories	Fresh waters	Terrestrial	Marine	All realms
Biological resource use	17	21	23	20
Agriculture	13	24	3	18
Urban development	10	12	12	11
Invasive species and pests	11	12	13	11
Pollution	19	5	14	11
Natural system modification	15	8	3	10
Climate change	6	6	13	7
No. species assessed	25 449	37 343	12 432	63 939
No. threatened species	5882	8682	1089	13 501
Mean no. threats per species	2.3	2.4	3.0	2.3

freshwater animals than to marine species (such as scleractinian corals). Some threatened species are associated with two realms (e.g. salmon or sturgeon that migrate between rivers and the sea; some amphibians and semiaquatic vertebrates) and are represented twice in the columns in Table 2.1, but they are unlikely to have influenced the relative rankings of threats or the finding that freshwater species appear to be disproportionately threatened.

The categorization used by the IUCN to encompass threats to biodiversity in all three realms, and which was followed in Table 2.1, does not completely align with the six general categories of threats to freshwater biodiversity set out in Chapter 1. For instance, dams and flow regulation are subsumed under the IUCN 'modifications of natural systems' category rather than appearing separately, whereas land-use change (e.g. by urbanization and conversion to agriculture), which also represents a modification of natural systems and influences inland waters through changes in the quality and quantity of runoff, is split over two categories. Moreover deforestation, another aspect of land-use change, is included within the IUCN 'biological resource use' category, which makes sense for land-based threat assessments but does not adequately reflect the implications of deforestation for inland waters. Nonetheless, the data in Table 2.1 serve to illustrate the relative endangerment of freshwater animals, the main categories of threat, and their tendency to act in combination rather than in isolation. Box 2.4 gives an example of how an evolutionarily significant population of freshwater fish has

---

Box 2.4 *Coho Salmon on the Brink in California*

It is not only large Eurasian salmonids that are at risk of extinction (see Box 2.3). In the New World, coho salmon (*Oncorhynchus kisutch*: Salmonidae) breed in drainages along the western coast of the United States. They are close to extinction in the California Central Coast region, where the population represents an evolutionarily significant unit of individuals reproductively isolated from other coho salmon, and was designated as federally endangered in 2005. Numbers of these fish returning from the sea to breed have declined from approximately 100 000 individuals in the 1940s to fewer than 10 000 in the 1990s, dropping to only 500–600 individuals in each of the three years leading up to 2010 (Miller, 2010). This succession of low returns was significant because the life span of these fishes is three years, and thus the population comprises only three year classes. The causes of decline are attributable to several interacting factors: dam construction, water extraction, climatic variability (especially drought), and fluctuating conditions in coastal waters, as well as land-use change and associated sedimentation of streams. Given their location at the southern end of the current geographical range of coho salmon, the central Californian population is highly vulnerable to warming associated with climate change.

The prognosis for survival of Californian coho appears bleak, but fishing for them has been proscribed since the 1990s, and there are ongoing efforts to restore in-stream habitat and establish riparian buffers. Release of young hatchery-reared animals also may have a role to play in supplementing wild populations, provided that excessive inbreeding during aquaculture is avoided by selecting and pairing fish in a manner that maximizes genetic diversity. Stocking with juveniles is supplemented by raising some captive-bred fish in fresh water and releasing them into streams as adults, which has the advantage that the offspring of these fish imprint on the stream where they hatched rather than on the salmon hatchery. This approach may be a viable method for translocating coho salmon and reestablishing them in streams where they have been wiped out (Miller, 2010).

dwindled in response to a combination of anthropogenic threats, and the options that might be available to reverse these declines.

What Is the Most Threatened Group of Freshwater Animals?

While unionids may be the most imperilled taxon in North America (see Box 2.2), the conservation status of these mussels in other parts of their global range – apart from Europe – has been incompletely documented. Among those taxa for which IUCN assessments are more comprehensive (Table 2.2), there are high levels of endangerment among freshwater reptiles and mammals, which exceed those of their terrestrial counterparts (e.g. Böhm *et al.*, 2013), as is also evident from the general trend for vertebrates in the LPI (see Fig. 2.1). Likewise, freshwater fishes and decapods are more threatened than those in the sea, although the marine assessments for these taxa are somewhat less comprehensive.

Amphibians – especially frogs (Anura) – are widely considered to be the most threatened major animal group and, certainly, the most threatened vertebrates globally (Warkentin *et al.*, 2009; Hof *et al.*, 2011). Almost 40% of anurans are classified by the IUCN as at risk (if near-threatened species are included), and only 39% of them are currently included in the 'Least Concern' category. One estimate of the recent global total of frog extinctions is roughly 200 species or 3.1% of all frogs (Alroy, 2015). This is four orders-of-magnitude higher than background rates due, in part, to infestation of frogs in Latin America by an invasive chytrid fungus (see Chapter 4), but at least another 6.9% of all frog species are projected to be lost within the next century, even if there is no acceleration in the growth of environmental threats (Alroy, 2015). This bleak projection for frogs takes no account of other amphibians that may be at risk, such as that described in Box 2.5. To such particular cases may be added the potential losses of salamanders that might be caused by the emerging pathogen *Batrachochytrium salamandrivorans*, a second amphibian-specific chytrid discovered in 2013 (see Chapter 4).

The Problematic Status of Species That Are Data Deficient

One drawback of using the IUCN Red List for determining levels of imperilment among taxa is the proportion categorized as data deficient

Table 2.2 Threatened freshwater animal species and, where relevant (in parentheses), their terrestrial (reptiles, mammals) or marine (fishes, decapods) counterparts, as recorded in the IUCN Red List (2017). Data are percentage of extinct and threatened (CR, EN and VU) species out of the total number assessed; the proportion of species classified as data deficient (DD) is shown also. Pearly mussels (Unionida) are the only freshwater bivalves that have been assessed; marine bivalve assessments remain at an initial stage. Fishes = Actinopterygii; Frogs = Anura; Decapods = crayfishes, freshwater crabs and shrimps.

	Fishes	Frogs	Reptiles	Mammals	Decapods	Unionida	Dragonflies
Number assessed	8230 (7320)	5796	448 (5359)	145 (5466)	2637 (254)	533	3208
Threatened species (%)	24 (5)	32	30 (21)	39 (23)	19 (1)	38	10
DD (%)	21 (15)	24	11 (17)	12 (13)	39 (25)	17	28

Box 2.5 *The Critically Endangered Axolotl*

While many amphibians dwindled to extinction before steps to conserve them were taken, some have benefitted from such attention although it represented little more than a last-ditch effort to preserve the few remaining wild individuals of a once-abundant species. A case in point is the axolotl or Mexican salamander (*Ambystoma mexicanum*: Ambystomatidae; CR): this neotenic salamander lives permanently in water and derives its name from Xolotl, the dog-headed Aztec god of death, lightning and monstrosities. It was formerly abundant in a network of interconnected lakes and waterways in the Tenochtitlan Basin at an elevation of 2290 m where Mexico City now stands. Axolotl abundance and habitat occupancy have been reduced by the combined and long-term effects of wetland drainage, urbanization, pollution and the presence of invasive fishes, as well as their long-standing exploitation for human food and as folk remedy. Wild individuals of axolotl are now confined to a few canals in southern Mexico City within the Ejidos de Xochimilco and San Gregorio Atlapulco Protected Natural Area established by the Mexican government in 1992, and since designated as a Ramsar site. Axolotl numbers have, nonetheless, continued to decline. Extensive surveys since 2004 have failed to yield any of these salamanders at locations where they were present formerly, although a few individuals have seemingly been offered for sale in markets (Zambrano *et al.*, 2010).

The axolotl is in no danger of global extinction due its popularity in the aquarium trade and as a model organism in scientific research. Captive breeding is straightforward but faces the constraint that the resulting recruits cannot be reintroduced into the highly compromised and degraded habitats in their native range around Mexico City. Population recovery attempts through captive breeding will need to be combined with habitat restoration and improvements in water quality. To address this, local community projects have been launched to revitalize the tradition of farming in wetlands using structures – called *chinampa* – that are essentially floating gardens. The trailing vegetation associated with *chinampa* provide habitat for the salamanders in the adjacent waters, while the cultivated plants take up dissolved nutrients and enhance water quality. An alternative approach would be translocation and introduction of axolotl to some suitable habitat beyond its original range, but, at present, enthusiasm for this option appears limited.

(DD) because there are insufficient data to permit sufficiently accurate assessments of population trends or extinction risk. Apart from mammals and birds, a significant proportion of species in many taxa fall into this grouping, representing up to 23% of all freshwater species assessed. Although a high proportion of amphibians are threatened, nearly one quarter of the more than 6536 species assessed by the IUCN are classified as DD: essentially, those that are too rare for population trends to be assessed adequately. When this 24% is added to the percentage of species known to be threatened (32%) and near-threatened (NT; a further 6%), a total of 62% of *all* amphibians (not only Anura) are either imperilled or too rare to assess. Global assessments for two invertebrate groups – dragonflies and freshwater crabs – suggest that while only 14% of dragonflies are threatened, the proportion rises to 40% if DD species are included (Clausnitzer *et al.*, 2009); recent evaluations (Table 2.2) put these numbers at 10% and 38%. The equivalent values for freshwater crabs are 16% threatened rising to 65% if DD species are included (Cumberlidge *et al.*, 2009), with the proportions changing slightly (to 15% and 77%) according to the latest assessments (IUCN, 2017). Among unionid mussels, 17% are in the DD category. It is probable that a proportion of the DD species in all of these groups are more at risk than the average non-threatened species, because a lack of information about their population trends is due to the fact that they are rare, or have restricted distributions, or both. To the DD species can be added a fraction of unevaluated species (status 'Not Evaluated'), which constitute the majority of animals globally, comprising most of the invertebrates as well as nearly half of the world's fishes.

While IUCN assessments reveal that amphibians are the most imperilled group of vertebrates, almost a quarter of them (1548) fall into the DD category. This fraction is an underestimate of the actual total number of DD species, as the number of amphibians described in recent years has been increasing steadily (see above), and more than 1100 await IUCN assessment. A more complete picture of the extent of global extinction risk of amphibians, or any IUCN-assessed group, could be gained if it were possible to predict which DD species are potentially threatened. Howard and Bickford (2014) made such a projection by modelling extinction risk of DD amphibians by combining global distribution patterns with species-trait data and environmental variables such as habitat loss. The models were trained using data for species classified as threatened by the IUCN. Almost two thirds (63%) of DD species were projected as likely to be at risk (i.e. CR, EN or VU), double the proportion of amphibians formally assessed by IUCN that fell into these categories. Based on this evidence, taxon-specific extinction risks are therefore probably considerably underestimated if DD species are not taken into account.

The results of modelled projections for DD species should not be used as a substitute for formal IUCN assessment, but could inform allocation of scarce conservation resources. Furthermore, any blanket assumption that DD species are not threatened would give rise to an unduly conservative perception of extinction risk in the face of ubiquitous and escalating threats to freshwater biodiversity (Vörösmarty *et al.*, 2010; Hof *et al.*, 2011). Howard and Bickford (2014) have shown that there is a considerable disparity between recognized and unrecognized extinction risk for amphibians. A similar discrepancy would very likely arise if their approach was extrapolated to other under-assessed taxa such as fishes.

If, as stated at the beginning of this chapter, we have already transgressed planetary boundaries for terrestrial and marine biodiversity loss, then – given the relative extent of endangerment of freshwater species in general, as well as frogs and pearly mussels in particular – it seems obvious we have far exceeded whatever margins would have been sustainable for freshwater biodiversity. Indeed, the extent of population declines and losses described above is probably a reliable indicator of the extent to which current practices are unsustainable, and demonstrate how human exploitation and impairment of inland waters have outpaced our best attempts at management (Dudgeon *et al.*, 2006; Ormerod *et al.*, 2010). Extinctions are likely to continue over the next few decades, regardless of actions taken now, resulting from a large unredeemed – and unquantified – extinction 'debt'. This could, for instance, be incurred by habitat fragmentation reducing the viability of populations that are now in the process of dwindling to extinction (Matthews & Marsh-Matthews, 2007; Strayer & Dudgeon, 2010).

Why Are Fresh Waters So Rich in Species?

A few freshwater species have large geographical ranges, but the insular nature of freshwater habitats has led to the evolution of many species with small geographical ranges, sometimes encompassing just a single lake or drainage basin (Strayer, 2006; Strayer & Dudgeon, 2010). This sets the context for generating diversity in fresh waters, and high levels of local endemism and species richness seem typical of several major groups, such as fishes that are almost as rich in fresh waters as they are in the expanses of the world's oceans (Carrete & Wiens, 2012). Among invertebrates, decapod crustaceans, molluscs and aquatic insects such as caddisflies and mayflies are also highly speciose (Balian *et al.*, 2008a,b). The tendency towards restricted ranges and local endemism results in considerable species turnover (= β diversity) between basins or catchments, especially in tropical latitudes that were not affected by glaciation during the last Ice Age (Leprieur *et al.*, 2011). An outstanding example is cichlid diversity in African Rift Valley lakes where hundreds of endemic species are found, with each lake supporting a unique combination of species most of which are found in only one of these large water bodies; a similar pattern is seen among cichlids in many smaller African lakes (Thieme *et al.*, 2005; see also table 3 in Dudgeon *et al.*, 2006). Cichlids are, however, not the only group of animals that have undergone speciation in these lakes; Lake Tanganyika offers an outstanding example of endemic diversity in a range of taxa (Box 2.6).

Local endemism and high species turnover reflect the limited ability of most freshwater species to disperse through terrestrial landscapes or to migrate from river to river along the coast. Moreover, the hierarchical arrangement of riverine habitats means that the populations they harbour are differentially connected to – or isolated from – each other depending on the vagaries of confluence patterns, stream gradients or the presence of natural barriers such as waterfalls. Geographical distance may appropriately reflect the degree of isolation among terrestrial habitats, whereas stream distance, which is defined as the shortest distance between two locations where distance is measured only along the stream network, is often much larger than straight-line distance and better reflects the degree of isolation among stream habitats. Headwater streams tend to be isolated habitats for fully aquatic species, even if they are in adjacent valleys and geographically proximate, because there can be large stream distances between them (e.g. Clarke *et al.*, 2008). Geographical distance is a more appropriate measure of isolation among lakes or other standing-water bodies, but the problem of overland dispersal remains. More generally, it seems obvious that small geographical ranges are associated with a higher extinction risk (e.g. Giam *et al.*, 2011), and this link probably explains why small-bodied freshwater fishes tend to have an increased chance of extinction compared to closely related larger species or to marine fishes of similar size (Olden *et al.*, 2007).

The main point is that because of their high species turnover, freshwater bodies tend not to be 'substitutable' with respect to their faunal assemblages (Revenga *et al.*, 2005; Dudgeon *et al.*, 2006), and this contributes to regional and global species richness. However, it is

Box 2.6 *Hyperdiverse Communities: The Example of Lake Tanganyika*

After Lake Baikal, Tanganyika is the second oldest lake in the world (having existed for around 20 million years), as is manifest in its rich endemic biodiversity. For instance, there are more than 250 species of cichlid fishes, almost all of which are endemic, and while this is fewer than the two other African great lakes, Tanganyika contains more endemic genera and they exhibit striking morphological diversity (Turner *et al.*, 2001; Meyer *et al.*, 2013), including specialized molluscivores and scale eaters. Almost all of these cichlids are benthic, and the majority live along the shoreline, but a number inhabit deep water with some descending to depths of 200 m. Lake Tanganyika cichlids occupy almost all imaginable trophic niches, and many colourful or remarkable species are sought after as aquarium fishes. Among non-cichlids, there has been a remarkable evolutionary radiation of *Mastacembelus* (Mastacembelidae) spiny eels, all but one of the 15 species endemic to Lake Tanganyika. In addition, there are several endemic catfish genera (mainly Claroteidae), a species flock of *Synodontis* catfishes (Mochokidae), and four endemic *Lates* species (Latidae), three of which are either threatened (*Lates angustifrons*, *L. microlepis*; EN) or at risk (*Lates mariae*; VU) from overfishing. As with the cichlids, the richness of all of these fish families is greater than other Rift Valley lakes (Day *et al.*, 2009; Brown *et al.*, 2010), and Lake Tanganyika contains the highest diversity of lacustrine catfishes on Earth (Peart *et al.*, 2014). Among the 10 *Synodontis* species (there is also a single non-endemic), at least one (*S. multipunctata*) is a brood parasite of mouthbrooding cichlids – a cuckoo-like habit once supposed to be confined to birds. The catfish eggs are incubated together with the eggs of the host, but hatch earlier and eat the cichlid eggs. In addition, the endemic *Synodontis* catfishes exhibit Müllerian mimicry (Wright, 2011): all have striking black spots on a paler body, fins with black bases and white margins, coupled with well-developed venom glands and pectoral spines. Piscivorous fishes such as *Lates* conditioned to avoid one species will avoid others, and it is significant that, unlike other Tanganyikan catfishes, *Synodontis* species in the lake are diurnal, presumably ensuring that their distinctive markings are clearly visible to predators.

Caenogastropod snails (mainly Paludomidae, but formerly classified within the Thiadidae) in Lake Tanganyika have radiated to a remarkable extent, numbering 18 endemic genera and approximately 80 species, with a wide range of habitat and substrate preferences. As in Lake Baikal (for details, see Chapter 6), many have a thalassoid (= marine-like) appearance with thickened, sculpted and ornamented shells, and some live only in deep water (Glaubrecht, 2008). Narrow habitat occupancy has put several of these paludomids at risk of extinction due to inshore sedimentation as well as urban and agricultural pollution: examples include *Bathanalia howesi*, *Hirthia littorina*,- *Potadomoides pelseneeri* (EN), *Reymondia tanganyicensis* and *Tanganyicia michelae* (VU). Empty shells of a large viviparid (*Neothauma tanganyicense*) endemic to the lake provide breeding sites for several species of lamprologine cichlids in which either of both sexes are sufficiently small to fit in gastropod shells, or the sexes display extreme size dimorphism with dwarf females (Koblmüller *et al.*, 2007).

Like the snails, microcrustaceans (ostracods and copepods) in Lake Tanganyika are diverse, and it is the only lake known to harbour an evolutionary radiation of decapods. In fact, there are two, involving parathelphusid crabs and atyid shrimps, each made up of about a dozen endemic species. Leeches and sponges in Lake Tanganyika are diverse but poorly known, and there is also an endemic jellyfish (*Limnocnida tanganyicae*: Olindiidae). Among reptiles, a water snake (*Lycodonomorphus bicolor*: Lamprophiidae) occurs nowhere else, and there is a Lake Tanganyika subspecies of the water cobra *Naja* (= *Boulengerina*) *annulata stormsi* (Elapidae); unfortunately, the slender-snouted crocodile (CR) has been virtually extirpated (see also Chapter 3). Together, the fishes and other endemic fauna may number as many as 600 species (Brown *et al.*, 2010), making Lake Tanganyika one of the most important hotspots of freshwater biodiversity on Earth.

There are particular challenges to managing this lake, because the catchment is shared by four countries (Burundi, Democratic Republic of Congo, Tanzania and Zambia) that have rapidly expanding populations, are undergoing high rates of land-use change and lack an integrated fisheries management policy. In addition, as will be seen in Box 7.3, the Lake Tanganyika ecosystem has been strongly influenced by climatic warming.

precisely because of this lack of substitutability that a local extinction event in a particular lake or river drainage can be equivalent to global extinction for that species. Moreover, the hierarchical architecture and/or isolation of fresh waters that contribute to richnesss (through evolution of endemism) also limit the rate at which recolonization proceeds following local extinction events that may be caused by droughts, contaminants and so on (Cook *et al.*, 2007; Lake *et al.*, 2007). Thus, the features generating freshwater biodiversity also contribute to its vulnerability in the face of the many interacting threats posed by human activities, especially individual highly diverse water bodies (such as the three African Rift Valley lakes; see Box 2.6), or particular types of freshwater habitat. For instance, the peatswamp forests of Southeast Asia have been undervalued as a habitat for rare and endangered mammal species, fishes and plants, but it is clear that they play an important role in maintaining regional biodiversity as well as atmospheric carbon balance (see Box 2.7). These habitats warrant conservation attention for a

---

Box 2.7 *Habitat Degradation and Its Significance: The Case of Asian Peatswamp Forest*

Overuse or contamination of fresh water has deleterious effects for humans and biodiversity, as does overexploitation of the aquatic animals – mainly fishes – that many people depend upon as a source of protein and nutrients (see Chapter 3). But there are other good reasons to protect freshwater habitats, as some of them contain major reserves of organic carbon. Soil carbon is vital in regulating climate and water supplies, and affects biodiversity – all of which are essential contributions to the provision of ecosystem services. There is more carbon in the soils of the Earth than there is in the atmosphere, and wetlands hold a disproportionate amount of that carbon. Despite occupying only around 5% of the Earth's land surface, wetlands contain between 20% and 30% of the global soil carbon, with their soils sometimes constituting up to 40% carbon by weight relative to 2% or less typical of agricultural soils (Nahlik & Fennessy, 2016); freshwater wetlands also contain more methane than coastal sites such as saltmarsh or mangrove. Drainage and other alteration of freshwater wetlands affects local hydrology but also leads to the release of carbon dioxide or methane to the atmosphere. Wetlands that are less affected by human activity are more efficient at storing carbon, and intact wetlands have an important influence on the global carbon balance and climatic warming.

Ombrogenous swamps, more commonly known as peatswamps, are one of the most important wetland carbon sinks globally (Page et al., 2011). These forests are usually flooded to depths >1 m during the wet season, but some remain inundated for much of the year. Peatswamps are underlain by acidic nutrient-poor soils, and accumulate layers of organic matter up to 20 m deep. They, and the 'blackwater' streams that drain them, are rich in humic acids, poorly buffered and low in dissolved oxygen, yet support many highly specialized aquatic animals adapted to low pH (Giam et al., 2012; Miettenen et al., 2012). Peatswamps in Southeast Asia represent around 60% of the global extent of this habitat, and ~10% of the world's peatlands. In 1970, they occupied an area of around $250\,000\,km^2$ – two thirds of that in Indonesia. This represented a carbon stock of ~65 Gt (or 65×10^9 t) that could be released when the swamps were cleared or drained and burned (Miettinen et al., 2012). The proportion of the remaining peatswamp forest cover in Peninsular Malaysia, Sumatra and Borneo fell from 77% to 36% between 1990 and 2010. The loss was associated primarily with conversion for oil-palm cultivation (see Box 8.8) and, secondarily, for wood-pulp plantations. At an estimated annual loss rate of 4.9% in 2011, all Asian peatswamps could be cleared by around 2030, with serious consequences for global carbon emissions (Miettinen et al., 2012) – to say nothing of potential consequences for biodiversity. Continued deforestation and conversion into managed land-cover types has meant that, by 2015, only 29% of the original peatswamp forest in Malaysia, Borneo and Sumatra remained (Miettinen et al., 2016). As a result, the region ceased to serve as an efficient carbon sink and became one of the top emitters of greenhouse gases. Cumulative carbon emissions since 1990 have been in the order of 2.5 Gt, mostly from peat oxidation under managed land cover, although peat fires also contributed (Miettinen et al., 2017). Regrettably, conversion of peatswamp for agriculture yields limed benefits: although easy to polderize, when drained, deep peat yields land with unproductive acid-sulphate soils.

The destruction of peatlands will jeopardize the existence of plant and animal species endemic to these habitats, including many fishes, and threaten terrestrial animals that have moved into peatswamp as their dryland habitats have been given over to agriculture (Giam *et al.*, 2012; Wilcove *et al.*, 2013). A limited selection of swamps in Peninsular Malaysia confirmed their richness, uncovering 48 fish species (including 15 stenotopic taxa and five new species) – almost 20% of the national total (Ng *et al.*, 1994). Among them were the obscure Chaudhuriidae (or earthworm eels) and the catfish genus *Encheloclarias* (Clariidae) that are largely confined to peatswamp, as well as a variety of Osphronemidae (some of which are threatened: e.g. *Betta persphone*; CR), and small cyprinids such as the tiny *Paedocypris* spp. If current rates of peatswamp conversion in Southeast Asia continue through 2050, as many as 16 fish species may become globally extinct, but if rates of habitat loss across the region are similar to those in the most rapidly deforested river basins, 77% (79 of 102 species) of stenotopic fish species could be driven to extinct, mostly in Borneo (Giam *et al.*, 2012). This figure would substantially increase the number of recorded extinctions of freshwater fishes globally. More generally, of more than 1000 threatened aquatic and terrestrial species in the tropics, 85% rely on carbon-rich alluvial or wetland habitats – a tendency observed in plants, mammals, reptiles, amphibians and crustaceans (Shiel *et al.*, 2016). Protecting peatswamp would both safeguard the biota and curtail atmospheric carbon emissions, suggesting that opportunities for biological conservation could be enhanced by protecting wetlands where the carbon content of soils is highest.

It is not only highly diverse peat-associated ecosystems in Asia that are at risk from anthropogenic change, but also their more depauperate European counterparts. Simulations suggest that, by 2051–2080, climate change and summer droughts could result in drying and almost complete loss of peat (blanket) bogs from the United Kingdom. These north-temperate ecosystems are defined by extensive deep peat soils over rolling terrain, and are associated with cool, wet climates and vegetation of low stature. Falling water tables will lead to reductions in abundance of tipulid larvae – the dominant invertebrates in blanket bogs – by as much as 80%. There would be knock-on effects on birds such as dunlin (*Calidris alpina*: Scolopacidae; up to 50% decline predicted) and European golden plover (*Pluvialis apricaria*: Charadriidae; 30% decline), which breed in these peatlands and depend on tipulids as food for chicks and adults (Carroll *et al.*, 2015). Reductions in food could drive declines in these ecosystem-specialist birds, increasing their risk of local extinction as the climate changes.

number of reasons: uniqueness in Asia; habitat for endemic or stenotopic species; contracting coverage and declining condition of remaining extent; high degree of threat; and extreme fragility. They are good candidates for designation as priority habitats of global conservation significance. Such recognition could be leveraged on the propensity for anthropogenically modified peatswamps to become nett emitters of carbon rather than serving as long-term sinks.

3 · *Overexploitation*

I began to wonder whether, as each fish died, the world was reduced in the amount of love that you might know for such a creature. Whether there was that much less beauty & wonder left to go around as each fish was hauled up in the net. And, if we kept on taking & plundering & killing, if the world kept on becoming more impoverished of love & wonder & beauty in consequence, what, in the end would be left? ... I imagined a world of the future as a barren sameness in which everyone had gorged so much fish that no more remained, & where Science knew absolutely every species & phylum & genus, but no-one knew love because it had disappeared along with the fish.

Richard Flanagan (2001: pp. 200–201), *Gould's Book of Fish*

The (Columbia white) sturgeon can grow twenty feet long, live for more than a hundred years, and weigh up to 1,800 lb. It has small myopic eyes, the streamlined build of a pike or a U-boat, with a tapering snout from which trail four whiskery barbules, and hose-like lips which it extends, concertina style, to savour likely morsels. . . . Pictures of trophy sturgeon, taken before strict size limits came into force, make their grinning captors look as if they should have been the fish's lunch.

Jonathan Raban (2010: p. 503) *Driving Home*

What and When?

Overexploitation ranks second in overall importance as a threat to freshwater biodiversity (Table 2.1). It chiefly affects vertebrates, mainly fishes, as well as reptiles, amphibians and a few mammals and birds, but some invertebrates are also subject to human depredation.

Evidence of large-scale early exploitation of freshwater fishes comes from excavation of a 9200-year-old Early Mesolithic settlement in Sweden, where there were signs of fermentation and storage of more than 60 t of fish. The catch was dominated by cyprinids, chiefly roach (*Rutilus rutilus*), although Eurasian perch (*Perca fluviatilis*: Percidae) and northern pike (*Esox lucius*: Esocidae) were also exploited, as were European eel (*Anguilla anguilla*), European smelt (*Osmerus eperlanus*: Osmeridae) and salmonids (Boethius, 2016). These findings imply a greater societal complexity and higher degree of sedentism than the foraging lifestyle assumed to be typical of the Mesolithic, and may be somewhat comparable to the Neolithic revolution in the Middle East that took place at roughly the same time. Whereas people in the latter group domesticated animals and harvested crops, the northern foragers exploited rivers and lakes; their reliance upon fishes allowed the development of sizeable communities with sophisticated lifestyles (Boethius, 2016). It would also have set the scene for overfishing. In Europe, at least, most of the fishes consumed up until the twelfth century came from inland waters. It was only in late medieval times that preservation technologies advanced sufficiently to permit transportation and consumption of large quantities of marine fishes any great distance from the coast. Prior to that, growing human populations placed ever-greater demands on freshwater fish stocks at the same time that deforestation and agricultural and urban expansion were degrading fish habitats and limiting yields (Hoffman, 2001).

Irrespective of when it first became apparent, there were signs of overfishing in Europe by around 1000 AD, when reductions in mean size and abundance of exploited species became evident. This occurred in parallel with stock declines attributed to a combination of siltation from intensive agriculture, increased nutrient loads, pollution from tanneries and other sources, proliferation of mill dams, and introduction of non-native species (Barrett *et al.*, 2004; Hoffman, 2005). Allowing for some differences in scale or intensity of action, these are much the same factors that threaten freshwater biodiversity today. The obvious degradation of European fresh waters gave rise to attempts to regulate or control fish catches, some originating almost 1000 years ago. These were intended initially for Atlantic salmon (*Salmo salar*: Salmonidae) (Plate 4), which proved to be extremely vulnerable to overexploitation during spawning runs when nets could be deployed across the entire width of natal rivers. Upstream migrations were also impeded by mill dams, and susceptibility to pollution reduced recruitment success in rivers that flowed through cities or areas of intensive agriculture. Early attempts to limit salmon catches in England and Scotland involved royal proclamations by the countries' respective kings dating back to the eleventh century, and later regulations that imposed closed seasons for fishing. Large-scale commercial fishing did not become a significant threat to Atlantic salmon until the eighteenth century when preservation and transport methods improved to an extent that, by 1820, fish consumption in growing urban centres had depleted many rivers across the United Kingdom. This trade, overfishing of large older fish, and exploitation of individuals that had never bred, in combination with the effects of water pollution from factories built during the Industrial Revolution, almost eliminated salmon from English rivers including the Thames and Mersey between the mid-1700s and 1850. Salmon had likewise all but disappeared from French rivers by the middle of the nineteenth century, but salmon runs into the continental interior of Germany continued – albeit diminished – until a *coup de grace* administered by twentieth-century water pollution (for details, see Montgomery, 2003). Only Scotland, Ireland, Iceland and Norway now retain substantial populations of Atlantic salmon. Catches from rivers along the Atlantic seaboard of North America were in decline by the 1850s. A century later there was little commercial capture of salmon, and wild *S. salar* has since declined to historic lows in North America (Limburg & Waldman, 2009). This is a tragic fate for the 'King of Fish' – a title Izaak Walton (*The Compleat Angler*, 1653) bestowed on the Atlantic salmon. Yet it is not atypical: instances of equally catastrophic declines of freshwater fishes in other parts of the world are provided later in this chapter.

Historical losses of salmon and diadromous fishes from North American and European rivers, as well as reductions in some other exploited freshwater animals, would have had profound ecological effects (e.g. Humphries & Winemiller, 2009); some of the consequences of such overexploitation will be described below (see also Box 3.10). The range of species that have been targets of consumptive use by humans is wide: it includes (as documented later in this chapter) many – non-piscine – vertebrates, principally reptiles and amphibians, well as mammals. They are fished, trapped or otherwise collected for food, but some – especially herpetofauna – are consumed as medicines or tonics, while crocodiles and mammals are hunted for their skins. The case of beavers is considered later in this chapter, but various otters (Mustelidae) – mainly the Eurasian otter (*Lutra lutra*; NT), the smooth-coater otter (*Lutrogale perspicillata*; VU) in Asia, as well as the North American river otter (*Lontra canadensis*) – have likewise been killed for their pelts, as well as for 'sport', or due to the belief that their feeding activities reduce fish catches. Neotropical otters have been extensively exploited also, as described below.

Waterbirds are subject to human depredation by shooting (especially ducks and geese), snaring or netting, either as subsistence foods or –

mainly in Europe and North America – for recreation (the practice of culling nuisance geese that damage crops is beyond the scope of this book). Growth rates of waterbird populations in Europe tend to be negatively affected if they are hunted (Jiguet *et al.*, 2012), and migrants can be susceptible to unrestricted exploitation in different parts of their ranges (e.g. Madsen *et al.*, 2015). Colonial breeders and waterbirds that nest on the ground (e.g. on sand bars in rivers) are vulnerable to people collecting eggs and chicks for food (Thewlis *et al.*, 1998), and in areas such as Dobruja (which includes the wetlands of the Danube delta) in Eastern Europe, the late nineteenth century was marked by great reductions in a wide variety of waterbirds due to unrestricted hunting of adults and overcollection of eggs. The Danube Delta is now a UNESCO Biosphere Reserve, Ramsar, and World Heritage Site. In many other parts of the world, loss and degradation of wetland habitat, rather than overexploitation, are primary causes of continuing waterbird decline. Nonetheless, where waterbirds are intensively hunted for subsistence or as an economic activity – both of which tend to target larger individuals – outcomes can include substantial declines in species richness and a shift in assemblage structure towards smaller species (Ramachandran *et al.*, 2017). As described below, a reduction in mean body size of overexploited stocks also occurs in animals as diverse as fishes and crocodilians, and this 'great shrinking' is a typical fingerprint of Anthropocene threats to freshwater biodiversity.

Among invertebrates, pearly mussels were subject to high levels of utilization for their shells during the last century (see Box 2.2), palaemonid shrimps (*Macrobrachium* spp.) support artisanal fisheries, and a few other taxa are subject to local exploitation as subsistence food or, even, delicacies. Invertebrates are, increasingly, collected from the wild and traded internationally as aquarium ornamentals, but have yet to rival fishes as the main focus of that industry. Mention should be made also of human exploitation of freshwater plants. Among them, arrowheads (*Sagittaria* spp.: Alismataceae), water-lilies (species of *Nymphaea* and *Nuphar*: Nymphaeaceae) and lotus (*Nelumbo nucifer*: Nelumbonaceae) form large edible tubers that are a common source of food in many parts of the world. Others such as watercress (*Nasturtium* [= *Rorippa*] *officinale*: Brassicaceae) and hornwort (*Ceratophyllum demersum*: Ceratophyllaceae) have been used for medicinal purposes. Such consumption does not represent a conservation threat to freshwater plants and, anyway, would likely be far less influential than habitat degradation. The Thai water onion (*Crinum thaianum*: Amaryllidaceae; EN) is an exception to this generalization: excessive collection as an ornamental plant for aquaria has substantially diminished its narrow geographical distribution.

How Large Is the Global Freshwater Capture Fishery?

Taking account of fishes only, and leaving aside aquaculture, the Food and Agriculture Organization of the United Nations reported total landings from inland waters of 11.9 million t in 2014 (FAO, 2016). This is 13% of the global capture fishery of 93.4 million t. Asia accounted for two thirds of the freshwater landings, with China, Myanmar, India, Bangladesh and Cambodia representing the top five ranked countries in terms of landings, while Africa (with Uganda ranked sixth) contributed around a quarter. Despite the presence of the world's largest river, South America accounted for only 5% of the total yield, and catches in Brazil have declined recently, apparently as a result of overfishing (FAO, 2016). Catches in Tanzania, ranked eighth globally, have fallen also.

The values of the absolute magnitude of freshwater landings are certainly underestimates, as much of the catch does not pass through formal markets, and subsistence fishing is not included (Beard *et al.*, 2011; Fluet-Chouinard *et al.*, 2018). Moreover, fishing can be either a full-time occupation or an informal activity, often highly diffuse and varying seasonally, which compounds the difficulty of obtaining accurate catch statistics. That data challenge needs to be placed into context. Freshwater fisheries are complex in terms of the variety of species involved, the many factors affecting population fluctuations of fishes, the range of capture gear employed, and the dispersed nature of the activity. The number of people involved is also substantial. At least 21 million people are engaged regularly in freshwater fisheries (FAO, 2014), over a third of the global total for capture fisheries, and most are small-scale operators concentrated in Asia (>84% of freshwater fishers) with Africa ranking second (~13%). Almost one third of the global fishing fleet operates in fresh waters, but most vessels are small.

The FAO have repeatedly emphasized (e.g. FAO, 2012, 2014, 2016) that their estimates of freshwater fish landings are subject to underreporting and certainly far lower than actual catches, and '... 'inland waters' remains the most difficult subsector for which to obtain reliable capture statistics' (FAO, 2014: p. 17). There has been a gradual increase in reported landings, which have risen 18% (from 10.1 million t) in 2007, and this trend is mainly attributable to Asian countries. It is very likely that '... inland waters are being overfished in many parts of the world . . .', and in countries that report high landings, such as China, '. . . a good proportion of inland catches comes from water bodies that are artificially restocked' (FAO, 2012: p. 8). In contrast to the contribution made by aquaculture, the natural production of freshwater fishes seems likely to be decreasing (Youn et al., 2014), a trend that is undoubtedly exacerbated by widespread deterioration in the condition of freshwater habitats (Vörösmarty et al., 2010). Unfortunately, '. . . without information on the status of the fish populations, it is difficult to manage such fisheries towards sustainability' (FAO, 2016: p. 45; see also Beard et al., 2011). To gauge the extent of underreporting of landings, Fluet-Chouinard et al. (2018) used household consumption surveys to estimate freshwater fish catches in 42 low- and middle income nations over an eight-year period. Collective consumption of wild-caught fishes was far greater than officially reported, and an extrapolation from the data collected suggested that the global freshwater catch of 10.3 million t in 2008 was more likely to have been 16.6 million t. If this estimate of underreporting is correct, it complicates use of catch statistics to determine the sustainability of exploitation of freshwater fisheries.

Analysis of long-term trends in freshwater landings that might uncover evidence of overexploitation at the global scale is not feasible because the FAO did not separate statistics on freshwater and marine fishes prior to the 1990s. There have been some recent improvements in data collection, but accounting typically does not occur at the level of individual species or particular fisheries. Until comprehensive data on freshwater fisheries become available, their contribution to human livelihoods through the economic and nutritional benefits that they bring will be undervalued, as is now plainly the case (Beard et al., 2011; Hall et al., 2013). This oversight is of critical concern because wild-caught fishes have been reported to be a rich source of protein and calcium that are crucial to human health (Youn et al., 2014), and in the Lower Mekong Basin, for example, fishes and other aquatic animals account for more than half of the animal protein consumed (see Box 3.6). Globally, freshwater fisheries have been estimated to provide the equivalent of all dietary animal protein for 158 million people, with poor and undernourished populations particularly reliant on inland capture fisheries compared with marine or aquaculture sources (McIntyre et al., 2016). Underreported landings alone may represent the equivalent of the total annual animal protein consumption of as many as 36.9 million people, contributing far more to food security than has been recognized previously (Fluet-Chouinard et al., 2018). Furthermore, a spatial coincidence between productive freshwater fisheries and low food security highlights the crucial role of rivers and lakes in providing locally sourced, low-cost protein. Put simply, there are many people in the world who do not eat much animal protein, and what they do eat is mostly freshwater fish.

Small freshwater fishes are also the primary source of lipids and essential micronutrients (calcium, iron, zinc, vitamin A) in many parts of the developing world (Kawarazuka & Béné, 2011; Hall et al., 2013). This fact is not generally recognized by decision makers who encourage habitat modifications (particularly dams for hydropower and water diversions for human use) that reduce fisheries productivity and undermine food security at local and regional levels. Wild fishes are more nutritious (higher protein and micronutrient content) than farmed individuals, even within the same species; thus a switch to consumption of farmed fish as wild catches decline (assuming that were practicable) would result in poorer diets (Youn et al., 2014). And not all fish species are equally substitutable: for instance, large carp are a good source of omega-3 fatty acids but not calcium, which is more readily obtained from smaller fish species with bones that can be chewed up and swallowed. The subject of human nutrition is beyond the scope of this book, but the need to ensure that freshwater capture fisheries are fully taken into account in decisions about water-resource management will require that their contribution to food security is reliably assessed, valued and communicated.

In addition to the many people who engage in freshwater fisheries to meet their livelihood needs, numerous others pursue recreational fishing, typically with a rod and line. Their captures may amount to as much as 12% of the total global yield of fresh- and saltwater fishes (FAO, 2012). The proportion of overall catch constituted by the recreational take will be far higher from inland waters than from the seas and might add 2 million t to the 11.9 million-t annual yield of freshwater capture fisheries. There is, however, no reliable estimate of the total magnitude of recreational catch. With increasing economic and industrial development, the importance of commercial freshwater fishing tends to decrease. Thus inland-water bodies in Europe and North America, where there are an estimated 85 million anglers, are increasingly or exclusively used by recreational fishers (FAO, 2012), with around 10% of the population of the industrialized world participating in recreational fisheries (Cowx et al., 2010). Such fisheries can be valuable: two river basins in Idaho and Wyoming with high-quality catch-and-release trout fisheries generated US$12–29 million in annual income and 341–851 jobs (Hughes, 2015).

Recreational fishing effort is generally highly targeted, typically species- and size-selective, and thus may modify population structure and assemblage composition. Potential impacts of recreational fishers can, however, be offset by their involvement in projects directed towards conservation of exploited species, especially where these are rare and charismatic. One example is replacement of an extractive sport fishery for taimen in Mongolia (see Box 2.3) by catch-and-release ecotourism; the income generated is used to incentivize local communities to take part in habitat protection and fish monitoring (for details, see Granek *et al.*, 2008).

As with recreational fishing, the capture of wild fishes and other animals for the international aquarium trade can also have local impacts on stocks of targeted species (see Box 3.1). It can, however, be a driver of economic development. Moreover, people who make a living collecting fish for sale to the aquarium trade have an interest in contributing to habitat preservation and likely avoid participation in occupations that damage the environment. There is also scope for involvement of aquarists in conservation efforts, as demonstrated by the establishment, in 2015, of a subgroup of the IUCN Freshwater Fish Specialist Group – the Home Aquarium Fish Subgroup – dedicated to conservation and wise management of wild populations of tropical fishes that are part of the aquarium trade (www.iucnffsg.org/about-ffsg/home-aquarium-fish-sub-group/).

Evidence of Overfishing: Effects on Anadromous Species Are Particularly Manifest

Despite the inadequate data base of landings and fishing effort over time, there are some evident trends in fish stocks, especially migratory fishes and diadromous species that travel between fresh waters and the sea. Among them, anadromous salmonids, sturgeons and clupeids have declined by as much as 98% from former levels of abundance in rivers on both sides of the Atlantic (Limburg & Waldman, 2009), and the fate of the Atlantic salmon in Europe – as described above – is emblematic of those declines. During the late nineteenth century, commercial fishing fuelled a booming export trade of canned Pacific salmon (*Oncorhynchus* spp.) to Europe and elsewhere; extensive use of gill nets, seines and fish wheels on natal rivers rapidly depleted populations. While such mass-catch fishing methods were outlawed in the early twentieth century, dam

Box 3.1 *The Fishery for Ornamental Species*

The global trade in ornamental aquarium fishes is a multi-billion-dollar industry, and while most animals traded are now captive bred in farms, some species are wild caught. A wide range of source countries in the tropics is involved. Aquarium fisheries have historically not been well studied and, although field data are limited, there is evidence that overexploitation can deplete stocks of species that are particularly sought-after. For instance, the bala shark (*Balantiocheilos melanopterus*: Cyprinidae; EN) was exploited to near-extinction in Sumatra and Borneo during the 1970s, and all individuals in the aquarium trade are now captive bred. (Although the causes are unclear, the related Indochinese species, *B. ambusticauda*, seems to have become extinct.) The Asian arowana (*Scleropages formosus*: Osteoglossidae; EN) (Plate 5) is the subject of a targeted fishery in Cambodia, even though commercial trade in the species is prohibited under Appendix I of CITES (the Convention on International Trade in Endangered Species), with high levels of exploitation driving some populations to local extinction (Rowley *et al.*, 2008). Collection of large mouthbrooding males, in order to obtain the offspring that they harbour, is a particularly unsustainable practice that has driven the species' decline. A growing legal trade of captive-bred and micro-chipped Asian arowana, combined with better enforcement of illegal practices, may help reduce exploitation of wild populations.

In India, at least 30 endemic species categorized as threatened in the IUCN Red List are in trade (Raghavan *et al.*, 2013), the most numerous being *Botia striata* (Botiidae; EN), *Carinotetraodon travancoricus* (Tetraodontidae; VU), *Puntius denisonii* and *P. chalakkudiensis* (Cyprinidae; both EN). Despite the fact that threatened species of conservation concern are routinely exported, India has yet to frame national legislation on freshwater aquarium trade. Unregulated collection of some endemic fishes, at least in some countries, is probably a more severe threat to biodiversity than hitherto recognized.

The Amazon supports a large fishery for ornamental species. This includes juveniles of the silver arowana (*Osteoglossum bicirrhosum*), which are collected in part as a substitute for its rarer Asian counterpart; the adults are an important food fish. There are signs that in areas where the ornamental fishery is most developed, overexploitation has depleted wild populations (Gerstner *et al.*, 2006) leading to, for example, regulation of the exports of freshwater stingrays

(Potamotrygonidae) from Brazil and Colombia; these fishes are now CITES Appendix-II listed. Controlled exploitation of ornamental species could be permitted in order to provide indigenous communities with economic opportunities and incentives for habitat preservation, and a managed fishery centred around the middle Rio Negro in Brazil and largely based on the cardinal tetra (*Paracheirodon axelrodi*: Characidae) has existed since the 1950s. Mass production of captive-bred tetras in Southeast Asia has led to a waning of the wild-capture of some ornamental species, and the loss of income has been only partly compensated by an increase in tourism-based sport fishing (Zehev et al., 2015). Nonetheless, the persistence of the fishery for ornamental species in some localities shows that the presence of marketable species can provide effective incentives for Amazonian communities to avoid practices that degrade the environment upon which the fishes depend, resulting in protection of the target species and the ecosystem. Arguably, then, the aquarium trade can sometimes be an effective instrument for poverty alleviation and preservation of remaining areas of biological importance that support endangered species. The introduction of an international certification programme for wild-caught aquarium fishes from sustainable sources could incentivize conservation efforts by those involved in the trade. There is also scope for involving hobbyists and public aquaria in *ex situ* conservation efforts for threatened fish species at national and international levels (Koldewey et al., 2013; see also Chapter 9).

It is not only fishes that are exploited for aquaria: there is a well-established ornamental trade in amphibians and turtles, and exports of freshwater molluscs and decapods (especially crayfishes) have been growing. Almost 60 mollusc species were recorded in a survey of dealers in Singapore, mainly snails in the families Neritidae, Thiaridae, Pachychilidae and Viviparidae, and a few pearly mussels (Ng et al., 2016). The most expensive species traded were those with the most restricted distribution, such as *Tylomelania* spp. and *Celetaia persculpta* (Pachychilidae) endemic to Lake Poso, Sulawesi. The rarity of these species, and some large colourful crayfishes (e.g. Lukhaup, 2015), tends to increase demand for them, driving further exploitation and population decline (Courchamp et al., 2006). This process, which is already plainly evident for freshwater turtles, has begun to impact Asian salamanders (as discussed in the context of overexploitation of Old-World herpetofauna).

construction along Pacific-coast rivers posed new challenges for salmon, and many runs failed to recover. Overfishing caused long-term declines in the Pacific salmon that migrated into the Columbia River, with annual landings plummeting from 20 million kilogrammes in the late nineteenth century to around half that level by the 1930s. Salmon must now contend with 14 mainstream dams that are among 31 multipurpose dams constituting the Federal Columbia River Power System and, despite the provision of fishways, landings are now less than 5% of peak catches. Salmon have disappeared from over one third of their former range in California and the Pacific Northwest, and stocks in the latter are probably less than 10% of pre-1850 levels. On the eastern side of the Pacific, overexploitation of migratory fishes is just as apparent where anadromous subspecies of the cherry or masu salmon (*O. masou*) in Japan have undergone marked declines with only a few remnant runs in Hokkaido. Montgomery (2003) has written an informative synthesis of the history of salmon overexploitation, including the particular case of the Columbia River, which provides considerable further detail.

While declines in the magnitude and geographical extent of salmon runs have ecological ramifications, which will be described together with the effects of overfishing on ecosystem functioning, they represent a reduction in a valuable ecosystem service: provision of food and animal protein for humans. This role is common to many anadromous fishes, arising from their (historic) vast abundances, the seasonal predictability of spawning migrations, and their ease of capture when swimming upstream. They thus had considerable importance for indigenous and (later) nonindigenous people in North America and elsewhere. Over-exploitation of runs that significantly reduced yields in subsequent years could cause hardship and food shortages. Thus, for example, Native American tribes in the Pacific North West (as well as equivalent peoples in the Eastern Pacific) observed rituals intended to ensure the perennial return of salmon. To the East, Native American settlements were usually located along rivers, and salmon was a major dietary item. Atlantic salmon also featured in Celtic and Norse mythology and are included in ancient cave paintings at Altamira in northern Spain. These, and many other examples (for details see Montgomery, 2003), attest to the import-ance of the ecosystem service provided by anadromous fishes, particularly salmon.

Of other anadromous species subject to overexploitation, the Yangtze (or obscure) pufferfish (*Takifugu fasciatus*: Tetraodontidae) sustained a commercially important fishery amounting to hundreds of tonnes annually along the Yangtze River until the 1960s but then dwindled to almost nothing (Turvey *et al.*, 2010). Also in the Yangtze, Reeve's shad (*Tenualosa reevesii*) was one of the most economically valuable species caught in the river; stocks of this clupeid collapsed in 1975 following after a gradual decline in the mean size of individuals captured due to the use of excessively fine-meshed nets (Blaber *et al.*, 2003; Wang, 2003). Reeve's shad no longer exists as a commercial species in the Yangtze and it is classified as nationally endangered in the *China Red Data Book*, with capture of juveniles prohibited (Wang *et al.*, 1998). As in the Yangtze, populations of Reeve's shad in the Zhujiang (Pearl River) once supported a lucrative fishery, but it declined by around 80% between the 1950s and early 1980s, and appears to have vanished subsequently (Wang, 2003); no shad at all were encountered during a recent three-year survey of fish larvae in the Zhujiang (Tan *et al.*, 2010). While the primary driver of collapse of these fisheries was overexploitation, this would not have been the sole cause, as described in Box 3.2.

In addition to Reeves shad in China, all other species of *Tenualosa* in Asia are highly susceptible to overfishing, a vulnerability that is amplified by dam construction and pollution of many rivers that they ascend to spawn. Thus, the demise of the overfished but once important hilsa shad fishery of the Ganges took place after completion of the Farakka Barrage blocked their migration route (Box 3.3; see also Chapter 1). Remnant stocks of hilsa in downstream Bangladesh and elsewhere are highly overfished. The longtail shad (*T. macrura*) and toli shad (*T. toli*) have declined also, while populations of the endemic Mekong or Laotian shad (*T. thibaudaeui*; VU), which is secondarily potamodromous rather than anadromous, have been greatly depleted by overfishing and are further threatened by dam construction (Blaber *et al.*, 2003; Vidthayanon, 2013). Similar – and sometimes dramatic – population declines of other anadromous clupeids, including the alewife (*Alosa pseudoharengus*) in eastern North America and the allis shad (*A. alosa*) in Europe, have also been attributed to overfishing (Limburg & Waldman, 2009). In addition, the eulachon (*Thaleichthys pacificus*: Osmeridae), which once migrated in vast numbers into streams of the Pacific Northwest, were subject to intensive commercial fishing, especially in the Columbia and Fraser Rivers. Catches of these smelt declined in the Fraser during the 1940s and 1950s, when the southernmost Californian euchalon populations were extirpated (NatureServe, 2013). Stocks in the Columbia and Fraser crashed in the 1990s, leading to the subsequent closure of the commercial fisheries, although subsistence fisheries for smelt continue.

Box 3.2 *Decline and Fall: The Yangtze Fishery*

The Yangtze River offers a model example of how overfishing can combine with other anthropogenic stressors and cause fisheries to collapse. It is China's longest and largest river, with a catchment of 1.8 million km^2 inhabited by more than 400 million people; unsurprisingly, much of the landscape is human dominated and the river – particularly the floodplain reach downstream of the giant Three Gorges Dam – is highly polluted. The channel is navigable for more than 2000 km, with commercial boat traffic numbering more than 200 000 large vessels, and intense human modification of the river through sand mining, levees, channelization and so on (for a summary, see Dudgeon, 2010) has degraded fish habitat and resulted in the extinction of the baiji (see Box 2.1).

The Yangtze fishery peaked in 1954 when the annual yield was at least 450 000 t, but catches from the river fell by half between 1950 and 1970, and declined subsequently to *c.* 130 000 t in 2000 (Chen *et al.*, 2004, 2009). Almost all (97%) of this yield came from the Yangtze below the Three Gorges Dam, and a considerable amount was derived from floodplain lakes (Fang *et al.*, 2006). The main fishery species in this part of the Yangtze are major carp: i.e. black carp (*Mylopharyngodon piceus*), grass carp (*Ctenopharyngodon idella*), silver carp (*Hypophthalmichthys* [= *Aristichthys*] *molitrix*) and bighead carp (*H. nobilis*). A considerable portion of the catch of these large cyprinids depends upon the well-established practice of stocking the river and floodplain lakes with cultured fry (Fu *et al.*, 2003). Despite this, yields of major carp, which constituted 80% of the overall catch in the 1980s, fell to only 35% of the total during the 1990s, with reductions from 90% in the 1960s to no more than 5% 30 years later in some localities (Chen *et al.*, 2004; Fang *et al.*, 2006). Annual catches of major carp downstream of the Three Gorges Dam declined by a further 50% after 2003 when the impoundment began to fill, and drifting eggs and larvae were reduced markedly (Xie *et al.*, 2007). This is a clear demonstration of the synergistic relationship between overfishing and anthropogenic modification of rivers. Much the same trend – driven by similar factors – has been seen in the Zhujiang (Pearl River) in southern China, where major carp made up almost 50% of fish larvae during the early 1980s, but scarcely a tenth of that 25 years later (Tan *et al.*, 2010). During the same period *Luciobrama microcephalus*

(DD), the sole representative of a monotypic genus of predatory carp, entirely disappeared from Zhujiang.

Among Yangtze fishes affected by overexploitation, declines of large-bodied species are especially notable. The fishery of both highly threatened sturgeon species ceased to be viable more than 30 years ago, although some capture of hatchery-reared and stocked Chinese sturgeon (*Acipenser sinensis*; CR) continued until recently (Plate 6). It has entirely disappeared from the Zhujiang River (Tan *et al.*, 2010), which historically supported a fishery for the Chinese sturgeon. Catches of this huge fish (up to 3 m in length) in the Yangtze were of greater magnitude than endemic Yangtze sturgeon, *A. dabryanus* (CR: Plate 7), which no longer seems to have a viable breeding population (Wei, 2010). Both sturgeons are now legally protected from fishing, as is the even larger Chinese paddlefish (*Psephurus gladius*: Polyodontidae; CR; Plate 8). The paddlefish, which was confined to the Yangtze, may well be extinct as there are no reports of any captures since 2003. It has not been caught in economically significant numbers for decades, the fishery having collapsed prior to the 1970s (Zhang *et al.*, 2009; Turvey *et al.*, 2010).

The Chinese sucker (*Myxocyprinus asiaticus*), another Yangtze endemic and the only catostomid found in the Old World, formerly contributed >10% of the catch in sections of the Yangtze above the Three Gorges Dam. Numbers have dwindled to such an extent that the Chinese sucker is now a protected species. As with many large riverine fishes, susceptibility to overfishing can be attributed to a combination of large body size (almost 1 m total length), long life (~25 years) and late maturity (requiring at least six years of growth), plus an obligatory breeding migration (Gao *et al.*, 2008). Smaller Yangtze fishes that have suffered from overexploitation include the Chinese longsnout catfish (*Leiocassis longirostris*: Bagridae) which, prized for its swim bladder, has almost disappeared from markets (Chen *et al.*, 2009). The anadromous Yangtze pufferfish and Reeve's shad (see main text) have likewise vanished as commercially valuable species. Overexploitation of these species has been facilitated by illegal fishing methods such as poisons, electricity and explosives, as well as excessively fine-meshed nets.

Box 3.3 *Multiple Stressors Exacerbate the Effects of Overfishing along the Ganges*

It is seldom the case that declines in fish stocks result from over-exploitation alone, or even in combination with just one other stressor (see also Chapter 2). Instead, as in the Yangtze and Zhujiang (Box 3.2), multiple factors acting in consort have reduced landings and changed the composition of catches from the Ganges. Overfishing, dam construction, channelization, water extraction, gross pollution from multiple sources, sand mining for construction and siltation have degraded the riverine environment, and allowed alien species to proliferate.

At Allahabad in the Ganges middle section, the composition of catches in the 1960s was dominated by large-bodied major carp (*Cirrhinus mrigala, Catla* [= *Gibelion*] *catla, Labeo calbasu* and *L. rohita*) in the 1960s, with large bagrid (*Aorichthys aor, A. seenghala*) and silurid catfishes (mainly *Wallago attu*) ranking second. Four decades later, yields had fallen by almost two thirds (Fig. 3.1), and 20% of catches

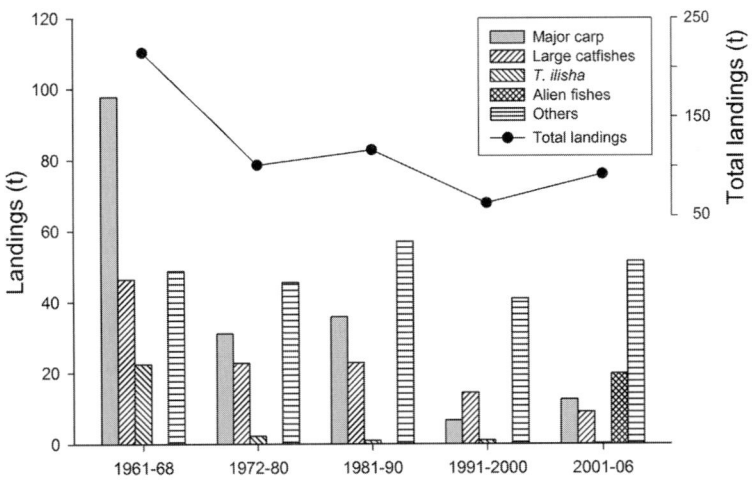

Fig. 3.1 Statistics from the Central Inland Fisheries Research Institute, Barrackpore (compiled by Vass *et al.*, 2010), show annual landings (t) of fishes from the Ganges at Allahabad. There is a long-term trend towards reductions in major (large-bodied) carp and catfishes, and a consequent decline in total landings, as well as a steep and sudden collapse of the hilsa shad fishery during the 1970s. A slight increase in total landings after 2000 was mainly attributable to the establishment of alien species, predominantly common carp.

consisted of alien species, principally common carp as well as Nile tilapia (see Chapter 4). Hilsa shad declined to only 1% of their previous abundance, in part attributable to completion of the Farakka Barrage in 1975, with annual landings at Patna, downstream from Allahabad reduced from 14 t in the 1960s to less than 0.1 t between 1986 and 1993 (Vass et al., 2010). Data on landings from Bhagalpur, closer to the mouth of the Ganges, are less comprehensive but reveal a similar pattern, with landings of major carp in the 1980s no more than a third of those in the 20 years earlier but smaller reductions in the yield of large catfishes. Representation of major carp in catches at Patna fell by 70% over the same period, and catches of catfishes halved.

Even further downstream in Bangladesh, overexploitation of open-access fisheries and the effects of dams and flow regulation that limits access to floodplains have led to a substantial long-term decline in landings of river fishes and a reduction in per-capita fish consumption (Craig et al., 2004). This is a substantial livelihood concern, not least because the country has an estimated 12 million people involved in fishing (Hossain et al., 2006). Conflicts over water (especially during the dry season) and attempts to control floods will likely further increase the extent of water engineering in Bangladesh to the detriment of capture fisheries.

While the Chinese paddlefish and Yangtze sturgeon may now be extinct, or almost so, elsewhere at least one migratory fish was driven to extinction almost 90 years ago as a result of overexploitation. The diadromous (apparently anadromous) upokororo or New Zealand grayling (*Prototroctes oxyrhynchus*: Retropinnidae) ascended rivers in vast shoals up until the nineteenth century. This smelt-like fish was caught by the Maori people, using a variety of techniques, and both the flesh and roe were eaten. Upokororo was still abundant at the time of European settlement of New Zealand in the 1860s, when it became subject to intense exploitation. Declines began within a decade and it soon became scarce; by the 1920s, upokororo were confined to a few rivers in remote areas (West et al., 2014). Habitat degradation and introduced brown trout probably contributed to the eventual demise of upokororo, which was last reported in 1930, but overfishing was the initial and primary driver of its extinction.

Overfishing has marked impacts on diadromous species in general, but is particularly detrimental to anadromous acipenserids, with 21 out of 25 sturgeon species classified as globally threatened (Jacoby *et al.*, 2015). Catadromous fishes, and especially angullids, are also highly vulnerable. Overexploitation was the main threat to the now critically endangered European eel, with immature and post-larval glass eels representing the major target. They were exported and reared to marketable size in aquaculture facilities in East Asia, until the trade was banned in 2010. Greater protection of temperate eels has, however, led to the appearance of glass eel fisheries for tropical *Anguilla* spp. in parts of Africa and Southeast Asia (Jacoby *et al.*, 2105). Notwithstanding the primary role of overexploitation in driving population declines of diadromous migrants, most are at risk from the combined effects of fishing, dams and habitat degradation, and – as mentioned above – fishery impacts on most freshwater species likewise tend to occur in combination with other stressors (reviewed by Allan *et al.*, 2005). Thus, the overfished Pacific salmon of the Columbia River must contend with multiple dams as well as degradation of in-stream habitat resulting from farming, logging and land-use change. In the Columbia drainage, however, as is generally the case elsewhere, overexploitation is the initial cause of population declines of fishes, and archaeological evidence from the Early Mesolithic in Europe (see above) plainly shows it to be the first insult in the historical progression.

Overexploitation of Lake Fishes

The vulnerability of fishes that move between fresh and salt water and are susceptible to exploitation within channel bottlenecks is mirrored by those that migrate from lakes into affluent rivers and streams for spawning, and they are fished intensively at river mouths or in places where they congregate upstream. Such fisheries are, for example, well developed around Lake Malawi, which is one of the world's largest and most ancient lakes, and have resulted in overexploitation of migratory species (Tweddle *et al.*, 2012). There, stocks of ntchila (*Labeo mesops*: Cyprinidae; EN) have declined to less than 1% of levels in the 1960s, with destructive fishing practices (use of fine-meshed nets and piscicides) also contributing to this population reduction. The same phenomenon of overexploitation of migrating ningu (*L. victorianus*: CR) during the 1960s led to the depletion of what had been the most important fishery along the rivers flowing into Lake Victoria. Land-use change and sedimentation of spawning grounds have led to further dwindling of populations, despite attempts to regulate fishing in affluent rivers during the spawning period.

A parallel situation of overexploitation is manifest in another ancient lake, and one that holds around one fifth of the Earth's unfrozen fresh water. The omul (*Coregonus migratorius*) is an endemic salmonid that is the mainstay of the commercial fishery of Lake Baikal. Adults undertake annual spawning runs into affluent rivers, particularly the 1590-km-long Selenga River. The fishery dates back to at least the seventeenth century, but landings peaked in the 1940s, and then crashed to such an extent that the fishery was closed between 1969 and 1982. That allowed some recovery of stocks, whereupon annual catch quotas were introduced (Smirnov *et al.*, 2012). However, populations and average body size declined steadily from the mid-1980s until recently and, in 2017, the commercial fishery was again closed for two years; more controversially, angling was also banned. Because both the flesh and the eggs of omul are prized, illegal fishing targets spawners, and prohibitions on capture (and closed seasons) are often ignored. This poaching also reflects the high levels of unemployment among residents around Lake Baikal, and the absence of large-scale farming in the region.

To return to Lake Malawi: due partly to its large size and great age, Lake Malawi has more fishes than any other lake – perhaps 1000 species, although only ~350 have been described. Cichlids make up the vast majority, almost all of which are endemic, but the lake also contains a unique genus of clariid catfish (a dozen species of *Bathyclarias*) and the monotypic sardine-like *Engraulicypris sardella* (Cyprinidae). However, even in this large water body – the word's ninth largest lake – fishes are subject to high levels of exploitation. More than one quarter (28%) of Lake Malawi cichlids assessed by the IUCN are regarded as threatened. Most are relatively sedentary but have slow growth and low fecundity – features that enhance their susceptibility to overfishing and hinder recovery of exploited populations. Populations of the most valuable food fishes in Lake Malawi, chambo or *Oreochromis* (*Nyasalapia*) spp., collapsed from a peak of ~17 000 t in 1982 to <3000 t in 1996, reflecting the intensity of fishing pressure from a rapidly growing riparian population (Kazembe & Makocho, 2004). Three of the formerly important chambo are now regarded as globally threatened (*O. karongae*, *O. lidole* and *O. squamipinnis*; EN). Fishing effort subsequently shifted to small haplochromine cichlids, known collectively as kambuzi, but their economic value was much less than the chambo fishery, and yields declined after an initial peak in exploitation. A shift to smaller species as large fishes become depleted is, as will become evident later in this chapter, a trend that is manifest in most overexploited communities. In Lake Tanganyika, for example, four large endemic species of *Lates* have been overexploited, whereupon catch effort shifted to small, pelagic clupeids. Subsequent declines in these fishes has, however, been linked to climatic warming rather than overexploitation (Cohen AS *et al.*, 2016; see Box 7.3).

There have been some attempts to manage the fishery of Lake Malawi. Around 10% of the area of Lake Niassa, the portion of Lake Malawi governed by Mozambique (Malawi and Tanzania are the other riparian nations), was declared a reserve in 2011 and is managed by the Ministry of Fisheries. It encompasses zones that have complete prohibition of fishing, and seasonal protection during spawning periods in others. Agreement with local communities was also reached on prohibition of fishing in affluent rivers during spawning runs of ntchila, and protection of spawning areas of chambo within the lake. Malawi has a long-established national park in the south of the lake, but it is relatively small, and fisheries management is only weakly developed.

Shifting Baselines May Result in Underestimates of Exploitation Rates

Extensive losses of anadromous fishes in drainages around the Atlantic, as well as from many other rivers and lakes worldwide, will certainly have taken place before any formal stock assessments. Thus, a contemporary observer might gain the mistaken impression that conditions in the immediate past reflected those prevailing in the intermediate and more distant past. This 'shifting baseline' syndrome results in a tendency to underestimate the extent of impacts (Humphries & Winemiller, 2009). For example, the Delaware River was the chief fishery for anadromous sturgeons along the Atlantic seaboard, but landings in 1901 were only 6% of their 1889 peak of more than 2000 t, showing how rapidly declines can occur. Gulf sturgeon (*Acipenser oxyrinchus*: NT) remain so scarce in the Delaware that it is not known whether any reproduction occurs there. Its eastern Atlantic counterpart, the (inappropriately named) common sturgeon (*A. sturio*; CR), was likewise an important commercial fish until the beginning of the twentieth century but, as its threat status indicates, few adults remain in the wild (Limburg & Waldman, 2009). For both of these species, any attempt to extrapolate from current levels

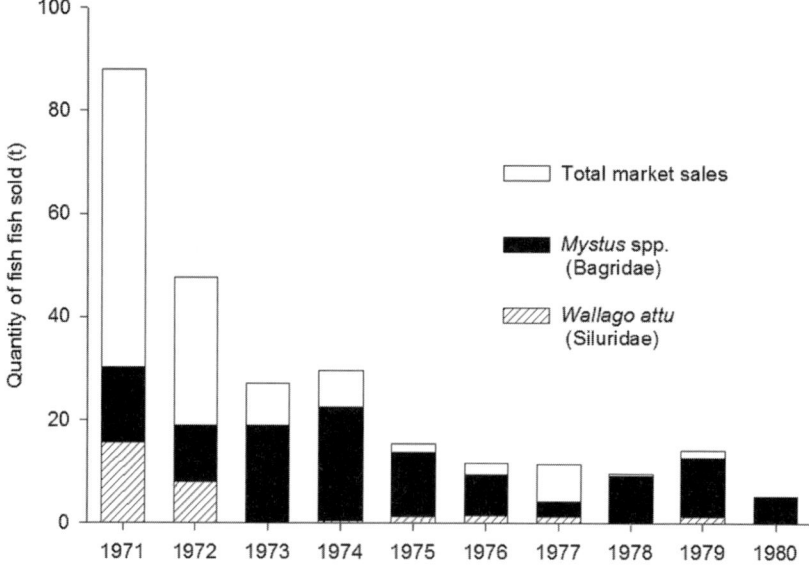

Fig. 3.2 Market sales of river fishes in Perak State, Malaysia, where the number of commercially important species fell from nine in 1971 to only three in 1980. The disappearance of large *Wallago attu* catfish is notable. Original data from Khoo et al. (1987).

of abundance to those in the nineteenth century would give rise to misleading conclusions. Even the earlier reference states may represent a 'shifted baseline' that differs significantly from the original population size. As Fig. 3.2 shows, a scientist assessing catch statistics of river fishes in 1980 would have gained a very different impression of the composition and magnitude of the fishable stock comparted to someone doing the same thing in 1971, and it is unlikely that these earlier values represent any sort of pristine initial baseline.

Shifting baselines are not just a matter of historic interest: the transition from the chambo to the kambuzi fishery in Lake Malawi took place over no more than 15 years during the 1980s and 1990s. Furthermore, even large and charismatic species exploited by fishers, such as the sturgeons mentioned above, can be affected by baseline shift within the span of a human generation; when these species are not encountered on a fairly regular basis, they are rapidly forgotten (Turvey *et al.*, 2010). This breakdown in expectation of what species should be present in fresh waters, and society's consequent absence of interest in their conservation, has been dubbed 'ecosocial anomie' (Limburg & Waldman, 2009). The loss of species and the services they provided, combined with dwindling interest or recollection as the baseline shifts, will tend to make it increasingly difficult to motivate stakeholders and secure funding for population management or habitat restoration.

Which Attributes Increase Vulnerability to Overfishing?

It will be evident from what has been said thus far that migration makes fishes susceptible to overexploitation as they travel through the estuarine bottleneck between river and sea or become concentrated within a relatively small space as they enter affluent rivers from lakes. Compared to marine species that constitute one or a few large and geographically widespread populations, most anadromous fishes comprise numerous relatively small populations that are specific to natal rivers, and this makes them particularly vulnerable to local extinction arising from a river-specific anthropogenic insult. The accumulation of such local losses across a number of natal rivers can lead to regional or even global extirpation,

Many, but not all migrants, are large-bodied. However, most freshwater fishes are small: around half of the fish species present in any inland-water body do not exceed 15 cm maximum length, and the majority of these are relatively abundant; fewer than 90% of species grow larger than 50 cm (Allan *et al.*, 2005). Thus, the majority of species constituting any fished assemblage are small, and body size has an important influence on species-specific vulnerability to overexploitation. Both the largest and smallest species of freshwater fishes appear to be at greatest risk of extinction globally, but the largest are more likely to be threatened by overfishing while threats to the smallest species are particular to local circumstances and the species concerned (Olden *et al.*, 2007).

Whenever large fish are targeted, their relatively smaller numbers and slower population growth – because they are generally long-lived and take years to attain sexual maturity – make them vulnerable to overexploitation (see Fig. 3.2). Some of these fishes are also apex predators, and energetic constraints place limits on their abundance; thus, as in the case of freshwater sharks and rays, they are prone to being overfished.

The Particular Vulnerability of Freshwater Elasmobranchs

Sharks and rays – the elasmobranchs or cartilaginous fishes – are familiar components of marine ecosystems worldwide, where many of them are apex predators. A number of them are associated with fresh waters (principally rivers) where they can be among the largest fishes, although they constitute only a small fraction of the number of species present. Some elasmobranchs are euryhaline and move between rivers and the sea, while others have been reported from fresh water only. These animals are interesting with respect to their biodiversity, as well as the biological attributes that make them vulnerable to overexploitation. Accordingly, they are worthy of comprehensive treatment here.

Among elasmobranchs that live in fresh water, or remain there for extended periods, the charismatic sawfishes (Pristidae), which can grow to more than 3 m in length, are the most well known. Of these, the largetooth sawfish (*Pristis perotteti*, CR) can breed in fresh water. It once abounded in Lake Nicaragua until it became the target of a dedicated commercial fishery for the meat, livers and fins; the rostrum or 'saw' was also valued as a curio or totem. Following establishment of the fishery in 1970, these large, long-lived fishes were quickly overexploited to such an extent that they were almost eliminated from Lake Nicaragua (Thorson, 1976a); the species is now nationally protected. Regardless of this local legislative measure, the largetooth sawfish is at grave risk from over-exploitation throughout its global range, and international trade in its body parts is proscribed under CITES. Other pristids likewise suffer from overfishing, especially for their fins, which are a preferred ingredient in shark-fin soup. The toothed rostrum also makes sawfishes prone to entanglement in nets and to bycatch (Martin, 2005). Their tendency to swim far up rivers increases the likelihood that sawfishes will be captured or encounter degraded habitat and other stressors. Such movements into fresh water are suspected to be associated with breeding (Peverell, 2005), although *P. microdon* (at least) seldom ventures into the sea and is suspected to be confined to inland waters and estuaries over its Indo-Pacific range (Thorburn & Morgan, 2005). Susceptibility of sawfishes to overexploitation is exacerbated by slow growth, late sexual maturity, and viviparous reproduction (specifically, lecithotrophic viviparity) with pro-duction of a small number of pups during each breeding episode. Con-sequently, pristids (and hence all Pristiformes) are categorized as critically endangered by the IUCN, and listed under CITES Appendix I.

Much about the ecology of sawfishes (which are more closely related to rays than to sharks) is incompletely understood; even their taxonomy may not yet be fully resolved. Seven species in two genera have been recognized at one time or another, although the validity of the six nominal species assigned to *Pristis* is uncertain. For instance, western and eastern hemisphere species that appear ecologically different (*P. microdon* and *P. perotteti*) have been treated as subpopulations of a single, widely distributed species (*P. pristis*), thereby reducing the species total for the family to five (see Faria *et al.*, 2013).

The bull shark (*Carcharhinus leucas*: Carcharhinidae), which attains more than 3 m in length, is the most wide-ranging of the euryhaline elasmobranchs. Most un-impounded rivers in tropical and subtropical latitudes are potential habitats for bull shark, which probably breeds in fresh water, and it has the potential to travel hundreds of kilometres upstream. Such movements brings the bull shark into proximity with people, making it more susceptible to anthropogenic impacts than counterpart species living offshore (Compagno & Cook, 1995; Martin, 2005); interactions with this large predator sometimes have unfortunate consequences for humans also. The bull shark is not usually targeted by freshwater fishers, although it is taken as bycatch, but became the subject of a specialized fishery in Lake Nicaragua, where it was formerly abundant. The Lake Nicaragua population was supposed variously to be an endemic species (known as *C. nicaraguensis*), or a land-locked population of *C. leucas*, but it now known that the resident lake sharks were augmented by individuals that migrated ~180 km up the Rio San Juan from the Atlantic Ocean (Thorson, 1976b). The bull shark was heavily exploited during the 1960s and 1970s after the Nicaraguan government granted a concession to a Japanese company that built a shark-fin processing plant beside Rio San Juan. An artisanal fishery also targeted bull shark for their fins and liver oil, capturing them as they transited the river. As with the large-toothed sawfish, the fishery ended following collapse of the bull shark population, whereupon protection of the species was mandated under national legislation. Irrespective of this local depletion of bull shark numbers, the near-threatened global conservation status of this elasmobranch is due largely to a multispecies marine fishery for shark fins, and not impacts arising from the occupation of freshwater habitats.

Overfishing has been implicated as one cause of the decline of 'true' freshwater – or river – sharks, *Glyphis* spp. Members of this obscure genus of carcharhinids are thought to breed in rivers in tropical Asia and Australia, although they have been recorded from coastal marine waters also (White *et al.*, 2015). They comprise the Ganges shark (*G. gangeticus*), Irrawaddy river shark (*G. siamensis*; represented by a single museum specimen) and the New Guinea river shark (*G. garricki*) – all of which are critically endangered – as well as the endangered speartooth shark (*G. glyphis*), and the little-known Borneo river shark (*G. fowlerae*).

Molecular data suggest that some of these species may be conspecific, while at least one other lineage within *Glyphis* remains undescribed (Li *et al.*, 2015). Reexamination of the morphological data and further work on the genus is surely warranted. Due to their extreme rarity, and despite the fact that they are apex predators reaching 2.5 m in length (White *et al.*, 2015), the ecology and habits of river sharks are almost undocumented. Like the rest of the carcharhinids (such as the bull shark), they probably exhibit placental viviparity.

In terms of species richness, tropical stingrays constitute the majority of elasmobranchs found in fresh water but, like the pristids and river sharks, their taxonomy is incompletely resolved and their diversity may have been underestimated (Martin, 2005). Freshwater stingrays comprise two families: the Potamotrygonidae (currently, 32 species in *Paratrygon*, *Plesiotrygon*, *Heliotrygon* and, predominately, *Potamotrygon*) in the Neotropics, and the Dasyatidae (seven *Dasyatis*, six *Himantura* and at least one *Pastinachus* – *P. sephen*) in the Indo-West Pacific and Africa. The dasyatids include both euryhaline and obligate freshwater species, in addition to many marine representatives, but the potamotrygonids are the only elasmobranch family entirely confined to fresh water. They occur in rivers flowing into the Atlantic or Caribbean, and some (e.g. *Potamotrygon leopoldi*) are limited to a single drainage. Dasyatids can be spectacularly large: *Himantura chaophraya* (= *Urogymnus polylepis*: Plate 9) exceeds 2 m in disc width and weighs up to 600 kg, making it one of the world's largest freshwater fishes. Potamotrygonids are generally smaller, although some species attain disc widths (or lengths) of more than 1 m.

Freshwater stingrays share the attributes of other elasmobranchs – i.e. slow growth, late maturation and viviparity with low fecundity – that are typical of strongly K-selected life histories, making them susceptible to overexploitation and limiting their tolerance of the other anthropogenic influences typical of riverine environments (Compagno & Cook, 1995; Martin, 2005). Although potamotrygonids would appear to be particularly vulnerable due to their restricted freshwater distributions, none of them is yet classified as globally threatened. However, 16 species are data deficient, and *Potamotrygon magdalenae*, endemic to two river basins in Colombia, is near-threatened. There is evidence that several potamotrygonids are being overfished in the Paraná River, where populations of only one out of six species remained relatively stable over a 10-year period (Lucifora *et al.*, 2017). In addition, at least six potamotrygonids, principally the Rio Negro endemic *Potamotrygon hystrix* and also *P. magdalenae*, are collected for the international aquarium trade. Possible impacts on wild stocks are inadequately documented, but, despite this, Brazil has taken the precaution of introducing export quotas for some species (Martin, 2005).

Dense human populations and rapid environmental change in Southeast Asia pose a threat to freshwater dasyatids. Five of them (*Himantura signifier, H. polylepis, H. oxyrhyncha, H. kittipongi* and *Dasyatis laosensis*) are globally endangered, while a sixth (*H. dalyensis*) is data deficient; *D. laosensis*, endemic to the Mekong River, is collected for the aquarium trade. One African species (*D.* [=*Urogymnus*] *ukpam*) is also endangered, and another (*D. garouaensis*) is classified as vulnerable. Given that virtually all freshwater or euryhaline stingrays are confined to the tropics where human populations are growing rapidly, it is likely that the high proportion of threatened species seen in Southeast Asia will soon be equalled in South America.

Have All the Bigger Freshwater Fishes Really Had Their Chips?

It is not only populations of freshwater elasmobranchs that – in significant part because of their size – frequently succumb to the impacts of overexploitation. Large fishes that are migratory, irrespective of whether or not they are diadromous, are susceptible to overexploitation as well as other human impacts and, where they are present, tend to be among the first members of any freshwater assemblage to exhibit population reductions. Their need to use different habitats at different times and stages of their life cycles makes such fishes vulnerable since it heightens the chances of them encountering fishing gear and contaminated or degraded habitats. The further – or more often – fishes migrate, the greater the risk. The vulnerability of large fishes more generally will be evident from the information given about the Yangtze in Box 3.2, where migratory paddlefish and sturgeons were susceptible to overexploitation, as were large catfishes in the Ganges (Box 3.3), but, in both rivers, other drivers of environmental change contributed to their decline also. But this has not been the case along the Mekong (Box 3.4), where the effects of overexploitation on large fishes, and their subsequent decline and endangerment, have not been confounded by other anthropogenic stressors.

The Mekong is a particularly informative example of the phenomenon of overfishing, since the river is relatively unpolluted and, until

Box 3.4 *Overexploitation of Mekong Megafishes*

The Mekong giant catfish (*Pangasianodon gigas*: Pangasiidae; CR) (Plate 10) serves as a typical example of the outcome of overexploitation of a large migratory fish – in this case, a potamodromous species that migrates within a single river basin. It can weigh up to 350 kg, although there are anecdotal records of even larger body sizes. This behemoth formerly supported a commercial fishery in the Lao People's Democratic Republic (henceforth, Laos) and Cambodia, dating back more than 100 years, for the oil that could be extracted from its fatty flesh. The catfish migrates throughout the Lower Mekong Basin, with records of occurrence from the delta region upstream to Yunnan Province in China. Stock decline attributable to overfishing were evident by the 1940s, and the fishery had waned greatly by the 1970s although some landings continued in Thailand during the 1990s. The latest assessment of the conservation status of *P. gigas* (Hogan, 2013a) included an estimate that the remaining stocks had declined by more than 80% in the preceding two decades. It has virtually disappeared from the Mekong in Thailand, where some licensed fishing of the species is still allowed, but is protected by law in Cambodia and Laos, although this may not protect it from being fished there. Moreover, such regulations do not prevent adults from being landed occasionally as part bycatch aimed at other fishes, and juvenile Mekong giant catfish resemble the young of other pangasiids taken as part of a multispecies fishery that targets small catfishes. This incidental mortality will continue as fishing effort increases throughout the Mekong Basin and gear becomes more efficient. Captive breeding of the Mekong giant catfish is possible, and survival of hatchery fish in reservoirs is good (although they do not breed in impoundments), whereas it appears that they do not survive when released into rivers. This may have the unintended benefit of limiting genetic introgression of wild populations with artificially bred stock. However, maintenance of a genetically representative captive population would be appropriate as *ex situ* insurance against possible extinction of the Mekong giant catfish in the wild.

Unfortunately, the ecology and habitat use of the Mekong giant catfish is poorly known – even its feeding habitats are not fully understood – and migration routes or spawning sites have yet to be adequately documented, thereby limiting attempts to manage this species. Even genetic data on the species are contradictory, although some findings suggest this catfish needs to be managed as a single panmictic population (Hogan, 2013a). A robust well-funded research programme that identifies key habitats used by the Mekong giant catfish, accompanied by establishment of protected areas to maintain them, will be needed to ensure its survival. Any such efforts may, however, be moot in the face of ongoing dam construction along the Mekong mainstream that will obstruct migration to the only known spawning site of this megafish (see Case Study: Damming The Mekong Commons in Chapter 5).

A second giant pangasiid, the 300-kg dog-eating catfish (*Pangasias sanitwongsei*; CR), may be approaching extinction also, as it has almost vanished from the Mekong; the only other river where it has been found is the Chao Phraya in Thailand. The giant carp (*Catlocarpio siamensis*: Cyprinidae; CR) (Plate 11) attains a similar maximum size to both pangasiid megafishes. It was once an important food fish but has become very rare in the Mekong and has been fished out of the adjacent Chao Phraya (Hogan, 2013b). The giant carp is now legally protected from fishing in all riparian nations of the Lower Mekong Basin and is the national fish of Cambodia.

Aside from near-disappearance of megafishes, there is other evidence that the Mekong is being overfished. There have been gradual declines in the catches and average size of long-distance potamodromous migrants, such as the seven-striped barb (*Probarbus jullieni*) and thicklipped barb (*P. labeamajor*: Cyprinidae; both EN) that can reach 1.5 m long and *c.* 70 kg, together with an increased proportion of smaller species in landings (Baird, 2006a). The seven-striped barb has been almost completely eliminated over much of its former Southeast Asian range, and is now probably restricted to the Mekong, whereas the thicklipped barb is a Mekong endemic. Numbers of the migratory silurid *Wallago attu* (NT), a large-bodied migrant, have also declined due to overfishing in the Mekong and throughout the rest of its extensive Asian range. Another declining Mekong species, the small-scale croaker, *Boesemania microlepis* (Sciaenidae; NT), is not migratory but is one of the most sought after (and hence highly priced) food fishes in the river; it can exceed 1 m in length, although most fished individuals landed are much smaller. Overexploitation has reduced populations to less than 20% of previous stock levels in Laos, despite a law prohibiting its capture during the breeding season or sale at any time of the year (Baird *et al.*, 2001). Despite declines throughout its range, local community conservation efforts in southern Laos (Lao People's Democratic Republic) that protect the deep pool habitats *B. microlepis* occupies during the dry season show signs of having benefitted this species (Baird, 2013). The Mekong giant salmon carp (*Aaptosyax grypus*; CR) – a piscivorous migratory cyprinid about which little is known – has also become extremely rare as a result of overfishing.

recently, lacked any mainstream dams. By contrast, in other large rivers, such as the Columbia and Delaware, the effects of overexploitation were exacerbated by dam construction or, in the Yangtze, Zhujiang and Ganges (Boxes 3.2 and 3.3) by a combination of pollution, dams and overfishing. Much the same remarks about the multiple causes of declining fisheries apply to the Thames and Rhine during the nineteenth and twentieth centuries. Nonetheless, in all such cases, the particular susceptibility of large fishes to overexploitation fishing is evident. Other examples of 'big-fish decline' include the mangar (*Luciobarbus escocinus*; VU) of the Euphrates and Tigris, which ranks second in body size among cyprinids to the giant carp; the Murray cod (*Maccullochella peeli*: Percichthyidae; CR) in Australia; two species of Asian taimen (Box 2.3); the golden mahseer (*Tor putitora*: Cyprinidae; EN) of the Himalaya; the black mahseer (*T. khudree*; EN) and humpbacked mahseer (*Hypselobarbus mussullah*; EN) of the Western Ghats; the poorly known goonch or giant devil catfish (*Bagarius yarrelli*: Sisoridae; NT) of southern Asia; and the widespread yet highly threatened freshwater sawfish (*P. pristis*; CR) mentioned in the account of freshwater elasmobranch vulnerability earlier in the chapter. The same trend towards overexploitation of large fishes is also manifest in the Amazon, as described in Box 3.5.

The overall picture of the global status of large or migratory freshwater fishes is bleak, with most species in decline and many close to extinction. They are subject to both a 'great thinning' in abundance and a 'great shrinking' in average size. These fishes may disappear within the next few decades, sooner in some cases, but their loss is not inevitable. The anadromous (some populations are amphidromous) 1-m long shortnose sturgeon (*Acipenser brevirostrum*; VU) has been protected by the US Endangered Species Act since it came into force in 1966. The Hudson River population of this fish has shown signs of recovery in recent decades. Numbers have increased by more than 400%, and adults tagged

Box 3.5 *Fishing Down the Amazon?*

Evidence is accumulating about the extent of overfishing in the Amazon. An expansion in total fishing effort and extent has been accompanied by a progressive depletion of large high-value species, and a shrinkage in the mean body size of the catch. As a result, the growing fishery has begun to take increasing numbers of smaller-bodied, more abundant species. Of the 18 species that now dominate fishery yields, four of the largest species are overexploited in at least part of the basin (Castello *et al.*, 2013): the migratory pimelodid catfishes *Brachyplatystoma filamentosum* (reaching 2.5 m in length; this genus includes the aptly named goliath catfishes), *B. vaillantii* and *Pseudoplatystoma* spp. (both 1 m), as well as the fruit-eating tambaqui (*Colossoma macropomum*: Serrasalmidae; 1 m). The huge dorado catfish (*B. rousseauxii*), which can grow to 3 m long, is likewise commercially important and overexploited. It is also vulnerable to construction of dams planned for the Amazon because it migrates further than any other potamodromous fish, travelling westward from the estuary up to the Andes (Barthem *et al.*, 2017). Spawning takes place in the head-waters, and larvae drift downstream to the freshwater plume in the estuary, which serves as a nursery area. After two to three years, juvenile dorado catfish begin to make their journey along the Amazon floodplain towards the headwaters, feeding and growing along the way; the round trip totals ~11 600 km. Goliath catfishes such as *B. juruense* and *B. platynemum* undertake similar long-distance migrations with the latter travelling almost as far as the dorado catfish. These distances surpass the maximum life cycle migration (6000 km) reported for anadromous Pacific salmon and are nearly as long as the combined freshwater and marine journeys of the European eel (Barthem *et al.*, 2017).

Apart from reductions in abundance of large catfishes, signs of fishing down the Amazon are evident also from declines of pirarucu (*Arapaima* spp.: Osteoglossidae). These air-breathing megafishes, which can grow more than 2 m long and weigh 200 kg (Plate 12), were the most important species caught in the Amazon during the early twentieth century, but, by 1975, populations had been depleted to such an extent that commercial trade was restricted by a listing on CITES Appendix II (Antunes *et al.*, 2016). Average body size of *Arapaima* also declined substantially. *Arapaima* are not long-distance migrants, but their need to surface periodically in order to gulp air makes them vulnerable to spear fishers in canoes, especially during the dry season when water levels are low. In addition, males engaged in parental care are susceptible when guarding nests along the margins of floodplain lakes and channels. The Brazilian government has taken steps to conserve *Arapaima* and other large fishes through introduction of minimal size limits and a four-month closed season during which fishers are paid a minimum salary. Colombia similarly proscribes fishing of *Arapaima* during the breeding season. Unfortunately, such regulations are poorly enforced (Parry *et al.*, 2014).

A recent survey (Castello *et al.*, 2015a) concluded that high levels of illegal exploitation had been the cause of local extinctions of *Arapaima* populations in 19% of Amazonian fishing communities. *Arapaima* were approaching extinction in 57% of localities and overexploited in a further 17%. Almost one quarter of fishers admitted to catching *Arapaima* despite being aware of local declines in their populations; fishing during the closed season was more prevalent in communities where *Arapaima* were already depleted. A minority of communities have imposed local management rules, including bans on gill nets and setting fishing quotas based on fish abundance, which have enhanced the abundance of *Arapaima*. Despite such efforts, and measures introduced by governments, one or more of the five known species of *Arapaima* may now be extinct (there is some uncertainty, because the taxonomy of the genus is not fully resolved). Continued overexploitation is driven by the ever-increasing commercial value of *Arapaima* as they becomes scarcer (see also Box 3.9). However, aquaculture production of *A. gigas* has been expanding and may help reduce the pressure on wild populations.

The cascading effects of the reductions of *Arapaima* and other predatory fishes in the Amazon are unknown. Some large pimelodid catfishes play a role in ichthyochory (i.e., seed dispersal by fishes) although the details of this mutualism await study (Correa *et al.*, 2015). Fortunately, fruit eating by the tambaqui and other large migratory serrasalmids (e.g. *Piaractus brachypomus* and *P. mesopotamicus*) (Plate 13) has received more research, revealing that these fishes serve both as seed predators and dispersers of ingested seeds, according to the plant species concerned. These serrasalmids may play a keystone role as dispersal agents because they are highly mobile and can swim long distances upstream, defaecating seeds in floodplain habitats that are suitable for germination after floodwaters recede (Anderson *et al.*,

2009). Large serrasalmids are important fishery species in the Amazon and Orinoco Basins, but their size renders them vulnerable to over-exploitation resulting in substantial reductions in abundance and changes in population age structure. It is precisely their size that makes them effective and important seed dispersers, and overfishing has the potential to disrupt a 70-million-year-old mutualism between plants and fishes (see review by Correa *et al.*, 2015). An additional threat to this mutualism, and Amazon fisheries as a whole, is climate change: an increased frequency of droughts has been implicated in reductions in catches and the average size of frugivorous tambaqui in Lago Catalão, a floodplain lake near the confluence of the Amazon and Rio Negro, where 15 years of landing data are available (Röpke *et al.*, 2017).

in 1979 have been encountered almost two decades later (Bain *et al.*, 2007). This recovery is important because the Hudson River population (~60 000 fish, mainly adults) is as large as all the remaining combined stocks of this species. More significantly, it shows that so long as it is protected from exploitation, the persistence and proliferation of a long-lived migratory fish can take place within a human-dominated ecosystem in close proximity to one of the world's major cities. Enforcement of prohibitions on fishing was a significant factor in ensuring population recovery of shortnose sturgeon, as was continued access to suitable spawning sites and improvement of water quality in the Hudson (following implementation of the 1972 Clean Water Act), despite con-siderable physical transformation of parts the river. Strategies employed to enhance populations of overexploited fishes have frequently used hatchery-reared individuals, but the 'protect-and-wait' approach to shortnose sturgeon recovery involved minimal interference with natural processes and aimed to restore and maintain habitat conditions that would sustain this fish. The Hudson River population has recovered sufficiently to qualify for delisting as a federally protected species (Bain *et al.*, 2007), and improvements in habitat conditions have also benefitted the western Atlantic sturgeon (*Acipenser oxyrinchus*), listed under the United States Endangered Species Act. Elsewhere, imposition of fishery restrictions, better water quality and restoration of riverine spawning areas have been associated with a rebound of stocks of lake sturgeon (*Acipenser fulvescens*; see below) in recent decades. The establishment of invasive dreissenid mussels in the Laurentian Great Lakes (see Chapter 4) and beyond may have facilitated the process of population recovery by these molluscivores.

Fishing Down the Food Chain: The 'Great Shrinking'

While large fishes, which are often highly valued species, are frequently targeted by fishers and are vulnerable to overexploitation, overfishing of an entire species-rich assemblage can occur. It has, for example, been well documented in the Laurentian Great Lakes (in a historical summary by Smith, 1968) and within Lake Superior in particular detail (Waters, 1987). A sequence of areas was gradually depleted in turn from the late nineteenth century when a number of species or stocks were successively overfished beginning, primarily, with the top predators. However, after 1940 or thereabouts, the combined effects of pollution and invasive species (see Chapter 4) became dominant influences upon the Great Lakes' fish community. One unusual aspect of this succession was the initial (pre-1850) depletion of large (up to 2-m long) lake sturgeon, which was targeted as a nuisance species that destroyed fishing gear; captured fish were discarded to rot. A dedicated commercial fishery for these freshwater giants developed later, eventually reducing populations of this – the largest Great Lakes' species – to such low levels that restrictions on capture were introduced.

In many areas of the tropics, especially Asia and Africa (as in the account of Lake Malawi given above), the mean size and age of the catch have declined progressively over time until, in some cases, the major part of the catch consists of fish within the first year of life. This is another aspect of the 'great shrinking' mentioned above; in this case, reflecting the average size of fishes in a mixed assemblage. In Lake Victoria, where the fishery was dominated by invasive Nile perch that had greatly reduced the abundance and diversity of native haplochromines (see Chapter 4), overfishing led to a reduction in average body size of Nile perch and other demographic changes that initiated a shift in gear from large-meshed nets to small hooks on long lines (Mkumbo & Marshall, 2015). In this particular instance, depleting the stocks of a large invasive piscivore probably delivered conservation benefits for haplochromines, but such is seldom the case.

Fishing pressures in South America appear not to have reached these levels seen in Africa and Asia, with some reports suggesting that catches still include large species with relatively few signs of depletion (Welcomme *et al.*, 2010). Overexploitation of fishes and other aquatic animals has nonetheless been a significant historical driver of population declines in the Amazon (Antunes *et al.*, 2016), and it is becoming increasingly apparent that fishing down at the community level is taking place (e.g. Castello *et al.*, 2013: see Box 3.5). Population reductions of five out of six species of freshwater stingrays in the middle and lower Paraná River (between 2005 and 2016) have also been attributed to overexploitation, with lowest relative abundance occurring at localities with the most intense fishing effort; larger species were most susceptible, with annual rates of decline as high as 25% (overall average 15%: Lucifora *et al.*, 2017).

Overexploitation not only represents a threat to large-bodied freshwater fishes, but it also threatens even the high-yielding multispecies fisheries of relatively pristine rivers. The Lower Mekong Basin (essentially the portion of the river drainage downstream of China) is the world's most productive freshwater fishery (Box 3.6), and while there is unambiguous evidence that megafishes and other large species have been overexploited (Box 3.4), smaller species are also being affected. While there are limitations upon the reliability of landing statistics, fishing effort along the river has increased leading to a rise in the aggregate catch but a decline in the catch per fisher (FAO, 2010). Thus, the catch from the Tonlé Sap Lake on the Mekong floodplain in Cambodia, which is the largest wetland in Southeast Asia, doubled between 1940 and 1995, but the number of fishers tripled over that period. Stability or even increases in overall catches from a multispecies fishery can also conceal the fact that individual species are being overexploited. Declines in the contribution of larger long-lived species (such as *Probarbus* spp. or the small-scale croaker: see Box 3.4) are offset by increased capture of smaller short-lived species as the community is fished down. In Tonlé Sap Lake, for example, the 120 000 t annual catch in 1940 consisted mainly of large fishes, while the 235 000 t landed in 1995 was almost exclusively small individuals (FAO, 2010). Under Cambodian law, the large-scale commercial fisheries of the lake were managed through a system of fishing lots, and these limited-access fisheries operated only during eight months of the year. Artisanal and subsistence fishers could only operate in open-access areas, but these areas were fished throughout the year. In 2012, Prime Minister Hun Sen abolished the commercial lot system, designating the entire lake as a common fishing area for local residents. While this was intended to help conserve fish stocks, the effect of opening access was to further deplete the fishery. More recent fishery protection measures have involved establishment of a fish conservation zone by local communities (at Plov

Box 3.6 *Mekong Fisheries: Richness and Importance*

The Mekong is highly significant in terms of the magnitude of its multispecies fishery, as well as its contribution to global biodiversity of fishes. The river hosts *c.* 850 freshwater fish species; the total rises to around 1100 if estuarine species and marine vagrants are included (Hortle, 2009). Based on these figures, the Mekong is among the top three rivers in the world in terms of fish species richness, along with the Amazon and the Congo, yet it is relatively small ranking 15th globally in terms of discharge, 16th in terms of length and 25th in terms of drainage-basin area (Dudgeon, 2011). If the estimate of 850 (or more) species is correct, the Mekong is actually ranked second, after the Amazon, in terms of species richness (although fish diversity in the Congo may have been underestimated) but would rank top according to species richness per unit area.

The Mekong fishery is of great importance to humans: the Lower Mekong Basin (downstream of China) is the world's largest freshwater capture fishery, yielding an estimated 2.2 million t annually (Hortle, 2007). It may amount to 2.5 million if freshwater shrimps, crabs, molluscs, frogs and water snakes are included, such that it approaches one quarter of the global total freshwater yield (see main text). A more recent assessment of the Lower Mekong yield by regional fisheries authorities was 2.3 million t for 2012 (So *et al.*, 2015), a significant increase on the total of 1.9 million t in 2010 and almost double that for 1996. The changes were attributed to improved data collection rather than greater fishing effort (So *et al.*, 2015), but the latest figure is remarkably similar to the 2.2 million-t estimate. To put these yields into broader context, 1 million t of fish is equivalent, in terms of animal protein, to 1.2 million large buffaloes or 16–17 million pigs (Welcomme *et al.*, 2010). The Lower Mekong capture fishery was valued at US$11 billion in 2015 first-sale prices (So *et al.*, 2015). This is far in excess of the 2012 estimate of $5.5 billion for *global* first-sale value of freshwater capture fisheries made by the FAO (2014), and much greater than an earlier valuation of the Lower Mekong that ranged from $2.2 to 3.9 billion first sale and $4.3–7.8 billion as fish products (Hortle, 2007). Regardless of this uncertainty about its economic value, the fishery is evidently vitally important on a local, regional and global scale.

Apart from its size, the Mekong fishery has three distinctive characteristics: first, it is based upon a large number of species; second, much of the catch (40–70%) is contributed by around 50 species that migrate within the river (Welcomme *et al.*, 2016). A third feature is involvement in fishing of ~40 million people (more than half of them women) on a small-scale, subsistence or *ad hoc* basis, with the catch contributing to family welfare and food security. In land-locked Laos, for example, 83% of households engage in capture fishery some of the time, with 90% of the catch derived from rivers and streams, and fishes provide 20% of animal protein consumed (FAO, 2010). Surveys of what is consumed suggest that the amount captured from the Lower Mekong Basin is considerably more than is reported in fish capture statistics (FAO, 2014). Fishes and other aquatic animals (frogs, shrimps, molluscs, etc.) account for more than half of the animal protein consumed (on average, 65 kg per person annually), ranging from around 50% in Laos to 80% in Cambodia, compared to a global mean of 17% (Hortle, 2007; So *et al.*, 2015).

Tuok, where the IUCN played an advisory role) so as to restrict fishing in important dry-season habitat.

The *dai* (stationary trawl) fishery of the Tonlé Sap River that targets smaller migratory species of short-lived cyprinids was established more than 140 years ago. The fishery is managed using a closed season together with effort (number of trawls) and gear size restrictions. It has been monitored since 1997. Initially, total annual yields (which fluctuate around 12 000 t) showed a strong correlation with the height of the yearly flood peak, but this relationship broke down after 2003. Catches in 2010 were unusually low despite high water levels, attributed to recruitment failure due to overfishing of adults in dry-season refuges (Sopha *et al.*, 2010), highlighting a need for effective overall management of the *dai* fishery. A recent and comprehensive analysis of the fishery over a 15-year (2000–2015) period took account of 116 species and found that the total biomass of landings had remained remarkably resilient, with an increase in the catch of smaller species compensating for declines in larger species that occupy higher trophic levels. The majority (78%) of species exhibited decreasing landings through time, with downward trends most apparent in medium to large-bodied species such as the formerly common cyprinid *Cirrhinus microlepis* (VU); a decrease in the individual weights and lengths of other fishes and even small species was notable also (Ngor *et al.*, 2018). The increasing reliance on small cyprinids (*Henicorhynchus* and *Labiobarbus* spp.) and shifts in community composition are evidence that the Tonlé Sap *dai* fishery is being over-exploited, and it shows obvious signs of body-size shrinkage due to being fished down. The growing human population of Cambodia and a shortage of livelihood options has been responsible for an expansion of catch effort and the pressure on stocks, since entry into fishing sector is free and the gear is affordable. The current formal management regime in Cambodia favours community-based co-management of fisheries, and dozens of such institutions have been established on the Tonlé Sap floodplain (Ngor *at al.*, 2018). It remains to be seen whether they can cooperate to protect and conserve key areas of habitat, and reduce overexploitation of juveniles and migrating adults in order to sustain the fishery.

The correlation of annual fishery yields with the extent of floodplain inundation (the flood pulse) – as seen along the Mekong – is a typical feature of large rivers (Welcomme *et al.*, 2010; Castello *et al.*, 2015b). Both high and low waters exert strong effects on recruitment, and the time lag in response to water levels in a particular year indicates the time taken for a year class to grow to a size at which it is exploited. One line of evidence that an assemblage is being fished down is a reduction in this lag time from (say) three years to a correlation between yields and flood height within the same year, indicating that the fishery is based on either very small fishes or is dominated by species that mature extremely rapidly (or both). The reduction in the average size of individuals and species caught as an assemblage is fished down may stimulate management towards a reduction in fishing effort or imposition of size-based restrictions. However, smaller fishes are favoured in certain African cuisines and Southeast Asian countries where they are salted and fermented to make fish sauce and, under those circumstances, continued fishing has the potential to deplete the entire assemblage. And, as fishes become scarcer, there can be a tendency to adopt more destructive practices in order to increase catches, involving the use of finer-meshed nets, application of pesticides (sometimes replacing natural phytochemicals) and fishing by electrocution (using power generators).

Effects of Overfishing on Ecosystem Functioning

Surprisingly little is known about the ecological consequences of over-fishing, aside from the direct effects on the target species (Allan *et al.*, 2005; Flecker *et al.*, 2010; Beard *et al.*, 2011). However, Pacific salmon are an exception to this generalization. The overexploitation of large, migratory fishes such as salmon can have significant consequences because of their potential to connect food webs in distant locations, either as a result of their own feeding activity or through transport and remineralization of nutrients, and sometimes both. Depletion of Pacific salmon runs along the western North American coast by overfishing, or their obstruction by dams, impairs the 'uphill' transfer of marine-derived nutrients that would otherwise have become available inland following the death and decay of semelparous adults after spawning. There is evidence that an even greater proportion of marine-derived nutrients is released during the upstream journey as excretory products and later, towards its end, as gametes (Tiegs *et al.*, 2011). Excretion supplies nitrogen and phosphorus in forms that are readily available for uptake by stream biofilms, whereas nutrients in carcasses have to be mineralized before they can be assimilated. Migrating adult salmon provide food for brown bear (*Ursus arctos*: Ursidae; Plate 14), and nutrients derived from the fishes (and the sea) are transported into riparia and beyond by bears, other piscivores and carrion feeders, or via subsurface flow and flooding, where they fertilize plants influencing growth and timber yields (Gende *et al.*, 2002). The bear–salmon relationship has been termed a keystone interaction, and the subsidy of marine-derived nutrients is an essential promoter of stream and riparian productivity (Helfield & Naiman, 2006). Chum (*Oncorhynchus keta*), which are the most widely distributed and abundant of the Pacific salmon, are the leading contributors to this nutrient flux. One estimate is that the nutrient flux (as phosphorus) contributed by all anadromous fishes, which was greatest at high latitudes in the northern hemisphere, is less than 4% of historic values (Doughty *et al.*, 2016). In some places where spawning runs have been eliminated, compensatory fertilization through addition of nutrients reverses the effects of the loss, restoring productivity formerly derived from the salmon-mediated marine subsidy (Hyatt *et al.*, 2004).

Although they have been far less studied than salmon (but see West *et al.*, 2010), the spawning migrations of smaller anadromous fishes (*Alosa* and *Tenualosa* herrings, osmerid smelt) must likewise transfer marine-derived nutrients and energy into fresh waters. Where these fishes are consumed by piscivorous fishes and birds (such as cormorants, gulls, mergansers and eagles), they subsidize the upper trophic levels of riverine food webs (Flecker *et al.*, 2010). These uphill transfers of material are lost (or diverted) where spawning runs have been overexploited or obstructed by dams. Such reductions in the magnitude of a subsidy are of a different kind to the reduction in the provision of an ecosystem process (seed dispersal) attributable to overexploitation of large frugivorous fishes in the Amazon (see Box 3.5).

More generally, larger freshwater species targeted by fishers tend to have disproportionate effects on nutrient recycling rates and the relative availability of nitrogen and phosphorus, although the effects on a particular element change according to the identities of the fishes involved (McIntyre et al., 2007). This means that the overexploitation to which large species are particularly vulnerable can result in the disruption of ecosystem functioning, even in species-rich communities. For example, detritivorous *Prochilodus mariae* (Prochilodontidae) is an abundant migratory fish in tropical South American rivers where it plays an important ecological role as a bioturbator through sediment ingestion and resuspension. Manipulations of its abundance altered carbon dynamics within streams in the Orinoco Basin, matching ecological changes evident in parts of its range where stocks of this fish had been reduced by overfishing (Taylor et al., 2006). Reductions in stocks of *P. mariae* are associated with substantial reductions in nitrogen remineralization, meaning that overexploitation directly compromises the high primary productivity that actually supports tropical freshwater fisheries (McIntyre et al., 2007). This finding is important because *P. mariae* and its close relatives constitute a major component of biomass and fishery yields in rivers across South America, and their mass migrations enhance ecosystem connectivity between different parts of drainage basins. Among these prochilodontids, *Semaprochilodus* spp. are of major commercial importance in the Amazon Basin. They move as juveniles from the productive floodplains where they were spawned into unproductive blackwater rivers. There, many of them become food for a range of piscivores, allowing these predators to maintain higher growth and population densities than could be supported by *in situ* secondary production in the absence of migrating prochilodontids (Winemiller & Jepsen, 2004; Flecker et al., 2010). There is no guarantee that fish communities have sufficient functional redundancy to guarantee that increases in abundance of underexploited species can compensate for changes in nutrient dynamics caused by depletion of larger fishes, thereby moderating the effects of overfishing or functional extinction. Indeed, the limited evidence at hand suggests that such compensation is unlikely (McIntyre et al., 2007).

Intense fishing, particularly when it is directed towards larger fishes, not only depletes populations and alters assemblage species composition, but it also changes the remaining fish. Typically, fish able to reproduce before being caught are smaller and have reached sexual maturity earlier than their conspecifics. If these characteristics are heritable, then the size and age of sexual maturity in overfished populations will decline from generation to generation, with associated genetic changes attributable to the strong selective pressure imposed by fishing. Such changes are well known to marine fishery biologists, and are consistent with the life-history adjustments that occur when fishes experience high levels of predation (e.g. Walsh & Reznick, 2008), and have been reported for freshwater species (Audzijonyte *et al.*, 2013). For example, overfishing of piscivorous Eurasian perch and zooplanktivorous whitefish (*Coregonus larvatus*) in Lake Constanz leads to life-history modifications and a shift towards smaller individuals which, in the case of the perch, tend to rely more heavily on plankton. In turn, these qualitative changes can have knock-on effects on food webs and ecosystem stability (Kuparinen *et al.*, 2016). When overfishing results in smaller fish that grow quickly and reach sexual maturity earlier, their feeding induces instability in the production of plankton. As a result, fish abundance fluctuates wildly from one year to the next, due to increased recruitment variability as well the inherent tendency for populations with faster growth rates to be more unstable. As will be evident from the ecological ramifications arising from depleted salmon runs, it is essential to consider indirect impacts on non-target species and the potential for ecosystem-level effects when the consequences of overfishing are assessed. In the case of frugivorous Amazonian serrasalmids (Box 3.5), overfishing selects for the poorest seed dispersers, and has indirect effects on trees and lianas that rely on ichthyochory (Anderson *et al.*, 2009).

Irrespective of the underlying details, depletion of large or ecologically dominant fishes can have potentially far-reaching direct and indirect effects on the nutrient dynamics and functioning of freshwater ecosystems. This means that overexploitation leading to reductions in fish abundance, especially in cases where they can be so great as to be equivalent to functional extinction, can matter a great deal. Such changes are of particular significance because yields of inland fisheries are highly correlated with fish species' richness globally, suggesting a link between biodiversity and stable, high-yielding fisheries (Brooks *et al.*, 2016). Richness remains an important and statistically significant predictor of yield once other macroecological drivers are controlled for, despite the widespread reliance of many fisheries on targeting just a few fish species. Other research suggests that fish richness and catches from rivers (only) are positively but not causally correlated, revealing that fishing pressure is most intense in rivers where potential impacts on biodiversity are highest (McIntyre *et al.*, 2016). Furthermore, much of the global catch comes from rivers that are already subject to high levels of anthropogenic stressors. Specifically, 89% of total catch worldwide comes from ecoregions that have above-average threat levels. Intensive fishing in rivers that are already degraded may undermine efforts to conserve biodiversity, and the syndrome of poverty, nutritional deficiency, fishery dependence, and extrinsic threats to river ecosystems underscores the high stakes for improving fisheries management (McIntyre *et al.*, 2016).

Fisheries Management

A number of models have been developed to assist in the assessment and management of freshwater fishery resources. Many of them have been derived from the literature on marine fisheries. While some models might be adequate for the management of single-species fisheries in large lakes, such as the Nile perch fishery in Lake Victoria (Mkumbo & Marshall, 2015; see also the account of this species in Chapter 4), they do not perform satisfactorily in the more diffuse multispecies and multi-gear fisheries of rivers and their floodplains (Welcomme *et al.*, 2010); the *dai* fishery of Tonlé Sap River is an outstanding example of such a fishery. However, even if there were appropriate and useful models, it would be difficult to implement the recommendations arising from them because of limited management and enforcement capacity in many countries, particularly over much of the tropics, and the often diffuse nature of the fishery. Open-access fisheries are vulnerable to overexploitation, and attempts to regulate fishing effort tend to involve restricting the number of people involved by limiting access. That approach is usually combined with one or more of the following: permanent no-take zones (Arantes & Freitas, 2016); closed areas or sanctuaries that protect nursery grounds, spawning habitats or dry-season refuges (e.g. Baird 2006b); closed seasons linked to breeding or migration; catch limits or quotas (e.g. Petersen *et al.*, 2016), including control of the number of fisher licences or fishing lots (as formerly applied in Tonlé Sap Lake); restrictions on types of fishing gear (e.g. Castello *et al.*, 2015a) and prohibition of fine-meshed nets and destructive fishing practices (poisons, electricity); and proscriptions on the size or species of fish that can be taken.

In many places, management involves a central authority responsible for controlling access or setting catch regulations that limit overall fishing effort. The disadvantages of this approach are that subsistence fishers, who are those most reliant on the resource, may be denied access. Community-based local management (or co-management) probably offers a more effective way of regulating diffuse, multispecies fisheries, especially those that involve different people, using a wide range of gear, at different times of the year. The efficacy of community management can be increased if it has government support and there is willingness to look beyond 'one-size-fits-all' solutions (see also Baird, 2006b; Hossain *et al.*, 2006). This is not to say that community management is a panacea. The poverty faced by small-scale fishers in the topics may prevent them from curbing overexploitation of a declining fishery because they lack economic or nutritional alternatives, and hence are less able to stop fishing than individuals from wealthier households (Cinner *et al.*, 2009). And, as mentioned elsewhere in connection with the *dai* fishery, entry into the fishing sector is cheap, and so capture effort will tend to grow when livelihood options are limited.

In cases where overexploitation induces genetic changes of fished stocks (see above), the ecosystem instability they cause cannot be remedied simply by reducing fishing intensity, and management can be challenging. Alarmingly, when the demographic consequences of life-history changes are combined with increased overall mortality and selective removal of larger individuals by overfishing, they create a positive feedback loop that further amplifies fluctuations in fish populations and the ecosystem as a whole (Kuparinen *et al.*, 2016). Such effects have led some scientists to argue that a new paradigm in fisheries management is needed whereby a moderate mortality from fishing is distributed across the widest possible range of fish species and sizes in an ecosystem with the aim of maintaining relative size and species composition (Garcia *et al.*, 2012). While the goal of this 'balanced harvesting' approach is laudable, its relevance and implementation have been questioned, especially in cases where rivers and lakes have been severely fished down (Tweddle *et al.*, 2012).

Recent research suggests that there is some scope to refine current knowledge of fish yields from inland waters. Findings from lakes worldwide reveal a strong positive relationship between chlorophyll concentrations, representing a measure of aquatic primary production, and fishery yields (Deines *et al.*, 2015); additional variation in the relationship can be ascribed to regional climate differences. Remotely sensed chlorophyll data could conceivably be used to predict catches from lakes at a global scale, or from locations where current estimates are highly uncertain. A similar model for estimating potential riverine fish catches as a function of discharge has been developed from yield data for 40 basins worldwide (McIntyre *et al.*, 2016). The use of such surrogate environmental data will be useful for modelling fish yields and could inform fisheries management in places where catch statistics are scant or lacking (Deines *et al.*, 2015). Differences between predicted yield and actual catches could be used as an initial indication of overexploitation, assuming the latter are not simply underreported, although underreporting, as mentioned above, is widespread.

The scarcity of reliable data on the status of fish populations in Asia, Africa and Latin America makes it difficult to manage such fisheries towards sustainability (Beard *et al.*, 2011). While global organizations such as the FAO have been working with various partners '... on developing robust and credible methods to address this issue ... recent plans have yet to prove successful, and a revised practical and cost-effective strategy is needed in order to assess accurately the state of inland fisheries at a global scale' (FAO, 2016: p. 45). 'Poor governance, management and practices, including illegal, unreported and unregulated fishing ... as well as poverty ... continue to be major obstacles to achieving sustainable fisheries' (FAO, 2016,: p. 81). In the context of such uncertainty, one thing is clear: fisheries management should be incorporated within the wider context of protection of freshwater biodiversity, particularly in places with the highest-yielding capture fisheries, such as the Mekong and Amazon. These water bodies also tend to be hotspots of freshwater biodiversity or, in the case of Lake Victoria prior to the introduction of predatory Nile perch (see Chapter 4), formerly harboured a highly speciose community. Protection of species diversity in inland waters, and the positive correlation between biodiversity and productive fisheries (Brooks *et al.*, 2016), could provide the basis for a win–win outcome for human food security and conservation.

Overexploitation of Other Freshwater Vertebrates: The Neotropics

While fishes may be the well-known freshwater group that are subject to overexploitation, they are far from the only animals to be thus affected. The historical Amazon 'fishery' of more than a century ago was not then, nor is it in fact now, solely constituted by fishes. Although *Arapaima* was the pre-eminent component of the catch, Amazonian manatee

(*Trichechus inunguis*: Trichechidae; up to 2.8 m body length) and podocnemidid turtles (*Podocnemis* spp.) ranked second and third in importance. Crocodilians such as the spectacled and black caimans (*Caiman crocodilus* and *Melanosuchus niger*: Alligatoridae) (Plate 15) also made up a significant component of the overall catch. The skins of these animals were exported to markets in the United States and Europe, continuing for about eight decades until the basin-wide collapse of exploited populations (Antunes *et al.*, 2016). Amazonian people tend to live along rivers so that aquatic and semiaquatic mammals and reptiles tend to be more vulnerable to open-access, unregulated hunting than their terrestrial counterparts. This was especially so during low-water periods or drought years when animals concentrated in large waterways became readily accessible to hunters.

The international trade had its origins in the collapse of the Amazonian rubber boom in 1912 brought about by competition from plantations in Malaysia. Enterprises involved in growing, processing and transporting rubber shifted to the export of animal hides. An estimate of the total magnitude of the trade between 1904 and 1969 (Antunes *et al.*, 2016) was that (in addition to vast numbers of terrestrial animals) it involved 1.9 million aquatic and semiaquatic mammals and 4.4 million black caiman, as well as other aquatic reptiles such as the giant South American river turtle (*Podocnemis expansa*). Amazonian manatee were killed for both meat and skins, while capybara (*Hydrochaerus hydrochaerus*: Caviidae), giant and Neotropical otters (*Peronera brasiliensis* and *Lontra longicaudata*; EN and NT respectively) were exploited for their hides and pelts. The trade was reduced following Brazil's ratification of CITES in 1975, when these four exploited mammals were listed in Appendix I granting them maximum protection; *P. expansa* was included in Appendix II. However, illegal trade persisted until the early 1990s when changing fashions in recipient countries and the declining popularity of fur eliminated the market for Amazonian animals. These changes were, however, too late to stave off population collapses. The giant otter had been driven to commercial extinction in the 1960s initiating a shift towards exploitation of the Neotropical otter, which largely replaced its larger relative in trade. Likewise, populations of black caiman had dwindled so much by the 1960s that the smaller spectacled caiman, previously ignored by hunters as it reaches only half the size of the black caiman, entered the market as a substitute. Amazonian manatee were hunted in huge quantities in the 1930s (peaking at 16 000 animals in 1938), but annual 'harvests' were less than 10% of that by the 1970s. The overexploitation of capybara did not occur until later, with numbers in trade increasing steadily before dropping dramatically in the mid-1960s, coincident with a steep rise in the price of their hides (Antunes *et al.*, 2016).

Although the intensity of twentieth century commercial exploitation has now been relieved, subsistence hunting continues and is largely tolerated under current legislation that protects indigenous livelihoods; this may be one reason why manatee, giant otter and black caiman have yet to fully reoccupy their former Amazonian ranges (Antunes *et al.*, 2016). In addition, new threats to some species have recently emerged, such as the utilization of Amazon river dolphins described in Box 3.7. The potential basin-wide cascading effects of overexploitation of these and other aquatic mammals and reptiles has received insufficient research attention, although depletion of megaherbivores such as manatee and capybara has been implicated in overgrowth of beds of aquatic and semiaquatic plants on Amazon floodplains (Castello *et al.*, 2013).

Elsewhere in South America, steps to control overexploitation of the Orinoco crocodile (*Crocodylus intermedius*: Crocodylidae; CR), the largest crocodilian in the region, resulted in it being granted legal protection in Colombia and Venezuela, and gaining CITES Appendix-I status, in the 1970s. Intensive hunting for hides during the nineteenth and twentieth centuries, which peaked in the 1930s and 1940s, have so reduced numbers of Orinoco crocodile that only around 1500 individuals remain in the wild, mostly in Venezuela. Numbers have been supplemented by a small reintroduction programme in Colombia where, since 2015, more than 40 captive-bred *C. intermedius* have been released in El Tuparro National Natural Park. Certain other Neotropical herpetofauna have been exploited to the extent that they are close to extinction but, as explained in Box 3.8, the motivation is quite unlike that underlying the killing of crocodilians.

Overexploitation of Old-World Herpetofauna

Although often cited as a cause of declines of herpetofauna such as crocodilians (as in the Amazonian caimans discussed above), the effects of overexploitation on other reptiles, and particularly on amphibians, have been poorly quantified (Schlaepfer *et al.*, 2005). There is a significant international trade in wild-capture frogs' legs from Asia to The United States and, especially, Europe, as well as a regional trade serving markets in Singapore, Hong Kong and Malaysia. The main supplying countries are India, Bangladesh and Indonesia but, in Indonesia at least,

Box 3.7 *The Amazon Piracatinga Fishery*

In January 2015, the Brazilian government implemented a five-year moratorium on the commercial fishing of piracatinga catfish (*Calophysus macropterus*: Pimelodidae); subsistence fishing will continue to be permitted (Franco *et al.*, 2016). The prohibition was introduced because, over the previous decade or more, fishers had been hunting the pink Amazon river dolphin, or boto (*Inia geoffrensis*: Iniidae; DD) (Plate 16) in order to obtain its flesh for bait; tucuxis (*Sotalia fluviatilis*: Delphinidae; DD), and black caiman were also used. The use of these baits, which should preferably contain a large amount of fat, reflects their attractiveness to the predatory and often necrophagous piracatinga, which are also known as vulture catfish. Fishing for piracatinga often involves enclosing an entire decomposing carcass within a submerged corral into which the catfish are lured; the bigger the bait, the larger the catch. The piracatinga fishery had been increasing in popularity because annual landings tended to peak soon after the start of the closed season for other fishes in Brazil. However, the initial stimulus for development of the piracatinga fishery was overexploitation and consequent declines in the abundance of a more sought-after pimelodid, the now-threatened capaz (*Pimelodus grosskopfii*; CR) – an especially popular food fish in Colombia. This represents yet another example of fishing down the Amazon River community, and the outcomes that arise from that process.

Anecdotal accounts suggest that as the piracatinga fishery grew, the abundance of boto and caiman declined. Reliable population-trend data for the post-2000 period are lacking, although the longer-term overexploitation of black caiman during the twentieth century is clear (Antunes *et al.*, 2016). Killing boto is illegal in Brazil, but the prohibition is seldom enforced. These dolphin are subject to a small amount of targeted killing for predator control by fishers, which may well have been the initial step towards their subsequent deployment as bait. Accidental bycatch and drowning of boto trapped in nets could also be a significant source of carcasses. Ending the commercial catch of piracatinga may remove much of the incentive to hunt dolphins since the use of vertebrates as bait is exclusive to this fishery (Franco *et al.*, 2016). The Brazilian moratorium does not cover the entire Amazon Basin, but it may encourage neighbouring countries where there are piracatinga fisheries, such as Colombia and Venezuela, to introduce similar legislation. A shift to the use of fish offal as bait could also be mandated if the commercial fishery for piracatinga was allowed to continue or resume.

Box 3.8 *The Titicaca Water Frog*

At around 1 kg average weight, the Titicaca water frog (*Telmatobius coleus*: Telmatobiidae; CR) is one of the largest frogs in the world. It is endemic to Lake Titicaca and its tributaries in Peru and Bolivia. Once common, this frog has been reduced to only a fraction of its former abundance. The primary cause can be attributed to overexploitation for food, in part because the frogs are believed to have aphrodisiac properties and can cure ailments; they are most often consumed after being skinned and blended into a 'smoothie' (*jugo de rana*) with honey, roots and herbs. They are also used in poultices. Secondary causes of frog decline are predation of tadpoles by introduced rainbow trout, and algal blooms associated with lake eutrophication (Icochea *et al.*, 2004). There is no evidence of a role for chytridiomycosis in the decline of the Titicaca water frog, but it may have been responsible for the apparent extinction of three *Telmatobius* species in Ecuador (Angulo, 2008).

Thus far, the Titicaca water frog has proved difficult to breed in captivity – the feat was not accomplished until 2010 – and there are as yet no restrictions limiting its collection from the wild. The high altitude of Lake Titicaca (surface elevation 3812 m), and hence the low oxygen content of its waters, has led to evolution of adaptations such as a high count of unusually small red blood cells and a low metabolic rate. However, the most obvious feature of the Titicaca water frog is possession of large, capillary-rich skin folds that function as gills. These enable the frog to remain under water without needing to surface for air; consequently, the lungs are greatly reduced and this species is wholly aquatic. Because of its excessive skin and wrinkled appearance, the Titicaca water frog has been dubbed the scrotum frog (as mirrored in its specific epithet), and this resemblance may have contributed to the unfortunate imputation of aphrodisiacal properties. Other Andean species of *Telmatobius* such as the Lake Junin giant frog, *T. macrostomus* (EN), are also threatened by overexploitation, as are their close relatives in the genus *Batrachophrynus*. However, as with many amphibians, long-term population declines of these frogs are attributable to a combination of interacting factors.

In October 2016, media reports of mass mortality of frogs in an affluent tributary of Lake Titicaca, which had become polluted by sewage and wastes from illegal gold mining, focused international attention on the Titicaca water frog. Some tangible action to protect it – and the lake as a whole – appeared forthcoming in January 2017 when the President Pablo Kuczynski announced the construction of 10 pollution treatment plants along rivers emptying into the Peruvian portion of Lake Titicaca, at an estimated cost of more than US$400 million.

the domestic demand is several times greater than the amounts exported (Warkentin *et al.*, 2009). This commercial exploitation involves tens of millions of frogs annually, mainly large-bodied dicroglossids (formerly placed in Ranidae), with the international trade in the order of 10 000 t frogs annually. Export of frogs from India was banned in 1987 because of environmental concerns, but this prohibition has not stopped the exploitation of wild populations since collection as a subsistence food has continued.

Data on species composition and total quantities of wild-capture frogs are generally lacking. Although such information is needed for management of stocks in frog-exporting countries, it would be difficult to collect because the body parts are skinned before export, and there is taxonomic uncertainty surrounding Asian frogs with large geographical ranges as they probably comprise cryptic-species complexes. In addition, demographic data on wild populations are scant to non-existent, so that population viability analyses under different offtake rates cannot be used to inform management. In terms of impacts on populations, there is little to rely on apart from anecdotal accounts of reductions in the abundance and range occupancy of large dicroglossids. Of perhaps greater concern is the collection and trade of Southeast Asian salamandrids for use as traditional medicine and to supply the ornamental trade (Rowley *et al.*, 2010; Sung & Fong, 2018). The scale of the trade in these amphibians has been largely unmonitored but, given the high prices that they can fetch in Europe and North America when sold as pets, some rarer – and hence more sought-after – salamanders are under considerable threat from overexploitation. Examples are species of *Paramesotriton* (including *Laotriton*) and *Tylototriton* that are generally range-restricted with small populations (e.g. *L. laoensis* and *T. vietnamensis*; both EN).

Humans in many parts of the world have traditionally eaten turtle meat and eggs, as in the case of podocnemidid turtles along the Amazon (see above), but the main consumers of turtles are now in East Asia (Japan, South Korea and, principally, China) where the meat and shells are considered to have medicinal value. Large numbers of many types of turtle are imported from all over the world, with 155 species recorded passing through Hong Kong alone (Cheung & Dudgeon, 2006); this represents 61% of the total global complement of chelonian species. Almost half of them (72 species) were globally threatened, and 77 species were listed (at that time) on CITES Appendix I or II. Most of the non-Asian species were traded as pets, but the bulk of the trade was in Asian turtles – mainly batagurids. Rapid growth of the Chinese economy during the 1990s was accompanied by increased demand for Asian turtles during the 1990s when overexploitation for international trade led to steep population declines. Imports into China came initially from Vietnam and Bangladesh and subsequently from Thailand and Indonesia. As wild populations declined, these countries began acquiring turtles from neighbouring countries and transshipping them. Thus turtles in India, Myanmar, Laos and Cambodia became subject to intensive collection pressures (for a summary, see van Dijk, 2000). The outcome was that, by 2003, more than 80% of the Asian turtle fauna, which consists of around 90 species, was categorized as threatened by the IUCN, with many of the others judged to be data deficient (and mostly very rare). Although more than 20 Asian turtles have since been added to CITES Appendix II, the rareness of some species, which are mainly those believed to bestow medicinal benefits, only adds to their market value and thereby incentivizes their illegal exploitation. Unless wild populations can be protected effectively, economic imperatives could result in some freshwater turtles being exploited to extinction (see Box 3.9).

One remarkable example of exploitation of wild animal populations involves seven homalopsine water snakes (mainly *Enhydris* spp.: Colubridae) from Tonlé Sap Lake, where market sales exceeding 8500 individuals per day have been recorded, and an annual exploitation rate in the range of 2.7 to 12.2 million snakes (mean = 6.9 million) – the actual number is probably towards the upper end of this range (Brooks *et al.*, 2008, 2010). This represents the greatest exploitation of any snake assemblage in the world, and seems most unlikely to be sustainable. The snakes are caught as food for humans but, principally, to sustain farmed crocodiles (mainly *Crocodylus siamensis*); the skins of large snakes are sold for export. Per-capita declines in catches of between 74% and 84% between 2001 and 2005 confirm that this activity is a conservation concern (Brooks *et al.*, 2010). The abundance of *Enhydris longicauda* (VU), the endemic Tonlé Sap water snake, which formerly made up 16% to 39% of the overall catch, has fallen to such an extent that the

Box 3.9 *The Economic Logic of Overexploitation*

One might anticipate that exploitation would cease as species become rarer because it is increasingly expensive to pursue species as they become harder to find. However, if the rarity of a species enhances its economic value, the incentive to exploit it remains or increases, even as their densities decline and they become vanishingly rare (Courchamp *et al.*, 2006). In population ecology, an Allee effect can be recognized as occurring when there is a positive relationship between the number of individuals in a population and their fitness. Exploitation of rare species can result in an anthropogenic Allee effect: rare species become ever-more valuable, therefore experiencing more exploitation and thereby becoming even rarer, with this positive feedback driving overexploitation until extinction. To look at this another way, speculators could profit from increasing rarity of species as this would increase their price, giving rise to a market in extinction (Silvertown, 2015). That market has driven declines of African elephants and rhinos as well as some freshwater ornamental species, such as the Asian arowana (see Box 3.1) and rare salamanders. It applies to freshwater turtles also. *Cuora trifasciata* (Bataguridae; CR) (Plate 17) was formerly distributed widely in southern China, Vietnam and Laos. Because it is thought to have cancer-curing properties, populations have been exploited to such an extent that only a few individuals remain in the wild in Hong Kong, southern China, where they are protected by law. The value of these animals has risen as they have become rarer, with media reports on a 2016 theft of 12 *C. trifasciata* turtles estimating their market prices at close to HK$1.5 million (~US$190 000).

species is threatened. Larger, low-fecundity species such as *E. bocourti* and *Homalopsis buccata* have also become increasingly rare. Snake capture is of relatively greater importance to the livelihoods of the least well-off people around Tonlé Sap, as it reduces their vulnerability to seasonal fluctuations in fish catches as well as longer-term reductions in landings from the lake (Brooks *et al.*, 2008).

Control of the commercial trade in snakes could begin with a closed season that corresponds to the main breeding season of the species caught and would likely be more acceptable to local communities and feasible to implement than an overall ban. In the absence of such measures, continued overexploitation of the Tonlé Sap snake assemblage may result in scarce and vulnerable species being driven to local or even – in the case of *E. longicauda* – global extinction, because relatively abundant species may continue to make generic hunting worthwhile. The same phenomenon would also make rare species in a multispecies river (or lake) fishery susceptible to loss and hence, as a driver of extinction, differs from the increase in exploitation intensity that occurs as valuable species become ever-more rare (see Box 3.9).

Most Old-World crocodilians have been overexploited historically for their skins, and the eggs of some species, such as the Indian gharial (Plate 18), mugger and false gharial (Table 3.1), are eaten also. As a result, they tend to be considerably more threatened than most of their Neotropical counterparts, although the Orinoco crocodile is an exception to this generalization (see above). Hunting for trophies, consumption of meat for medicinal reasons, or killing crocodiles because they are thought to compete with fishers have also contributed to declines in abundance to the extent that some species are now critically endangered in the wild (Table 3.1). As with other examples of overexploitation, such as the shift from black caiman to spectacled caiman in the Amazon, some 'fishing down' of Old-World crocodilians is evident, with populations of the slender-snouted crocodile being hunted for their meat and skins only after the commercial extinction of the larger Nile crocodile within their sympatric African ranges (Shirley, 2014).

Restrictions on international trade established by CITES have reduced the exploitation of some crocodilians, resulting in an improvement in the conservation status of *Crocodylus porosus* (it was downgraded from EN to VU by the IUCN in 1990, and to LC in 1996), but this is far from the case for most species. Recent habitat loss and degradation has resulted in continued declines of some species, with depleted populations of the Indian gharial, the false gharial (the skin of which has limited value), and the Philippines and Siamese crocodiles confined to fragments of their original ranges. The Chinese alligator is almost extinct in the wild, with its habitat in the lower Yangtze having been almost entirely transformed, but thousands of individuals exist in captivity as is also true for the Siamese crocodile. In that case, however, the animals are farmed for their skins and many captive animals are hybrids (or backcrosses) with *C. porosus* and the introduced Cuban crocodile (*C. rhombifer*; CR), which is highly threatened in its Neotropical range.

Given the large number of crocodilians held in captivity, release of individuals to supplement or re-establish wild populations seems feasible.

Table 3.1 *The conservation status of Old World crocodilians. The desert crocodile, which was discovered relatively recently (see Hekkala et al., 2011), has not been included. Threatened crocodilians have been historically depleted by hunting for their skins. CITES Appendix I includes endangered species, trade in which is normally prohibited; Appendix II includes threatened species, trade in which is controlled by permits. For two species, the listing depends on the geographical origins of the animals traded.*

Species	Range	IUCN status	CITES Appendix
Chinese alligator (*Alligator sinensis*)	Lower Yangtze River only	CR	I
Australian freshwater crocodile (*Crocodylus johnsoni*)	Northern Australia	LC	II
Philippines crocodile (*C. mindorensis*)	The Philippines; now confined to three isolated portions of its former range	CR	I
Nile crocodile (*C. niloticus*)	Widespread in Africa, including Madagascar	LC	I/II
New Guinea crocodile (*C. novaeguineae*)	Island of New Guinea	LC	II
The mugger (*C. palustris*)	Indian subcontinent and Iran; extinct over large parts of its former range (Bangladesh, Bhutan, Myanmar)	VU	I
Fresh- and saltwater crocodile (*C. porosus*)	Widespread from Australia to Southeast Asia, Bangladesh and the Philippines, but nowhere abundant north of New Guinea	LC	I/II
Siamese crocodile (*C. siamensis*)	Formerly widespread in Southeast Asia; now confined to Cambodia, but a few animals in Kalimantan, Thailand, Laos and (possibly) Vietnam	CR	I
Indian gharial or gavial (*Gavialis gangenticus*)	Indian subcontinent (excluding Sri Lanka); extinct in Myanmar	CR	I
Slender-snouted crocodile (*Mecistops cataphractus*)	Two isolated populations in central and West Africa	CR	I
False gharial (*Tomistoma schleglii*)	Borneo, Sumatra, Peninsula Malaysia and, possibly, Java	VU	I

LC = lower risk/least concern.

Some stocking of mugger populations was undertaken in India during the 1980s and 1990s and, more recently, has been carried out on a small scale for the Philippines and Siamese crocodiles, and also – with some limited success – for the Chinese alligator. However, apart from concerns over hybridization of farmed crocodiles, many animals held in captivity have limited value for conservation purposes because their original provenance cannot be determined. This may matter less for a species with a restricted range, such as the Chinese alligator, than for a once widely distributed species like the Siamese crocodile, which is now represented in the wild by scattered remnant populations.

The Consequences of Overexploiting a Keystone Mammal

While instances of overexploitation of aquatic mammals in the Neotropics are given in the previous section, and Old-World otters have long suffered from human depredations as mentioned close to the start of this chapter, overexploitation of beavers has had more far-reaching implications for ecosystem structure and functioning. The two extant species of beaver (Castoridae) are the second and third largest rodents in the world (after the capybara), with adults of both reaching around 20 kg in weight and, sometimes, 30 kg or more. They are ecologically similar, representing biogeographical equivalents. The Eurasian (or European) beaver (*Castor fiber*), which is the slightly smaller species, was hunted to near-extinction over much of its range for fur, meat and castroeum (secretions from scent sacs at the base of the tail used as an ingredient in perfumes and medicines). By 1900, only a few relict populations of perhaps little more than one thousand individuals remained. Similar depredations reduced populations of the North American Beaver (*C. canadensis*) (Plate 19); although they began somewhat later (early seventeenth century), beaver had been virtually exterminated across the United States and between Canada and Mexico by the mid-nineteenth century. Populations recovered to some extent in the twentieth century following legal protection, but translocations and reintroductions also played a role. Eurasian beaver populations have likewise increased substantially and, since the 1920s, there have been separate reintroductions to 15 European countries.

What were the ecological consequences of overexploitation of beavers? Both North American and Eurasian beavers are keystone species and ecological engineers that build dams and associated lodges. These structures alter river flow and sedimentation regimes, create wetlands and backwaters, and change food and habitat availability for other animals (Wright *et al.*, 2002; Pollock *et al.*, 2003; Rosell *et al.*, 2005). The purpose of the dam is generally to maintain a stable water level so that the entrance to the lodge (or den) remains underwater and can be accessed while remaining safe from predators. The North American beaver makes larger dams than the Eurasian species, and so tends to have more marked effects on landscapes and biodiversity. These are mediated through hydrogeomorphic processes, such as reduced bank erosion and stream incision, increased sediment trapping, and changes in stream flow and ground-water replenishment, leading to greater aquatic habitat hetero-geneity, provision of rearing, overwintering habitat and flow refuges for fishes, and benefits for amphibian populations and invertebrate produc-tion (Wright *et al.*, 2002; Kemp *et al.*, 2011). Aquatic animal biomass may be two to five times greater in streams with beaver ponds than those without (Naiman *et al.*, 1988), with abundance of water birds increasing by as much as 75 times (McKinstry *et al.*, 2001). There has been great interest in the possible effects of beaver dams on salmonids in western North America, but these turn out to be generally positive, with increased growth, abundance or body size in streams with beaver ponds (Pollock *et al.*, 2003). There is no substantial evidence that beaver dams obstruct movement or migration by trout and salmon (Kemp *et al.*, 2011). Despite such findings, there remains a view that the North American beaver is a nuisance animal whose activities increase the risk and the extent of flooding, damage trees and reduce timber yields, and even deplete fish stocks. Such concerns have led to the establishment of large-scale beaver control programmes in American states, such as New York, South Carolina, Nebraska and Utah.

The Eurasian beaver has similarly positive effects on biodiversity to its North American congener: among the more surprising consequences are increases in vespertilionid bats along streams due to the greater quantities of emerging aquatic insects, and provision of open foraging areas among tree gaps or over pond surfaces (Ciechanowski *et al.*, 2011). Reintroduc-tion of Eurasian beaver into degraded pastureland streams in Scotland showed the ecological benefits these animals bring: channels with beaver dams had higher levels of organic-matter retention and aquatic plant biomass, and beaver activities helped bring about in-stream habitat res-toration (Law *et al.*, 2016). Landscape-scale changes included higher biodiversity of macroinvertebrates, greater nutrient retention and flood attenuation. Even in places where beavers do not build dams, as in lentic habitats with stable water levels, they can enhance biodiversity and modify wetland characteristics through selective consumption of aquatic macrophytes and other plants (Law *et al.*, 2014; for more information, see Rosell *et al.*, 2005). In addition to building dams, other beavers coppice trees, and their activities are associated with expanded riparian zones and overall enhancement of environmental heterogeneity which, in turn, has positive effects on biodiversity. For this reason, beaver reintroductions have not invariably been directed only towards increasing their populations, but also towards a range of environmental management objectives such as restoration of degraded aquatic or riparian habitat, and even providing opportunities for ecotourism (Rosell *et al.*, 2005; Kemp *et al.*, 2011).

More generally, beaver dams in low-gradient rivers enhance formation of wetland complexes and braided stream channels where sediment deposition and storage is enhanced. Additional beaver-induced sedimentation below dams is induced by a process of positive feedback because the braided channels increase potential stream length, presenting further opportunities for dam construction thereby increasing the area occupied by beaver ponds and the amount of sediment trapped. And because beavers substantially influence floodplain aggradation (by as much as 0.5 cm annually), channel complexity within and among floodplains decreases significantly as beaver populations decline (Polvi & Wohl, 2012).

Beavers not only act as geomorphic ecological engineers but also play an intermediary role in trophic cascades between predators, large browsing herbivores and their plant forage species. Reintroduction of the North American beaver to forests in Yellowstone Park resulted in an increase of their populations only after the grey wolf (*Canis lupus*) had been reintroduced. These predators reduced elk (*Cervus elephas*: Cervidae) abundance thereby allowing a recovery of riparian stands of willows (*Salix* spp.: Salicaceae) – a major food of beaver. Subsequent beneficial ecological changes along Yellowstone streams arose from the activities of resurgent beavers (Ripple & Beschta, 2012). Through their action as ecological engineers and keystone species, beavers offer another example of the shifting baseline. Inspection of a beaver-free landscape would give a very incomplete picture of its historic aspect in the presence of beavers.

The ecological influence of extant beavers may, however, pale into insignificance when compared to those attributable to *Castoroides ohioensis* and *C. leiseyorum* – enormous Pleistocene beavers, reaching up to 100 kg and 2.5 m long, which lived in North America until the end of the Ice Age around 12 000 years ago. Unfortunately, we know nothing about the ecology of these particular mega-beavers, which were the last representatives of the now-extinct Castoroidinae. There is, however, good evidence of woodcutting and lodge-building by species of the genus *Dipoides*, a precursor of *Castoroides*, and it is now supposed that tree-exploitation and swimming appeared within a single Holarctic beaver lineage during the Miocene, roughly 24 million years ago (Rybczynski, 2007, 2008). It is therefore likely that *Castoroides* spp. had ecological effects on a much larger scale than produced by *Castor* spp. in recent history. More importantly, if various beavers have been engineering landscapes for millions of years – as seems to have been the case – the ecological consequences would greater and far more profound than if the behaviour had evolved only a few thousand years ago. We do not know what caused the extinction of *Castoroides* spp., but the extinction of Pleistocene megafauna in North America took place soon after the arrival of *Homo sapiens*; the role of humans in their demise is set out in the overkill hypothesis (Burney & Flannery, 2005). If that notion is correct, we may need to reassess our perception of the amount of landscape engineering attributable to beavers – and hence the extent of the historic baseline shift that took place after they were eliminated by humans.

Overexploitation of Invertebrates

In terms of the scale, geographical extent and number of species involved, invertebrates have suffered comparatively less from overexploitation than fishes and other vertebrates. There has, however, been some targeted fishery of crayfishes or yabbies (species of *Astacopsis*, *Cherax* and *Euastacus*: Parastacidae) as human food in Australia and Papua New Guinea, thereby reducing their numbers (Strayer, 2006). Although proscriptions on collection of certain species have been introduced, the Tasmanian giant freshwater lobster (*A. gouldi*; EN) is still subject to illegal fishing, while some rare crayfishes (e.g. *E. sulcatus*; VU, *C. leckii*; CR, and recently described *C. pulcher*) are also targeted for the ornamental trade (see Box 3.1; Plate 20). Elsewhere Malagasy endemic crayfishes (seven *Astacoides* spp.) are collected for food, and the largest of them, *A. betsileoensis* (VU), has undergone substantial population declines.

Some of the larger species of parathelphusid and potamid crabs are eaten in Asia, as are pseudothelphusids in South America, but this occurs on a far smaller scale than the exploitation of freshwater shrimps and prawns in tropical rivers. The most important components of the catch are palaemonids, primarily *Macrobrachium* spp., which are heavily exploited in Asia and parts of South America where they are the target of subsistence and commercial fisheries. Some small atyids (*Caridina* spp.) are taken also, and often eaten dried or fermented, while certain rare or colourful species are subject to potential overexploitation to satisfy the demands of aquarists. There is a huge literature on *Macrobrachium*, but it focuses on the aquaculture of large commercially valuable species (*M. rosenbergii* and a few others) and pays scant regard to the status of wild populations. Formal catch statistics for river shrimp are generally lacking, but there are anecdotal accounts of population declines (and increased market prices) due to overexploitation, illegal or destructive fishing practices, and intensive collection of juveniles for aquaculture. In rivers such as the Ganges, pollution has doubtless worsened the effects of overexploitation on shrimp populations – just as it has done for the fishes (Box 3.3) – and, in many places, dams have blocked the breeding migrations that are an obligatory part of the life cycle of many species of *Macrobrachium*. Of the 239 species of *Macrobrachium* that have been assessed by the IUCN, only a handful are classified as threatened (a great many more are data deficient) but, in some of these instances, overexploitation has contributed to their declines.

In addition to decapods, other freshwater invertebrates, mainly molluscs and a few insects, are collected for food. For instance, large belostomatid bugs (*Lethocerus indicus*) are regarded as a delicacy in Southeast Asia, where they are sometimes even farmed commercially, while dytiscid beetles (*Cybister* spp.) are snacked upon in southern China. However, there is no published information on the extent to which this practice impacts wild populations. Adults of planktonic *Chaoborus edulis* midges that emerge from Lakes Malawi and Victoria in vast swarms are cooked in patties and eaten by people dwelling along the shores, but this consumption likely has negligible effect on insect numbers. The same seems certain to have been the case for alkali flies *(Ephydra hians*: Ephydridae), the pupae of which were historically important as a food and protein source for a population of Paiute native Americans (called Kutzadika'a or Kucadikadi) that lived around hypersaline Mono Lake in California (see Box 6.4). However, at least one gastropod shows clear signs of overexploitation: the viviparid *Margarya melanioides* (EN), formerly a major commercial species in Dianchi Lake and other inland waters of Yunnan Province (China), has experienced a substantial reduction in range and population size due to collection for food and traditional Chinese medicine, although environmental degradation probably also played a role in its decline. During the 1940s, this snail and other members of the genus was widely distributed across plateau lakes in Yunnan and dominated molluscan assemblages, but shrank to less than 10% of former abundance after the 1990s; *M. melanoides* is now present in only two of eight lakes where it occurred formerly (Song *et al.*, 2013).

The pearly mussels (mostly unionids and a few margaritiferids) are one instance of a group of freshwater invertebrates where there is clear evidence of depletion by human exploitation (see also Box 2.2). This took place principally in Europe and North America due to demand for their nacreous shells and/or pearls but, in other parts of the world, unionids represent an important subsistence food. In the United States, these mussels formed the basis of a substantial pearl industry beginning in the 1850s; around 10 species were involved, and as pearls were present in as few as one mussel in a 1000, there was much mortality for little gain. Eventually, populations became overexploited and insufficient to sustain the industry, which collapsed in the 1900s (Humphries & Winemiller, 2009). Beginning in 1890, a wider variety of mussels were collected and their shells used for button manufacture, but less than 20 years later, many larger species had declined and attention had shifted to smaller species. Some harvests seem astounding: in 1913 alone, more than 13 million kg of shells were removed from living mussels in Illinois, and 100 million mussels were taken from a single 73-ha bed in the Mississippi River (Strayer, 2006). In parts of the lower Yangtze Basin in China, a pearl 'industry' continues, based on culture of a few relatively hardy species (mainly *Hyriopsis cumingii*, but *Cristaria plicata* and *Sinanodonta woodiana* have been used) yielding virtually all of the global supply of freshwater pearls.

Pearly mussels continue to be collected as subsistence food in parts of Asia and Africa (where the Irinidae are endemic), but no longer experience high levels of exploitation in the United States due, in large part, to the replacement of mussel-shell buttons by plastic substitutes in the mid-twentieth century. Nonetheless, overexploitation devastated the mussel fauna to such an extent that it has yet to recover. A return to pre-nineteenth century levels of abundance is, at any rate, unlikely due to habitat degradation, competition with invasive dreissenid mussels, and other contemporary threats facing North American unionids (see Box 2.2). Large-scale historic declines of many unionid species would have given rise to conditions whereby – as with large river fishes – recollections of their former abundance became subject to baseline shift. There would also have been associated changes in ecosystem functioning (Box 3.10).

Box 3.10 *Ecological Consequences of Pearly-Mussel Depletion*

Pearly mussels provide a range of ecological functions in freshwater ecosystems (reviewed by Vaughn, 2010, 2018). Living mussels and their spent shells offer or improve habitat for other organisms. In addition, an abundant and diverse mussel community, at historical population densities, could have filtered the entire contents of the water column of streams, at least at times of low flow. Overexploitation leading to loss of a substantial biomass of pearly mussels must have had a significant impact on food webs and transport or transformation of suspended organic matter, phytoplankton and so on. Their filter-feeding activity links benthic and pelagic compartments by transferring energy and nutrients from the water column to the sediment, biodepositing organic matter (as pseudofaeces) and excreting nutrients, thereby stimulating both algal and macroinvertebrate production. Thus, reductions in the abundance of unionids would reduce the extent of bentho-pelagic coupling. However, individual species do not play identical roles in nutrient cycling and are not necessarily substitutable. Pearly mussels are thermoconformers, and different species have disparate thermal optima for filtering and nutrient excretion. Thus, the interaction of species composition and temperature results in strong species–identity effects. This means that reductions in total biomass as well as changes in pearly mussel composition caused by overexploitation can have ecosystem-level consequences. These higher-level outcomes have some parallels with the profound changes wrought to food webs following the establishment of invasive dreissenid mussels in North America (see Chapter 4), although in that case the cause was an addition rather than a reduction of filter-feeding bivalves.

Mussels may regulate ecosystem services in places where their filtering activity enhances water clarity, such that high densities of unionids might remove enough phytoplankton to reduce the effects of eutrophication leading to 'biological oligotrophication'. This is of value to humans, in addition to the ecological effects on food webs mentioned above, since it increases water clarity and recreational value, and reduces the costs of water treatment, especially where the water is used for drinking. Chowdhury *et al.* (2016) reported that densities of two species of pearly mussels in polluted Dhanmondi Lake in the centre of Dhaka, Bangladesh, were sufficiently high to generate filtration rates with the potential to clear the near-shore waters within

24 hours, in spite of excessive nutrient levels favouring algal blooms. Parts of the lake with abundant mussels (mostly *Lamellidens marginalis*) supported a diverse macroinvertebrate fauna, suggesting that mussels acted as ecological engineers promoting localized biodiversity. Loss of mussels from this lake, and the associated ecosystem service resulting from biofiltration, would be costly to replace through other means (see also Strayer, 2017).

Reductions in ecosystem services have been directly linked to mussel losses in the Kiamichi River, Oklahoma (Vaughn, 2018). Over a 20-year period, drought-induced changes in flows and poor water management from a tributary reservoir caused declines in mussel biomass. Laboratory-derived measurements of filtration rates and other physiological traits combined with field estimates of species-specific mussel biomass revealed that biofiltration, nitrogen and phosphorus cycling, and nitrogen, phosphorus and carbon storage provided by mussels declined almost 60%, representing a substantial loss of ecosystem services. While this reduction was not a direct result of overexploitation, the overall effects would be similar.

Aside from the freshwater invertebrates collected to meet the demands of the aquarium trade (see Box 3.1), or exploited for human food, one other example is noteworthy. Large numbers of the medicinal leech, *Hirudo medicinalis* (Hirudinidae), were taken in Europe during the eighteenth and nineteenth centuries when the practice of blood-letting became particularly prevalent. Although overexploitation drove local population declines of this leech, habitat degradation is probably the major factor contributing to its current near-threatened conservation status (Utevsky *et al.*, 2014).

A Future for Freshwater Fisheries?

It is very clear that human exploitation of species for food or other products has significant impacts on freshwater biodiversity, leading to reductions in population size, altered age structure and shifts in the relative abundance of species that can have consequences for ecosystem functioning and service provision. As stressed repeatedly, overexploitation seldom acts in isolation, but its effects can often be direct and obvious. Although overexploitation is often thought of as something that mainly affects fishes, as the summary in Table 3.2 shows, a much wider range of animals can be involved.

But, to conclude this chapter, the focus returns to fishes. Fishing itself is usually the primary anthropogenic driver of the stock status of *marine* fish and their fisheries. Improving the management and governance of fisheries can thus serve the dual objectives of maintaining human food security and conservation of species that are being overexploited. In contrast, and with a few exceptions – principally involving large species and relatively pristine lakes or rivers – external anthropogenic stressors often pose a greater threat to freshwater fishes and other animals exploited by humans than the exploitation itself. This is especially true for river fisheries, which face a combination of threats from dam construction, water abstraction, land-use changes, pollution and other drivers described in Chapter 1. Overexploitation is often only one of a suite of interacting factors that need to be addressed if stocks are to be managed sustainably. A prerequisite for such management will be enhancing appreciation of the importance of fisheries and wild-capture as a food-supply system that underpins human well-being, especially in countries where people are most dependent on such fisheries to meet their nutritional requirements. Unfortunately, it is exactly these nations where fisheries may be most at risk in future. The inadequate consideration of fisheries in plans for water-resource and hydropower development threatens human communities and biodiversity in the Mekong and Amazon basins, as well as many smaller subsistence fisheries worldwide.

Even in the absence of grandiose water-engineering schemes, the security of freshwater fisheries may be at risk. A coarse-scale analysis of the effects of climate change on fish yields globally shows that the countries that are most vulnerable are the poorest or least developed with populations that are highly reliant on fish as a source of dietary protein (Allison *et al.*, 2009). They include nations in eastern and central Africa (e.g. Uganda and Malawi) where fisheries landings are derived largely from lakes (e.g. Chilwa and Chad, the Rift Valley lakes) and yields have declined (as described from Lake Malawi), as well as countries such as DR Congo, Bangladesh and Cambodia where river fishes are relatively important. All have limited capacity to adapt to climate-driven change. Although the precise impacts of climate change for particular fisheries are uncertain, and may act in a direct or indirect fashion to affect catches (Cochrane *et al.*, 2009), they will most likely augment the pressures of overexploitation and amplify the interacting anthropogenic

Table 3.2 *Summary of the effects of overexploitation on freshwater animals, indicating the range of taxa involved and variety of uses to which they are put*

Reductions of fish stocks:
  — Initially impacts larger or long-lived, late-maturing species, resulting in fishing down the food chain and exploitation of smaller, faster-maturing species
  — Migratory species, especially anadromous fishes, are especially susceptible to overexploitation because the timings and routes of migration facilitate a highly targeted fishery
  — The use of destructive fishing practices, such as poisons, electricity and fine-meshed nets, drive further overexploitation and are deployed as large fish become increasingly scarce
  — Changes in species composition can have implications for nutrient dynamics, material transfers and food-web subsides

Reductions of amphibians, water snakes, river birds, pearly mussels and other invertebrates:
  — Mostly exploited as a source of food, especially in Asia, where the largest freshwater snake 'fishery' in the world occurs at Tonlé Sap Lake, Cambodia
  — Birds that nest colonially in floodplain or riparian forest, or on sand bars in rivers, are vulnerable to collection of eggs or nestlings for food
  — Most frogs are exploited for food, but some amphibians are collected for use as tonics or medicines
  — Pearly mussels are exploited for food and formerly on an industrial scale for their nacreous shells and pearls; reduction in diversity and biomass could have effects on river food webs through reductions in bentho-pelagic coupling
  — Some other invertebrates (decapods, caenogastropod snails) exploited for food, as are some aquatic plants, but use of other invertebrates typically minor

Reductions of mammals, crocodiles and turtles:
  — Some exploitation for food, but other valuable products include the hides of crocodiles and shells or flesh of turtles that are used in traditional Chinese medicine; increasing scarcity of target species drives up their value and stimulates further exploitation
  — Growing prosperity of China has led to import of turtles from all parts of globe (especially other parts of Asia) to supply demand for medicines or tonics
  — Populations of aquatic mammals depleted by hunting in Amazon, Europe, North America and elsewhere for hides and pelts with peak exploitation in nineteenth and twentieth centuries; international market now greatly reduced

Collection for global ornamental (aquarium) trade:
  — May affect some fishes and herpetofauna, and a few aquatic plants, as rare or wild-caught specimens can fetch high prices, but many fishes formerly collected are now cultured on industrial scales
  — Increasing exploitation of crayfishes, as well as atyid shrimps and tropical snails

influences upon fishes and other freshwater species used by humans (Beatty *et al.*, 2014). Conventional fisheries management and the science underpinning it have tended to focus on target fish populations and are thus greatly hindered by a scarcity of data on the status of freshwater fish stocks in many parts of the world (as described above in the section on the magnitude of the global freshwater catch). And they seldom incorporate the impacts of other human activities and environmental drivers such as climate change. A more sustainable model for fisheries management would need a broadened scope to include conservation of freshwater biodiversity in general, while explicitly taking into account human livelihoods, as well the value of fisheries and their social and economic contributions to sustainability. Formulation of such a model would have to be accompanied by finding some means to integrate fisheries management into cross-sectoral national or regional governance structures. The food security of millions of people and the preservation of a substantial proportion of global vertebrate diversity hinge on achieving success in this matter.

4 · *Alien Species and Their Effects*

... what havoc the introduction of any new beast of prey must cause in a country, before the instincts of the aborigines become adapted to the stranger's craft or power.
Charles Darwin (1839: p. 478), Journal of Researches

... man is everywhere a disturbing agent. Wherever he plants his foot, the harmonies of nature are turned to discords.
George Perkins Marsh (1864: p. 36), Man and Nature

'Bad Biodiversity': What Are Alien Species?

To begin, a note on terminology: an exotic species is one originating in a different country or region, whereas one that is non-native (or nonindigenous) may originate within the same region or country but not occur naturally in a particular lake or river basin until colonization is facilitated by humans. Both are treated herein under the general category of alien species. An invasive species is an alien (exotic or introduced) species that either gives rise to ecological, economic, health or other concerns as a result of its establishment and spread, or has the potential to do so. The latter qualification is added because the impacts of invaders may remain undetected for some time, because they are subtle or because no one has investigated them. In addition, there may be a time lag between the arrival of an invader and the occurrence of any impacts arising from its establishment. This, incidentally, implies that native biotas may be experiencing an 'invasion debt' whereby the effects of recently arrived aliens have yet to be felt (Strayer *et al.*, 2006). For such reasons, I have opted in this chapter to deal with alien species in general, rather than restricting the account to invasives that have a clearly demonstrated ecological or socioeconomic impact. The main focus will be the harm or potential harm wrought by alien species on freshwater biodiversity. Except in a few cases where aliens provide habitat for indigenous species, or feed upon other invasives and hence have the potential to limit their numbers and activities, there are scarcely any instances where aliens have had positive effects on biodiversity. The benefits to humans that can be derived from non-native species used in aquaculture are, however, manifest.

Whether through deliberate or accidental introductions, alien species become established and may often spread to the detriment of native biodiversity, representing a leading cause of extinctions (reviewed by Loehle & Eschenbach, 2012; Bellard *et al.*, 2016). Over a half a century ago, Charles Elton (1958) wrote that the Earth was undergoing a great historical convulsion in its flora and fauna, likening the establishment and spread of invasive species to 'ecological explosions'. In fresh waters, as in other realms, such species have direct impacts on native species through predation and disease transmission, as well as by competition and hybridization. Aliens also exert indirect influences on receiving communities by altering energy flow and food-web architecture. Both processes are reflected in a global trend, at least among river fish communities, for the proportion of endangered native species within a drainage basin to be correlated with the percentage of alien species that have become established (Leprieur *et al.*, 2008). Furthermore, that percentage is strongly related to the amount of economic activity in the basin, and hence the demand for imported goods, as well as fluxes of alien species via the aquarium trade, sport fishing, aquaculture and, more recently, facilitated by the internet and social media (e.g. Sung & Fong, 2018). It is thus not surprising that, at the global scale, greater numbers of species of alien freshwater fishes have become established in areas with higher human population densities (Dawson *et al.*, 2017). Stocking of non-native species has long been a cornerstone of fisheries management, and such fishes represent a significant proportion of landings by recreational anglers, but they may be detrimental to native counterparts, particularly through predation. For instance, depredation of native species over wide geographical scales is attributable to the introduction of northern-hemisphere salmonids to inland waters of the southern hemisphere (reviewed by McDowall, 2006; see also McIntosh *et al.*, 2012).

This chapter will begin by describing some of the transformations brought about by piscivorous fishes stocked in lakes, but it is important to stress while fishes are widely stocked in inland waters, and transported internationally for aquaculture or as ornamental species for aquaria, they are by no means the only potential invaders of fresh waters. The aquarium trade has a long history of transporting plants and snails, and other invertebrates, into regions where they are not native and where they may be deliberately or accidentally introduced (Duggan, 2010). The global trade in some non-fish taxa is remarkable: at least 120 crayfish species have been reported from aquaria, around 20% of the total number of species globally (Chucholl, 2013), although only a fraction of these are commonly exported. Molluscs too are an increasingly important component of this ornamental trade (Ng *et al.*, 2016). 'Hitchhiking' snails that are transported along with more desirable species (or on aquarium plants), but which are overlooked because of their small size and cryptic appearance, frequently have greater potential to become established outside their native range than ornamental gastropods and bivalves (Patoka *et al.*, 2017).

Piscivores and Other Predators in Lakes

One of the earliest comprehensive documentations of the impacts of a non-native tropical fish dealt with the introduction of piscivorous peacock bass (*Chicla ocellaris*: Cichlidae) from the Amazon to man-made Gatún Lake in the Panama Canal (Zaret & Paine, 1973). They included the loss of six species of formerly abundant native fishes, and dramatic reductions in almost all populations of secondary consumers, with consequences for trophic levels both higher and lower than that occupied by the peacock bass. The largemouth bass (*Micropterus salmoides*: Centrarchidae) offers another example of the profound changes to a tropical lake that can be wrought by a single introduced fish species. This North American native eliminated more than half of the indigenous fishes in Lake Atitlán, Guatemala, and it contributed to the global extinction of the endemic Atitlán (or giant) grebe (*Podilymbus gigas*: Podicipedidae) by depleting the fishes and freshwater crabs (various Pseudothelphusidae) that constituted its principal food. The extinction of the confamilial Alaotra grebe (*Tachybaptus rufolavatus*) was likewise subsequent to the introduction of largemouth bass to Aloatra Lake, Madagascar, where this grebe was endemic. The predatory blotched snakehead (*Channa maculata*: Channidae) from China has also been implicated in the disappearance of this waterbird. Habitat degradation and entanglement in fishing nets seemingly played an auxiliary role in population declines of both grebe species (BirdLife International, 2014).

In naturally fish-free lakes, the introduction of piscivores can have adverse consequences for alternative prey: for example, salmonids – such as *Salvelinus* spp. (various species of charr or trout) – can deplete zooplankton (copepods and, especially, Cladocera) causing changes in energy flow through food webs in temperate lakes. Amphibian larvae in formerly fish-free lakes are severely affected by introductions of widely stocked species such as rainbow trout (*Oncorhynchus mykiss*) because they lack adaptations to evade or avoid these novel predators (Knapp, 2005). For instance, declines of the mountain yellow-legged frog (*Rana sierrae*: Ranidae; EN), endemic to the Sierra Nevada Range of the western United States, began coincident with the aerial 'planting' of fish fingerlings into remote high-altitude (>2000 m elevation) lakes during the 1950s, and had knock-on effects also on terrestrial snakes that preyed on the frogs. Experimental removal of trout from some of these mountain lakes led to the recovery of their imperilled frog populations (Knapp *et al.*, 2007), and subsequent reductions in trout abundance (as aerial planting has ceased) have been associated with further increases in frog abundance (Knapp *et al.*, 2016).

Introductions of predators can have especially grave consequences for biodiversity if they involve lakes that are large and ancient or if, for whatever reason, they contain significant numbers of endemic species. For example, the combined effects of predation by rainbow trout, introduced in the 1940s, and pejerrey (*Odontesthes* [= *Basilichthys*] *bonariensis*: Atherinopsidae) that invaded in 1955, led to marked reductions in abundance of several members of a species flock of *Orestias* killifishes (Cyprinodontidae) in Lake Titicaca on the Altiplano Plateau of Peru and Bolivia. More than half of this flock of ~25 species is endemic to Lake Titicaca, which at 8372 km^2 is South America's largest lake, and they exhibit considerable diversity of form and habit. *Orestias cuvieri* (Plate 21), the largest of these endemics, is thought to be extinct (Harrison & Stiassny, 1999), and several others are regarded as threatened. Elsewhere, the introduction of predatory snakehead gudgeon (*Giuris margaritacea*, formerly known under a variety of synonyms: Eleotridae) to Lake Lanao in the Philippines eliminated all but two of an endemic species flock of 18 or more *Puntius* barbs (Capuli & Froese, 1999). Of these, *Puntius lindog* (Cyprinidae; VU) formerly constituted the bulk of fisheries landings from Lake Lanao, but now comprises <0.01% of fish biomass (Ismail *et al.*, 2014). The losses of endemic species from Lakes Titicaca and Lanao, although tragic, pale into relative insignificance when compared with the best-known and, arguably, most disastrous instance of the depletion of such fishes by an alien piscivore. It took place in Lake Victoria, East Africa, as described in detail later in this chapter.

The profound and far-reaching consequences for lake ecology that can result from the introduction of predatory fishes are examples of what has been dubbed the 'Frankenstein effect' (Moyle *et al.*, 1986), whereby well-intentioned introductions or stock-enhancement efforts give rise to unanticipated and detrimental outcomes. It is not only fishes that can induce such effects. Predatory mysid (or opossum) shrimps (in the genera *Hemimysis*, *Limnomysis* and *Mysis*) introduced to lakes in Europe and North America were intended to exploit zooplankton and serve as a food for salmonids, thereby increasing their production (Spencer *et al.*, 1991). The outcomes were unexpected, and sometimes even involved declines in abundance and productivity of the very fish that mysids were intended to benefit. What had been overlooked is the fact that mysids are largely nocturnal, and hence can evade predatory fishes. They are able to deplete the largest zooplankton, particularly Cladocera, which are important food of pelagic fishes such as the Arctic charr (*Salvelinus alpinus*) of lakes in Canada and Scandinavia. In other instances, the impacts of alien mysids have been more complex, involving both positive and negative interspecific interactions with fishes, and giving rise to trophic cascades that extend beyond the land–water boundary. Such outcomes are exemplified by Flathead Lake in Montana (see Box 4.1).

Box 4.1 *An Introduced Mysid Reshaped the Ecology of Flathead Lake*

The ecology of Flathead Lake, which at $510\,\mathrm{km}^2$ is the largest lake in the western United States, has been transformed as a result of changes wrought by *Mysis diluviana* (formerly *M. relicta*) after they became established in 1981 (for details, see Spencer *et al.*, 1991; updated by Ellis *et al.*, 2011). The mysid was far from the only alien species in Flathead Lake. Although it was habitat to 10 native fish species, a further 14 non-native fishes were introduced between 1890 and 1920, but it was not until the arrival of *M. diluviana* that major changes in lake ecology were detected. The mysid was stocked to enhance sport-fishing opportunities by supplementing the food supply of kokanee salmon (*Oncorhynchus nerka*: these fish are land-locked populations of the anadromous sockeye salmon). Nocturnal *M. diluviana* evaded the depredations of the visually hunting kokanee, instead competing with this zooplanktivore for copepods and cladoceran prey. The mysid did, however, serve as a benthic food resource for juvenile lake trout (*Salvelinus namaycush*), a non-native species that had been introduced at the beginning of the twentieth century, releasing a recruitment bottleneck for this piscivore allowing them to increase in abundance and extirpate kokanee. For the same reason, the non-native Lake whitefish (*Coregonus clupeaformis*: Salmonidae), introduced from the Laurentian Great Lakes in 1909, also became more numerous, although their impacts on kokanee were relatively minor as they are smaller and less voracious than lake trout. The establishment of *M. diluviana* thus had the indirect effect of facilitating predation on kokanee and led to the ecological dominance of Flathead Lake by lake trout. Kokanee were themselves not native to Flathead Lake: after introduction in 1916, they thrived and largely supplanted indigenous westslope cutthroat trout (*Oncorhynchus clarkii lewisi*) as the main sport fish. Both kokanee and native bull trout (*Salvelinus confluentus*; VU) are now imperilled by predatory lake trout (Ellis *et al.*, 2011).

The establishment of alien species has altered energy flow within Flathead Lake from the original food web based on pelagic prey (zooplankton → westslope cutthroat trout → bull trout), through a transitional phase during the mid-twentieth century following fish introductions (zooplankton → kokanee and small lake whitefish → bull trout), to one involving a benthic or profundal pathway (zooplankton and benthos → mysids and small lake whitefish → lake whitefish and small lake trout → large lake trout) brought about by mysid facilitation of lake trout (Fig. 4.1). The resultant changes in energy supply to the upper trophic levels of the lake food web led to declines in bald eagle (*Haliaeetus leucocephalus*: Accipitridae), gulls (*Larus* spp.: Laridae) and grizzly bear (*Ursus arctos*: Ursidae) that all fed on kokanee during their spawning migrations up affluent tributary streams of Flathead Lake within Glacier National Park. The collapse of spawning runs and declines in associated wildlife also diminished the number of tourists visiting the national park. The effects of this introduced mysid were thus sufficiently far-reaching to compromise the provision of ecosystem services to humans.

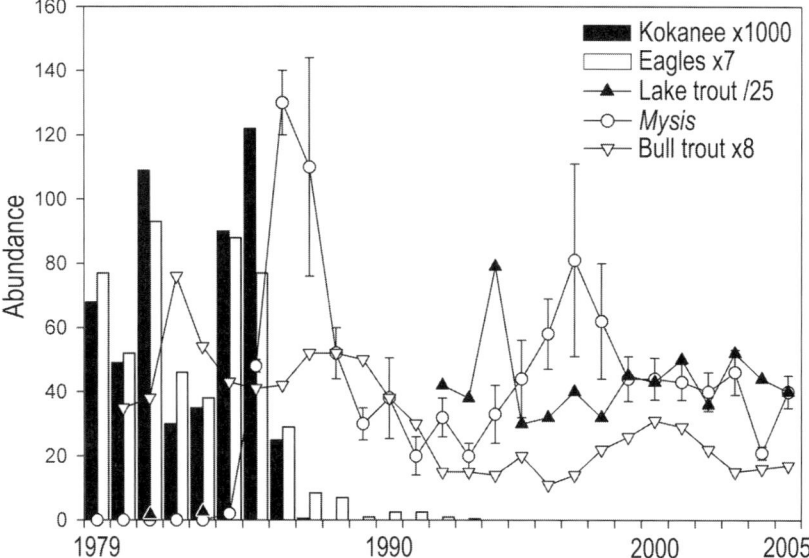

Fig. 4.1 Abundance history of kokanee, eagles, *Mysis* (mean numbers per square metre with 95% confidence intervals), lake trout and bull trout in Flathead Lake. Note that abundance of lake trout in 1981 and 1983 was near zero, and they were not sampled again until 1992. Redrawn from Ellis et al. (2011), where information on sampling methods is given; used with permission.

Diverse Impacts of Aliens: Hybridization and Competition

While alien predators evidently have considerable potential to impact native species or ecosystems, not all changes attributable to aliens are directly (or, in the mysid example in Box 4.1, indirectly) mediated by predation. Competition for food or space may also play a role, in addition to transmission of pathogens to native species, thereby causing their decline. These effects are often aggravated by land-use change and forest clearance that increase sedimentation and are usually associated with stream warming that may be detrimental to native biota. In addition, hybridization with a relatively eurytopic invasive species may swamp the gene pools of more narrowly distributed native species. Introduced trout provide instructive examples of both hybridization and competition.

Salmonids – particularly trout – are among the most successful fish invaders, and were priority species for translocation because of their popularity as sport fishes. Brown trout (*Salmo trutta*) was first introduced from Europe to the United States in 1883 – they had arrived in Australia almost 20 years earlier in 1864, and reached New Zealand by 1865. Rainbow trout from the United States became established in the southern hemisphere during the 1890s, arriving in Europe not long after and, since 1874, has spread across the North American continent far beyond its native range along the Pacific coast. Rainbow trout is now the most widely introduced salmonid globally, but it is not well suited to the European climate where it has fared less well than its European counterpart has done in North America. Brown trout have largely replaced the smaller indigenous brook trout (*Salvelinus fontinalis*) in streams of the American Midwest (for details, see Moyle *et al.*, 1986). Various populations of the native cutthroat trout (*Oncorhynchus clarkii*), which constitutes 14 separate subspecies, are at significant risk of extinction because they readily hybridize with rainbow trout in its expanded North American range. First-generation hybrids between rainbows and westslope cutthroats in the Rocky Mountains have lower fitness – but higher fecundity – than purebred parents, and the reproductive success of these hybrids may result in westslope cutthroats being interbred to extinction (Muhlfeld *et al.*, 2009). Hybridization with rainbow trout was likewise a contributing cause of extinction of the Alvord cutthroat (*O. clarkii alvordi*) in Oregon and Nevada during the mid-twentieth century. Rainbow trout also compete with cutthroats, as does the lake trout that has been stocked in many lakes well beyond its native range (they have even reached New Zealand), and such competition has been implicated in a dramatic decline in populations of Yellowstone cutthroat trout (*O. clarkii bouvierii*). The long-term prospects of the various subspecies of cutthroat within the United States is complicated further by the incidence of some misguided translocations that have led to hybridization among subspecies that were formerly geographically isolated.

Both hybridization and competition with rainbow trout threaten Gila trout (*O. gilae*; EN) in the southwestern United States, where captive breeding and aggressive stocking has been adopted as a means of attempting to boost its abundance and range. Hybridization with rainbow trout is likewise the major threat faced by the critically endangered Apache trout (*O. apache*) – the state fish of Arizona; it also interbreeds with stocked cutthroat trout. Other effects of translocated salmonids are evident in Eurasian lakes, where it has recently become clear that whitefish (*Coregonus* spp.: Coregoninae, a salmonid subfamily) represent a larger and more species-rich taxon than was previously supposed, and where translocations of competitively dominant species (such as the vendace, *C. albula*) could give rise to hybridization with native congeners or be otherwise detrimental to species such as *C. lavaretus* and *C. danneri* (both VU) that are endemic to one or a few lakes. Brown trout, reared in hatcheries in northern Europe and stocked widely for angling in streams further south, also pose a threat to the population integrity of native congeners with which they readily hybridize, including the Rhône trout (*Salmo rhodanensis*; DD), the marbled trout (*S. marmoratus*) of the Balkans, and the Zeta trout (*S. taleri*: DD) of Montenegro.

Competitive effects of invasive fishes, which may be accompanied by hybridization with some or all of the affected native species, are especially well reported in the Salmonidae but not confined to this family. Clariid catfishes are affected also: the broadhead catfish (*Clarias macrocephalus*) in Southeast Asia is undergoing genetic introgression due to interbreeding with the African sharptooth catfish (*C. gariepinus*) that has become widely established in the region after escaping from aquaculture facilities. The Esocidae provide a second example: piscivorous northern pike is prized by anglers, and stocked in Europe beyond its native range, where it poses a threat to southern pike (*E. cisalpinus* [= *flaviae*]; DD) through hybridization (Lucentini *et al.*, 2011). Northern pike also hybridize with muskellunge (*E. masquinongy*) in parts of the United States where introductions have brought these species into contact, although predation of native fishes by northern pike is a matter for greater concern than this hybridization. Among the Cyprinodontidae, the Pecos pupfish (*Cyprinodon pecosensis*; VU) has a limited native range within the southwestern United States and is threated by hybridization with the sheepshead minnow (*C. variegatus*), a bait fish commonly used by anglers that has been introduced into the Pecos River from the east coast. Another bait fish, the red shiner (*Cyprinella* [= *Notropis*] *lutrensis*: Cyprinidae) likewise hybridizes with congeners in its introduced range, including the blacktail shiner (*C. venusta*) and some of its geographically restricted subspecies.

The mahseers (all formerly grouped under *Torr* spp.: Plate 22) are a group of around 16 species of large-bodied, southern-Asian cyprinids that are prized sport fishes and of great cultural significance in Indian literature and religious scriptures. The impacts of overfishing and habitat degradation have led some species to be classified as threatened by the IUCN, but there is much ambiguity over the systematics and distribution of many members of the genus. Further confusion arises from the fact that hybridization can occur if previously isolated species (or morphotypes) are brought into contact. The results can be seen from recent evidence that the humpbacked mahseer, an inadequately described *Tor* sp. endemic to the River Cauvery in the Western Ghats, declined after the river was stocked with hatchery-reared juveniles of another mahseer morphotype – probably a hybrid – during the 1980s. Thirty years later, the humpbacked mahseer is close to extinction due to the combined effects of hybridization and competition with the introduced form (Pinder *et al.*, 2015), as well as a general deterioration in habitat quality. Added to the taxonomic confusion around mahseers, it is evident that the account of the humpbacked mahseer (*Hypselobarbus* [= *Tor*] *mussullah*; EN) in the IUCN Red List (Dahanukar & Raghavan, 2011) does not refer to the same species (or morphotype) as the humpbacked mahseer of the Cauvery (that has since been described as *Tor remadevii*), although both are endemic to the Western Ghats.

Most studies of the conservation implications of hybridization between native and alien fishes focus on the potential for introduced alleles to overwhelm native genomes. While hybridization with one or more native species can occur after the arrival of an invasive species, hybridization between invasive species is also a possibility, and it can exacerbate impacts on native species. This is because established populations of invasives often exhibit limited genetic variability resulting from founder effects or bottlenecks, and hybridization can increase genetic and phenotypic diversity within the invasive species. Such hybridization has been reported in several families: for instance, between cyprinids such as roach and common bream (*Abramis brama*) in Ireland (Hayden *et al.*, 2014), and between introduced silver carp and bighead carp in the United States (Lamer *et al.*, 2010). It has also been recorded between the monkey goby (*Neogobius fluviatilis*: Gobiidae) and round goby (*N. melanostomus*) in the River Rhine (Lindner *et al.*, 2013), and between invasive *Oreochromis* (Cichlidae) tilapias that are widespread in the tropics and subtropics (Plate 23). This brief list is by no means exhaustive. While there are other instances where introductions of fishes have led to hybridization, genetic introgression and harm to indigenous populations, the point has been sufficiently made herein. Hybridization within other – non-fish – taxa has been reported less frequently, but some examples are given in Box 4.2.

As well as, or in addition to, hybridization, competition with invasive species is detrimental to native species in a range of taxa. For instance, populations of three species of *Yunnanilus* loaches (Balitoridae) endemic to a karstic endorheic basin on the Yunnan Plateau in China declined following the introduction of cyprinid competitors. Two other

Box 4.2 *Hybridization between Non-fishes*

Hybridization between native and alien species taxa occurs in taxa other than fish. The Japanese giant salamander (*Andrias japonicus*: Cryptobranchidae) is affected in parts of its native range by hybridization with the Chinese giant salamander (*A. davidianus*) that became established in rivers around Kyoto after being imported as a gourmet delicacy between the 1950s and 1970s (www.nies.go.jp/biodiversity/invasive/DB/detail/40250e.html). These salamanders are the two largest amphibians in the world, but the Chinese species reaches 1.8 m in length, compared to the 1.5 m of its Japanese counterpart; the size differential allows *A. davidianus* males to monopolize females of both species during mating, and to occupy the best nest sites. This competition results in introgressive hybridization with the native salamander, and *A. japonicus* is now globally near-threatened. While the Chinese giant salamander is regarded as an invasive nuisance in Japan, it is critically endangered at the global scale and protected by law in its native range; international trade in this species is also proscribed under CITES (Appendix 1).

Another example of non-fish introgressive hybridization involves the invasive North American ruddy duck (*Oxyura jamaicensis*: Anatidae) which interbreeds with the white-headed duck (*O. leucocephala*; EN) in Europe. The widely domesticated mallard (*Anas platyrhynchos*: Anatidae) also readily hybridizes with native ducks in various parts of its invaded range, including the New Zealand grey (or black) duck (*A. superciliosa*) and the Hawaiian duck (*A. wyvilliana*, EN). Among invertebrates, there is evidence for hybridization and introgression between the pleurocerid snails *Pleurocera* (= *Elimnia*) *virginica* and *P. livescens*. These allopatric molluscs came into contact after the opening of the Erie Canal connecting the Hudson River with the Great Lakes during the nineteenth century. This allowed *E. virginica* to move from the Atlantic Slope drainage into Lake Ontario and the interior basin streams that are the habitat of *E. livescens* (Bianchi *et al.*, 1994). Further examples of threats posed by hybridization and introgression are given by Rhymer & Simberloff (1996).

As with fishes, there are instances where the hybridization associated with invasions occurs among two or more aliens. Several species of willows (*Salix* spp.) and their cultivars have spread from their origins in Europe and Asia to Australia, New Zealand, South Africa and parts of the New World. There, they displace native trees and shrubs and come to dominate riparian zones; reduce in-stream flows because of their high water use; and alter organic-matter dynamics by changing the quantities and quality of allochthonous litter entering streams (Serra et al., 2013). A few species have been naturalized in Australia and, while some can spread by vegetative means (becoming established from twigs or stem fragments), the most widely distributed, S. cinerea, readily hybridizes with other species; S. babylonica breeds with willows introduced from New Zealand that are themselves hybrids based on S. matsudana. The extent of hybridization is such that many of the invaders cannot be reliably identified to species. Interbreeding is facilitated by changes in flowering phenology in response to the novel environment, thereby removing pre-reproductive barriers that would have prevented hybridization within the native range.

Hybridization that arises from the establishment of alien species can have ecological effects on species not involved directly in the hybridization. For instance, hybridization between the barred tiger salamander (Ambystoma tigrinum mavortium) and the threatened native California tiger salamander (A. californiense; VU) has proceeded over several decades following the introduction of the former from Texas where it is sold as live bait for anglers. Hybrid salamanders are now widespread in California. Larvae have a tendency to develop a cannibalistic morph that is the apex predator in seasonal ponds (Ryan et al., 2009). They dramatically reduce survival of California tiger salamanders and other, more common, amphibians such as the Pacific chorus frog (Pseudacris regilla: Hylidae) and the California newt (Taricha torosa: Salamandridae). Demonstrated impacts of the hybrid salamanders on abundant species do not bode well for the threatened and endemic California red-legged frog (Rana draytonii; VU) or the Santa Cruz long-toed salamander (Ambystoma macrodactylum croceum), which are both listed under the federal Endangered Species Act.

Yunnanilus endemic to Dianchi Lake, also in Yunnan, have apparently become extinct due, at least in part, to competition with top-mouth gudgeon (Pseudorasbora parva: Cyprinidae) introduced from elsewhere in China (Kottelat & Chu, 1988). One of these loaches (Y. nigromaculatus) fed exclusively on atyid shrimps, particularly Sinodina (= Caridina) *gregoriana*, and an additional contributing factor in its decline was likely reductions in atyid populations due to competition with the considerably larger palaemonid shrimp, *Macrobrachium nipponense*, introduced to Dianchi Lake. An array of fishes has been introduced to this lake and, in combination with eutrophication, have had major consequences for the community structure of native species (Ye *et al.*, 2015; Ding *et al.*, 2017). Elsewhere, competition among other decapod crustaceans came about after establishment of the western North American signal crayfish (*Pacifastacus leniusculus*: Astacidae) in Europe, where the relatively large-bodied invader displaced native crayfishes (Dunn, 2012); they also fed on juveniles of the European crayfish (*Astacus astacus*). Prior to their arrival in Europe translocations of signal crayfish within the United States during the nineteenth century, and those of the red swamp crayfish (*Procambrus clarkii*: Cambaridae) (Plate 24), together contributed to the global extinction of the sooty crayfish (*Pacifastacus nigrescens*) in California. Red swamp crayfish, which have been widely introduced outside the United Sates, have also been implicated in declines and extirpations of populations of the California newt, because they are resistant to the tetrodotoxin that protects newt eggs from native predators. The rusty crayfish (*Orconectes rusticum*: Cambaridae) is likewise invasive in parts of North America beyond its native range but has yet to spread more widely.

The red swamp crayfish is an important aquaculture species and the most widely introduced crayfish species, occurring also in the aquarium trade. It is tolerant of a wide range of conditions and has established feral populations outside its native range in the New World as well as in Europe, Asia and East Africa. Stocking of red swamp crayfish has even been undertaken in Kenya in an attempt to control snail vectors of schistosomiasis (Lodge *et al.*, 2005), although this seems an unwise strategy in view of the circumstantial evidence that this animal was responsible for the elimination of native plants from Lake Naivasha (Smart *et al.*, 2002; see also Box 4.3). The congeneric marbled crayfish (*P.* 'marmorkrebs' *fallax*), which recently begun to be traded internationally, originates in the southern United States but has been found in the wild in Madagascar and parts of Europe (Martin *et al.*, 2010). Marbled crayfish are triploid parthenogenetic descendants of the sexually reproducing *P. fallax* from North America; they have obscure origins in the aquarium trade and do not exist naturally in the wild – except where they have been introduced. As they are capable of asexual reproduction, marbled crayfish have been described as the 'perfect invader' (Jones *et al.*, 2008), posing a threat to endemic Malagasy parastacids such as *Astacoides betsileoensis* and *A. hobbsi* (both VU).

Box 4.3 *Of Carp and Crayfish: Shifts in Ecologically Dominant Invaders of Lake Naivasha*

Lake Naivasha in Kenya offers a remarkable example of a water body in which changes in ecology reflect the dominant role of a succession of alien species that became established after anthropogenic alteration of conditions within the lake (reviewed by Gherardi *et al.*, 2011). This shallow (3–6 m deep; ~140 km^2) equatorial lake was designated a Ramsar site in 1995 in recognition of its important bird biodiversity, but, since then, the condition of the lake and fringing swamps has deteriorated as a result of land-use change within the catchment and, especially, the presence of alien species. The former reflects the huge expansion of horticulture and floriculture around Lake Naivasha during the last three decades, and a more than 10-fold increase in human population densities. The consequences have been eutrophication of the lake and increased sedimentation. Furthermore, although lake area and depth vary considerably reflecting fluctuations in rainfall, water consumption within the catchment has risen markedly and lake levels are only about one third of former values. The combined effects of reductions in water quantity and quality within Lake Naivasha probably helped facilitate the establishment of alien species.

At least 23 species have been accidentally or deliberately introduced to Lake Naivasha, and 14 of them are well established (for a complete list, see Gherardi *et al.*, 2011). Alien species dominate each trophic level of the lake food web except the top carnivores, which are all indigenous birds. The red swamp crayfish, stocked in 1970, and common carp (*Cyprinus carpio*: Cyprinidae), accidentally introduced and first reported from the lake in 2001, are the most important of these aliens. During different periods, they each act(ed) as ecologically dominant species in the lake. The omnivorous crayfish fed on native aquatic plants, eradicating them within 15 years of arrival; crayfish also decreased water clarity through bioturbation. Lake-bed seed banks provided the basis for some recovery of macrophytes after crayfish numbers declined due to a combination of overexploitation, water-level reduction and disease, and there were signs that a 'boom-and-bust' cycle between the plants and their herbivore was developing. However, this relationship was complicated by the establishment of alien plant species that were not eaten by crayfish, and the subsequent arrival of common carp. The carp fed on and competed with red swamp crayfish, suppressing their numbers and supplanting them as an ecologically dominant species in Lake Naivasha. Like crayfish, common carp uproot and consume submerged macrophytes (e.g. Weber & Brown, 2009; see also the section on Asian carp in the United States), but, contrary to expectations, regeneration of native macrophytes (e.g. *Potamogeton pectinatus*: Potamogetonaceae) took place despite carp establishment, covering substantial parts of the lake area (Britton *et al.*, 2007). Common carp also displaced the fishes that constituted most of the catch from the lake: the largemouth bass, introduced from North America as a sport fish, and two non-native African tilapias (*Tilapia zillii* and *Oreochromis leucostictus*: Cichlidae).

By 2008, common carp made up more than 90% of the total fish catch from Lake Naivasha, amounting to more than 9000 kg annually and providing a significant economic benefit to the surrounding inhabitants. The catch, as well as that derived from the former fishery for red swamp crayfish, is a significant provisioning ecosystem service derived from alien species that would not have existed prior to their arrival. This is because, unusually for a tropical lake and probably due to historic drying events, Lake Naivasha contained only one native fish: an endemic, but undescribed, species of *Aplocheilichthys*. This minnow-sized poeciliid could not have sustained a significant fishery and was apparently eliminated by largemouth bass. One impetus for the introduction of the red swamp crayfish was to provide food for bass, in addition to establishing an export fishery based on this crustacean (Gherardi *et al.*, 2011). However, the establishment of common carp thwarted both ambitions. It may be too early in the invasion of Lake Naivasha to fully assess the ecological impact of common carp, especially if their population continues to increase. And in view of the impacts of carp on a well-established fishery of naturalized tilapias, this alien introduction should not be seen as a panacea for declining fisheries in other degraded Kenyan lakes (Britton *et al.*, 2007).

Alien Pathogens and Their Impacts

Aliens have the capacity to devastate native biodiversity through the diseases they carry with them and transmit to indigenous species not previously exposed to such pathogens. A salutary example is the spread of crayfish plague in Europe, following the introduction of signal crayfish from the United States. This pathogen – an oomycete fungus, *Aphanomyces astaci* – devastated native astacids such as the white-clawed crayfish (*Austropotamobius pallipes*: Astacidae; EN), whereas the signal crayfish is resistant to the disease. The red swamp crayfish is likewise unaffected and can transmit the pathogen, which is a matter for concern given its global spread and the fact that all non-North American crayfishes are susceptible to the fungus. In fact, of around 120 crayfish species in trade, 105 species originate from the New World and are likely to be potential vectors of crayfish plague (Chucholl, 2013). White spot syndrome virus (*Whispovirus* sp.: Nimaviridae) can also be spread by American crayfishes to their European counterparts, although the virus probably originated in China and spread to the United States during the 1990s via export of infected peneaid shrimps.

A range of pathogens is transmitted by fishes: species implicated include the top-mouth gudgeon, which has spread widely through central Asia into Europe from its origin in China (Gozlan *et al.*, 2005), and is a vector of the rosette agent *Sphaerothecum destruens* (Rhinosporideaceae), an intracellular protistan parasite that infects a wide variety of native fish species with chronic or fatal effects. This protist has been implicated in declines of cyprinids and brown trout in Europe and mass mortalities of salmonids in the United States (Andreou *et al.*, 2012). The spread of *Myxobolus cerebralis* (Myxobolidae) serves as a second example of an invasive pathogen transmitted by non-native fish species; it causes fatal whirling disease of young salmonids. *Myxobolus cerebralis* was introduced from Europe to the United States along with infected brown trout which, due to a common evolutionary history with the pathogen, suffers relatively mild effects. However, outside Europe, rainbow trout and steelhead (anadromous rainbow trout) proved particularly susceptible to its ravages, and this pathogen has caused population declines of mountain whitefish (*Prosopium williamsoni*), as well as bull, brook and various cutthroat trout subspecies (Gilbert & Granath, 2003). Some – such as Yellowstone cutthroat – are already threatened by other factors (see section on Diverse Impacts of Aliens: Hybridization and Competition), and whirling disease may be contributing to further declines (Koel *et al.*, 2006). The spread of *M. cerebralis* was made easier by the fact that the intermediate host required to complete its life cycle is a tubificid worm (*Tubifex tubifex*) with an almost cosmopolitan global distribution. In addition, some piscivorous birds (such as great blue heron, *Ardea herodias*: Ardeidae) that eat infected fishes concentrate viable myxospores in their faeces and disseminate them to new habitats (Koel *et al.*, 2010). Whirling disease has been spread to countries (probably via infected trout) as far afield as South Africa and New Zealand. Thus far, the effects of the pathogen have been less damaging than in the United States, because of the absence of native salmonid hosts.

As well as serving as a vector for pathogenic microorganisms, alien species may transmit metazoan parasites that harm native species. The swimbladder nematode (*Anguillicoloides crassus*: Dranuculidae) is an endo-parasitic worm accidentally introduced to Europe along with Japanese eels (*Anguilla japonica*) translocated from Asia during the 1980s for food and aquaculture purposes. The nematode readily infested the European eel; it spread widely due to an ability to develop in a range of intermediate hosts (usually copepods) and because stocking of infected fish into waterways led to declines in the vitality of native eel populations (Didžiulis, 2013). In addition to spreading introduced parasites, alien species can also facilitate native parasites. The invasive faucet snail *Bythinia tentaculata* (Bythiniidae) from Eurasia is able to serve as a novel intermediate host for two trematode parasites *Sphaeridiotrema globulus* (Psilostomatidae) and *Cyathocotyle bushiensis* (Cyathocotylidae) in North America, allowing them to expand their range. The snail also transmits a European trematode (*Leyogonimus polyoon*: Lecithodendriidae) introduced after it had become established in North America. The parasites use native waterfowl as final hosts, and mass mortalities of birds may result from eating infested snails (Herrmann & Sorensen, 2009).

Diseases of freshwater animals can spread via human agencies without the need for a non-native vector such as a crayfish or fish or snail, and such instances must therefore represent alien invasions. Pre-eminent among these is a fungus of African origin that causes chytridiomycosis (*Batracho-chytrium dendrobatidis*: Chytridiomycota). This pathogen has played a role in global amphibian declines and is responsible for unprecedented species losses among Anura within the last 25 years, often aggravating the impact of other threat factors, as for yellow-legged frogs depleted by rainbow trout introduced into mountain lakes. Chytridiomycosis has impacted more than 200 anuran species and threatens many more (Wake & Vredenburg 2008; Rödder *et al.*, 2009), especially frogs in the Neotropics (Lips *et al.*, 2006; Skerratt *et al.*, 2007). In this case, non-native vector species may have a role to play in pathogen spread as the invasive

American bullfrog (*Lithobates catesbeiana*: Ranidae), which is resistant to the effects of the fungus, can transmit it to other amphibians. Recent research has shown that frogs in Africa (Cameroon), thought to be unaffected by chrytrid fungus, are declining in the same way as has been reported elsewhere, and unpredictable temperature fluctuations associated with global warming appear to increase the severity of these chytrid-related die-offs (Hirschfeld *et al.*, 2016).

A second pathogenic fungus of the same genus (*B. salamandrivorans*) that affects European salamanders (i.e. the order Urodela, but not frogs and toads) was discovered in 2013 (Martel *et al.*, 2014). *Batrachochytrium salamandrivorans* is thought to have been imported with Asian salamanders, which are traded internationally as pets, but are unaffected by the pathogen as a result of their shared 30-million year evolutionary history (Martel *et al.*, 2014). Non-Asian urodeles are highly susceptible to this fungus, which has been spreading rapidly in Europe (Feldmeier *et al.*, 2016) and already caused rapid declines in populations of the fire salamander (*Salamandra salamandra*: Salamandridae) in The Netherlands. Its possible transmission from imported Asian salamandrids, such as the widely-traded fire-bellied newt (*Cynops orientalis*), into wild populations in North America could have grave consequences (Ducatelle *et al.*, 2014), given that the Holarctic region is the global centre of salamander diversity (Salamandridae plus, especially, Plethodontidae), containing two thirds of all urodele species.

In addition to infection by chytrid fungi, amphibians worldwide are susceptible to at least three species of *Ranavirus* (Iridoviridae) that cause haemorrhagic diseases. These pathogens can cause mass mortality across multiple species in communities where they become established (Gray *et al.*, 2009; Price *et al.*, 2014). As with chytridiomycosis, one route of disease spread is the international trade in amphibians and consequent transport of *Ranavirus*-infected individuals; the American bullfrog, which is not especially susceptible to this pathogen, is a known vector (Gray *et al.*, 2009). *Ranavirus* occurrence is exacerbated by the presence of non-native fishes, which tend to increase amphibian deaths attributed to this pathogen (North *et al.*, 2015); humans moving frog spawn between garden ponds have also been implicated in the spread of the virus. (Price *et al.*, 2016). Other examples of invasive pathogens and parasites are given in Table 4.1.

A peculiar outcome of disease transmission by an alien species involves hydrilla (*Hydrilla verticillata*: Hydrocharitaceae), a now-cosmopolitan sub-merged macrophyte that became established in the inland waters of North America from Asia during the 1960s. The disease, avian vacuolar

Table 4.1 *A summary of the variety of freshwater taxa (aquatic or semiaquatic, but not estuarine) represented among alien invasive species globally, with an indication of their ecological niche or mode of interaction. The list is not intended to be exhaustive but to illustrate the range of taxa that have become established outside their natural ranges. Elaboration on the impacts of some of these species is given in the text also. Additional examples of especially pernicious aliens, and their impacts on native biodiversity, are given in Table 4.2. Data compiled from the Global Invasive Species Database (ISSG, 2015), supplemented by Dudgeon (1999), Leuven et al. (2009); Simberloff & Rejmánek (2011) and Francis (2012).*

Pathogenic microbes:

- *Sphaerothecum destruens* (Rhinosporideaceae): rosette agent causing disease of fishes
- Tilapia lake virus disease (Orthomyxovididae): TiLV or syncytial hepatitis of tilapia; emergent disease causing mass mortalities in aquaculture
- Cyprinid herpes virus 3 (Alloherpesviridae): CyHV-3 or koi herpes virus; host specific, causing necrosis and mass mortalities of common carp
- *Whispovirus* sp. (Nimaviridae): white spot syndrome virus; lethal pathogen spread by invasive American crayfishes to European counterparts
- *Ranavirus* spp. (Iridoviridae): cause ranavirosis, a virulent haemorrhagic disease of amphibians
- *Novirhabdovirus* spp. (Rhabdoviridae): causes highly infectious haemorrhagic septicaemia in fishes
- *Vesiculovirus carpio* (Rhabdoviridae): causes haemorrhagic and contagious spring viraemia of carp and other fishes
- *Piscirickettsia salmonis* (Piscirickettsiaceae): muskie pox; bacterial pathogen of salmonid fish
- *Aphanomyces astaci* (Leptolegniaceae): fungus causing crayfish plague
- *Batrachochytrium dendrobatidis* (Chytridiomycota): fungal skin disease of amphibians (chytridiomycosis)
- *Batrachochytrium salamandrivorans* (Chytridiomycota): fungal skin disease of urodeles
- *Myxobolus cerebralis* (Myxobolidae): myxosporean parasite; causes whirling disease of salmonids

Algae and cyanobacteria:

- *Cylindrospermopsis raciborskii* (Nostocaceae): toxic filamentous cyanobacterium linked to fish kills in native range; fixes nitrogen
- *Gonyostomum semen* (Raphidophyceae): planktonic alga; causes blooms in acidified lakes
- *Didymosphenia geminata* (Gomphonemataceae): rock snot; diatom causing epilithic and epiphytic blooms
- *Spirogyra fluviatilis* (Zygnemataceae): filamentous green algae causing blooms that smother benthos (e.g. in Lake Baikal)

(cont.)

Table 4.1 (*cont.*)

Macrophytes:

- *Azolla pinnata* and *A. filiculoides* (Azollaceae): water velvet; small floating ferns; nitrogen fixers
- *Hygrophila polysperma* (Acanthaceae): Indian swamp weed; amphibious perennial herb
- *Alisma plantago-aquatica* (Alismataceae): European water-plantain; emergent wetland perennial
- *Sagittaria sagittifolia* (Alismataceae): arrowhead; emergent herbaceous perennial
- *Alternanthera philoxeroides* (Amaranthaceae): alligator weed; amphibious perennial herb forming floating mats
- *Heracleum mantegazzianum* (Apiaceae): giant hogweed; tall perennial riparian herb with toxic sap
- *Pistia stratiotes* (Araceae): water lettuce; floating perennial
- *Eupatorium cannabinum* (Asteraceae): hemp-agrimony; perennial herb in riparia and wetlands
- *Gymnocoronis spilanthoides* (Asteraceae): Senegal tea plant; emergent or floating perennial herb
- *Impatiens gladulifera* (Balsaminacea): Himalayan balsam; annual riparian herb
- *Nasturtium* (= *Rorippa*) *officinale* (Brassicaceae): watercress; perennial herb with floating stems
- *Butomus umbellatus* (Butomaceae): flowering rush; emergent wetland perennial
- *Cabomba caroliniana* (Cabombaceae): Carolina fanwort; submerged perennial
- *Ceratophyllum demersum* (Ceratophyllaceae): hornwort; submerged perennial
- *Ipomoea aquatica* (Convolvulaceae): water spinach; perennial herb, floating stems and leaves
- *Crassula helmsii* (Crassulaceae): New Zealand pygmyweed; amphibious succulent perennial herb
- *Mimosa pigra* (Fabaceae): bashful plant; perennial shrub in wetlands, riparia and on floodplains
- *Acacia mearnsii* (Fabaceae): black wattle; dominant riparian tree; depletes in-stream water supplies
- *Myriophyllum* spp. (Haloragaceae): parrot-feather or water milfoil; submerged perennial
- *Egeria densa* (Hydrocharitaceae): Brazilian elodea; submerged perennial
- *Elodea canadensis* (Hydrocharitacae): common elodea; submerged perennial; *E. nuttallii* is invasive also
- *Hydrilla verticillata* (Hydrocharitaceae): hydrilla; submerged perennial
- *Hydrocharis morsus-ranae* (Hydrocharitaceae): European frogbit; floating annual herb
- *Lagarosiphon major* (Hydrocharitaceae): African elodea; submerged perennial; displaces invasive *Elodea* spp.
- *Vallisneria* spp. (Hydrocharitaceae): eelgrass; submerged perennial herb
- *Landoltia punctata* (Lemnaceae): spotted duckweed; very small, floating plant

Table 4.1 (*cont.*)

- *Utricularia gibba* (Lentibulariaceae): humped bladderwort; carnivorous perennial herb in bogs
- *Lythrum salicaria* (Lythraceae): purple loosestrife: perennial wetland herb
- *Trapa natans* (Lythraceae): water chestnut; submerged annual with floating leaves
- *Nymphoides peltatum* (Menyanthaceae): fringed water-lily; submerged perennial, floating leaves
- *Melaleuca quinquenervia* (Myrtaceae): broad-leaved paperbark; wetland perennial tree
- *Ludwigia peruviana* (Onagraceae): water primrose; wetland perennial, forms floating mats
- *Limnophila sessiliflora* (Plantaginaceae): Asian ambulia; submerged perennial
- *Arundo donax* (Poaceae): giant reed; tall emergent perennial grass, dominates riparia and wetlands
- *Phragmites australis* (Poaceae): common reed; tall emergent perennial grass in lakes and wetlands
- *Zizania latifolia* (Poaceae): Manchurian wild rice; perennial aquatic grass in wetlands and riparia
- *Eichhornia crassipes* (Pontederiaceae): water hyacinth; floating perennial forming dense mats
- *Fallopia* (= *Reynoutria*) *japonica* (Polygonaceae): Japanese knotweed; tall herbaceous perennial that dominates riparia
- *Potamogeton perfoliatus* (Potamogetonaceae): clasping-leaf pondweed; submerged perennial
- *Salix* spp. and their hybrids (Salicaceae); willows; shrub or small trees in riparia or shallow water; e.g. *S. babylonica*, *S. nigra* and *S. cinerea*
- *Typha latifolia* (Typhaceae): bulrush; tall, emergent perennial herb; cattail (*T. angustifolia*) is invasive also

Mollusca:

- *Bellamya chinensis* (Viviparidae): Chinese mystery snail; eats algae and fine detritus; ovoviviparous (= live-bearing)
- *Marisa cornuarietis* (Ampullariidae): giant ramshorn snail; omnivorous, chiefly herbivorous, may eat pulmonate eggs
- *Pomacea canaliculata* and *P. insularum* (Ampullariidae): golden apple snail; chiefly herbivorous, but eats snail and amphibian eggs
- *Bythinia tentaculata* (Bythiniidae): faucet snail; eats periphyton and biofilm, also filter feeds; trematode host
- *Pyrgophorus platyrachis* (Cochliopidae): Florida serrate crownsnail; eats periphyton and biofilm; ovoviviparous (Singapore only)
- *Potamopyrgus antipodarum* (Hydrobiidae): New Zealand mud snail; eats periphyton and detritus
- *Lithoglyphus naticoides* (Lithoglyphidae): gravel snail; eats algae; host of parasitic trematodes

(*cont.*)

Table 4.1 (*cont.*)

- *Mieniplotia* (= *Thiara*) *scabra* (Thiaridae): parthenogenetic and ovoviviparous family with high invasion potential; eats periphyton and detritus
- *Tarebia granifera* (Thiaridae): host of parasitic trematodes; eats periphyton and detritus, monopolizing these resources
- *Melanoides tuberculata* (Thiaridae): host of parasitic trematodes; eats periphyton and detritus, monopolizing these resources
- *Aplexa marmorata* (Physidae): eats periphyton and biofilm
- *Physella* (= *Physa*) *acuta* (Physidae): tadpole snail; eats periphyton and biofilm
- *Pseudosuccinea* (= *Lymnaea*) *columella* (Lymnaeidae): mimic *Lymnaea*; host of parasitic trematodes; eats periphyton and biofilm
- *Radix rubiginosa* (Lymnaeidae): host of parasitic trematodes; eats mainly periphyton and biofilm
- *Biomphalaria straminea* (Planorbidae): host of blood flukes; eats periphyton and biofilm
- *Gyraulus chinensis* (Planorbidae); eats periphyton and biofilm
- *Menetus dilatatus* (Planorbidae); eats periphyton and biofilm
- *Dreissena polymorpha* and *D. rostriformis bugensis* (Dreissenidae): zebra and quagga mussels; filter feeders, biofoulers
- *Limnoperna fortunei* (Mytilidae): golden mussel; filter feeder, biofouler
- *Sinanodonta woodiana* (Unionidae): Chinese pond mussel; benthic filter feeder
- *Corbicula fluminea* (Cyrenidae): Asian clam; benthic filter feeder; *C. fluminalis* has also invaded parts of Europe
- *Musculium transversum* (Sphaeriidae); fingernail clam; pollution-tolerant benthic filter feeder, can attain high densities

Crustacea:

- *Acrtodiaptomus dorsalis* (Diaptomidae): calanoid copepod; consumes phytoplankton, displacing native calanoids
- *Pseudodiaptomus forbesi* (Pseudodiaptomidae): calanoid copepod; consumes phytoplankton
- *Eurytemora 'affinis'* (Temoridae): calanoid copepod species complex; consume phytoplankton
- *Limnoithona tetraspina* (Oithonidae): cyclopoid copepod; consumes phytoplankton
- *Lernaea cyprinacea* (Lernaeidae): anchor worm; copepod ectoparasite of fishes
- *Bythotrephes longimanus* (Cercopagididae): spiny water flea; predatory on other zooplankton
- *Cercopagis pengoi* (Cercopagididae): fish-hook water flea; predatory on other zooplankton
- *Daphnia lumholtzi* and *D. pulex* (Daphniidae): cladoceran grazers, alter phytoplankton species composition
- *Hemimysis anomla* (Mysidae): bloody-red opossum shrimp; omnivorous and predatory, reduces algal and zooplankton populations
- *Limnomysis benedeni* (Mysidae): opossum shrimp; omnivorous but highly predatory; dominates assemblages, alters food webs

Table 4.1 (*cont.*)

- *Mysis diluviana* (Mysidae): opossum shrimp; predatory on zooplankton; depletes zooplankton affecting upper trophic levels
- *Proasellus coxalis* and *P. meridianus* (Asellidae): freshwater isopods; eat algae & detritus
- *Jaera istri* (Janiridae): freshwater isopod; eats algae & detritus; can dominate stony benthic assemblages
- *Crangonyx pseudogracilis* (Crangonyctidae): freshwater amphipod; eats algae and detritus
- *Dikerogammarus* spp. (Gammaridae): including killer and demon shrimp; omnivorous benthic amphipods, predatory tendencies
- *Echinogammarus* spp. (Gammaridae): omnivorous benthic amphipods, predatory tendencies
- *Gammarus tigrinus* and *G. lacustris* (Gammaridae): omnivorous benthic amphipods; can reach high densities and dominate communities
- *Gmelinoides fasciatus* (Gammaridae): omnivorous benthic amphipod; can reach high densities and dominate communities
- *Orchestia cavimana* (Talitridae): aquatic and semi-terrestrial amphipod; eats algae and detritus
- *Chelicorophium curvispinum* (Corophiidae): Caspian mud shrimp; filter-feeding gallery builder, alters habitat and food availability
- *Macrobrachium nipponense* (Palaemonidae): Oriental river prawn; omnivorous
- *Atyaephyra desmarestii* (Atyidae): shrimp that eats periphyton and detritus
- *Astacus leptodactylus* (Astacidae): narrow-clawed crayfish; omnivore, chiefly carnivorous
- *Pacifastacus leniusculus* (Astacidae): signal crayfish; omnivore; carries crayfish plague
- *Orconectes rusticum* and *O. virilis* (Cambaridae): rusty and virile (or northern) crayfishes; omnivores
- *Procambrus clarkii* (Cambaridae): Red swamp crayfish; omnivore; carries crayfish plague; *P. fallax* is invasive also (see text)
- *Cherax quadricarinatus* and *C. destructor* (Parastacidae): Australian red-claw crayfish and yabby; omnivores; carry crayfish plague

Insecta:

- *Cloeon smaeleni* (Baetidae): larvae of this mayfly are mainly algivorous
- *Synaptonecta issa* (Micronectidae): Asian water boatman; eats algae and fine detritus
- *Hydropsyche bulgaromanorum* (Hydropychidae): filter-feeding caddisfly; occupies hard substrates in streams
- *Parapoynx diminutalis* (Crambidae): larvae of this aquatic moth eat macrophytes
- *Aedes aegypti* and *A. albopictus* (Culicidae): yellow fever mosquito and Asian tiger mosquito; disease vectors
- *Anopheles quadrimaculatus* (Culicidae): common malaria mosquito; disease vector
- *Culex quinquefasciatus* (Culicidae): southern house mosquito; disease vector
- *Ochlerotatus japonicus* (Culicidae): Asian rock pool mosquito; disease vector

(*cont.*)

Table 4.1 (*cont.*)

Other invertebrates:

- *Cordylophora caspia* (Cnidaria: Clavidae): small colonial hydroid that eats zooplankton; biofouler
- *Craspedacusta sowerbyi* (Cnidaria: Olindiidae): freshwater jellyfish; hydromedusan that eats zooplankton
- *Girardia* (= *Dugesia*) *tigrina* (Platyhelminthes: Dugesiidae): triclad turbellarian; eats small arthropods and other invertebrates
- *Dendrocoelum romanodanubiale* (Platyhelminthes: Dendrocoelidae): triclad turbellarian; eats small arthropods and other invertebrates
- *Centrocestus formosanus* (Platyhelminthes: Heterophyidae): Asian gill trematode; impacts fish intermediate hosts
- *Leyogonimus polyoon* (Platyhelminthes: Lecithodendriidae): causes mass mortalities of waterfowl; uses the invasive faucet snail as a host
- *Nitzschia sturionis* (Platyhelminthes: Capsalidae): sturgeon gill trematode; infests gills and causes skin lesions in *Acipenser* spp.
- *Bothriocephalus acheilognathi* (Cestoda: Bothryocephalidae): Asian tapeworm; internal parasite of fishes; introduced with grass carp
- *Anguillicoloides crassus* (Nematoda: Dranuculidae): swimbladder nematode; endoparasite of eels in Europe and North America; Asian origin
- *Pectinatella magnifica* (Bryozoa: Pectinatellidae): colonial bryozoan; filter feeder; large gelatinous colonies usually attached to hard surfaces
- *Pomphorhynchus tereticollis* (Acanthocephala: Pomphorhynchidae): 'thorny worm' fish parasite; uses gammarid intermediate hosts
- *Branchiura sowerbyi* (Annelida: Tubificidae): deposit-feeding oligochaete; dense aggregations in organic-rich sediments
- *Quistadrilus* (=*Peloscolex*) *multisetosus* (Annelida: Tubificidae): deposit-feeding oligochaete; dense aggregations in organic-rich sediments
- *Hypania invalida* (Annelida: Ampharetidae): freshwater polychaete; filter and deposit feeder; dense aggregations dominate benthic substrata
- *Helobdella europaea* (Annelida: Glossiphoniidae): predatory and scavenging leech, eats molluscs and other macroinvertebrates
- *Hirudinaria manillensis* (Annelidae: Hirudinidae): blood-feeding leech; invasion probably facilitated by transport for medical purposes
- *Barbronia weberi* (Annelida: Salifidae): predatory leech, eats molluscs and other macroinvertebrates
- *Caspiobdella* (= *Piscicola*) *fadejewi* (Annelida: Piscicolidae): leech ectoparasitic on fishes
- *Caspihalacarus hyrcanus danubialis* (Acarina: Halacaridae): euryhaline water mite; micropredator

Fish:

- *Petromyzon marinus* (Petromyzontidae): sea lamprey; haematophagus fish ectoparasite
- *Potamotrygon motoro* (Potamotrygonidae; DD): ocellate river stingray; a live-bearing predator (established in Singapore only)

Table 4.1 (*cont.*)

- *Amia calva* (Amiidae): bowfin; piscivorous, also eats crayfishes and amphibians
- *Scleropages formosus* (Osteroglossidae; EN): Asian arowana; a mouthbrooding predator (established in Singapore only)
- *Heterotis niloticus* (Arapaimidae): African arowana; predator, chiefly of invertebrates (introduced in parts of Africa and Madagascar only)
- *Chitala ornata* (Notopteridae): clown knife fish; predator, chiefly piscivorous
- *Alosa pseudoharengus* (Clupeidae): alewife; planktivore
- *Osmerus mordax* (Osmeridae): rainbow smelt; eat zooplankton and small fishes; *Hypomesus olidus* is invasive also
- *Odontesthes* (= *Basilichthys*) *bonariensis* (Atherinopsidae): pejerrey; predatory (important only in Lake Titicaca)
- *Salmo trutta* (Salmonidae): brown trout; predator
- *Salvelinus fontinalis* and *S. namaycush* (Salmonidae): brook and lake trout; predators
- *Oncorhynchus mykiss* (Salmonidae): rainbow trout; predator
- *Coregonus clupeaformis* and *C. albula* (Salmonidae): whitefish; eat zooplankton and small fishes; may hybridize with congeners
- *Colossoma macropomum* (Serrasalmidae): tambaqui; omnivorous, eats fruit
- *Piaractus brachypomus* (Serrasalmidae): red-bellied pacu or pirapatinga; omnivorous, chiefly phytophagous (perhaps New Guinea only)
- *Barbonymus gonionotus* (Cyprinidae): Java (or silver) barb; omnivore
- *Carassius auratus* (Cyprinidae): goldfish; omnivore
- *Catla catla* (Cyprinidae): catla; omnivore
- *Cirrhinus cirrhosis* (Cyprinidae): white carp; omnivore, widely introduced but threatened (VU) in native range
- *Ctenopharyngodon idella* (Cyprinidae): grass carp; phytophagous
- *Cyprinella lutrensis* (Cyprinidae): red shiner; omnivore, eats fish eggs; may hybridize with certain conspecifics
- *Cyprinus carpio* (Cyprinidae): common carp; omnivore
- *Hypophthalmichthys nobilis* and *H. molitrix* (Cyprinidae): bighead and silver carps; planktivorous
- *Labeo rohita* (Cyprinidae): rohu; omnivorous, chiefly herbivorous and planktivorous
- *Mylopharyngodon piceus* (Cyprinidae): black carp; chiefly molluscivorous
- *Pseudorasbora parva* (Cyprinidae): top-mouth gudgeon; omnivore, vector of fish pathogens
- *Rutilus rutilus* (Cyprinidae): roach; omnivore.
- *Tinca tinca* (Cyprinidae): tench; omnivore, chiefly predatory
- *Misgurnus anguillicaudatus* (Cobitidae): Japanese weatherfish; predator
- *Silurus glanis* (Siluridae): wels catfish; voracious predator (of fish, amphibians, etc.)
- *Ameiurus melas* and *A. nebulosus* (Ictaluridae): black and brown bullheads; predatory
- *Ictalurus punctatus* (Ictaluridae): channel catfish; omnivorous, chiefly predatory
- *Pylodictis olivaris* (Ictaluridae): flathead catfish; predator, chiefly piscivorous

(*cont.*)

Table 4.1 (*cont.*)

– *Hoplosternum littorale* (Callichthyidae): atipa; omnivorous, chiefly predatory (perhaps Florida only)
– *Pterygoplichthys* (= *Liposarcus*) spp. (Loricariidae): sailfin catfishes (e.g. *P. pardalis*); omnivorous, chiefly algae and plants
– *Pangasias hypophthalamus* (= *sutchi*) (Pangasiidae): striped or iridescent shark catfish; omnivorous, larger individuals predatory
– *Clarias gariepinus* (Clariidae): sharptooth catfish; omnivorous, chiefly predatory
– *Esox lucius* (Esocidae): northern pike; predator (chiefly piscivorous); may hybridize with native congeners
– *Fundulus zebrinus* (Fundulidae): plains killifish; omnivorous
– *Cyprinodon variegatus* (Cyprinodontidae): sheepshead minnow; omnivorous; hybridizes with certain native congeners
– *Gambusia affinis* and *G. holbrooki* (Poeciliidae): mosquito fishes; omnivorous, chiefly predatory
– *Poecilia reticulata* (Poeciliidae): guppy; omnivorous
– *Xiphophorus helleri* (Poeciliidae): swordtail; omnivorous; *X. variatus* is invasive also
– *Phalloceros caudimaculatus* (Poeciliidae): speckled mosquito fish; omnivorous
– *Monopterus albus* (Synbranchidae): Asian swamp eel; generalist predator, eats fishes and amphibians
– *Macrognathus siamensis* (Mastacembelidae): peacock spiny eel; eats macroinvertebrates (perhaps Florida only)
– *Morone americana* (Moronidae): white perch; predator (especially of fish eggs)
– *Gymnocephalus cernuus* (Percidae): Eurasian ruffe; benthivorous predator
– *Perca fluviatilis* (Percidae): Eurasian or redfin perch; generalist predator
– *Sander lucioperca* (Percidae): pike perch or zander; chiefly piscivorous
– *Lepomis* spp. (Centrarchidae): sunfishes; omnivorous, but some (e.g. bluegill) chiefly predatory
– *Micropterus dolomieu* (Centrarchidae): smallmouth bass; generalist predator, eats macroinvertebrates, fishes and amphibians
– *Chicla ocellaris* (Cichlidae): peacock bass; predator (mainly piscivorous)
– *Cichlasoma urophthalmus* (Cichlidae): Mayan cichlid; predator; certain other *Cichlasoma* spp. have established local populations
– *Acarichthys heckelii* (Cichlidae): threadfin acara; omnivorous, but mainly eats macroinvertebrates (thus far, Singapore only)
– *Parachromis managuensis* (Cichlidae): jaguar guapote; predator (mainly piscivorous)
– *Tilapia mariae*, *T. rendalli*, *T. sparmanii* and *T. zilli* (Cichlidae): tilapias (term also applied to *Oreochromis* spp.); omnivores and mouthbrooders
– *Prochilodus* spp. (Procholodontidae): *P. lineatus* and *P. argenteus* have established feral populations from aquaculture; omnivores
– *Lates niloticus* (Latidae): Nile perch; predator (piscivorous)
– *Parambassis* spp. (Ambassidae): glass perchlets; *P. ranga* and *P. siamensis* have established feral populations
– *Hyporhamphus intermedius* (Hemiramphidae): Asian pencil halfbeak; zooplankton feeder; euryhaline; invasive in some Chinese lakes

Table 4.1 (*cont.*)

- *Giuris margaritacea* (Eleotridae): snakehead gudgeon; generalist predator (perhaps important in Lake Lanao only)
- *Perccottus glenii* (Odontobutidae): Amur sleeper; generalist predator
- *Neogobius melanostomus* (Gobiidae): round goby; generalist predator; the monkey goby (*N. fluviatilis*) is invasive also
- *Ponticola kessleri* (Gobiidae): bighead goby; generalist predator
- *Proterorhinus semilunaris* (Gobiidae): tubenose goby; generalist predator
- *Salangichthys tangkahkeii* (Salangidae): Taihu icefish; eats zooplankton and fish fry; invasive in some degraded Chinese lakes
- *Anabas testudineus* (Anabantidae): climbing perch; omnivorous, chiefly predatory
- *Betta splendens* (Osphronemidae): Siamese fighting fish; generalist predator; potential hybridization with other *Betta* spp.
- *Osphronemus gouramy* (Osphronemidae): giant gourami; omnivorous, eats aquatic plants
- *Trichogaster trichopterus* and *T. pectoralis* (Osphronemidae): three-spot and snakeskin gouramis; omnivores, predatory tendencies
- *Channa* (= *Ophiocephalus*) spp. (Channidae): snakeheads; predators (piscivorous)

Amphibians and reptiles:

- *Andrias davidianus* (Cryptobranchidae): Chinese giant salamander; large-bodied predator
- *Duttaphrynus [= Bufo] melanostictus* (Bufonidae): Asian common toad; predatory adults eat other amphibians; toxic
- *Kalula pulchra* (Microhylidae): Asian painted frog or banded bullfrog; highly eurytopic, but mainly myrmecophagous
- *Xenopus laevis* (Pipidae): African clawed frog; predator of other amphibians and fishes; vector of chytridiomycosis
- *Caiman crocodilus* (Alligatoridae): spectacled caiman; predatory adults
- *Varanus niloticus* (Varanidae): Nile monitor lizard; large semiaquatic predator (in Florida only)

Birds:

- *Anas platyrhynchos* (Anatidae): mallard; omnivorous, transmits avian influenza virus H5N1; hybridizes readily with native ducks
- *Branta canadensis* (Anatidae): Canada goose; omnivorous, chiefly herbivorous
- *Cygnus olor* (Anatidae): mute swan; omnivorous (chiefly aquatic macrophytes)
- *Oxyura jamaicensis* (Anatidae): ruddy duck; omnivorous (chiefly benthic macroinvertebrates); hybridizes with certain congeners

Mammals:

- *Castor canadensis* (Castoridae): North American beaver; herbivorous, damages trees
- *Ondatra zibethicus* (Cricetidae): muskrat; primarily herbivorous, semiaquatic rodent
- *Myocaster coypus* (Myocastoridae): coypu or nutria; primarily herbivorous, semiaquatic rodent
- *Neovison vison* (Mustelidae): American mink; semiaquatic predator

myelinopathy, is associated with lesions in the brains of birds, and has been spreading since it was first detected in 1994. Bald eagle and American coot (*Fulica americana*: Rallidae) are principally affected, but mallard are also susceptible. Bird deaths from vacuolar myelinopathy tend to occur in habitats with extensive growths of hydrilla but only where these plants have dense colonies of a neurotoxin-producing cyanobacteria *Aetokthonos hydrillicola* (Hapalosiphonaceae) growing on the underside of their leaves (Wilde *et al.*, 2014). American coot and other waterfowl eating hydrilla are affected by the neurotoxin and become sickened, with the vacuolar myelinopathy being passed to bald eagles that feed on impaired coot (Fischer *et al.*, 2006). Thus, the invasive hydrilla facilitates the poisoning 'disease' of herbivorous waterfowl caused by a cyanobacterial neurotoxin, with consequential secondary intoxication of apex terrestrial predators. Triploid (and sterile) Asian grass carp are a potential control agent of hydrilla but are susceptible to the cyanobacterial neurotoxin, although they do not accumulate it in their tissues and hence may not pose a major risk to piscivorous birds (Haynie *et al.*, 2013).

Direct and Indirect Impacts of Aliens and Their Role As Ecological Engineers

Not all alien species will affect native biodiversity deleteriously but, as the examples given thus far show, many do. Invasive species are a leading cause of animal extinctions, second only to habitat loss or degradation. One estimate is that 20% of all recent species extinctions were directly caused by alien species (Clavero & García-Berthou, 2005), and a fifth of freshwater fish species categorized as threatened by the IUCN are at risk from alien species (Olden *et al.*, 2007); that proportion falls to 11% if all freshwater species (not only fishes) are considered (see Table 2.1). A study of 12 invasive species (11 fishes and a crayfish) in six large European countries reported ecological effects in an average of 69% of instances where establishment was successful, ranging from a low of 36% in Spain to 100% in Germany (García-Berthou *et al.*, 2005). Impacts on native species can be direct and hence relatively obvious – often via predation, hybridization or pathogenic infestations – or by way of trophic cascades and other indirect interactions such as the spread of disease (and the role played by hydrilla; see previous section). Instances of such indirect outcomes can be seen from the effects of stocked fishes on apex predators in lacustrine habitats such as Gatún Lake and Flathead Lake (Box 4.1), and the example of Lake Naivasha (Box 4.3) shows how ecological conditions within a water body can change in response to a transition in the dominant alien species. The establishment of North American beaver, and at least one other semiaquatic rodent, outside their native ranges (Box 4.4) shows that aliens can bring about change through ecological engineering.

One example of an invader-mediated cascading interaction became apparent in Japan following the establishment of rainbow trout. They usurped the 'rain' of terrestrial insects that fell into streams from the surrounding forest canopy, causing Dolly Varden charr (*Salvelinus malma*) to shift their diet to incorporate benthic insects that grazed periphyton. The consequences were a reduction in emergence of adult aquatic insects and a substantial – almost 70% – decrease in the abundance of riparian spiders (Baxter *et al.*, 2004). In this case, the invader intercepted land-to-water energy subsidies and indirectly reduced the magnitude of water-to-land transfer, affecting the interconnection and flow of resources between streams and the surrounding forest.

Some invasive species act as ecological engineers, affecting ecosystem characteristics directly without them being mediated by alterations in native populations or communities. Such engineering effects can then have a strong influence on biodiversity. Changes to *Nothofagus* (Nothofagaceae) beech forests and stream hydrology, with consequential effects on benthic and riparian assemblages, that followed the introduction of North American beaver to Patagonia in 1946 are a good example of this (Anderson & Rosemond, 2007; see also Rosell *et al.*, 2005; Chapter 7). The omnivorous habits of crayfishes combined with their relatively large size (especially signal and rusty crayfishes) can result in major changes to the lakes they invade (e.g. Smart *et al.*, 2002; Rosenthal *et al.*, 2006). These are usually manifest in a tendency to eliminate macrophytes, as in Lake Naivasha (see Box 4.3), thereby simplifying habitat structure and causing declines of macroinvertebrates, and a substantial increase in the relative importance of planktonic primary producers. The feeding activities of introduced signal crayfish also increases nutrient recycling in lakes, which stimulates near-shore primary production (Wittmann *et al.*, 2013). Conversely, removal of invasive rusty crayfish from lakes can result in rapid recovery of macrophytes, fishes and native virile crayfish (*Orconectes virilis*) populations, as well as some – although not all – macroinvertebrates (Hansen *et al.*, 2013). Declines in snail abundance attributable to rusty crayfish occur far more quickly than recovery following release

Box 4.4 *Invasive Mammals: Semiaquatic Rodents*

Beavers are keystone species and ecological engineers: overexploitation of their populations has had significant ecological consequences, whereas reintroduction to areas where they have been historically overexploited has been beneficial and facilitated habitat restoration (see Chapter 3). However, the introduction of North American beaver to Tierra del Fuego, well outside their native range, and their subsequent spread through the Patagonian wilderness, has had significant impacts on stream flows and forest ecology that have been deemed detrimental (Anderson & Rosemond, 2007), leading to attempts by the Argentinian and Chilean authorities to control their populations. Other semiaquatic rodents are native across parts of Europe, Asia and the Americas: they include capybara (two species of *Hydrochoerus*, predominately *H. hydrochaeris*), muskrat (*Ondatra zibethicus*: Cricetidae) and coypu (*Myocastor coypusa*: Myocastoridae). Feral capybara occur in Florida, but they have yet to be confirmed as established or invasive. In contrast, muskrat and coypu are cited by Elton (1958) as being responsible for 'ecological explosions' in their expanded ranges, and the impacts of these rodents offer an informative contrast with beavers.

Muskrat (up to 2 kg) are native to the wetlands of Canada and the United States, where they are frequently sympatric with the much-larger North American beaver, and have been introduced to parts of South America, Europe and Palaearctic Asia (reviewed by Skyrienė & Paulauskas, 2012). Like beaver, they were trapped for their pelts in their native range and for that reason were introduced to Europe in the early twentieth century, subsequently spreading eastward. Muskrat live in family groups and maintain a semi-permanent, often dome-like, den with an underwater entrance. They maintain open-water swimming routes and can thus create habitat for other species on a similar, albeit smaller, scale to beavers. In Sweden, introduced muskrat is considered a keystone species, because it increases the extent of open-water habitat for water birds (Danell, 1996). Conversely, muskrat is regarded as a pest in the Netherlands, Germany and elsewhere, because their burrows damage dykes and drainage structures. Muskrat feed predominately on aquatic macrophytes such as reed (*Phragmites communis*) and cattail (*Typha angustifolia*: Typhaceae), and affect vegetation composition to such an extent that they may reduce feeding opportunities for European water vole (*Arvicola amphibious:* Cricetidae). To supplement their vegetarian diet, muskrat eat crustaceans and molluscs, and are reported to have caused declines in populations of pearly mussels, such as *Margaritifera margaritifera* (CR) along the River Lutter in Germany (Box 5.12), and *Unio crassus* (EN) across Central Europe (Lopes-Lima *et al.*, 2017). Muskrat predation can alter the age-structure and species composition of unionid assemblages (Owen *et al.*, 2011), and this can be detrimental to European bitterling (*Rhodeus amarus* [= *sericeus*]: Cyprinidae) that depend on certain pearly mussels as repositories for eggs and larvae (Skyrienė & Paulauskas, 2012; see also the account of *Sinanodonta woodiana* given in the section on ecological impacts of molluscan invaders).

Coypu (5–9 kg) have become established in Europe, North America (where they are known as nutria), and a few localities in Asia and Africa. They are mainly herbivorous and were intentionally introduced to Florida as an agent of biological control for invasive aquatic macrophytes. Coypu are voracious feeders that not only reduce plant biomass but alter wetland characteristics and degrade habitat for other animals. In Japan, coypu have been implicated in the extinction of a dragonfly, and local declines of two bitterling species via depredation of their unionid egg repositories. Such 'ecological explosions' have led to their designation as one of the 100 'world's worst' invasive species by the IUCN (ISSG, 2015). Unlike beavers, coypu reduce rather than increase landscape heterogeneity within their invaded range, while both muskrat and coypu have detrimental effects on native animals. In parts of the United States where they overlap, coypu tend to suppress native populations of muskrat, because both species eat similar foods and the gluttonous coypu has a substantial size advantage.

Capybara have the distinction of being the largest rodents in the world, weighing more than 50 kg. They are gregarious grazers that eat grasses and aquatic macrophytes, feeding both in water and on land, with a dental structure that allows them to crop short-statured plants down to ground level. The native range of capybara is Central and South America where they are largely confined to floodplains and waterlogged areas. Seasonal changes on floodplains necessitate ecological flexibility, and capybara breeding is synchronized with the annual flood cycle to coincide with maximum productivity of grasses (Pacini & Harper, 2008). Because they are the largest herbivorous species over much of their range, capybara grazing can have significant effects on vegetation stature and composition. This would be a matter of concern if feral individuals in Florida establish breeding populations. Effects of capybara on other species might include facilitation, as reported within their native range: insectivorous birds such as the cattle tyrant (*Machetornis rixosus*: Tyrannidae), as well as ardeids and jacanids, frequently associate with capybara using them as 'beaters' to flush insects from vegetation; other birds groom parasites from the rodents' hides (Tomazzoni *et al.*, 2005).

from crayfish predation (Kreps *et al.*, 2012), indicating that recovery from the impacts of invasives may be a lengthy process. Indeed, re-establishment of macrophytes in lakes where invasive rusty crayfish have been reduced or eliminated entirely is not always possible without human intervention (i.e. manual planting) as seed banks become increasingly depauperate over time (Rosenthal *et al.*, 2006). Thus, by acting as an ecological engineer, the rusty crayfish can switch a lake to an alternative state that is maintained even after the crayfish have been removed, unless measures are taken to restore the lake to its initial condition.

Less conspicuous but, nonetheless, far-reaching impacts have been associated with invasions of South American sailfin (or armoured) catfishes of the genus *Pterygoplichthys* (Loricariidae) that are common in the aquarium trade. They have become established in Mexico, Puerto Rico, Hawaii, and the southern United States (Florida), as well as Taiwan and parts of Southeast Asia where at least three species seem to have become established (Page & Hall, 2006). These large herbivorous catfishes change habitat conditions by burrowing and bioturbation, with other impacts that include depletion of algae and competition with native species that can result in detrimental effects on fisheries (Capps & Flecker, 2013a). *Pterygolichthys* attain densities that can be over two orders of magnitude greater than that of entire native fish assemblages; they form large aggregations during the day, dispersing to graze at night. In nutrient-limited Mexican streams, they generate biogeochemical hotspots of elevated nutrients (nitrogen and phosphorus) during daylight hours (Capps & Flecker, 2013b). This comes about because loricariids have a high requirement for phosphorus, and thus are net sinks of this nutrient, whereas they remineralize nitrogen through excretion. Much of the phosphorus in invaded streams becomes accumulated within the high loricariid biomass, so these fishes have the potential to alter ecosystem function by constraining processes such as primary production and decomposition that depend on phosphorus availability (Capps & Flecker, 2013a, 2013b).

What Environmental Factors Facilitate Establishment of Aliens?

Regardless of the taxon concerned, establishment of alien species has the potential to supplement the various physical and chemical transformations of freshwater ecosystems arising from human activities that threaten native biodiversity (see Chapter 1). In part, this is because invasions tend to be more successful in habitats that are already modified or degraded, where flow regimes or water quality (or both) may no longer suit native species, allowing adaptable species from elsewhere to colonize (Strayer, 2010). For instance, many invasive aquatic macrophytes tend to proliferate under nutrient-rich conditions where they can out-compete native species, although reduction of nutrient loading may reverse that advantage. Perhaps for this reason, aliens rank second only to habitat transformation – in its many forms – as a threat to freshwater biodiversity in general, particularly fishes (Millennium Ecosystem Assessment, 2005; see also Olden et al., 2007). A recent analysis found that human populations were strong predictors of richness of alien fishes (Dawson et al., 2017), reflecting both the greater number of introductions in more densely populated regions, and the fact that such places would, inevitably, be those where freshwater habitats would be highly modified or degraded.

Overexploitation of many freshwater vertebrates, particularly fishes, which is globally pervasive but inadequately monitored (see Chapter 3), might also be predicted to enhance invasion success. Depletion of native species would be accompanied by reductions in predation or competition (or both) and an increase in available niche space within the recipient community. However, predicting the combined impacts of exploitation and invasion upon freshwater biodiversity is by no means simple, as shown by a recent study of 23 Japanese lakes (Matsusaki & Kadoya, 2015). Fishery yields (in terms of catch per unit effort) declined in most lakes regardless of whether a 10-, 20- or 30-year time span was considered, and best-fitting models of this decline included species richness and, especially, functional diversity of alien piscivorous fishes in each lake, rather than fishing effort or capacity. Aliens tended to have greater impacts in smaller lakes, causing reductions in the magnitude and temporal stability of catches. Increased functional diversity of alien piscivores probably increased food-chain length and induced trophic cascades (Matsusaki & Kadoya, 2015). Evidently, overexploitation of native species is not a prerequisite for establishment of alien species, particularly when the invaders are predatory, as evinced by the numerous examples of successful invasion of native fish communities by piscivores given elsewhere in this chapter. Nonetheless, in many instances, proliferation of invasive species follows overexploitation of natives which, in turn, is often accompanied by other anthropogenic stressors. This was evidently the case in Inlé Lake where a variety of factors are implicated in the endangerment of endemic fishes (see Box 4.5).

Human-modified habitats present new opportunities for establishment of invaders. For example, non-native species are 2.4 to 300 times more likely to occur in impounded rivers and reservoirs than in natural lakes (Johnson et al., 2008). Impoundments frequently support multiple invaders and act as 'stepping stones' that facilitate spread of alien species by acting as a source for colonization of downstream reaches or natural water bodies nearby. This process tends to homogenize faunas, particularly where transformation of lotic habitats by reservoir construction results in the replacement of local riverine faunas by invasive lentic species that may have a near-cosmopolitan distribution (Rahel, 2002). Furthermore, there are no indications that the rate of anthropogenic transformation of freshwater ecosystems is slowing. Human-induced climate change also increases the vulnerability of natural lakes and other freshwaters to invasive species (Wiedner et al., 2007; Rahel & Olden, 2008: see also Box 7.7). For example, water temperatures in Lake Tahoe, a deep (501 m) subalpine lake in the United States Sierra Nevada, rose sufficiently during the second half of the twentieth century to allow the establishment (in 2002) and subsequent proliferation of the Asian clam, *Corbicula fluminea* (Cyrenidae, formerly Corbiculidae). Within less than a decade, it reached densities exceeding $10\,000\ \mathrm{m}^{-2}$ in some parts of the lake and became the dominant benthic taxon in terms of abundance and biomass. The growth and reproduction of Asian clam in Lake Tahoe is still limited by temperature, but further warming during this century will allow it to become even more widespread and abundant (for projections, see Wittmann et al., 2013). Since rising temperatures enhance filter-feeding by *C. fluminea*, a future Lake Tahoe dominated by dense clam populations feeding at an

Box 4.5 *Invasive Fishes Interacting with Other Factors Drive Declines of Inlé Lake Endemics*

Inlé (=Inlay) Lake, the second largest freshwater body in Myanmar, is situated at an altitude of 880 m in the Shan Hills. Its relative isolation and considerable age (1.5 million years) have resulted in local speciation of cyrinids, loaches, spiny eels and freshwater snails; the lake is also rich in freshwater macrophytes. Inlé Lake was added to the UNESCO Man & the Biosphere global network of reserves in 2015 in recognition of its importance as a site for migratory waterbirds (82 species, including the endangered eastern sarus crane [*Antigone antigone*: Gruidae; VU] which nests there) and as habitat for around 35 native fishes. Heritage value also accrues from the distinctive local culture of the Intha people who plant fruit and vegetable gardens on floating mats of elephant grass (*Saccharum spotaneum*). Fifteen of the Inlé Lake fishes are endemics, and some are threatened, including the monotypic cyprinid *Sawbwa resplendens* (EN), which is remarkable for lacking scales. The danionine genus *Inlecypris*, which includes *I. auropurpurea* (EN), may also be confined to the lake, but there is dispute over its taxonomic status (some view the genus as synonymous with the more widespread *Devario*). Two other endemics – predatory *Systomus compressiformis* (Cyrinidae; CR) and *Silurus burmanensis* (Siluridae; DD) – that have not been recorded in more than 20 years may be extinct. A likely cause is interaction with 17 non-native species that have become established in the lake, including highly invasive mosquito fish (see main text), tilapias, grass carp and common carp (Kano *et al.*, 2016a). The last of these has begun to hybridize with endemic *Cyprinus intha* (EN), formerly the main fishery species, and probably also competes with it. Deforestation of the catchment and siltation have caused Lake Inlé to shrink by around one third since the mid-twentieth century, and increasing settlement and intensifying tourist development along the shores has contributed to eutrophication and pollution. Invasive water hyacinth (see *Eichhornia crassipes*: Pontoderiaceae; Box 4.12) has proliferated in the lake also, reducing the extent of open water. A long-term reduction in the duration of the monsoon period, which has been attributed to climate change, is an additional driver of deteriorating lake conditions. No-take fish reserves have been established to reduce the effects of overexploitation due a switch from traditional fishing methods to widespread use of monofilament gill nets and illegal electrofishing. However, such sanctuaries are unlikely to benefit endemic Lake Inlé species affected by invasive fishes.

accelerated rate would be characterized by reduced diversity and bio-mass of phytoplankton, lower primary productivity and scarcity of food for pelagic planktivores. Box 4.6 describes additional instances of the role of anthropogenic transformation in facilitating establishment of aliens in Lake Tahoe.

Box 4.6 *Invasion of Lake Tahoe Is Facilitated by Anthropogenic Change*

Human activities affecting Lake Tahoe have led to declines in water clarity attributable to eutrophication and increased sedimentation. The lake was formerly highly oligotrophic with clear waters that allowed penetration of ultraviolet light and limited the establishment of new species. However, when bluegill sunfish (*Lepomis macrochirus*: Centrarchidae) were introduced to Lake Tahoe in the late-1980s, reduced transparency together with warmer temperatures permitted the survival of their UV-sensitive larvae, and these fish have been able to maintain populations at near-shore sites that receive high loadings of particulates from terrestrial sources (Wittmann *et al.*, 2013). Other fishes were intentionally stocked in Lake Tahoe during the early twentieth century, including rainbow, brown and lake trout. By 1939, predation and hybridization with other trout, combined with overfishing and siltation of influent spawning streams had led to the local extirpation of indigenous Lahontan cutthroat trout (*Oncorhynchus clarki henshawi*), one of three federally endangered cutthroat subspecies and the state fish of Nevada. As in Flathead Lake (Box 4.1), non-native kokanee salmon were stocked (beginning in 1945) and, two decades later, *Mysis diluviana* shrimp were introduced. The arrival of these shrimp corresponded with shifts in the diet of kokanee and lake trout and dramatic reductions in the abundance of native zooplanktivores. Signal crayfish were also stocked several times in the late nineteenth and early twentieth century, intended as food for both fishes and humans. Whatever predation or exploitation took place, they were not sufficient to prevent the crayfish population from increasing to well over 200 million, and a concomitant decline in abundance and diversity of aquatic plants and native macroinvertebrates. The result was a complete transformation of the Lake Tahoe food web. Further information can be found in Whittman *et al.* (2013), including an account of invasion history of aquatic macrophytes in the lake.

Biotic Homogenization

The scale and intensity of some impacts associated with invasive species remains to be seen as recently established aliens (such as some of those discussed later in this chapter) are still expanding and have yet to occupy the full extent of their potential ranges (Strayer *et al.*, 2006). In addition, their impacts can vary over time, having both acute (causing mortality) and chronic sublethal effects (as through hybridization). Introductions or invasions are likely to increase rather than lessen in future as globalization of trade and increased ease of travel facilitate deliberate or accidental spread of non-native species, and so the 'great mixing' of the Earth's biotas continues. Furthermore, there no indications that the rate of anthropogenic transformation of freshwater ecosystems is slowing, and it is apparent that such changes tend to facilitate invaders. Aliens have already become pervasive in many places: one need only point to the ubiquity of salmonids in temperate latitudes of the southern hemisphere, or records dating back 25 years of the establishment of alien fishes in more than 90% of the major lakes of Indonesia (Giesen, 1994). A gradual homogenization of the Earth's biota is taking place as invasive species come to dominate habitats that they would not have been able to colonize in the absence of some human involvement. Thus, once-distinct groups of organisms in different parts of the world (or even different drainage basins) come to be more similar as natural biogeographical boundaries become irrelevant in the face of deliberate or unintentional translocation of species by humans, and the continuing physical transformation and simplification of freshwater habits (e.g. Leprieur *et al.*, 2008). As an alternative to the notion of the Anthropocene epoch, the current era of mixing could well be labelled the Homogenocene.

The growing uniformity of biotas is well illustrated by a comparison of the fish fauna of the United States in the past (before the effects of European settlement were manifest) with that of the (near) present day (Rahel, 2002). Calculations of similarity in species complement between all 1128 possible pairwise combinations of the 89 coterminous states showed that pairs of states averaged 15 more species in common than they did in the past. Remarkably, 89 pairs of states that historically had no species in common had come to share an average of 25 species. The main cause of this homogenization was deliberate introductions of fishes for sport or aquaculture purposes by management agencies. The overall biodiversity gains and losses are evident at the scale of the United States:

at least 39 species have been added to the fish fauna, all of which were already common in other parts of the world, while 19 species found nowhere else have become globally extinct (Rahel, 2002). While the number of species gained has exceeded the losses, all the gains involve common species, but the losses are endemics that contributed to inter-regional diversity. While both additions and subtractions contribute to biotic homogenization, the addition of alien species appears the stronger driver. However, the relative importance of the latter might grow if an as-yet unpaid debt associated with aliens leads to further global extinctions. The same phenomenon is apparent in some Chinese lakes where the combined effects of new introductions and extirpation of endemic species have reduced taxonomic dissimilarity among lakes through time, tending to increase regional homogenization (Ding *et al.*, 2017). Thus, the 'great mixing' proceeds.

Who Are the Invaders and What Are Their Attributes?

There are no complete global inventories of the extent to which non-native species have established themselves in fresh water beyond their original range. However, based on the records that are available, the variety of taxa involved is remarkable, ranging from microbes and macrophytic plants through invertebrates to reptiles, amphibians and mammals (Table 4.1), although the list consists of a highly non-random assortment of taxa. For instance, in contrast to the number of invasive freshwater flowering plants (32 macrophyte families listed in Tables 4.1 and 4.2), it is surprising that few algae seem to have become successful invaders (but see Sheath & Vis, 2013). All but one of the most successful mammalian invaders are rodents (see Box 4.4). The exception is the American mink (*Neovison vison*: Mustelidae), a semiaquatic predator that has colonized Europe, northern Asia and Patagonia, displacing European mink (*Mustela lutreola*; CR) and Eurasian otter wherever it overlaps with them. It is also a highly effective predator of European water vole, depleting their populations within its introduced range. Among other mammals, hog deer (*Hyelaphus* [= *Axis*] *porcinus*: Cervidae; EN) are threatened within their original Asian range where they inhabit flood-plains and wet grassland, but viewed as a pest in Australia where they damage wetlands and riparian vegetation; hog deer are also present in Texas and Florida. A population of hippo has become established in the Magdelena River, Colombia, originating from a private menagerie belonging to Pablo Escobar of the Medellín drug cartel. The animals

Table 4.2 *Examples of alien invasive species in fresh water and their primary modes of impact on native biodiversity. Unless indicated otherwise, data were compiled from the Global Invasive Species Database (ISSG, 2015) where all the species included here are categorized among the 100 of the 'world's worst' invaders.*

Species name	Taxon	Origin	Region invaded	Mode of impact	Native species affected
Salvinia molesta (water fern; *S. minima* is invasive also)	Salviniaceae	Brazil	Much of tropics and subtropics (including United States); ornamental plant	Fast-growing floating fern that covers water surface, depleting light and oxygen in water column	Submerged plants, fishes and other aquatic animals
Tamarix ramosissima (salt cedar)	Tamaricaceae	Temperate Asia	Arid regions: western United States, South Africa, Australia, Argentina; ornamental and landscaping plant	Shrub that dominates riparian zones in arid climates; depletes ground water and in-stream flow; displaces native riparian species and alters detrital inputs	Aquatic and riparian biodiversity; food-web architecture; causes substantial economic losses (Zavaleta, 2000)
Eriocheir sinensis (Chinese mitten crab)	Grapsidae	China	Widespread in Europe, also California; via ballast water	Modifies habitat by burrowing activity; migratory (catadromous) with significant biomass transport; omnivorous, may alter food webs	Macrophytes and macroinvertebrates; may be local depletion during migrations
Oreochromis spp. (tilapias; major	Cichlidae	Africa	Throughout tropics and subtropics; major	Competition and displacement of native	Fishes, invertebrates, macrophytes

(cont.)

Table 4.2 (*cont.*)

Species name	Taxon	Origin	Region invaded	Mode of impact	Native species affected
O. aureus, *O. mossambicus* and *O. niloticus*)			aquaculture species; *O. mossambicus* may be the world's most introduced fish	species; deplete phytoplankton and macrophytes; omnivorous feeding alters ecosystem structure; mouthbrooding habit likely facilitates invasion; *O. niloticus* may hybridize with congeners	
Clarias batrachus (walking catfish)	Clariidae	Southeast Asia	Other parts of Asia, southeastern United States; aquaculture species	Voracious predator, but will eat almost anything; competes for food with native fishes; can travel overland facilitating spread in invaded areas	Fishes, tadpoles, probably macroinvertebrates
Micropterus salmoides (largemouth bass)	Centrarchidae	North America	Almost worldwide; introduced as sport fish	Voracious predator of smaller fishes, etc.; may displace native predators	Fishes, amphibians, crayfishes and macroinvertebrates
Rhinella (= *Bufo*) *marina* (cane toad)	Bufonidae	Central and South America	Caribbean islands, Hawaii, Australia, Pacific islands	Larvae grow fast and produce pheromones that harm other	Amphibians; terrestrial invertebrates; impacts on terrestrial predators

Species	Family	Native range	Introduced range	Impacts	Diet
				tadpoles; eggs poisonous; adults large-bodied and highly toxic that can eat invertebrates and other amphibians; all life stages can poison predators that try to eat them	(especially reptiles) but varies with species and over time (Shine, 2010)
Lithobates (= *Rana*) *catesbeiana* (American bullfrog)	Ranidae	Eastern North America and Mexico	Much of Asia and South America, parts of western Europe and western United States; traded as food	Large size facilitates predation, competition and displacement of native species; vector of the fungus causing chytridiomycosis	Amphibians; many other small vertebrates and large invertebrates
Trachemys scripta elegans (red-eared slider)	Emydidae	Eastern United States and Mexico	Over 70 countries, including much of topics and subtropics, via aquarium trade	Competition and displacement of native species; omnivore, but strong predaceous tendencies; more study on ecosystem-level effects needed	Turtles, tadpoles, macroinvertebrates, macrophytes

had been left to roam after Escobar was shot in 1993; they now constitute the largest herd outside Africa. If the population continues to grow and spread, their ecological effects – as in their homeland (see Pennisi, 2014) – could be substantial.

Considering only the freshwater macroinvertebrate invaders (including zooplankton: Tables 4.1 and 4.2), it is evident that they are co-dominated by crustaceans (20 families), among which crayfishes (three families) are conspicuous, and molluscs (both snails and bivalves: 15 families). This is certainly not a random assortment of taxa, given that freshwater macroinvertebrate assemblages are often constituted by insects. Only five insect families are listed, compared to 14 other invertebrate families (or 49 including crustaceans or molluscs). The observation that the taxonomic composition of aliens is decidedly idiosyncratic is not new and was apparent from a survey of Holarctic invaders (Karatayev et al., 2009), which also revealed that aliens were relatively tolerant of compromised water quality and able to withstand moderate amounts of organic pollution (see also Ye et al., 2015). This capacity would permit their establishment in eutrophic or contaminated waters, and allow them to exploit sites vacated by native species that are more sensitive to environmental conditions. The underrepresentation of aquatic insects among invaders (and, perhaps, that of amphibians relative to, say, fishes) could be a consequence of their amphibiotic life cycle, which demands that both the aquatic and the terrestrial environments are suitable for the invader. Moreover, the adults must be able to locate a mate on land at the time and place of emergence. This probably means that a much higher initial number of invading individuals is needed to bring about population establishment of aquatic insects than would be the case for taxa that spend their entire lives in water. It could account for the fact that crustaceans and molluscs, which complete their life cycle in water and lack a terrestrial adult stage, are well represented among invaders.

In terms of trophic guilds, filter- and suspension-feeding species occur more frequently among invaders than among native macroinvertebrate assemblages (Karatayev et al., 2009). Invasions therefore tend to shift communities towards greater numbers of filter feeders, enhancing bentho-pelagic coupling with consequences for food-web architecture and energy flow. Some of the most highly invasive alien species are those that occupy relatively unusual ecological niches: for instance, there are only three species of byssate bivalves (two Dreissenidae and one Mytilidae) in fresh waters, and all are invasive. Similarly, tube-dwelling amphipods, such as the euryhaline Caspian mudshrimp (*Chelicorophium curvispinum*: Corophiidae), do not have freshwater equivalents, and can become dominant species in invaded water bodies (e.g. Van der Velde, 1994).

Some invasive species have become spectacularly successful beyond their native ranges. For example, *Gambusia affinis* and *G. holbrooki* (Poeciliidae) have been introduced worldwide from eastern North America for the purposes of mosquito control, in most instances becoming invasive after their initial introduction. As a result, these two mosquito fishes – present on every continent except Antarctica – are collectively the most abundant, widespread freshwater fish in the world (Pyke, 2008), although some contend that this honour should be bestowed upon the common carp. Mosquito fishes and some other poeciliids (see below) are highly invasive because they are live-bearers (i.e. they are ovoviviparous), which means that a single fertilized female can found a new population (Deacon *et al.*, 2011). Moreover, although *Gambusia* spp. are small, they are more pugnacious than other poeciliid invaders, and their impacts include predation and physical injuries to denizens of invaded water bodies, as well as induction of pelagic trophic cascades and shifts in nutrient dynamics. To quote a review by Pyke (2008: p. 184): '. . . rigorous experimental studies have almost unanimously demonstrated negative impacts of *Gambusia* on a wide range of aquatic animals including aquatic invertebrates, amphibians, and other fish species' (see also Karraker *et al.*, 2010). The impacts of *G. affinis* appear to vary with sex, and populations dominated by females are more damaging to native biota because females are larger than males and have higher feeding rates (Fryxell *et al.*, 2015). Mosquito fishes tolerate a wide range of temperature, salinity and water quality, and can live in near-desert conditions where few other aquatic species can survive. They even pose a threat to the red-finned blue-eye (*Scaturiginichthys vermeilipinnis*; Pseudomugilidae; CR), one of the world's rarest fishes endemic to a single complex of artesian springs in the Australian outback. Blue-eye have been displaced from a number of sites they occupied prior to the arrival of *G. holbrooki*. Conservation efforts have involved drilling bores to source underground water, and translocating blue-eye to newly created wetlands free from mosquito fish (www.bushheritage.org.au/species/red-fin-blue-eye).

The tolerance of *Gambusia* spp. allows them to become more pernicious invaders than other poeciliids such as guppies (*Poecilia reticulata*) or swordtails and their relatives (*Xiphophorus helleri* and *X. variatus*) that have become established in – and are largely confined to – tropical latitudes. The guppy has nonetheless spread to 69 countries because, like *Gambusia* spp., it has been deliberately introduced to control mosquitos (Deacon et al., 2011). Guppies have similar detrimental effects to *Gambusia* spp. when they invade water bodies where diversity is naturally low (e.g. Hawaii, some Caribbean islands), but they appear less influential in high-diversity continental fresh waters (El-Sabaawi et al., 2016).

Few freshwater invasives have enjoyed the same global-scale success as *Gambusia* spp., but one very different species, the freshwater jellyfish *Craspedacusta sowerbyi* (Olindiidae), has spread from its native range in China to such an extent that it now occurs almost worldwide (Dumont, 1994). The likely mode of transport is association of the polyp stage of the jellyfish life cycle with ornamental aquatic plants. Both sexual and asexual reproduction are possible, but the latter appears common within its invaded range, as populations are frequently comprised of one sex only. Fortunately, the ecological effects of this invader appear relatively benign, although the capacity for asexual reproduction must surely have facilitated range expansion.

Ovoviviparity seems to have contributed to the success of invaders among the poeciliids and thiarid snails; the latter are also parthenogenic so that – as with *Craspedecusta* jellyfish – any single individual can start a new population if conditions are suitable. Parental care by mouthbrooding cichlids (such as *Tilapia* and *Oreochromis* spp.: Plate 23) would likewise favour population establishment if the first immigrant arrived with a mouthful of eggs or larvae. Again, it is notable that poeciliids and mouth-brooders have an aspect of their niche – in this case, related to breeding – that is fairly unusual among other freshwater fishes, and which probably makes them highly effective invaders. Variability or divergence in life history is also associated with invasion success in certain fishes (e.g. *Gambusia holbrooki* and the round goby), with intraspecific differences in growth, maturation, and reproductive investment facilitating adaptation to novel conditions at invasion fronts (Alcaraz & García-Berthou, 2007; Kornis et al., 2017). Among freshwater macrophytes, many invasive species have the twin attributes of an ability to reproduce vegetatively (species of Hydrocharitaceae, for example) and rapid growth rates, allowing them to displace native plants. Traits related to competitive ability are positively associated with the likelihood of successful invasions by non-native macro-phytes, and such aliens can be characterized as '. . . species producing soft leaves faster' (Lukács et al., 2017: p. 950), tending also to have short life cycles and rapid growth rates with lower investment structural tissues.

In addition to poeciliids and cichlids, some of the other 40 fish families included in Tables 4.1 and 4.2 have representatives that have been especially successful as invaders, particularly the Cyprinidae and Salmonidae; a few species of Centrarchidae and Gobiidae have also become widespread. Most are habitat generalists or species found in river backwaters within their native range (e.g. brown and rainbow trout, largemouth bass) and thus are able to live in both lotic and lentic habitats within their invaded areas. Some are rather generalist feeders, or are able to adjust their diets to whatever food is available, but others are more specialized, and predatory fishes seem to do well as invaders – especially in lakes (see section dealing with piscivores in lakes) that may have a vacant apex-predator niche. While the establishment of large-bodied predators would be expected to have rather predictable effects on receiving ecosystems, there can be surprises. Alien wels catfish (*Silurus glanis*: Siluridae), one of the world's largest freshwater fish, beach themselves in order to capture pigeons (*Columbia livia*: Columbidae), dragging the birds into the water before swallowing them (Cucherousset *et al.*, 2012). The behaviour, which is analogous to the intentional beaching of predatory marine mammals, has not been reported in the original range of this piscivorous catfish, and some individuals must have modified their behaviour to forage on novel prey in new environments. More generally, behavioural flexibility may increase the chance of alien species becoming established, but also leads to unexpected interactions with native biodiversity.

Despite the number and variety of taxa listed in Tables 4.1 and 4.2, there are surely countless unrecorded records of alien establishment (Strayer, 2010), especially for more inconspicuous species that do not cause any nuisance or obvious ecological impact, and particularly in the tropics where knowledge of much of the native biota is incomplete (Balian *et al.*, 2008a). Invasion by non-pathogenic microbes are certain to have been overlooked and underestimated (Litchman, 2010), although Sukenik *et al.* (2012) have documented the recent spread of two genera of Nostocales (cyanobacteria), *Cylindrospermopsis* and *Aphanizomenon*, among temperate and subtropical lakes and reservoirs. The invasions became apparent only because of declines in water quality attributable to the toxic compounds produced by these Nostocales. In that regard, it is revealing that the most well-known invasive aquatic insect, the Asian tiger mosquito (*Aedes albopictus*: Culicidae), is one that is of direct concern to humans, both as a nuisance biter and viral vector; moreover, as Table 4.1 shows, the majority of reported insect invaders are culicids. Among them, the southern house mosquito (*Culex quinquefasciatus*) is a vector of the avian malaria parasite (*Plasmodium relictum*: Plasmodiidae) that caused declines in the abundance of many endemic species of terrestrial birds following its establishment in Hawaii. The yellow fever mosquito (*A. aegypti*) that originated in Africa now has a pantropical distribution, so it – and the *Flavivirus* (Flaviviridae) it transmits – have been historical invaders; they probably spread to the Neotropics during the seventeenth century by way of the slave trade. The nematode parasite (*Oncocera volvulus*: Onchoceridae) responsible for river blindness (onchocerciasis) in West Africa arrived in Brazil and Central America via the same route. In that case, however, the African blackfly (Simuliidae) vector did not make the journey, and the nematodes were transmitted from infected humans to new hosts via the bites of at least three indigenous species of *Simulium* (Gustavsen *et al.*, 2011). Interestingly, one Afrotropical mayfly, the baetid *Cloeon smaeleni*, has become established in Brazil (Salles *et al.*, 2014), showing that other aquatic insects, apart from culicids, have been able to invade the New World.

Aliens are now so pervasive in some countries – New Zealand and Chile are good examples – that their non-native origins have been forgotten; habitat management and conservation plans have been developed to maintain fisheries based on introduced salmonids such as brown and rainbow trout. This approach may contribute to the maintenance of ecosystem heath and incidentally protect some native biodiversity, but it has led to the elimination of small native fishes – principally galaxioids in both countries – from much of their former habitat, confining them to places where salmonids are unable to colonize or fail to thrive. Irrespective of these impacts, there would be strong opposition to attempts to remove salmonids from places where they have become established. Many stakeholders view them as providing valuable ecosystem services, and the basis of profitable sport fisheries that cannot be replaced by native species (Becker *et al.*, 2007). To offer another example from a different taxon, most attention lavished on invasive watercress in the United States is directed towards its horticulture rather than control, probably because it is widely cultivated in its native European range. The perceived value or utility of alien species within their native range often prompts their deliberate introduction elsewhere without due consideration of the potential ecological impacts of their translocation.

Case Studies

A range of case studies of alien invasions are given in the rest of this chapter to illustrate the variety of mechanisms of impact, the diversity and complexity of outcomes, the scale and importance of the consequences of establishment, and the diversity of taxa involved. I will deal primarily with those aliens that pose significant threats to native freshwater biodiversity, but the species discussed represent only a small selection of the potential examples – both in terms of taxa included and geographical coverage – that could have been chosen. Summary data intended to demonstrate the variety of invaders and their effects on native biota, intended to supplement the text of this chapter, are shown in Tables 4.1 and 4.2. Francis (2012) provides a comprehensive species-specific global overview of some freshwater invasive species, and an encyclopaedic account of biological invasions of all types can be found in Simberloff & Rejmánek (2011).

I will begin by spotlighting some of the larger species of Asian carp that have become successful invaders in the United States, and describe some attempts to manage these invasions, the extents and consequences of which have yet to play out fully. This initial focus on the United States reflects the amount of scientific information available on invaders and their effects, whereas in many parts of the world – especially the tropics – invasions are less well documented, and evidence of impacts is often anecdotal. The chapter will continue with accounts of a range of invaders (mainly fishes and bivalves) in the Laurentian Great Lakes, which will be supplemented by some more general remarks about the effects of invasive molluscs more generally. The Great Lakes example is of particular import given that these inland 'seas' hold around 20% of the Earth's surface fresh water, with a combined coastline more than 10 000 km-long. The multiple invasions of the Great Lakes also provides an opportunity to discuss interactions among invaders in receiving ecosystems, and to explain how human actions and transformation of water bodies facilitate the spread of alien species within and between continents. A further case study deals with one of the great African lakes, Lake Victoria, where the arrival of a large-bodied non-native fish had devastating consequences. Finally, the focus will shift from lakes to describe, more briefly, the impacts of aliens in large rivers.

Asian Carp in the United States

Asian carp have been widely introduced in the United States and elsewhere, giving rise to concern about ecological and other impacts. The first arrival was the common carp introduced from Europe during the late nineteenth century, and deliberately stocked in rivers and lakes for food and angling; ornamental varieties (such as mirrored carp and, especially, koi) have also been imported and occasionally escape into the wild. Common carp is highly eurytopic and is established in all states south of the Canadian border (Nico *et al.*, 2014a), and in so many other countries that it is present on all continents except Antarctica. The Global Invasive Species Database (ISSG, 2015) ranks common carp as the third most frequently introduced species in the world, and it probably ranks second – after *Gambusia* spp. – among the invasive fishes. In the United States, as in all places where it has become established (see Box 4.3), common carp exerts strong effects as an ecosystem engineer by uprooting or damaging macrophytes and disturbing the many species associated with them. These effects (reviewed by Weber & Brown, 2009; see also Hicks *et al.*, 2012) are largely attributable to the behaviour of this benthonic omnivore. Common carp feed by roiling: sucking up sediments, retaining food items and expelling the unwanted reside. This behaviour stirs up or re-suspends bottom sediments thereby increasing turbidity which, in combination with carp excretion, can lead to higher nutrient concentrations and the potential for algal blooms. Impacts of common carp on invertebrates and other fishes can be manifest through habitat alteration, predation (e.g. of fish eggs or juveniles, or by eating zooplankton that limit planktonic algae) and, perhaps, competition with ecologically similar native species, as well as hybridization with Old-World congeners. However, untangling cause and effect can be difficult as carp often thrive in habitats that have been channelized or otherwise altered by human activities where native species are already in decline.

Common carp can have quite large-scale effects: extensive macrophyte beds in Utah Lake experienced a dramatic reduction after carp introduction in the mid-nineteenth century; this habitat degradation triggered a decline in the now critically endangered June sucker (*Chasmistes liorus*: Catostomidae), in part because the decline in plant cover deprived juveniles of a structural refuge against predation by carp and other introduced fishes (Nico *et al.*, 2014a). In other water bodies where common carp have been established for decades, their effects can be difficult to establish with certainty (Weber & Brown, 2011), but there is circumstantial evidence of declines in fishes and also for impacts on waterfowl via depletion of macrophytes. Throughout their invaded range, common carp serve as an example of the profound unintended consequences that can arise from the addition of a single non-native species (see also Box 4.3). Unfortunately, it is possible to point to many other such instances in an increasingly globalized world.

More recent introductions of Asian carp to the United States since the 1960s involve an ecologically diverse array of large-bodied species each reaching 1 m or more in length. The earliest was the plant-eating grass carp that was stocked to control nuisance aquatic macrophytes, although in China, the United States and elsewhere, it also serves as a food fish and is extensively cultured. Grass carp have become widespread occurring in most American states, and even in Lake Michigan, although a widespread prohibition on stocking anything but sterile triploid individuals for weed control has – thus far – helped limit the establishment of breeding populations in the Laurentian Great Lakes (Nico *et al.*, 2014b). The ability of grass carp to regulate nuisance macrophytes, sometimes eliminating them entirely, has made it the most consistently successful means of biological control of floating and submerged aquatic plants globally. However, declines in the extent of native (and non-native) macrophytes are associated with indirect impacts on other aquatic biota through reductions in habitat structure and availability of substrates and refuges for invertebrates and small fishes, as well as increased frequency of algal blooms. Introductions of grass carp also affect bird populations, and can reduce the abundance of phytophagous waterfowl such as American coot, gadwall (*Anas stepera*) and other ducks (Nico *et al.*, 2014b).

During the 1970s, three more species of Asian carp were imported: molluscivorous black carp, as well as bighead and silver carp, which are both filter feeders. Black carp have been used in fish farms to control snails that transmit parasites, and most of these fish are sterile individuals produced through artificial cultivation (Nico & Neilson, 2014). Nonetheless, because of their predatory habits and large body size (up to 1.8 m long and possibly weighing more than 100 kg), concern has been raised about the possibility that escaped black carp might reproduce in the wild, and there is anecdotal evidence of population establishment in the southern Mississippi. It is now illegal to import black carp or transport those already present across state lines (Nico & Neilson, 2014). Black carp have pharyngeal teeth adapted to grind the shells of snails and unionid mussels, and their ecological impacts on endangered molluscs (see Box 2.2) could be substantial.

Bighead and silver carp were initially deployed in fish ponds and reservoirs in the United States to control algal blooms. They subsist mainly on plankton sieved from the water with comb-like gill rakers, those of the silver carp being adapted to collect algae and cyanobacteria as small as 4 μm in diameter (Nico & Fuller, 2014a). Both species can deplete zooplankton populations, thereby competing for food with native fishes, and their regulating effect on phytoplankton has implications for food-web structure and energy flow (Garvey, 2012). Unlike the black carp, bighead and silver carp species are well established in the wild in the United States, as well as in many other countries worldwide, although they have yet to rival the invasive success of the common carp. The efficiency of bighead and silver carp as filter feeders suggests that their continued spread will have substantial ecological ramifications some of which have yet to be fully anticipated (Garvey, 2012; Nico & Fuller, 2014a, 2104b). Bighead carp is the more widespread of the two species: it is farmed in the southern United States (and dozens of other countries), but numerous wild populations have been established further north along the Mississippi to Illinois (Nico & Fuller, 2014b). Legislation prohibits import and possession of bighead carp in some states, but it is a popular food fish among Asian communities, and often transported and sold live. This, combined with the use of juveniles as bait fish, and the tendency for live fish to be released during Buddhist ceremonies, has facilitated the spread of bighead carp throughout the United States. Silver carp are no longer used in aquaculture and are a less popular food fish than bighead carp, but have nevertheless established breeding populations along the Mississippi and Missouri Rivers northward towards Lake Michigan. Silver carp have become notorious for their inclination to leap from the water when startled or disturbed by boat traffic. Given that they can reach 1 m long and weigh 50 kg, airborne silver carp can pose a significant threat of injury to humans.

Bighead and silver carp are poised to invade Lake Michigan, and thence the other Laurentian Great Lakes, via the Chicago Sanitary and Ship Canal. It was built in 1910 to connect Lake Michigan and the Mississippi thereby flushing sewage downriver to St Louis and providing Chicago with improved water quality and other services of waste disposal and navigation, albeit at the expense of downstream degradation. For several decades the canal was too polluted to permit fish passage. Better water quality in recent years has allowed fish to pass through the canal, and there is evidence from sampling of environmental DNA that suggests bighead and silver carp have already entered Lake Michigan (Jerde *et al.*, 2013). Based on their environmental tolerances in their native distributions, all invasive Asian carp species now in the United States could survive in northern Canada, and should find the Great Lakes habitable (Mandrak & Cudmore, 2010).

An electric barrier, installed in the Chicago Canal in 2002, and upgraded twice since then, is intended to prevent transit of Asian carp

(Garvey, 2012), and, to date, there is no firm evidence of bighead or silver carp establishment by that route in any of the Great Lakes. However, such a situation could change if there were power outages that reduced barrier effectiveness. Moreover, an electric barrier will provide no protection from invasives transported in ballast water or biofoulers on ship hulls. The US Congress introduced bills during 2011 and 2012 intended to prevent invasion of the Great Lakes by Asian carp and tasked the US Army Corps of Engineers to supplement the electric barrier and find a means of preventing carp passage between Lake Michigan and the Mississippi drainage. At the time of writing, public debate was ongoing about the desirability of damming the canal to prevent an invasion of the Great Lakes, but this would disrupt the transportation services afforded by the century-old canal and have long-term economic implications (Garvey, 2012). The effectiveness of such a proposed barrier, and the consequential reinstatement of the natural separation between the Mississippi and the Great Lakes, will depend upon it being completed before Asian carp become established in Lake Michigan. There were media reports of the capture of a single silver carp beyond the electric barrier in June 2017 (not far from where one bighead carp was caught in 2010), suggesting that such establishment might occur in the not-too-distant future. There would be serious ecological consequences if Asian carp gained access to the Great Lakes, not least for the US$4.5 billion recreational and commercial fishery of Lake Michigan – although one forecast suggests impacts on commercial fish landings in Lake Erie, where bighead carp have been introduced but appear not to have become established, might be minor (Wittman *et al.*, 2015). However, given that the Great Lakes hold around one fifth of the surface fresh water on Earth, and are a source of drinking water for around 40 million people, extreme care is needed to prevent the establishment of – or the subsequent need to control – Asian carp. However, this is not to imply that these lakes have remained free from alien species thus far. The recent history of the Great Lakes has been dominated by such invaders and the ecological fall out from their establishment and attempts to limit their numbers.

The Laurentian Great Lakes: An Alien Melting Pot

Fishes

The role of a barrier in preventing the entry of Asian carp from the Mississippi into the Laurentian Great Lakes – Ontario, Erie, Michigan,

Huron and Superior – draws attention to the fact that there is but one connection between these huge water bodies and river drainages further south. This strategy of using barriers to fragment linked fresh water bodies clearly has potential utility when it comes to preventing the spread of invasive alien species into habitats where they might cause ecological harm, or spread of disease, or result in hybridization between (for instance) hatchery or wild fish stocks (for details, see Rahel, 2013). The use of fragmentation as a management tool has been applied within another context in the Great Lakes, directed towards the sea lamprey (*Petromyzon marinus*: Petromyzontidae), which, in spite of its name, thrives in fresh water. The sea lamprey first appeared in Lake Ontario in 1835, perhaps having gained access via the Erie Canal, completed in 1825, which connects to the Hudson River and hence to the Atlantic. It was then able to invade the other four Great Lakes from Lake Ontario via a man-made connection with Lake Erie – the Welland Canal – that allowed passage around Niagara Falls; the lakes were completely colonized by 1938. The sea lamprey is a haematophagus ectoparasite of fishes, far larger and more aggressive than the four native lampreys in the Great Lakes – only two of which are parasitic. It had devastating effects on native species such as predatory lake trout (*Salvelinus namaycush*) that had no prior exposure to – or adaptations to withstand – such an enemy; sea lamprey thus triggered one of Elton's 'ecological explosions'. Lake Michigan populations of lake trout collapsed entirely within five years of sea lamprey arrival, although they had experienced declines prior to that due to overexploitation by commercial fishermen during the 1930s and 1940s. Sea lamprey were likewise the main cause of the extinction of the deep-water cisco (*Coregonus johannae*: Salmonidae), which was endemic to Lakes Huron and Michigan, as well as the shortjaw cisco (*C. reighardi*) endemic to Lakes Huron, Michigan and Ontario. Lampricides have been deployed since the 1950s to kill lamprey larvae in the affluent streams where the adults spawn, but the toxins had undesirable effects on a number of non-target species. However, a substantial degree of control of sea lamprey populations has been brought about by construction of low-head dams that block access to spawning tributaries flowing into the Great Lakes, thereby breaking the connectivity migrating lamprey need to complete their life cycle. Dam design is a crucial element in such fragmentation management since, ideally, it should impede only the progress of the pest species and allow others to pass unhindered.

Among other alien fishes in the Great Lakes, the alewife invaded from the Atlantic Ocean and Lake Ontario via the same putative route as the sea lamprey, reaching Lake Michigan by 1949. It became highly successful at the expense of its zooplanktivorous native counterparts, especially various Coregoninae that have a similar niche to the alewife and which, in the case of the bloater (*Coregonus hoyi*; VU), were also subject to its depredations. Alewife themselves were subject to little control by resident predators such as lake trout that had been depleted by overfishing and the ravages of sea lamprey. By the mid-1960s, alewife dominated fish biomass in the Great Lakes, especially Lakes Michigan and Huron, and landings of more economically desirable native species had dwindled. However, alewife experienced summer mass mortalities, perhaps because of thermal stress, and their accumulated corpses degraded water quality; windrows of dead fish along shorelines caused a public nuisance, necessitating expensive clean-up operations. Alewife populations began to come under control in the late 1960s following the introduction of first coho then chinook salmon (*Oncorhynchus tshawytscha*) from the North American west coast; they fed on alewife, and were much less susceptible than lake trout to sea lamprey. The two salmon had the added benefit of serving as sport fishes. Other salmonids were introduced subsequently, allowing establishment of a lucrative sport fishery based upon predation of alewife. Populations of native species that suffered from competition with the alewife had the potential to recover, the extent and duration of which depending on the extent to which they were subject to predations by introduced salmonids. Paradoxically, although it was regarded as an undesirable invader in the Great Lakes, the alewife is a species of conservation concern in parts of the United States, where anadromous populations have declined due to dams that prevent upstream migrants from gaining access to breeding sites. In places where they were once regarded as 'passenger pigeons of the sea', numbers have plummeted by >99%, and some states have banned fishing of alewife (Limburgh & Waldman, 2009).

The Great Lakes are host to around 35 established alien fish species (Mandrak & Cudmore, 2010; other authors give somewhat lower estimates of invader establishment). Changes in the fish fauna have included global extinction of two ciscoes mentioned above, as well as a subspecies of the blue walleye (*Sander vitreus glaucus*: Percidae) endemic to Lakes Ontario and Erie. It suffered from intense competition with the alien rainbow smelt (*Osmerus mordax*) – a zooplanktivore that also eats hatchlings of native fishes, including ciscoes. (Dams on affluent streams that impede the upstream passage of breeding sea lamprey will also obstruct rainbow smelt and help reduce their recruitment.) The loss of the blue walleye has particular import because it was a commercially valuable fishery species

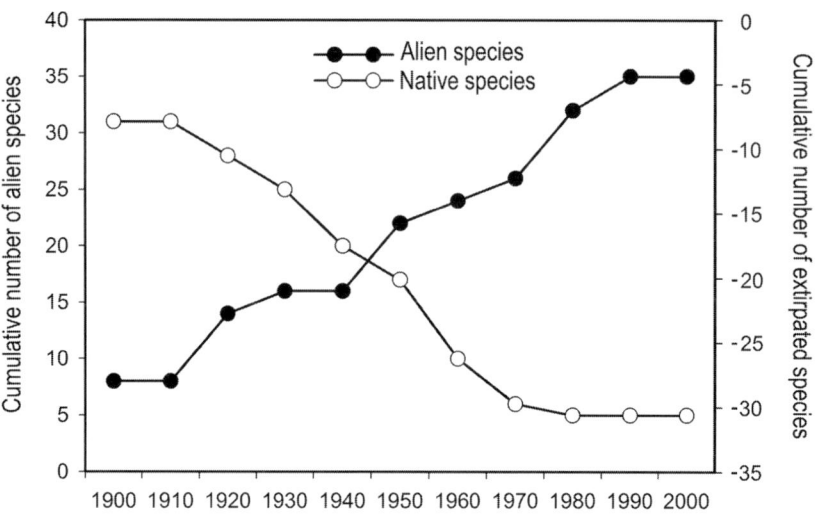

Fig. 4.2 Combined cumulative change in the number of established alien fish species and the number of extirpated fish species in the Laurentian Great Lakes during each decade of the twentieth century . Redrawn from Mandrak & Cudmore (2010) with permission.

until the 1950s; the synergistic effects of exploitation and competition were probably responsible for reducing populations to levels from which the species could not recover. In addition to the three endemic taxa that have become extinct, a further 14 non-endemic species have been extirpated, lowering the species total for the Great Lakes from 169 to 148; a further 82 species have declined so significantly that they should be regarded as threatened (Mandrak & Cudmore, 2010). It is not possible to ascribe all changes in the fish community to interactions with aliens, since other drivers were likely to have been influential for some species, but the association between the timing of alien establishment and species loss during the last century (Fig. 4.2) is very plain.

Dreissenid Mussels

Aside from fishes, another notable invader of the Great Lakes is the euryhaline zebra mussel (*Dreissena polymorpha*: Dreissenidae) (Plate 25) that spread from its native range in the Black and Caspian Seas to colonize several European countries during the nineteenth and twentieth centuries, reaching North America in the late 1980s (Benson *et al.*, 2014a). The route of introduction to the Great Lakes was likely transportation of its planktonic larvae in ballast water of ships, perhaps those entering Lake Ontario through the St Lawrence Seaway, with subsequent spread to the other four lakes via the Welland Canal. The earliest records of zebra mussel are from 1988 in Lake Saint Clair, situated on the Saint Clair River that flows from Lake Huron to Lake Erie. Within two years, zebra mussels had spread throughout the Great Lakes except Lake Superior, where calcium levels are low enough to limit shell formation. Their invaded range has since extended further afield, passing along some of the same routes as bighead carp, albeit in the opposite direction. Zebra mussels now occur in the Mississippi, Missouri, Colorado, Hudson and other rivers in almost 30 states (Benson *et al.*, 2014a). They are highly efficient suspension feeders and biofoulers that attach by byssal threads to hard surfaces forming dense beds. This biofouling habitat has very substantial economic implications due to the obstruction of pipes and interference with municipal water supplies and electric power generation. One estimate puts the annual cost of freshwater biofouling in North America in excess of US$72 million out of a global total of approximately US$277 million (Nakano & Strayer, 2014); other economic costs of alien species can be substantial also (Box 4.7).

Box 4.7 *Economic Impacts of Freshwater Aliens in North America*

The introduction of the zebra mussel could probably have been prevented by spending US$0.32 million in risk assessments and control measures (Leung *et al.*, 2002), far less than the annual economic costs of invasion or the sum spent each year in managing this invader in the United States. This clearly demonstrates the economic imperative for prevention of establishment rather than control or eradication of invaders. To give another example, the United States and Canada together spend around US$15 million each year in attempts to control sea lamprey within the Great Lakes. The economic impacts of this and other invasive species on the multibillion-dollar lake fishery are very large, amounting to US$800 million annually for the sport-fishing industry alone (Pagnucco *et al.*, 2015).

Even on a smaller scale, the economic costs of alien species can be substantial. A predatory zooplankter, the spiny water flea (*Bythotrephes longimanus*: Cercopagididae), invaded the Laurentian Great Lakes from Eurasia in the 1980s and spread subsequently to other water bodies, reaching Lake Mendota (Wisconsin) in 2009. There it reached extremely high densities, reducing the abundance of other Cladocera that feed as grazers – principally *Daphnia pulicaria* – thereby allowing algae to proliferate in this eutrophic lake; as a result, water clarity declined by nearly 1 m (Walsh *et al.*, 2016). Household surveys of local residents had shown that they were willing to pay a total of US$140 million ($640 per household) to improve water clarity of Lake Mendota by 1 m, a sum around the same order of magnitude – US$86.5 to 163 million – as the cost of restoring the lake to pre-invasion levels of clarity through a 70% reduction in phosphorus loading. Although representing only one aspect of the impacts of spiny water flea on Lake Mendota, these cost estimates illustrate the potential harm caused by a single invasive species within a single lake affecting a single ecosystem service – maintenance of water clarity, which enhances recreational use of the lake. The effects of invasions on biodiversity or ecosystem services are rarely measured in monetary terms useful to decision makers, but this example shows that the costs of preventing invasions are far less than those associated with restoring or compensating for damage to ecosystems. This is especially so if the alien species goes on to invade other water bodies beyond the initial site of establishment, because the costs of secondary spread aggregated across many locations will accumulate rapidly. Compounding these costs across multiple invasive species would surely suggest that the collective economic impact of freshwater invasives represents a colossal sum. And, as with many alien species, there are no options for directed control or eradication of the spiny water flea after it has become established.

Soon after the arrival of the zebra mussel, a second invasive bivalve from southeastern Europe appeared in the Great Lakes. The quagga mussel (*Dreissena rostriformis bugensis*) arrived by the same presumed mode as the zebra mussel, but did not begin to spread widely until after the turn of the century (Benson *et al.*, 2014b). The two congeners are ecologically similar, but the quagga mussel appears more suited to its new geographical range and has replaced the zebra mussel as the predominant invasive species, especially in deeper water (Bunnell *et al.*, 2014; Jones & Ricciardi, 2014). The quagga mussel is less dependent on hard substrates for attachment and is able to use a wider range of habitats than the zebra mussel, leading to a substantial recent increase in the total number and biomass of invasive mussels throughout much of the Great Lakes, apart from calcium-poor Lake Superior (Benson *et al.*, 2014b).

Zebra and quagga mussels both act as ecosystem engineers, so the ecological ramifications of their invasion have been substantial (Ward & Ricciardi, 2010; Nakano & Strayer, 2014). They include depletion of phytoplankton, a shift in stocks of organic matter from the water column to the benthos, and greater water clarity that stimulates growth of invasive plants such as Eurasian water milfoil (*Myriophyllum spicatum*: Haloragaceae) and, especially, filamentous *Cladophora* algae on the lake bed. Yellow perch (*Perca flavescens*: Percidae), which is commercially valuable as a food fish and serves as an important forage species for Great Lake salmonids, appear to have benefitted from these ecological changes, as has the native smallmouth bass (*Micropterus dolomieu*: Centrarchidae) that feeds on zebra mussel. On the other hand, depletion of phytoplankton by mussels disrupts pelagic food chains and reduces production of planktivorous fishes such as the lake whitefish, a commercially important species that also suffers from the depredations of sea lamprey (it is the same whitefish species introduced into Flathead Lake: Box 4.1). Stocks of other salmonids and alewife have also been affected. These findings are consistent with long-term research in the Hudson River estuary (Strayer *et al.*, 2004) where establishment of zebra mussel narrowed the distribution and led to a 28% reduction in abundance of pelagic zooplanktivorous fishes, but a near doubling of populations of shore-dwelling zoobenthivorous species. In the Great Lakes and beyond, dreissenid invasions have evidently given rise to both 'winners' and 'losers' among native species.

Dense populations of dreissenid mussels can exert both direct, top-down control on phytoplankton in the Great Lakes by removing them through filtration, altering community composition, and possibly pre-cipitating blooms of toxic cyanobacteria (*Microcystis*). Indirect, bottom-up control can take place through the influence of dreissenids on nutrient dynamics: specifically, by sequestering phosphorus that other-wise would be available to support growth of phytoplankton. The latter has been posited to occur in Lake Huron, while top-down control may be occurring in Lake Ontario, with both top-down and bottom-up limitation of phytoplankton by dreissenids (mainly quagga mussels) in Lake Michigan (Bunnell *et al.*, 2014). Additional effects correlated with the establishment of invasive mussels are huge mats of decaying

Cladophora that are washed up along the shores of Lake Michigan following seasonal die-backs, and are accompanied by an increase in the frequency and severity of botulism type-E outbreaks in birds, caused by *Clostridium botulinum* bacteria. There have also been reductions in populations of the formerly widespread and dominant benthic amphipod *Diporeia* spp. (Pontoporeiidae), which has been replaced by dreissenids. The underlying cause is uncertain, and may involve competition for food, but the nett result is that the benthic community has become a major energy sink rather than a pathway to upper trophic levels (Nalepa *et al.*, 2009). The near-disappearance of *Diporeia*, a readily available food source rich in energy, could have long-term implications for the biomass and structure of benthivorous fish assemblages within the Great Lakes.

Not all of the changes described above can be directly linked to the establishment of invasive dreissenids in Lake Michigan, and their effects on some indigenous species may be positive in instances where dense beds of these biofoulers provide microhabitat for other macroinvertebrates (see, for example, Ward & Ricciardi, 2007). However, their impacts on certain large invertebrates can be devastating. Zebra and quagga mussels settle on the shells of native Unionida (pearly mussels: see Box 2.2), competing with them for food, interfering with movement and siphonal feeding, and overgrowing or smothering the host individual (Strayer, 1999). Such biofouling (or, more technically, epizootic colonization) of pearly mussels began with their almost total elimination from Lake Erie, and has resulted in near-complete disappearance from former habitat in much of the Great Lakes, as well as great reductions in many rivers in the United States invaded by dreissenids (Schloesser *et al.*, 1996; Ricciardi *et al.*, 1998). All of the 23 species of Unionida within the Great Lakes drainage have been affected by dreissenid mussels, mainly zebra mussels which arrived first and expanded their range earlier. Populations of native mussels were extirpated from lakes and rivers four to eight years after the arrival of zebra mussels (Ricciardi *et al.*, 1998), representing a 10-fold increase in rates of population decline entirely attributable to this invader. The replacement of so many native mussel species by one or two alien species represents an extreme case of biotic homogenization (Rahel, 2002). While invasions by zebra and quagga mussels have – rightly – received considerable attention, the proliferation of an invasive biofouling invertebrate, mainly in the southern hemisphere, has garnered much less attention (see Box 4.8).

Box 4.8 *The Golden Mussel: An Asian Invader of the Neotropics*

Like its marine counterparts, and invasive dreissenids, the golden mussel (*Limnoperna fortunei*: Mytilidae) attaches to hard surfaces with byssal threads, building up dense mats of animals. Since native biofouling equivalents are lacking, the golden mussel establishes – in just the same fashion as dreissenids in North America – an entirely novel ecological niche in the freshwater habitats it invades. The golden mussel colonized reservoirs in Hong Kong from southern China via water-supply pipelines in the late 1960s (Morton, 1975); this was the first record of its potential as an invasive species. It subsequently appeared in Taiwan, South Korea and Japan. The golden mussel was then transported – most probably in ship ballast water – to the Río de la Plata in Argentina in around 1990 from where, just over a decade later and facilitated by transport on boat hulls, it spread throughout the Río de la Plata-Paraná–Uruguay drainage basin to Paraguay, Brazil, Uruguay and Bolivia (Darrigran & Damborenea, 2005).

The high population densities attained by golden mussels, combined with their high mass-specific filtration rate, give them the potential to deplete stocks of phytoplankton in standing waters, thereby reducing primary production and changing food-web architecture (Boltovskoy *et al.*, 2009), although their effects in rivers are probably much smaller (Cataldo *et al.*, 2012). They also cause a nuisance by clogging pipes and water intakes of power plants. These impacts are broadly analogous to those attributed to dreissenid mussels (but see Boltovskoy & Correa, 2015). Golden mussels may also be beneficial to some of the native biota, in much the same fashion as dreissenids. Mats of this biofouler can increase invertebrate abundance and diversity by providing complex physical habitat (e.g. Sylvester *et al.*, 2007). Their veliger larvae serve as a food for larvae of native fishes (Paolucci *et al.*, 2010), while juveniles and adults are reported to be eaten by more than 50 fish species (Boltovskoy & Correa, 2015). The golden mussel has considerable scope for further range expansion into North America and, perhaps, Europe. It could invade regions that are too warm to support dreissenids and, based on the climates of the countries the golden mussel has invaded in Asia, it might even be able to co-occur with dreissenids in some northern latitudes. In all likelihood, the spread of the golden mussel is far from over.

Other Invaders and Contemporaneous Ecosystem Changes

The ecology of the Great Lakes ecosystem has been greatly destabilized by successive waves of invasive and introduced species, and associated extirpations or reductions of native. Further changes can be anticipated, not least if Asian carp gain passage into Lake Michigan and beyond. Given that invasions in disturbed habitats seem to have a higher chance of success than those in undisturbed ecosystems, some of the susceptibility of the Great Lakes to establishment of aliens may be ascribed to the effects of pollution during the last century and before. Deterioration in water quality began with modern settlement and resultant land-use changes, bringing pollution from both non-point and point sources such as pulp and paper mills. By the 1920s, deoxygenation events began to occur in Lake Erie fuelling fears that the lake was 'dying'. Additional signs of deterioration in the waters of this and the other lakes became more apparent during subsequent decades due largely to excess phosphorus draining from croplands. Matters began to improve in 1972, when Canada and the United States signed the Great Lakes Water Quality Agreement (which was revised and strengthened in 1978 and 1987) and began to introduce measures intended to restore water quality, with an emphasis on phosphorus reduction. Overall, these efforts have met with success: phosphorus concentrations in Lakes Superior, Michigan and Huron have generally been reduced to levels that fall within acceptable limits; those in Ontario and Erie have shown signs of improvement, although frequently have been at or around the maximum set by the Agreement. Total phosphorus inputs trended downward in all of the lakes but Superior between 1978 and 2008, but some of the declines in phosphorus loading within Lakes Michigan and Huron over this period may be attributable to the influence of dreissenids on internal nutrient cycling (Bunnell *et al.*, 2014). There is current uncertainty regarding continued federal funding for efforts to maintain water quality within the Great Lakes, which places the progress made in restoration at risk, and is no small matter given the number of people that depend on them for drinking water. In addition, proposed funding cuts will put at risk continued maintenance of the electric barrier in the Chicago Sanitary and Ship Canal, leaving the Asian carp free to make an entrance.

There is another source of concern over Great Lakes water quality. There are signs that microcystin-producing cyanobacteria in Lake Erie became more sensitive to phosphorus over the period 2001–2013, increasing the Lake's susceptibility to toxic blooms, meaning that targets for reductions of this nutrient will need to become more stringent (Obenour *et al.*, 2014). A possible cause is depletion of algal phytoplankton by the burgeoning populations of dreissenids, especially quagga mussels. They do not consume toxic cyanobacteria, but by changing the composition of the pelagic primary producers, the mussels may have induced a shift in the phosphorus cycle of Lake Erie (see also Vanderploeg *et al.*, 2002; Bunnell *et al.*, 2014). This possibility receives support from a growing body of evidence that changes in nutrient cycling and ecosystem resilience mediated by cyanobacteria affect lakes across a gradient of nutrient concentrations (reviewed by Cottingham *et al.*, 2015). Fluctuations in the incidence and magnitude of algal blooms, and the size of deoxygenated 'dead zone' in Lake Erie, as modelled over a 28-year period, are not only affected by phosphorus concentrations, but also by the magnitude of river inflows, and hence rainfall, as well as wind conditions that influence water-column mixing (Zhou *et al.*, 2015). The interactive effect of multiple stressors on Lake Erie, and the Great Lakes ecosystem as a whole, is obvious: both control of alien mussels and compensation for a changing climate (with more frequent droughts and extreme precipitation events) will need to be taken into account in the development of nutrient-management strategies. Furthermore, there is evidence that global warming facilitates range expansions by some invasive toxic cyanobacteria (e.g. Wiedner *et al.*, 2007).

This account of Great Lakes' aliens is by no means comprehensive: even with respect to fishes, it contains no mention of the round goby (but see Box 4.9) or the tubenose goby (*Proterorhinus semilunaris*), nor the Eurasian ruffe (*Gymnocephalus cernuus*: Percidae) that appears to compete with the native yellow perch. Likewise, invasive crustaceans have not been considered. They include cladocerans such as the spiny water flea (mentioned in Box 4.7 in connection with Lake Mendoza) and fish-hook water flea (*Cercopagis pengoi*: Cercopagididae) that feed on other zooplankton affecting their community structure (e.g. Vanderploeg *et al.*, 2002); two predatory benthic amphipods (*Echinogammarus ischnus* and so-called killer shrimp, *Dikerogammarus villosus*: Gammaridae) that replace native amphipods and attack other benthic invertebrates; and the bloody-red mysid shrimp (*Hemimysis anomala*) that arrived in 2006 (Kipp *et al.*, 2013). The colonial hydroid *Cordylophora caspia* (Clavidae) has also become established in Lake Michigan. Suffice to say that all nine of these alien fishes and invertebrates share the same Ponto-Caspian origins as zebra and quagga mussels, and have complex interactions with them and

Box 4.9 *Interactions among Aliens Can Permit Forecasts of Invasion Success*

Several of the same Ponto-Caspian species that have wrought profound changes on the Great Lakes ecosystem have reached the United Kingdom, where they have been forecast to precipitate a similar process of invasional meltdown. As many as 23 species have been identified as potential invaders (Gallardo & Aldridge, 2015): six of them – including the zebra and quagga mussels, killer shrimp and the congeneric demon shrimp (*Dikerogammarus haemobaphes*) – are already present and expected to spread further. Another four – *Echinogammarus ischnus*, a mysid shrimp (*Limnomysis benedeni*), an isopod (*Jaera istri*: Janiridae), and a freshwater polychaete (*Hypania invalida*: Ampharetidae) – are also likely to have arrived in the United Kingdom although they have not, as yet, been detected. Indeed, establishment of the quagga mussel was confirmed only subsequent to Gallardo and Aldridge (2015: p. 7) predicting that '. . . the likelihood of its introduction is very high'. Forecasts about the incidence of aliens can be made because, typically, the first set of invaders facilitate establishment by other species from the same biogeographical region as their shared evolutionary history under similar environmental conditions allows them to coexist, and frequently results in commensalism. For instance, the zebra mussel provides food (as pseudofaeces) and shelter (among mats of shells) for killer shrimp, which have body markings that resemble the stripes of the mussels (Gallardo & Aldridge, 2015).

A successful invasion by zebra mussels can also be used to predict the likely establishment of other Ponto-Caspian species that feed on them. For instance, zebra mussels may have facilitated establishment of the round goby, a major predator of this bivalve in the Caspian Sea, as well as that of the hydroid, *Cordylophora caspia*, which consumes their planktonic larvae. This trio of invasive species is common to both the Great Lakes and the United Kingdom. Another interaction between Ponto-Caspian invasive species has been reported in the Danube, where killer shrimp displace native *Gammarus pulex* (Gammaridae) from refuges among stones thereby making them vulnerable to predation by the round goby (Beggel *et al.*, 2016). The goby thus facilitates the replacement of native gammarids by invasive counterparts, and this phenomenon may well occur in the United Kingdom where the same combination of invasives and native species are present.

Interspecific facilitation among alien species is a key component of the invasional meltdown hypothesis. Evaluation of this process as a means of accounting for successful invasions, relative to other alternative invasion-related hypotheses, reveals that facilitation has high explanatory power (supported in 77% of empirical tests) regardless of whether examples from the marine, terrestrial or freshwater realms are considered (Jeschke *et al.*, 2012). The absence of natural enemies is another probable cause of invader success within novel ranges, and possession of some competitive or predatory advantage that is new to the resident community greatly enhances the chances of successful establishment and concomitant ecological impacts. Relative to these three explanations, most other hypotheses proposed to account for and predict biological invasions perform relatively poorly (Jeschke *et al.*, 2012). Even the notion that ecosystems with high biodiversity are more resistant to invaders than species-poor systems (originating with Elton, 1958) gains little support. In a global analysis of patterns of river fish diversity, Leprieur *et al.* (2008) failed to find the expected negative correlation between native and non-native richness that would bolster this biotic resistance hypothesis. Instead, facilitation among alien species with common origins seems to be a more useful tool for predicting the likely success of future invasions.

other species within the Great Lakes (e.g. Haynes *et al.*, 2005). The faucet snail (mentioned above in connection with parasite transmission to waterfowl) also has a Eurasian origin, and has become established in Lakes Erie, Ontario and Michigan. Because it supplements grazing with filter feeding, this snail displaces North American caenogastropod grazers such as *Pleurocera virginica* (see also Box 4.2) and *P. livescens* through competition for food (Kipp *et al.*, 2014).

In addition to invasive animals, the Great Lakes are host to alien plants: for example, European frogbit (*Hydrocharis morsus-ranae*: Hydrocharitaceae) forms large floating mats that can reduce light and oxygen in the underlying water, with local implications for pelagic production. More generally, invasions have transformed food-web architecture within the Great Lakes, thereby changing toxic substance transfer routes and the accumulation of pollutants in consumer biomass (for an account of these matters, see Vanderploeg *et al.*, 2002). A recent invasion of euryhaline copepods in the genus *Euryemora* via the St Lawrence Seaway raises new concerns for the ecology of the Great Lakes: these zooplankters are hosts of pathogenic *Vibrio* bacteria that could have wide-ranging impacts on community composition and energy flow (Cordell, 2012).

As will be readily apparent from the variety of taxa involved and the continuing process of invasion, the number of non-native species that have become established in the Laurentian Great Lakes cannot be determined precisely, but one estimate puts the total at a minimum of 182 (Strayer, 2010). Around 20% of these species have been harmful ecologically. There is significant likelihood that other Ponto-Caspian species that have not thus far arrived in the Great Lakes will become established eventually, as changes to the system wrought by other aliens will facilitate their invasion in a process termed 'invasional meltdown' (Ricciardi & Rasmussen, 1998; see also Gallardo & Aldridge, 2015), which predicts that the presence of alien species enables invasion by additional species, increasing both their likelihood of establishment and their impacts on native species and ecosystem processes. Thus knowledge of interactions among invasive species in the Great Lakes might allow predictions about what might happen elsewhere (see Box 4.9). However, this matter is by no means simple, since the impacts of invaders can depend upon their abundance: for instance, round gobies reduce the growth rates of native fishes in tributaries of Lake Michigan, but the effects diminish at high densities due to interference competition among the invaders (Kornis *et al.*, 2014). One implication of this is that invasive species have greater impacts at moderate densities than when they are abundant.

The variety and potential complexity of interactions between native and alien species in the Great Lakes serve as a salutatory lesson in profound and far-reaching ecological consequences of the establishment of alien species, representing what could well be considered *the* textbook example of invasional meltdown (*sensu* Simberloff & Von Holle, 1999), and also of biotic homogenization (*sensu* Rahel, 2002) involving taxa from two different continents. It is, however, by no means always the case that interactions among aliens involve facilitation. If the invaders have different origins, or feed at different trophic levels, agonistic exchanges that are detrimental to one or more parties are far more likely. Some examples are given in Box 4.10.

The Great Lakes case study clearly demonstrates the limited ability of humans to control or influence the outcome of invasions once they had

Box 4.10 *Agonistic Interactions among Invaders*

While an invasive alien species may displace native species, they some-times colonize ecosystems already dominated by previous invaders that are functionally similar to the new arrival. The ecological consequences of such so-called over-invasions, and the competitive interactions between invasive species, have received little study (Russell *et al.*, 2014), but instances where established invaders are ousted by subsequent arrivals have been documented. Invasion of the lower River Rhine by Caspian mud shrimp during the 1980s was associated with substantial declines in populations of alien zebra mussel and native hydropsychid caddisflies, probably due to intense interspecific competition for hard substrata (van der Velde *et al.*, 1994). Likewise, the alien gammarid shrimp *Gammarus tigrinus* from North America, one of the few freshwater invasive species that has spread eastward across the Atlantic, which competed with indi-genous *G. pulex* in the Rhine, was itself displaced by the more predatory *Echinogammarus ischnus* and, particularly, *Dikerogammarus villosus* after these gammarids arrived from the Ponto-Caspian region (Leuven *et al.*, 2009). Among macrophytes, North American *Elodea canadensis* (Hydrocharitacae) tends to be replaced by its American congener *E. nuttallii* in European water bodies where both species are invasive, and especially at sites that are eutrophic; the American species is sometimes supplanted by African elodea, *Lagarosiphon major*, which has also become established in Eurasia (www.cabi.org/isc/datasheet/20761).

The effects of one alien species can be exacerbated by a later arrival if they serve as a food source for the initial invader. For instance, the abundance of American mink in Europe is enhanced by the presence of red swamp crayfish, and there is a positive correlation between mink densities and the proportion of crayfish in their diets (Melero *et al.*, 2014). The alien trophic subsidy also increases the resilience of mink to population control, which has important consequences for European mink and otter that are excluded by the invader (e.g. Santulli *et al.*, 2014). American mink have also caused declines in populations of alien muskrat in parts of Europe, which is perhaps not unexpected given that this rodent is a favoured food of this mink within its native range.

Although not planned, the mink–muskrat interaction is not dissimilar to instances where a non-native control agent is introduced with the intention of limiting the abundance of a second alien species that shares the same geographical origin. Typically, this involves introduction of a specialist insect herbivore, usually a beetle, to control an aquatic plant. Examples include the effective control of alligator weed (*Alternanthera philoxeroides*: Amaranthaceae) over large parts of its invaded global range by a chrysomelid beetle (*Agasicles hygrophila*), and similar success with water fern (*Salvinia molesta*: Salviniaceae; Table 4.2) following widespread introductions of the weevil *Cyrtobagus salviniae* (Curculionidae). Some control of nuisance water hyacinth has been achieved by two other weevils (*Neochetina* spp.: Curculionidae) in East Africa. In all three cases, the beetles and the plants shared a South American home range, and the specialized feeding habits of the herbivore restricted it to the host plant obviating any risk to native species (see Winston *et al.*, 2014).

In addition to the occurrence of trophic interactions between aliens that have the same geographical origins, complex and entirely novel interactions between invaders with different origins can come about. For example, the non-native parasitic acanthocephalan *Pomphor-hynchus tereticollis* (Pomphorynchidae) has become established in the Rhine, despite the absence of its usual hosts. There, it uses the Ponto-Caspian round goby as its second (paratenic) host and killer shrimp as an intermediate host; this new arrangement is favourable to the parasite because killer shrimp (along with three other invasive amphipods) comprise the main food of the goby (Emde *et al.*, 2012). The bighead goby (*Pontiola kessleri*), which shares the same geographical origin as the round goby, likewise hosts non-native acanthocephalan parasites in the Danube and the Rhine (Ondračková *et al.*, 2009), the new relationship again depending upon the ready availability of a food supply of non-native gammarids. Such interactions between sympatric invaders, whether they involve replacement or consumption of one alien species by another, establishment of new host–parasite relationships, or facilitation (Box 4.9) may well be more common than has been supposed thus far.

taken place, as well the necessity for repeated management interventions. Another noteworthy feature of the Great Lakes invasions is that they were mostly unidirectional, with aliens travelling westward to North America although there is a smaller reciprocal wave of eastward invasions (affecting the Thames and Rhine, as described later in this chapter). The mechanism underlying these westward invasions, and the role of human activities in aiding the spread of aliens within Europe and beyond, will be elaborated in the next section.

Hydraulic Engineering, Navigation and Ponto-Caspian Invaders of the Great Lakes

One might well ask why Ponto-Caspian invaders have been able to spread through Europe as far afield as the Great Lakes of North America. A large part of the answer is that they can be transported in the ballast water of commercial vessels that traverse the Atlantic Ocean, with subsequent passage into Lake Ontario and thence the other Great Lakes via the St Lawrence Seaway (opened in 1959) and the Welland Canal (completed considerably earlier) that links Lakes Erie and Ontario. These two man-made connections are augmented by a network of canals built in northeastern North America during the late nineteenth century, which effectively removed barriers to dispersal of a variety of freshwater species into the Great Lakes. These waterways have ecological parallels with links among major European rivers intended to permit navigation, such as the Rhône–Saône–Seine and the Rhine–Rhône Canals, which permitted the formerly isolated freshwater faunas of southern, northern and western Europe to disperse and interact. In addition, the opening of the Danube–Main–Rhine Canal in 1992 joined two separate drainages and, by creating a navigation route between the North Sea and Black Sea, allowed further biotic homogenization (Leuven *et al.*, 2009). Shipping and transport in ballast water along these waterways provided an effective mechanism for the spread of invasives allowing some Ponto-Caspian species to extend their ranges throughout Europe and – via transoceanic vessels operating from river-mouth ports – even further westward into the Laurentian Great Lakes. This transoceanic shipping probably served as the primary means of entry of over one third of invasive species established in the Great Lakes (Pagnucco *et al.*, 2015).

Within the Ponto-Caspian region, the endorheic Caspian Sea is the likely source of most freshwater invaders. It receives large quantities of fresh water from the River Volga so it has a lower salinity (particularly in the north) than the Black Sea or the Sea of Azov, although these three 'seas' are no longer isolated and navigation between them is possible. The Volga-Don Canal plus a network of other waterways serve as connections along which aquatic species can gain passage, between the Caspian and the Baltic Sea to the north, and westward from the Caspian via the Black Sea into the Mediterranean. In short, two centuries of expansion of navigation canals and their connection to rivers such as the Rhine have linked the previously isolated waters of the Caspian, Azov, Black, Mediterranean, Baltic, North and White Seas and the Atlantic Ocean, allowing mixing of some of the more adaptable constituents of their fauna (Leuven *et al.*, 2009). Further re-assortment of the aquatic fauna may result from a planned Sino-Czech scheme to build a Y-shaped waterway connecting Poland and the Czech Republic to Slovakia and Austria, thereby linking the Danube, Oder and Elde basins. More globally, a host of water-transfer schemes are planned or under construction (Shumilova *et al.*, 2018), and seem certain to further the 'great mixing' of freshwater biotas.

New invasions of the Great Lakes by alien species transported in ballast water from Europe has been effectively prevented since 2006, when regulations were introduced that required ships to discharge fresh water from their ballast tanks and fill them with salt water prior to entering the St. Lawrence Seaway (Pagnucco *et al.*, 2015). These quarantine measures appear to have been effective as there have been no new ballast-mediated invasions since they were introduced. That is not to say that invasions of the Great Lakes will now cease. Asian carps continue to pose a direct threat that requires no human mediation. Furthermore, most alien fish present in the Great Lakes were individuals conveyed to the region for sale as live bait, or as stock for aquaculture or enhancement of angling (Mandrak & Cudmore, 2010), and such transport continues. Climatic warming will also facilitate northward migration and colonization by southern species that are presently constrained because of a thermal barrier to their survival. Higher water temperatures in winter would also permit the further spread of invasives that have, thus far, only penetrated the southernmost Great Lakes (Erie and Michigan). Species of particular concern include the Asian clam and red swamp crayfish that – as described earlier in this chapter – are known to have significant impacts in places where they have become abundant in the United States, and they have been recorded in Michigan State. Significantly, proliferation of the Asian clam in Lake Tahoe was favoured by warming temperatures and other anthropogenic changes (see Box 4.6). Although not yet established in the Great Lakes region, the highly predatory northern snakehead (*Channa argus*) is also a threat, and escapes (or releases) from aquaria have resulted in populations establishing further south. Climatic conditions in its native Asian range suggest the fish has considerable capacity for northward expansion of its invaded range. Evidently, the Great Lakes remain vulnerable to further ecological and economic disruptions from alien species; these will be exacerbated by interactions with anthropogenic stressors that have been increasing in frequency and spatial extent (Pagnucco *et al.*, 2015).

The Fish That Ate Lake Victoria

While the Laurentian Great Lakes provide a dramatic illustration of the effects of alien species in fully one fifth of the habitat available for freshwater biodiversity worldwide, this is far from the only instance of where invasion results in profound changes in lake biota. Earlier in this chapter, reference was made to the havoc wrought by piscivorous fishes such as peacock bass (Gatún Lake), largemouth bass (Lake Atitlán) and lake trout (Flathead Lake: Box 4.1). However, in terms of the numbers of species affected, the invasion of Lake Victoria in East Africa by Nile perch (*Lates niloticus*: Latidae) represents a particularly egregious example of the damage that can be wrought by a non-native predator, resulting in the rapid extinction of ~200 endemic fishes, mainly haplochromine cichlids (see Kaufman, 1992; Lowe-McConnell, 1993; Verschuren *et al.*, 2002).

Impacts on Haplochromine Cichlids

The disaster unfolded over a considerable spatial scale: Lake Victoria is the world's largest tropical lake in terms of surface area (68 800 km^2) and second only to Lake Superior; in terms of water volume, it ranks 9th globally, and its coastline (4830 km) is around half the combined length of shores of the Laurentian Great Lakes. Lake Victoria was host to around 600 native fish species. More than 500 of them were haplochromines and, although estimates of total richness vary, this species flock represents perhaps the most dramatic example of adaptive radiation in any vertebrate taxon (Seehausen *et al.*, 1997; Verschuren *et al.*, 2002), resulting in the evolution of a great variety of feeding modes (Lowe-McConnell, 1993). There is uncertainty about the full extent of this radiation, and hence the precise number of species, because many of them had disappeared from the lake before being formally described. This precludes making an accurate estimate of what has been lost because there is no complete inventory of what lived in Lake Victoria prior to human modification of this ecosystem.

Nile perch – one of the 100 'world's worst' invasive species (ISSG, 2015) – was deliberately introduced into Lake Victoria in 1962 in an attempt to increase the value of the fishery; earlier surreptitious stocking may have taken place during the 1950s (for more history and background, see Pringle, 2005). The rationale behind the stocking was to shift the fishery of the lake from one based predominately on small native fishes to a commercially more valuable tonnage of the larger Nile perch. After a slow start (the reasons for this initial delay are not known), Nile perch proliferated through the late 1970s and early 1980s until, by the end of the 1980s, they dominated fish stocks (Table 4.3) – an event coinciding with the loss of many native haplochromines. The likelihood of such a catastrophic outcome was foreseeable: the Nile perch is a piscivore reaching almost 2 m long and 200 kg in weight, with no ecological equivalent in Lake Victoria. As with invasive sea lamprey in the Laurentian Great Lakes, the Lake Victoria haplochromines were ill-adapted to cope with such a formidable foe. The consequent loss of diversity has resulted in major shifts in energy flow and food-web structure, affecting ecosystem-service provision, fisheries yields and capture practices in the riparian countries of Kenya, Tanzania and Uganda where more than 30 million people depend on Lake Victoria (Kaufman, 1992; Goldschmidt *et al.*, 1993).

Synergies between invasion and habitat degradation in Lake Victoria also have been at play – again analogous to the Great Lakes scenario – because human population growth and land-use change within the drainage basin added nutrients and sediments to the Lake during the period following Nile perch introduction, altering phytoplankton composition, increasing its biomass and, hence, reducing water clarity (Hecky *et al.*, 2010). Impoverishment of fish communities was thus likely due to the combined effects of predation, eutrophication and deoxygenation of deep-water areas of the lake, as well as overfishing, although predation was probably the major driver for most species.

The situation remains unstable: changes in fishing gear and more intense exploitation of Nile perch since the 1990s has reduced the extent of top-down control by this predator, and – once more paralleling the Great Lakes scenario – further alterations in the ecology of Lake Victoria are taking place. Land-use change and siltation of the lake began around a century ago, so some ecological changes to Lake Victoria antedate Nile perch introduction. However, industrialization and urbanization along the lake shores that accompanied the boom in Nile perch fishing led to further deterioration in water quality, as well as some unexpected outcomes (Box 4.11; see also Hecky *et al.*, 2010). The introduction of new fishing regulations that would further focus capture efforts on Nile perch, deliberately depleting them in order to enhance the persistence of native

Table 4.3 *Species composition of Lake Victoria fish catches by percentage weight; all landings except those from 2006, which involved the use of a combination of fishing methods, are trawl catches only. Compiled from data in Katunzi et al. (2010).*

Fish taxa	1969	1985	1989	1995	1996	1999	2001	2006
Haplochromines	82.1	69.1	0	0	3.4	6.7	3.9	6.3
Oreochromis niloticus	0.5	<0.1	0.8	0.8	1.1	1.2	1.4	6.5
Lates niloticus	0.2	23.8	94.6	96.5	91.3	89.8	93.5	85.4
Others	17.3	7.1	4.6	2.7	4.2	2.3	1.2	1.7

Box 4.11 *Lake Victoria: Changing the Drivers of Diversity*

Some unexpected anthropogenically driven reductions in species richness within Lake Victoria have taken place because visually mediated sexual selection for male breeding colouration maintains reproductive isolation in many haplochromines, and increasing turbidity associated with eutrophication and siltation interferes with colour vision and mate recognition (Seehausen *et al.*, 1997; see also Selz *et al.*, 2014). This has had important effects on stenotopic rock-dwelling cichlids that dwell along the lake shoreline, of which there are about 200 species. Although they are seldom eaten by Nile perch, their richness has declined because sexual isolation between species has begun to break down: turbidity reduces the strength of sexual selection, leading to widespread hybridization. Dull colouration, few colour morphs, and low species diversity of fishes can be linked to reduced clarity of Lake Victoria, and eutrophication has thus destroyed both the mechanism of diversification and that which maintains species diversity (Seehausen *et al.*, 1997).

species, would be desirable but requires a significant shift away from the current open-access fishery in the lake. Both this, and action to improve water quality, would need coordinated action by the riparian nations. They could be brought about through the offices of the Lake Victoria Fisheries Organization and the Lake Victoria Basin Commission, which are both institutions of the intergovernmental East Africa Community headquartered in Tanzania. Such actions will require significant enhancement of management capabilities and infrastructure, as well as national and intergovernmental willingness to act.

There have been some signs of recovery of certain haplochromines during recent years, belying – at least to a small extent – some bleak initial forecasts for the Lake Victoria endemics (e.g. Kaufman, 1992). Among these, *Haplochromis* (= *Yssichromis*) *pyrrhocephalus* (formerly classified VU) nearly vanished during the 1980s, but numbers recovered coincident with the intense fishing of Nile perch in the 1990s and have risen to such an extent that it has since become abundant. However, the population recovery is by no means a return to pre-invasion conditions. Reductions in water clarity and oxygen concentrations associated with increasing eutrophication have driven morphological changes in *H. pyrrhocephalus*. Individuals making up current populations have smaller heads and eyes but expanded gill surface areas (64% greater) relative to their progenitors two decades earlier, and there have been morphological changes in the mouthparts consistent with a shift in the types of zoo-plankton eaten (Witte *et al.*, 2008). Population recovery by *H. pyrrhoce-phalus*, as well as a handful of other haplochromines with similar features (van Rijssel & Witte, 2013), has been facilitated by a combination of phenotypic plasticity and adaptation that allowed these fishes to exploit the novel conditions of post-invasion Lake Victoria – a 'no-analogue' ecosystem (*sensu* Strayer, 2010). The establishment of additional invasives (see Box 4.12) may have inhibited population recovery of some native species and, anyway, would have the effect of moving the lake further from its original state. It remains to be seen how many other species will be able to recover by taking advantage of changed conditions, or whether the increased incidence of hybridization in a turbid, eutrophic lake (Box 4.11) will result in the assembly of a novel assemblage comprised of a smaller number of haplochromine species.

Box 4.12 *Water Hyacinth in Lake Victoria*

A new invader arrived while the ecology of Lake Victoria was changing during the late 1980s. Water hyacinth is a floating plant native to the Amazon Basin that has become a pernicious nuisance in tropical and subtropical latitudes worldwide. It forms dense mats on the water surface that impede small fishing boats and other vessels, reduce light penetration and algal production, and may even bring about anoxic conditions in the underlying water column. Increasing nutrient loads during the 1990s favoured water hyacinth in Lake Victoria, and its coverage reached a maximum (over $170\,km^2$) in 1998. For reasons that are not altogether clear, the extent of mats then declined and have since fluctuated in extent; introduction of water-hyacinth weevils (see Box 4.10) in 1997 may have played a small role in controlling this weed (Albright *et al.*, 2004). While direct impacts of water hyacinth on fish diversity in Lake Victoria have not been demonstrated, unfavourable conditions prevailing beneath dense mats is unlikely to have favoured native species. Water hyacinth has also been a nuisance species in Lake Naivasha, the second-largest lake in Kenya (see Box 4.3); it covered large areas of the water surface prior to the introduction of weevils that imposed control on further expansion (Gherardi *et al.*, 2011).

Impacts on Non-haplochromines

The predatory impacts of Nile perch in Lake Victoria have not been confined to haplochromines, and not all declines in non-haplochromines were driven solely by predation. For instance, in 1969 the 'others' category in Table 4.3 included the Singidia tilapia (*Oreochromis esculentus*; CR: 3.2% of the total catch) and the Victoria tilapia (*O. variabilis*; CR) – endemic non-haplochromine cichlids that were formerly major herbivores in Lake Victoria. Both are now virtually extinct in the main water body, but persist in a few satellite lakes around the margins. While predation by Nile perch reduced populations of these endemic tilapias, they were affected also by overfishing as they were larger and more sought after than the smaller haplochromines. They also suffered from competition with Nile tilapia (*O. niloticus*) and *Tilapia zilli*, which were introduced to Lake Victoria to compensate for the decline in the fishery as the stocks of endemic tilapias were overexploited. Nile tilapia now ranks second only to Nile perch in terms of fish biomass in Lake Victoria. Great dietary flexibility enhances the success of this fish as an invasive species (Nico *et al.*, 2014c), and has contributed to making Nile tilapia one of the most widely cultured fish in the world.

Other non-haplochromines formerly abundant in Lake Victoria were mainly catfishes such as the African sharptooth catfish, the butter catfish (*Schilbe intermedius*: Schilbeidae) and the piscivorous Sudan catfish (*Bagrus docmak*: Bagridae); all occur widely beyond Lake Victoria. Nile perch have depleted populations of these catfishes through predation, and competition for haplochromine prey may also have played a role in the case of the Sudan catfish. Although subject to the depredations of an alien piscivore in Lake Victoria, the African sharptooth catfish has been widely introduced outside Africa as an aquaculture species, where it has potential to become invasive and impact native fishes (Vitule *et al.*, 2006), in some cases through hybridization (see discussion of hybridization impacts of aliens above). A fourth catfish, the Lake Victoria squeaker (*Synodontis victoriae*; NT), is mainly confined to the lake where its numbers have been reduced by Nile perch, but deoxygenation of deep-water habitat associated with progressive eutrophication of the lake may be a secondary contributory factor in its decline.

By devastating a host of native fish species, and particularly haplochromines that occupied a wide range of trophic niches, the Nile perch has had a major simplifying effect on the food web of Lake Victoria (Box 4.13). Depletion of those haplochromines that were primary consumers and detritus feeders (formerly around 55% of the total biomass) is thought to have enhanced rates of eutrophication and contributed to algal blooms (Goldschmidt *et al.*, 1993). However, some fish species that are not significant dietary items for Nile perch have flourished. Thus, the native mukene minnow or Lake Victoria sardine (*Rastrineobola argentea*: Cyprinidae) has become the dominant plankton feeder as competition from around two dozen species of zooplanktivorous haplochromines diminished. Sold dried, it is now a major component of the lake fishery, along with Nile perch and Nile tilapia. The atyid shrimp *Caridina nilotica* has thrived also with increases beginning in the late 1980s (Kaufman, 1992), as it seemingly replaced the detritivorous haplochromines and perhaps benefitted also from the cascading effect of reductions in the abundance of haplochromines that ate macroinvertebrates (Goldschmidt *et al.*, 1993).

Box 4.13 *Summary: Trophic Replacements and Ecological Simplification of Lake Victoria*

Changes in Lake Victoria since the 1960s have resulted in at least four ecological replacements: Nile perch replaced piscivorous haplochromines and catfishes; Nile tilapia replaced endemic Singidia and Victoria tilapias; the plankton-feeding mukene minnow replaced zooplanktivorous haplochromines; and *C. nilotica* shrimp replaced detritivorous haplochromines.

These replacements were accompanied by step-by-step reductions in species diversity resulting in an eventual simplification of food-web structure to a system dominated by only four species, with profound and irreversible consequences for the ecology of Lake Victoria. In some parts of the lake, the food web consists principally of shrimp that are eaten by small Nile perch which, in turn, are cannibalized by larger conspecifics. Elsewhere, however, the distorted biomass distribution among trophic levels that accompanied the Nile perch boom has returned to its pre-perch state, although community structure has been irretrievably altered (Downing *et al.*, 2012).

Lessons Learned?

One lesson from the dramatic changes wrought upon Lake Victoria, and the substantial changes that invaders have brought about in the Laurentian Great Lakes, is that high levels of diversity do not necessarily confer resilience upon ecosystems or allow them to resist invasions, despite longstanding views that this should be generally the case (e.g. Elton, 1958; Naeem & Li, 1997). The loss of much of the remarkable diversity of Lake Victoria could hardly have represented a clearer demonstration of the lack of any strong relationship between richness and resistance. At an aggregate level, Lake Victoria has proved to be resilient in terms of biomass distribution among trophic levels (Downing *et al.*, 2012), but community structure has not recovered and, in all likelihood, never will. The lake now supports much less biodiversity than it did during the first half of the twentieth century, with many fewer native species and a concomitant reduction in feeding groups. Limited resilience in terms of community structure may well be typical of instances where ecosystems are affected by multiple stressors, as in Lake Victoria – a situation that will surely become increasingly common in the world's fresh waters.

The examples of Lake Victoria and the Great Lakes also demonstrate that biodiversity in fresh waters, and particularly lakes, is highly susceptible to impacts from aliens, as is also the case for native species on oceanic islands (Millennium Ecosystem Assessment, 2005; Bellard *et al.*, 2016). This might be attributable to the insular nature of both types of environment, where species evolve in isolation making them vulnerable to novel competitors or predators to which they have had no prior exposure. The resulting interactions will be intensified by the bounded nature of lakes and oceanic islands, which limit the opportunities for natives to evade alien invaders. The notion that islands are susceptible to impacts from invasive species because they tend to be relatively species-poor and hence have low biotic resistance is well established in the literature, but has rather little empirical support (Jeschke *et al.*, 2012). Certainly, it cannot be relevant to Lake Victoria with its remarkably high indigenous biodiversity. Such richness did not confer any biotic resistance to the novel Nile perch 'super-predator', which was thus able to wreak ecological and evolutionary havoc.

Diverse Ecological Impacts of Molluscan Invaders

The imperilment of native pearly mussels in the Laurentian Great Lakes has parallels over large parts of the United States where the threat posed by biofouling zebra and quagga mussels is exacerbated by competition for food and space with another alien bivalve, the Asian clam; this clam has been mentioned in the context of its invasion of Lake Tahoe but has become far more widespread, occurring in 35 states. Unlike dreissenids, the Asian clam does not foul the shells of pearly mussels. However, it can attain sufficiently high densities in benthic sediments to displace native mussels as well as sphaeriid clams. The invader also depletes phytoplankton and serves as a sink for calcium thereby limiting its availability to other molluscs – even zebra and quagga mussels. They have even been reported to ingest unionid sperm and larvae, but the mechanisms of interaction with native species are not fully understood (Strayer, 1999). Like dreissenids, the Asian clam clogs pipes and is a major pest in water-supply systems and cooling plants. The original route of invasion has not been determined, but the clam is a source of human food in southern China and could have been brought into the United States by immigrants during the 1920s and 1930s. The Asian clam has also become established in some European rivers, where its detrimental effects on pearly mussels have been similar to those in the United States (see review by Sousa *et al.*, 2008).

As well as their direct impacts on pearly mussels and effects on lacustrine food webs (see above), dreissenids influence other native species indirectly. For instance, zebra mussels affect the European bitterling, because unionids serve as an egg repository for this fish (see also Box 4.4) as well as for the many other bitterlings found in Asia (i.e. around 70 species within the Acheilognathinae). Female bitterlings use a long ovipositor to deposit their eggs within the mantle cavities of unionid mussels by way of the exhalant siphon. The spawning process is completed by the male shedding sperm that is sucked into the inhalant siphon and fertilizes the eggs. Spawning does not occur in the absence of a unionid host, which serves as a shelter both for the bitterling eggs and the hatchlings during their first weeks of development. Experimental studies have shown that zebra mussels reduce the number of eggs of European bitterling inside *Unio pictorum* hosts and, above a critical density of biofoulers, the eggs and larvae do not survive (Vrtílek & Reichard, 2012). European bitterling females do not discriminate between fouled and unfouled mussels, even though oviposition into a fouled host results in partial or complete reproductive failure. The spread of zebra mussels across Europe has, presumably, been too recent to have resulted in any behavioural adaptation allowing females to distinguish between hosts according to their degree of fouling. The proliferation of zebra mussels in Europe is also bad for bitterling because the abundance of unionid hosts is reduced. Certain species of bitterling, such as *Tanakia tanago* (VU) in Japan, utilize a single species of pearly mussel only, which greatly limits their breeding opportunities and overall susceptibility to any factors (such as displacement by invaders) that might be detrimental to the unionid host.

As explained in Box 2.2, unionid mussels have glochidia larvae that are fish ectoparasites, and the relationship between larva and host is often highly specific involving one or a few host species. Surprisingly, since European bitterling must use native unionids as egg repositories, they do not usually serve as hosts for their glochidia, and could even be regarded as parasites of these mussels (Reichard *et al.*, 2006). A neat reversal of this relationship has come to light following the invasion of Eastern Europe by a unionid from China, *Sinanodonta woodiana*; the likely mode of invasion (on more than one occasion given the number of localities involved) is by way of glochidia larvae attached to introduced Asian carp. Outside Europe, *S. woodiana* is also invasive in parts of Asia outside its native range and, because it is eaten by humans, range extension has probably been facilitated by people moving it around (Zieritz *et al.*, 2016). European bitterling readily oviposit within *S. woodiana* but, unlike native European unionids, the potential host expels the fish eggs and the parasitism is unsuccessful. *Sinanodonta woodiana* also differs from other unionids in that their glochidia can successfully parasitize a wide range of fish hosts (Dudgeon & Morton, 1984), including European bitterling (Reichard *et al.*, 2012), although bitterlings in its Asian home range (with ancient sympatry) are poor hosts (Douda *et al.*, 2017).

There are at least two possible mechanisms by which *S. woodiana* could impact European bitterling. First, if it spreads through the range of European bitterling – which it could very well do since this fish serves as a glochidial host – the unsuitability of *S. woodiana* as an egg repository will reduce bitterling recruitment success if it does not learn to avoid the alien mussel. Second, if proliferating *S. woodiana* displace European unionids – and there is evidence of their potential to do so (Douda *et al.*, 2012) – they would reduce the availability of egg-hosts for European bitterling, with substantial implications for their populations, particularly in localities where the remaining native unionids are fouled by zebra mussels.

Among other invasive molluscs, herbivorous snails can have large impacts on ecosystems to which they have been introduced. The golden apple snail (*Pomacea canaliculata*: Ampullariidae; *P. insularum* is also present in some locations) originates in Brazil, but has spread widely through southeastern Asia as an escape from aquaculture; they also occur in the aquarium trade (Hayes *et al.*, 2008). There appear to have been multiple introductions of these snails in Asia as a potential food for humans although, in view of their current abundance, gourmands have limited enthusiasm for them. Golden apple snails attain high densities in the wetlands they invade, feeding voraciously on a wide range of aquatic plants, and causing significant reductions in rice yields. One estimate is that losses to rice crops caused by these alien snails exceed US$425 million annually in the Philippines alone (Naylor, 1996). Golden apple snails can eliminate macrophytes almost entirely from wetlands leading to changes in water-column nutrients, increases in phytoplankton and transformation of food webs (Carlsson *et al.*, 2004). Consumer biomass becomes concentrated in the snail invaders, and secondary production estimates for these animals are among the highest of any freshwater invertebrate (Kwong *et al.*, 2010). Golden apple snails probably represent a trophic 'dead end' in invaded ecosystems in Asia since they are not commonly eaten by native predators, although there is some evidence the introduced red swamp crayfish limits the abundance of these snails in Japan (Yamanishi *et al.*, 2012). In circumstances where they become abundant, these molluscan invaders act as ecological engineers – and in that sense are similar to beavers (see Chapter 3) – precipitating regime shifts that cannot be restored to the original condition without snail removal.

Although the predominant dietary item of golden apple snails is macrophytes, they also eat other snails (especially pulmonates) and their eggs (Kwong *et al.*, 2009) and so supplant them in invaded areas. In Hong Kong, for example, they have virtually eliminated the schistosome vector *Biomphalaria straminea* (Planorbidae) – itself an invasive species from central and South America – that was formerly abundant and widespread (Yipp, 1990). Predation on pulmonates by the golden apple snail have been attributed to another Neotropical ampullariid, the giant ramshorn snail (*Marisa cornuarietis*) that is invasive in the Caribbean and southern United States (United States Geological Survey, 2014). Alarmingly, golden apple snails also prey on the eggs of several species of amphibians (Karraker & Dudgeon, 2014), including the Asian common toad (*Duttaphrynus* [= *Bufo*] *melanostictus*: Bufonidae) – also an invasive species (Kolby, 2014) – that are generally avoided by other egg predators (e.g. Karraker *et al.*, 2010). As a result, there seems to be a trend towards reduced frog abundance in wetlands where golden apple snails have become well established (Karraker & Dudgeon, 2014). This direct effect of a herbivorous alien snail on amphibians has potentially serious implications for biodiversity given the extent of the established range of golden apple snails in southern Asia, which supports a rich diversity of anurans, as well as the potential for this species, which is among the 100 'world's worst' invaders, to spread even further. There are particular grounds for concern given that effective biological control methods for golden apple snails are lacking, although, as noted above, they are sometimes eaten by non-native crayfish.

Aliens in Large Rivers

As has been the case in lakes, rivers have been subject to multiple invasions. Impounded sections are particularly susceptible to establishment of aliens because conditions are unsuitable for indigenous fauna adapted to lotic conditions, giving rise to vacant niches that can be occupied by new arrivals (Pool *et al.*, 2010). And, as mentioned earlier, impoundments themselves can also serve as stepping stones that facilitate spread of invaders through drainage networks (see Johnson *et al.*, 2008). Furthermore, reductions in discharge due to aggressive water extraction, and changed flow and temperature regimes downstream of dams, also tend to favour non-native species (Bunn & Arthington, 2002). Reservoirs also offer an opportunity to stock species for recreational, sport or commercial fishing, accelerating the arrival of potential invaders.

Regardless of the route of invasion, rivers at a global scale now contain a substantial representation of alien species within some taxa, especially fishes and certain molluscs. The homogenization of faunas has been especially pronounced in the United States (Rahel, 2002; see the overview of biotic homogenization earlier in this chapter) and in the rivers of southern and western Europe where over one quarter of fish species are non-native species (Leprieur *et al.*, 2008). The Seine, which once would have had around 30 native fishes, now has 46, only 24 of which are native, demonstrating that, as in the United States, species additions apparently contribute more to faunal homogenization than losses (for a historical review of fish invasions in Europe, see Copp *et al.*, 2005). Examples from both regions are given in the next two sections of this chapter. Homogenization is proceeding more slowly outside of the Nearctic and Palaearctic realms, although the trend towards decreased uniqueness of fish biodiversity associated with individual river basins is apparent within each biogeographical realm and at the global scale (Villéger *et al.*, 2011).

Invasives in the Colorado River

The Colorado has been repeatedly dammed and diverted, notably by the Glen Canyon and Hoover Dams, which are the largest of several dams on the river mainstream; there are also dozens of dams affecting most major tributaries within the basin. They have altered thermal and flow regimes profoundly, and the post-impoundment Colorado has been subject to multiple invasions by nuisance species such as water hyacinth, zebra mussel, Asian clam, virile and red swamp crayfishes, common carp, rainbow and brown trout, *Micropterus* spp., *Lepomis* spp., *Oreochromis* spp., mosquito fish and American bullfrog (from the east coast). Over 100 non-native fish species have been introduced to the Colorado at one time or another, and at least 72 of them are established, far more than the 49 indigenous species (42 considered endemic) that previously lived there (Blinn & Poff, 2005). These aliens have become most prevalent in parts of the river most affected by habitat alteration (Tyus & Saunders, 2000). Notably, the lower ~400-km section of the Colorado has multiple impoundments, and, in some years, no river flow reaches the Gulf of California. This portion of the river now has a fish fauna comprising more than 40 species, almost all non-native, whereas the native fish fauna of the entire lower basin formerly numbered only 35 species (Blinn & Poff, 2005). Small cyprinodontid fishes that were formerly widespread, such as Quitobaquito pupfish (*Cyprinodon eremus*; EN) and desert pupfish (*C. macularius*; VU), persist only in disjunct populations in Arizona and California because poeciliids (especially mosquito fish) have made their original environments uninhabitable. Both pupfishes are categorized as nationally endangered, and the desert pupfish has been the subject of *ex situ* conservation initiatives involving translocation to *Gambusia*-free habitats (Minckley, 1995).

Further upstream, 14 native fishes are known from the Colorado above Glen Canyon Dam (Valdez *et al.*, 2001). Some are large or long-lived species that are endemic or near-endemic to the basin, such as the critically endangered razorback sucker (*Xyrauchen texanus*: Catostomidae), the sole representative of a monotypic genus, and the Colorado pikeminnow (*Ptychocheilus lucius*; VU), which is the largest cyprinid in North America growing to more than 1.5 m long. Other cyprinids include the bonytail chub (*Gila elegans*; CR), humpback chub (*G. cypha*; EN) and roundtail chub (*G. robusta*; NT). These are all primarily mainstream species (so-called big-river fishes) adapted to turbid fast-flowing conditions, and are threatened by the combined effects of loss of riverine habitat to dams, consequent changes in habitat characteristics and food supply, as well as by interactions with alien species. At least 60 non-native fishes have been introduced to the upper Colorado, resulting in establishment of 19 species of predators; consequently, alien species generally far outnumber the natives in any given river reach (e.g. the Grand Canyon: Schmidt *et al.*, 1998). Predation by aliens is thought to have been largely responsible for local extinctions (or elimination of particular populations) of natives after habitat alteration had reduced their abundance (Tyus & Saunders, 2000). A further hazard to some Colorado fishes is infestation by introduced parasitic tapeworms and copepods (see Table 4.1) that spread from aliens to native hosts.

Apart from the big-river fishes, the Apache trout inhabits the upper Colorado. Formerly found mainly in headwater streams, it now occupies only a fraction of its former range in the basin. And, as mentioned earlier, Apache trout are susceptible to hybridization with introduced rainbow and cutthroat trout. Also at risk from alien fishes are endemic or near-endemic cyprinids confined to small tributaries and springs in the upper basin that are susceptible to invasion; for example, the Moapa dace (*Moapa coriacea*; CR), the desert dace (*Eremichthys acros*; VU) and the, now extinct, Las Vegas dace (*Rhinichthys deaconi*). Three other native fishes have also entirely disappeared from the Colorado, but 40 are considered threatened, and the rate at which they are jeopardized has been increasing rapidly, particularly within the lower basin (Blinn & Poff, 2005). The endemic Las Vegas leopard frog (*Lithobates* [= *Rana*] *fisheri*) is also now extinct, attributable – at least in part – to the arrival of congeneric American bullfrog (see Table 4.2).

A highly specific threat to some Colorado River fishes is posed by the introduced white sucker (*Catostomus commersoni*) from the eastern United States. It hybridizes readily with the flannelmouth sucker (*Catostomus latipinnis*) and, less frequently, with the bluehead sucker (*C. discobolus*). Hybrids between white and flannelmouth suckers have facilitated introgression between the two native suckers, which were formerly reproductively isolated, producing individuals with contributions from all three genomes (McDonald *et al.*, 2008). Thus, the white sucker is not only hybridizing with natives, as can often happen when introductions bring once allopatric congeners together (in salmonids, for example), but has also breached the genetic barriers to interbreeding among previously distinct sympatric natives.

Several Colorado endemics or near endemics are currently listed under the United States Endangered Species Act, including the Colorado pikeminnow, humpback and bonytail chub, razorback sucker, Moapa and desert dace, and Apache trout. Wild populations of some species are supplemented by stocking with hatchery-bred individuals, while others such as the humpback chub are the focus of *ex situ* conservation efforts. The Apache trout and flannelmouth sucker have been the subject of translocations, and a stocked population of the latter has been able to persist in the lower Colorado after its initial elimination. However, major and costly programmes initially planned for the flannelmouth sucker and humpback chub have been scaled back over time to '... reduced and arguably insufficient efforts' (Minckley, 1995: p. 303).

Among other conservation measures, environmental water releases have also been used to mimic the flow dynamics of an un-impounded upper Colorado in an attempt to promote native fishes and disadvantage aliens (Schmidt *et al.*, 1998). The first instance was an artificial flood, involving several days of continuous water releases from the Glen Canyon Dam implemented in 1996 (for a fuller account, see Patten *et al.*, 2001). However, a before-and-after investigation (Valdez *et al.*, 2001) concluded that the flood caused only short-term reductions in some alien fishes, but supplemented that appropriately timed water releases of large magnitude may offer potential as a tool for reducing populations of non-natives. A later study, which took account of three post-1995 flood releases from the Glen Canyon Dam (Melis *et al.*, 2012), suggested that floods increased the survival of alien rainbow trout, and enhanced both sandbar habitat and macroinvertebrate food availability. In particular, the floods caused an increase in the abundance of the non-native benthic amphipod *Gammarus lacustris*, which is a major food of both rainbow trout and endangered humpback chub. Regular spring flooding might sustain greater trout abundance, and benefit the sport fishery, but would pose a competition and predation risk to humpback chub (Melis *et al.*, 2012). This finding, which was clearly contrary to one initial intention of the management intervention, reflects a growing concern in the environmental-flow literature that engineered artificial flooding may fail to achieve – or even be detrimental to – desired ecological outcomes (Bond *et al.*, 2014; but see Box 5.9). Studies of the effects of adjusting the rates (but not the volume) of water release from Glen Canyon Dam on downstream channel dynamics are projected to continue until 2020.

While habitat alteration and the establishment of alien species have been highly detrimental to endemic and near-endemic Colorado fishes, the total fish stock of the river has been augmented substantially by the non-natives. The regulated river reaches below the Glen Canyon Dam attract large numbers of breeding and overwintering waterbirds that were rare previously, with abundance tending to be higher in the sections where reductions in turbidity – and consequential shifts towards auto-trophy – have been greatest (Stevens et al., 1997a). They number around 60 species, including waders, shorebirds and piscivorous raptors, but are dominated by dabbling (herbivorous) and diving (predatory) waterfowl (Anseriformes). These native birds are making extensive use of alien species as food: for example, populations of great blue heron have increased due to the ready availability of prey in the form of non-native trout (Blinn & Poff, 2005). Dam tailwaters that support these and other cool-water species provide recreational opportunities prized by anglers, but offer little habitat for native fishes.

Dynamics of Invasions in Other Large Rivers

Like the Colorado, the River Rhine has been profoundly affected by humans. Its flow has been altered by hydraulic engineering and habitat modification to allow navigation by large vessels, which themselves induce waves that affect river-margin habitats. The water quality has gone through a cycle of deterioration to a state of serious pollution during the twentieth century followed by a period of recovery as attempts to clean up the river bore fruit. The improvement in water quality allowed establishment of invasive species that could withstand physical conditions within the modified river channel, and 45 species of non-native freshwater macroinvertebrates (more than 10% of the total), mostly crustaceans and molluscs, are present in the Rhine (Leuven et al., 2009). Transport by shipping and dispersal along canals are primary means of invasion as shown by the fact that there are greater numbers of alien species in the sections of the Rhine contiguous with its delta (42 species), where sea ports are located, and along the upper Rhine adjacent to the connection with the Danube–Main–Rhine Canal (37 species). This is confirmed by the geographical origins of the alien species: 44% are Ponto-Caspian and a further 27% are North American. Ballast-water transport of aliens has thus occurred in two directions, both westward to North America, and especially the Laurentian Great Lakes, as well as eastward from North America to Europe.

Elsewhere in Europe, the River Thames hosts at least 96 non-native species (Jackson & Grey, 2012), to which must be added the zebra mussel that arrived in 2014. Around half were introduced intentionally, and 40% of them originate from eastern North America reflecting the frequency of ship passages to the port of London. In terms of rates of discovery of new species, the Thames (averaging 1.04 species per year: Jackson & Grey, 2012) ranks second only to the Laurentian Great Lakes (1.52: Pagnucco *et al.*, 2015) and is slightly ahead of the Hudson River in the United States (1.0: Mills *et al.*, 1996), which has more invasive species in total (113) than the Thames. Ports for trans-Atlantic shipping are clearly implicated as factors enabling these high rates of invasion, although the rates themselves are sensitive to the specific period over which the averaging takes place. For instance, rates of discovery in the Thames were greatest during the 1970s and could be correlated with high levels of ship traffic, whereas they have declined in the Laurentian Great Lakes over the last 20 years due to effective control of ballast-water transport and release of aliens. Such regulations should be applied more widely, supplemented by legislation or other controls to limit the transport and release of known or potentially invasive species.

Are rivers less susceptible than lakes to the effects of invasion by aliens? The extent of invasional meltdown or species loss attributable to aliens that become established in lakes might suggest that the answer to this question is 'yes'. Why might this be? For strictly aquatic animals, such as fishes and (most) freshwater crustaceans that cannot disperse overland or along the coast, river basins flowing into the sea may not be at equilibrium with regard to the balance between extinction rates (attributable to natural events) and colonization of new species from neighbouring river basins. If so, river basins will be unsaturated with species and, possibly, more susceptible to the establishment of aliens because there will be vacant niches. Moreover, low levels of species packing (or unoccupied ecological space) would reduce the intensity of interspecific competition between natives and new arrivals. Species introductions might therefore have smaller impacts on native assemblages than would be expected if the community was saturated (Leprieur *et al.*, 2009), as might be the case in a biodiverse ancient great lake, unless the invader happens, for example, to be a rapacious piscivore.

A Utilitarian Case for Species Introductions?

While there are many examples of the profoundly damaging effects of invasive species on recipient freshwater ecosystems, there are nonetheless those who champion the introduction of non-native species in particular circumstances. The expansion of inland fisheries in Sri Lanka following the establishment of tilapias, mainly *Oreochromis mossambicus*, is an example of the benefits alien species can have on human livelihoods, and these reservoir fisheries are among the most productive in the world. In a review of the freshwater ecology of Sri Lanka, Fernando (2000) opined that there was no evidence that the introduction of tilapias had adversely affected native species. In part, this is because of the paucity of natural lakes on the island, and hence the absence of an indigenous fauna of lacustrine fishes. Fernando concluded that the drawbacks of tilapias were relatively minor compared to their contribution to fisheries and aquaculture. His view presents a remarkable contrast to the disquiet about alien species felt by many researchers in, for example, North America and Europe. This could be indicative of some disparity in attitudes toward management of fresh waters in different regions: maintaining human livelihoods is a paramount consideration in (say) Asia, irrespective of conservation concerns, whereas the need to protect native biodiversity (and, perhaps, the far smaller proportion of people depending on artisanal fisheries) influences decisions made in North America and Europe.

Is some reevaluation in order, given that certain alien species can be valuable sources of human food? Numerous species have been repeatedly and deliberately introduced to parts of the world outside their native ranges for aquaculture. Furthermore, given that the preservation of near-pristine freshwater environments is no longer a realistic option in many parts of the world, would the enhancement of ecosystem-service provision by introducing alien species be warranted? The introduction of sport fishes to boost recreational angling in rivers and lakes can be seen as one such case. In southern South America, chinook salmon (*Oncorhynchus tshawytscha*), and anadromous rainbow trout and sea-run brown trout support a tourist industry based on angling (Becker *et al.*, 2007); chinook and trout established in New Zealand supply the same recreational service.

Some authors have argued for a nuanced view, recognizing that alien species have potential value under certain circumstances (Gozlan, 2008; but see Vitule *et al.*, 2009), and might even contribute to achieving conservation goals by providing habitat or performing desirable ecosystem functions (Schlaepfer *et al.*, 2011). For instance, invasive hydrilla and the reed *Phragmites australis* (Poaceae) in North America provide as much or more waterfowl habitat and nitrogen-retention capacity than native wetland plant species, although hydrilla tends to compromise habitat quality for fishes (Hershner & Havens, 2008). Others strongly disagree with the contention that alien species should be used for conservation ends or to enhance ecosystem services (Vitule *et al.*, 2009, 2012). Even notorious invaders such as the zebra mussel may provide some benefits: their filtering activity may have the potential to control algal blooms (Elliott *et al.*, 2008). Provision of this service can scarcely be invoked as a benefit that offsets the ecological and economic damage caused by this mussel (Nakano & Strayer, 2014). There is, however, one noteworthy instance of a large-scale, multispecies introduction of non-native species that had the specific objective of augmenting ecosystem services in order to enhance human food security. It took place in the Sepik River, Papua New Guinea, and is worthy of some consideration. A summary is given here: for a fuller account see Dudgeon & Smith (2006, and references therein).

Aliens in the River Sepik

The 1100-km-long River Sepik is the largest in Papua New Guinea in terms of area drained (78 000 km^2). The floodplain fishery has a low yield, around 10% that of comparable tropical rivers, and catches on a per-capita basis are small. Fish yields from upland tributaries are even less. The low yields reflect, in part, the general poverty of freshwater fishes in New Guinea, and the distinctive composition of the Sepik fauna. With the exception of invasive common carp, Cyprinidae, which are diverse on the islands to the west and mainland Asia, are entirely lacking from New Guinea. Instead, the fauna is made up of diadromous species (in the Anguillidae, Lutjanidae and Megalopidae), plus permanent inhabitants of freshwater derived from marine ancestors (Ambassidae, Apogonidae, Ariidae, Eleotridae, Hemirhamphidae, Melanotaeniidae and Theraponidae). However, compared to other rivers on the island, the Sepik is species-poor: by comparison, the Fly River, which drains southern Papua New Guinea and is slightly smaller than the Sepik, supports around twice the number of native fishes. The poverty of freshwater fishes in the Sepik reflects the recent geological history of northern Papua New Guinea, and the formation of the river basin in what was a recently uplifted (<6000 years ago) intermontane, marine trough. Apparently, there has been insufficient time for the evolution of a suite of Sepik fish species capable of fully utilizing lowland floodplain habitats, and most species appear rather unspecialized, although the presence of a pelagic, filter-feeding, fork-tailed catfish (*Arius nox*) of the primarily marine Ariidae is noteworthy.

Despite its low yield, the Sepik supports a locally important subsistence fishery. The catchment has the highest population density of any of the island's river systems, and the population is growing rapidly. Nutritional surveys in Papua New Guinea have indicated a high incidence of malnourishment among children, protein deficiency being a major problem. Given that the Sepik fish fauna is distinctive because of what is absent, rather than what is present, the introduction of alien species could increase fish production and protein supply for humans with minimal impacts on indigenous fishes, provided that the species chosen for introduction would occupy underutilized or vacant niches. The analogous situation in Sri Lanka was the niche filled by introduced tilapia that had been left vacant in the absence of indigenous fishes adapted to lacustrine conditions in reservoirs. After weighing the potential threats to native species and habitats against the benefits of improved fish stocks and human livelihoods, in 1984 authorities in Papua New Guinea took the decision to initiate a project intended to enhance food security along the Sepik through the introduction of non-native fishes.

Part of the rationale for fish introductions was realization that enhancement was clearly possible. About half of the fishery yield from the Sepik in the 1980s was African *Oreochromis mossambicus* tilapia, which had been introduced to Papua New Guinea in 1954 and subsequently appeared in the Sepik. Common carp was also an important component of catches and, in some parts of the Sepik, constituted as much as one third of landings. Another reason was the recognition that capricious introductions of freshwater fishes have occurred in New Guinea, and would continue in the absence of introductions planned in the context of sustainable development. For example, Climbing perch (*Anabas testudineus*: Anabantidae) spread into Papua New Guinea in 1976, soon after introduction to neighbouring Irian Jaya. Possession of accessory breathing organs allows it to survive for some time out of water and travel overland to colonize new habitats by 'walking' on its spiny, hinged opercula.

The National Fisheries Authority of Papua New Guinea contracted the FAO and the United Nations Development Programme (UNDP) to advise on the introduction of a suite of alien species to the Sepik under the aegis of two projects: the Sepik River Stock Enhancement Project (SRSEP; 1987–1993) and Fisheries Improvement by Stocking at High Altitudes for Inland Development (FISHAID; 1993–1997). Introductions took place following evaluation of potential benefits and impacts, some very preliminary ecological studies, and with reference to the International Council for the Exploration of the Sea and the European Inland Fishery Advisory Commission (EIFAC) codes of purposeful movement of aquatic organisms. Among the requirements of the EIFAC codes is that quarantine, containment, monitoring and reporting programmes should be implemented. Fishes to be introduced to the Sepik were imported as eggs or fry and quarantined prior to introduction; unfortunately, the requirement to monitor and report outcomes was largely disregarded. Selection of species was based on their potential to fill vacant niches and thus, presumably, minimization of potential impacts on indigenous fishes.

Tilapia rendalli was the first species imported and introduced to the Sepik. Java barb (*Barbonymus* [= *Puntius*] *gonionotus*) was the second, and was stocked in the lower course of the Sepik, while three other cyprinids – the copper mahseer (*Neolissochilus hexagonolepis*; NT), snowtrout (*Schizothorax richardsonii*; VU) and golden mahseer – were introduced in fast-flowing stony tributaries at various altitudes. It is ironic that three of the introduced species are threatened within their home range of India, where the snowtrout is displaced by brown and rainbow trout that have been widely introduced in Himalayan streams (Vishwanath, 2010). In addition to these Old-World species, a serrasalmid *Piaractus brachypomus* (= *Colossoma bidens*) (Plate 13), and the prochilodontid *Prochilodus argenteus* (= *P. margravii*) from the Neotropics were also stocked in floodplain lakes and the lowland reaches. Introduction of the cyprinid *Bangana* (= *Sinilabeo*) *dero* from India was also proposed but had not been undertaken by the end of the FISHAID project, while translocation within New Guinea of the introduced snakeskin gourami (*Trichogaster pectoralis*) and giant gourami (*Osphronemus gouramy*: Osphronemidae) to the Sepik was also mooted but not undertaken on a large scale.

Details of the fate of fish introductions along the Sepik are sketchy. *Tilapia rendalli*, Java barb, *Piaractus brachypomus* and *Prochilodus argentus* have become established in the floodplain, and *T. rendalii* is also present in upland streams together with snowtrout and golden mahseer. All six species are being fished and are sold in markets thereby making some contribution to enhanced food security, but fishery yields have not been quantified. The National Fisheries Authority has undertaken no monitoring in the Sepik since 1997, despite the fact that this is a requirement under the EIFAC codes of practice (see above), nor did it request FAO or UNDP assistance in developing or undertaking such a programme. It is surprising that post-introduction monitoring was not undertaken, given the history of accidental introduction of non-native species in New Guinea. Waterway clogging by the South American water fern (see Table 4.2) was a major problem in the middle and lower Sepik during the 1970s and early 1980s until the plant was brought under control by the introduction of a host specific herbivorous weevil (see Box 4.10). Water hyacinth is also present.

In the absence of any formal monitoring programme, it is not possible to determine the extent of the spread of introduced fishes with the Sepik, and nor can their contribution to food security or impacts on native species – if any – be properly assessed. However, limited observations in tributary streams suggest that the impacts of some aliens may be locally severe. For instance, although initially stocked into an upland region with hardly any native fishes, the golden mahseer has spread downstream to lower altitudes. It attains a very large size – up to 80 kg and 2.5 m long – and is piscivorous as an adult. A small amount of time-series data from one site (2001 to 2004) showed that catches were increasingly dominated by alien species (see Dudgeon & Smith, 2006), and that this change coincided with a decline in the richness and abundance of native fishes from nine species to only two (comprising <1% of all fish caught). The dominant alien was the omnivorous copper mahseer, although golden mahseer, which may well have been feeding on native fishes, was present also.

The lack of post-introduction monitoring limits our ability to learn from the Sepik example. The species introduced to the Sepik were representatives of families not represented within New Guinea, making it difficult to predict how they might interact within recipient ecosystems and whether they would displace or even extirpate native species. There is some evidence that introduced species which are confamilial or congeneric with indigenous species are more likely to become established and potentially invasive because they share with their native relatives traits that pre-adapt them to their new environment (Hulme, 2003). This tendency is exemplified by the strong impacts that introduced trout can have on native salmonids, as described earlier in this chapter. However, none of the fishes now established in the Sepik is closely related to native species that inhabit the river.

In the case of the Sepik, one might ask whether the livelihood benefits to humans accruing from fish introductions might outweigh the known or suspected damaging effects of aliens on native fishes. Certain ecosystems may be regarded as of aesthetic and/or conservation value because biogeographical accidents have resulted in unique communities. The Sepik is an obvious instance, and that might have been a justification for attempting to maintain it in a pristine state – disregarding, of course, the common carp and *Oreochromis mossambicus* tilapia that had become established. While the later introductions involved selection of species that seemed likely to occupy vacant niches in the Sepik, there was no subsequent check on the consequences of these decisions. The benign and apparently beneficial effects of tilapia introductions in Sri Lanka, where the aliens were largely confined to man-made lakes, is an entirely different matter to fish introductions into a large interconnected river system such as the Sepik, because species do not remain confined to the locations where they were introduced. Even if fishes had been selected for introduction on the basis that they appeared to occupy a niche left vacant by indigenous species (such as the upper tributaries), there could be little assurance that displacement of natives would not take place. Although informed by scientific considerations, the choice of whether or not to deliberately stock alien species in the Sepik was, ultimately, a political decision made by the Papua New Guinea government within a socioeconomic context where conservation concerns were far from paramount. So little basic science had been undertaken on the ecology of the Sepik that it was not possible to discern, then or now, or even partially, the results of this giant experiment. Hence, we can learn little from it, beyond the likelihood that – when attempting to predict the likely ecological impacts of alien species – it is best to adopt the precautionary principle.

Invasion is Forever

Alien invasions – especially those involving several species – create 'no-analogue' ecosystems in a Homogenocene world (*sensu* Strayer, 2010); such novel ecosystems will require different management strategies to those scientists are familiar with, and present serious challenges for the preservation of native biodiversity. Ongoing climate change (see Chapter 7) and human population growth, along with further expansion of international trade and travel, will exacerbate matters by providing more opportunities for the transport of aliens to fresh waters where conditions are altering to become less and less like those to which the indigenous species are adapted, and are more likely to favour the establishment of arriving aliens. Furthermore, control measures for invasive species are frequently inefficient or ineffective, and may be harmful to native species. Elimination of invaders from small water bodies through the use of herbicides, molluscides, piscicides and the like is feasible – especially if it is feasible to temporarily remove species of conservation concern. It may also be possible in situations where the size and structure of a water body permit complete removal of aliens by trapping or netting them – for example, in small alpine lakes. Such interventions can result in population recoveries by affected native species (e.g. Knapp *et al.*, 2007, 2016), but this is not always the case especially in cases where the invaders have had a lengthy residence time (e.g. Knapp & Sarnelle, 2008). In contrast, complex riverine habitats offer many refuges and opportunities for invasive species to evade or escape control, and the chances of successful eradication are low. For most practical purposes, and in the majority of instances, one might well conclude that – as with extinction – invasion is forever. In such circumstances, biologists might well turn their attention to investigating the evolutionary response of native species to invaders, in the search for management interventions that enhance rates of adaptation or persistence of the former.

Clearly the most effective means of limiting the impacts of invasive species on native biodiversity is to prevent their arrival. This would involve restriction on imports, particularly widely traded ornamental species such as fishes and crayfishes that are known to be potentially invasive. Counties such as Australia already have strict proscriptions on what can be imported, while the EU 'List of Invasive Alien Species of Union Concern' (EU Regulation No 1143/2014) could, if adequately enforced, provide a basis for prevention of further introductions and spread alien species. More generally, there is a need to draw up national or regional 'black lists' of known invasive species whose importation is proscribed in law (the species categorized as among the 'world's worst' invaders in the Global Invasive Species Database (ISSG, 2015) represent a good starting point), and 'green lists' of species with no known potential risk would be desirable, with risk assessments based on the precautionary principle undertaken to screen candidate species (for an example, see Leung & Dudgeon, 2008). The investment needed to put in place necessary measures to prevent the import or export of invasive alien species will surely result in high potential gains in terms of offsetting ecological damage that they could wreak in future.

One possible means of controlling invaders after they have arrived is to introduce a highly specific predator or pathogen that might reduce their abundance, as in the case of weevils that have had some success in limiting floating nuisance weeds (see Box 4.10) or, more famously, the myxoma virus that was brought to bear on burgeoning rabbit populations in Australia. Perhaps encouraged by that successful example of biological control, the Australian Department of Agriculture and Water

Resources announced in 2016 their intention to develop a National Carp Control Plan involving introduction of a strain of the herpes virus – cyprinid herpes virus 3 (CyHV-3: carp or koi herpes virus) – into rivers and wetlands at an estimated cost of around US$11 million. The virus has only been observed as naturally occurring in common carp, and infections that have been reported in countries where these fish are invasive are usually fatal (Thresher *et al.*, 2018). Further research and risk assessments will need to determine the susceptibility of other aquatic vertebrates prior to deployment but, if all goes according to plan, introductions of CyHV-3 could begin in the near future.

It is necessary to conclude this chapter on a note of realism. The potential ecosystem services derived from alien species in Sri Lankan reservoirs, in the Sepik River, in Lake Naivahsa and in other inland waters worldwide where aliens are fished suggest that the maintenance of pristine ecosystems might not invariably be the most desirable goal nor, in an increasingly populous, Anthropocene – or Homogenocene – Earth, will it always be achievable. While protection of near-pristine habitats from alien species, and all habitats from the most perniciously invasive species, is certainly warranted, it will be challenging to sustain the 'fortress' model of conservation in many places. It may be wise to develop strategies for the management of damaged or degraded aquatic ecosystems, which may contain some alien species, or 'bad' biodiversity, with the overall aim of maximizing the persistence of 'good' biodiversity. This is not a new idea. Elton (1958: p. 155) opined that '. . . conservation should mean the keeping or putting in the landscape of the greatest possible ecological variety . . . and provided the native species have their place, I have no reason why the reconstitution of animal communities to make them rich and interesting and stable should not include a careful selection of exotic forms, especially as many of these are in any case going to arrive in due course . . .'. This could represent the most achievable outcome for many freshwater habitats in a rapidly changing world.

5 · *River Regulation*
Impacts and Mitigation

Nature to be commanded must be obeyed.
> Francis Bacon (1620), *Novo Organum Scientiarum*

I don't know why we keep building these fucking dams. Not only do they cause environmental and social disasters, they, with very few exceptions, all fail to do what they were supposed to do in the first place. Look at the Amazon, where they've all silted up. What is the reaction to that? They're going to build another eighty of them. . . . We must have beaver genes or something.
> Douglas Adams, interview with the *Daily Nexus*, April 2001

From each dam, lines of pylons marched every which way over the bare hills, like gangs of thieves making off with their swag, robbing the Columbia of its life in order to sell corporate farmers electricity at the laughable price of $1.50 a megawatt.
> Jonathan Raban (2010: p. 503), *Driving Home*

Ecological Impacts of Flow Regulation

Anthropogenic alteration of river flows, lake water levels, and the duration and extent of wetland or floodplain inundation (often collectively referred to as flow regulation) is a key contributor to freshwater habitat degradation and species decline, with manifold direct and indirect effects on populations, communities and food-web architecture. These can arise from an insufficiency of water as well as changes in flow that result in there being too much or too little water at particular times or seasons. Due to the absolute scarcity of fresh water (see Chapter 1), there is an inevitable trade-off between human appropriation of water and that available to sustain freshwater ecosystems including the water perceived to be 'wasted' water if it flows to the sea or to a lake or river reach that lies downstream of an administrative boundary. While human appropriation tends to reduce the quantities of water in rivers and lakes, much effort has been directed toward increasing the predictability of flow and reducing variability. This is understandable, since humans view flow variability as undesirable or – in extreme cases when associated with floods and droughts – disastrous. One outcome has been the widespread practice of storing water in an impoundment behind a dam and releasing it downstream when needed. Water storage occurs for a variety of reasons, but mainly for irrigation or domestic consumption, hydropower generation or maintenance of sufficient depth for vessel traffic; sometimes more than one of these functions can be achieved by the same facility. The downstream effects vary, even where regulation is intended to serve a single purpose only; for instance, the consequences of extraction of water for irrigation will depend upon the amount and timing of the demand. In general, however, downstream discharge will be most affected during natural low-flow periods, which are the times of least rainfall and hence greatest demand for irrigation water. The amplitude of flood flows may also be reduced due to water capture within empty impoundments.

The extent of river regulation should not be underestimated: a survey of almost 3000 sites across the United States found that the magnitudes of mean annual (1980–2007) minimum and maximum flows were altered in 86% of streams assessed (Carlisle *et al.*, 2011); unsurprisingly, the alteration becomes more severe in arid regions of the United States and elsewhere (e.g. Kingsford & Thomas, 2004). Downstream changes are usually much greater when water is withdrawn from irrigation than for domestic consumption because the volumes involved in the former are usually much larger, and the agricultural sector by far the largest consumer of water globally. Hydropower dams typically cause little or no change to downstream flows if they are generating baseload power, as inflow to the impoundment is balanced by outflow. By contrast, facilities that generate power to meet peak-load demand tend to increase water releases at the start of the generating period and decrease them abruptly at the end. Regulation for navigation enhances predictability in flow conditions ensuring that a minimum flow volume is always present within the channel. Thus, seasonal declines in discharge tend to be reduced or disappear entirely; any intermittency in flow duration tends, likewise, to be eliminated. In addition, obstacles to vessel movement, such as snags or log jams that can be important feeding or resting sites for fishes, are cleared from river channels. The influence of such practices on salmon, for example, have been well described by Montgomery (2003), and there is a large technical literature on the importance and management of large woody debris in streams (e.g. Lassettre & Kondolf, 2012; Roni *et al.*, 2015).

The multifarious impacts of river regulation and dams have been extensively documented in the primary technical literature and in numerous text books (e.g. Allan & Castillo, 2007), and are also explained well in an excellent volume written for non-specialists (McCully, 2001). In the case of dams, these vary depending on the size of the structure and its location along the river courses. Many of these impacts do not require further elaboration here but are presented in summary form in Table 5.1. The most obvious is a change in downstream flow patterns, including a reduction in volume, and shifts in both the amplitude and timing of peak flows, and a 'flattening' of the fluvial disturbance needed to rejuvenate habitat (Carlisle *et al.*, 2011). Such changes could involve smoothing the hydrograph and reducing both the amplitude and frequency of peaks or, in the case of some hydropower dams, producing unnaturally frequent flow fluctuations. Even slight modifications to the natural flow regime have significant consequences for the structure of riparian plant communities for instance, which become more simplified with increasing flow alteration (Tonkin *et al.*, 2018). The effects of flood reduction are particularly marked, but drought and flow homogenization both result in greater simplification of riparian communities than increased flooding. Regardless of the particular details, a fundamental reason why regulation of flow is problematic is that rivers are dynamic systems, exhibiting variability in discharge, inundation or other aspects of the water regime. River regulation results in a general shift in flow conditions away from those to which the aquatic biota are adapted, and the cues that initiate or terminate life-cycle events may change or even disappear entirely. And, because floods tend to enhance the robustness and resilience of ecological networks that are drivers of diversity in riverine ecosystems, reductions in flood magnitude or frequency can be particularly damaging.

The extent of dam building or 'artificialization' (*sensu* Meybeck, 2003) of the world's rivers has been such that by the early 21st century, most major river basins were fragmented by large dams (i.e. those exceeding 15 m in height), and in a number of basins, the combined storage capacity of dams far exceeded mean annual flows: for instance, roughly four times that of the River Volta and the Colorado. Globally, the total water volume trapped behind dams is over five times annual river runoff, and in their normal, 'near full' operational state, these impoundments have been estimated to control about 15% of surface runoff (Nilsson *et al.*, 2005). This storage of water results in 'river ageing' that can be up to one year and more for large impoundments such as Lake Nasser on the Nile, or in reservoir cascades along Russian and Ukrainian rivers (the Dniepr, Dnestr,

Table 5.1 *Ecological impacts of various types of river regulation*

Dam construction markedly alters flow conditions to which riverine biota are adapted
- Changed flow upstream of dam; impoundment of standing water replaces section of flowing river, with consequent shifts towards taxa favouring lentic conditions
- Altered flow downstream; natural flow regime replaced by pattern of water release determined by dam operations; in extreme cases, downstream flows may cease entirely for periods as dam (re)fills
- Diminished downstream flow favours taxa with preference for fine-grained substrates and slow-moving currents; taxa with rheophilic traits decline
- Creates barriers to movement of organisms and material
- Physicochemical characteristics of water (dissolved oxygen, temperature, sediment loads) up- and downstream of dam altered
- Trophic conditions altered in impounded section (e.g. increased autochthony); changes composition of transported material downstream of dam (e.g. more phytoplankton; less leaf litter)
- Overall, shifts in conditions may favour invasive species; often at the expense of native biota

Channelization
- River flow characteristics altered by channel straightening and constraints of 'hard' concretized banks; increased rates of runoff in engineered channel
- Snags and log jams cleared, combined with a general reduction of in-stream heterogeneity that tends to make habitats more uniform
- Levees or barriers prevent exchange of water with – and inundation of – floodplain
- In extreme cases, natural habitat entirely destroyed as river channel replaced by channel with concrete sides and base

Flow reduction due to water abstraction
- Overabstraction of water for irrigation or other human needs reduces flows and, in extreme conditions, may result in dewatering downstream
- Unused irrigation water returned to river channel will be of reduced quality due to contaminants, nutrients or salinization

Water transfers between drainage basins
- Change flow conditions in contributing and recipient rivers, and may alter water chemistry of the latter; allow exchanges in biota thereby facilitating invasive species

Don and Volga), whereas water residence within their un-impounded channels would have been a few weeks only (Meybeck, 2003).

Much of the water stored behind dams is eventually diverted and used for irrigation and does not flow directly to the sea. Irrigated agriculture is responsible for production of 40% of food on 20% of the land area.

Perhaps a third or more of this is based on water drawn from large dams into canals and hence into networks of smaller channels. The proportion rises to more than half in China, and likewise in India and Pakistan (Box 5.1); the effects on downstream flows are predictable. The Yellow River (or Huang He) did not flow to the sea in more than 20 of the years between 1972 and 1997; the period of annual dryness in the lower course lasted for one or two weeks in the 1970s and 1980s, increasing to more than 200 days in 1997, with hundreds of kilometers of river bed left dry. By that time, reservoir capacity had grown so much that control of the entire flow of the Yellow River became possible, and integration of water releases from multiple dams along the river network could ensure that the mainstream contained some water throughout its course. This could be regarded as an example of an environmental water allocation (see Chapter 1), implemented on a grand scale, but most of the water was released in the form of an annual flood. It was intended to flush the river, scouring accumulated sediments and pollutants from the lower course of the Yellow River, which reportedly had detrimental consequences in the delta and coastal region (Wang *et al.*, 2017).

Among other over-abstracted rivers, the Colorado River seldom flows to its delta in the Gulf of California (see also Box 3.5), and the Rio Grande has been similarly diminished. Dams, barrages and irrigation in the Murray-Darling basin have reduced river flows to the sea to around one quarter of former levels, but this varies considerably between years and can fall to virtually zero. Little to none of the water of the River Jordan reaches the Dead Sea, so its level is falling. The failure of the Amu Darya and Syr Darya to replenish the Aral Sea, and the consequences thereof, are described in Chapter 6, where additional examples of the outcomes of excessive appropriation of water on lakes and wetlands are presented. This chapter focuses on the effects of various types of flow regulation on rivers.

Even where the channel is not dewatered, dams can cause marked shifts in the physical conditions in downstream river reaches. Temporal variability in temperature, as well as sediment and nutrient fluxes, which can occur on diurnal, seasonal and inter-annual timescales, are pervasive in unregulated rivers, and interact with flow regime. Construction of dams can have profound effects on downstream temperatures (reviewed by Poole & Berman, 2001): for example, water that is drawn from the upper layers of water in impoundments tends to be warmer than water typical of an un-impounded river. The converse is true of water drawn from the lower layers, especially if the reservoir is stratified with a well-developed thermocline, and such water often also has a far more stable temperature than would be normal for river water (Box 5.2).

Box 5.1 *Overabstraction and Linking of Rivers in South Asia*

The Indus basin offers one example of the profound effects of humans on runoff and the global water system. It contains the largest contiguous irrigation network in the world, and portions have their origins more than 5000 years ago during the Harappan civilization. Much of the surface drainage of the basins has been altered by large and small dams, barrages, canals and smaller channels such that only around one eighth of the $\sim$400 km^3 of rain water that falls on the basin annually flows down the Indus, with the rest captured for irrigation and returned to the atmosphere by evapotranspiration. In turn, this influences local temperatures with broader-scale effects upon the patterns of air circulation, precipitation and timing of the south Asian monsoon (Saeed *et al.*, 2009). The Indus is now one of the world's most modified rivers: its dry-season flow generally fails to reach Karachi situated near the coast, and the river delta and its wetlands have shrunk dramatically. Nonetheless, despite human appropriation of much of the flow of the Indus, the region still suffers from devastating floods such as took place in 2010 and, again, the following year.

The ecological impacts of flow regulation in the Indus are evident from dams that impose high levels of obstruction upon movement of the many diadromous fishes in the River (Reidy Liermann *et al.*, 2012). However, it is even more clearly reflected in the relationship between the distribution of the Indus river dolphin (*Platanista gangetica minor*: Platanistidae; EN) and the construction of irrigation barrages (i.e. gated diversion dams). Between 1870 – before the first dam was built – and 2007, both the extent and rate of range reduction of this endemic dolphin subspecies within the Indus drainage were correlated with the extent of flow alteration and river engineering: specifically, construction of 20 barrages between 1886 and 1971. Barrages and the consequent depletion of river flows truncated the former contiguous distribution of this dolphin into 17 segregated sections; only six of them contained dolphins. Thus, by the 1990s, these animals occupied no more than 20% of their former range (Braulik *et al.*, 2014). Reduced dry-season river discharge due to water abstraction was the principal factor that explained dolphin range decline, influencing the spatial pattern of persistence, the temporal pattern of extirpation of isolated populations, and the speed of extirpation after habitat fragmentation. In addition, dolphins that pass through regulatory gates into irrigation canals during the wet season become entrapped and may be stranded when the canals dry up. Pollution would have made a secondary contribution to population declines, and more than 90% of industrial and municipal effluents enter rivers untreated. Rapid human population growth will increase rates of water abstraction and pollution within Pakistan, and does not augur well for the long-term future of the Indus river dolphin. Preservation of this dolphin will require management that focuses on restoring both the timing and duration of flood pulses, and maintaining critical minimum flows in the dry season (Braulik *et al.*, 2014). But even if the amount of water needed could be ascertained, implementing such a flow allocation would be challenging given the broader context of water scarcity in what is largely a desert nation. And Indian plans for a huge dam on the Jehlum River, a major tributary of the Indus, seem likely to further alter flows and reduce water availability downstream. This may pose challenges to The Indus Waters Treaty 1960 under which water sharing between Pakistan and India has – on the whole – been accomplished successfully.

Elsewhere in India, dams on the Ganges have reduced water flow into Bangladesh, where the river is known as the Padma, resulting in saline intrusion and impacts on agriculture. The Krishna, India's fourth largest river, and the Teesta, which flows through Sikkim into Bangladesh, are similarly overabstracted. One possible solution to dewatering of river channels is a grand scheme to link the major rivers of India to allow water transfers from the wettest to the driest regions, which could have additional benefits in terms of hydropower generation and flood control. Northern tributaries of the Brahmaputra and Ganges, and rivers flowing from Western Ghats could be dammed and the impounded water sent to drought-prone parts of the country in the south and west. Feasibility studies by the Indian government were initiated in 2005. Anticipated to take at least 15 years to complete, the entire project involving around 30 linkages has never formally been initiated, although a canal connecting the Godavari River with the over-abstracted Krishna River was completed in 2015. Concerns about the potential impacts of linking river drainages on biodiversity and fisheries have been raised by Lakra *et al.* (2011), who give more details of this grandiose scheme.

Box 5.2 *Thermal Impacts of Dams along the Colorado River*

Dams can change the thermal regimes of downstream channels through the selective release of hypolimnetic (cold) or epilimnetic (warm) water from thermally stratified impoundments with detrimental consequences for the lotic biota (Olden & Naiman, 2010), irrespective of the extent to which the natural flow patterns have been mimicked or retained. For example, deep-release of cool hypolimnetic water from Lake Powell behind the Glen Canyon Dam lead to year-round uniform temperatures of 9°C in the Colorado River downstream of Lake Powell; it has been estimated that it would require 930 km of downstream flow for the Colorado to recover its natural temperature (Stevens *et al.*, 1997b), but this is not possible given the presence of other dams along the river. An additional impact of the Glen Canyon Dam is a reduction in downstream turbidity arising from the deposition of sediments within Lake Powell, and the Colorado delivered very little sediment or water downstream during the filling phase of the lake (1964–1981), and, normally, the channel is nearly dry as it nears the Gulf of California (Blinn & Poff, 2005). These profound changes, in combination with the cooler temperatures, allowed alien fishes such as trout to proliferate in the river at the expense of natives that were adapted to the turbid conditions and warmer water of the un-impounded Colorado. The changes and increased clarity of the water was, however, beneficial to waterfowl in some regulated reaches of the river sections (Stevens *et al.*, 1997a; see also Chapter 4). The prevention or mitigation of thermal degradation downstream of dams has received little attention from researchers relative to the effects of changed flow. Better characterization of the seasonality and variability of stream temperatures is needed to identify the 'manageable' components of the temperature regime below impoundments (Olden & Naiman, 2010).

Stratification of impoundments is often accompanied by hypoxic conditions or even deoxygenation of deeper waters, and the downstream release of oxygen-poor hypolimnetic waters can harm aquatic biota (e.g. Allan & Castillo, 2007). The chemistry of such waters may be distinctive also, with higher acidity and nutrient loads than water from the surface layers, while release of water from the euphotic epilimnion may contribute phytoplankton to river reaches downstream. Dams trap suspended sediments and bed load, so that water released downstream tends to erode river beds and stream banks until the normal sediment load has been restored. Buildup of inorganic sediment in the reservoir bed will reduce storage capacity and shorten its operating lifetime; it also profoundly influences the supply of material to deltas and coastal waters, especially if a given drainage network has many dams or a mainstream cascade of impoundments (see Box 5.3).

Box 5.3 *The Great Sediment Trap*

The role of dams in sediment sequestration at a global scale has been well established (Syvitski & Milliman, 2007), resulting in at least a 15% global average decline in annual sediment delivery to coastal zones during the twenieth century (Syvitski & Kettner, 2011), much since the 1950s, and a perhaps 25–30% reduction in sediment flux relative to that within un-impounded drainage basins. In some rivers, such as the Colorado, more than 90% of the sediment flux is trapped, and this rises to almost 98% for the Nile. The nutrients transported with this sediment (phosphorus, iron, manganese, etc.) play a vital role in production of coastal waters, and their capture within impoundments has important implications for marine pelagic food webs. However, the situation is complicated by anthropogenic addition of nitrogen and phosphorus to rivers, and – in the case of the Nile – inputs from agriculture and urbanization balance the dramatic loss of natural river nutrients that occurred after construction of the Aswan High Dam (Meybeck, 2003).

The effects of dams on sediment flux have been particularly notable in some parts of Asia: the number of large dams (>15 m tall) in China, for instance, increased from eight in 1950 to 18 600, more than half of the world's total, by 1982 and has continued to grow since (Syvitski & Kettner, 2011). One outcome has been a substantial reduction since the 1980s in the rate of aggradation with consequential shrinkage of deltas of the Zhujiang and Yangtze in China, as well as the Yellow River mentioned earlier. East Asian rivers such as the Chao Phraya and the Mekong have been affected also (Syvitski *et al.*, 2009). Subsidence and compaction of the Chao Phraya delta contributed to the 2011 flooding of Bangkok during which much of the city was submerged for months. This was a repeat of less serious flooding of the Chao Phraya that took place in 2007–2008, when extensive areas of the Ganges-Brahmaputra, Godavari, Krishna, Irrawaddy and Mekong deltas were also inundated causing thousands of human fatalities.

While changes in sediment transport may not have direct impacts on riverine biodiversity, there are likely to be strong indirect effects via the dynamics of erosion and deposition and hence habitat creation and changes in the characteristics of riverbed substrata, as well as greater water clarity (Box 5.2), and so on. Sequestration of transported organic particles behind dams will also have implications for the supply of food to consumers downstream, although there may be cases where the loss of this allochthonous material is partially compensated by downstream release of impounded surface water containing phytoplankton. Of particular direct relevance to biodiversity is the impact of dams as obstacles to the up- and downstream movement of animals, most especially migratory fishes (Reidy Liermann et al., 2012), although decapod crustaceans are also important in some rivers. Dams and flow alteration also limit hydrochoric dispersal of plants and hence affect the structuring of riparian and wetland communities (e.g. Nilsson et al., 2010; Tonkin et al., 2018). The impacts of dams on longitudinal connectivity are pre-eminent, but lateral exchanges of rivers with floodplains and riparia are affected also as dams will tend to reduce the amplitude of seasonal flow fluctuations and the inundation of low-lying land. Channelization and construction of levees similarly degrade floodplains by constraining or severing their connection with the river channel, thereby preventing migration and reproduction of species that are adapted to and dependent on seasonal flooding (Lytle & Poff, 2004).

The Big Dam Boom

Natural barriers such as waterfalls can fragment populations and even lead to speciation in upstream reaches (Dias et al., 2013), but they allow downstream passage of organisms and material. As will be evident from the preceding sections, the construction of man-made barriers has quite different ecological consequences to those arising from ancient geological events. This dam-induced fragmentation is also of a different order of prevalence. In 2005, more than half (172 out of 292) of the world's major rivers were affected by fragmentation (Nilsson et al., 2005); Reidy Liermann et al. (2012) present similar data on dam obstruction for 397 ecoregions. Since then, the situation has worsened significantly. Globally, we are in the midst of a boom in dam construction arising, in part, from attempts to increase the share of energy generated from non-fossil sources. A 2015 estimate was that, worldwide, 3700 large (at least 1 MW installed capacity) hydropower dams were under construction or planned (Zarfl *et al.*, 2015). Combining those statistics on future dams with a data set of ~6400 existing large dams, Grill *et al.* (2015) projected their cumulative impacts on fragmentation (a discharge-weighted component of river length) and flow volume (the proportion of annual discharge withheld in reservoirs) of rivers globally. That assessment, at a finer spatial scale than earlier work (Nilsson *et al.*, 2005), revealed that 48% of river volume was moderately to severely impacted by either flow regulation, or fragmentation or both. The proportion would almost double to 93% in 2030 if all dams under construction or planned were completed, with huge dams in the Amazon Basin making a substantial contribution to the increase (see Box 5.4).

After construction of planned dams within the Amazon Basin, total reservoir storage volume would increase almost 40% from 5759 km^3 in 2010 to 8007 km^3 in 2030 (Grill *et al.*, 2015). The 59% of all river basins now containing large dams would rise to 65%, and 25 of the 120 large rivers formerly classified as free-flowing (by Nilsson *et al.*, 2005) – primarily in South America – would be fragmented. Even this dramatic expansion in generating capacity would result in only a slight rise in the contribution of hydropower to global energy production, from 16% in 2011 to 18% by 2040, because of the concurrent projected increase in global energy demand (Zarfl *et al.*, 2015). Conversely, if dam construction ceased and there was no increase in electricity generating capacity, the contribution of hydropower to global energy production would fall to 12%. Even with the damming of most of the world's great rivers, it is evident that hydropower lacks the potential to come close to substituting for fossil fuels or other non-renewable energy sources.

Projections of the changes in global hydropower generating capacity, and the effects of future dams on river ecology, are subject to a degree of uncertainty, not least because ambitious plans for some pharaonic schemes may never come to completion (see Box 5.5). Furthermore, Grill *et al.* (2015) projected the impacts of large hydropower dams only; if sufficient data had been available to take account of smaller dams that impound water for irrigation and flood control, the extent of flow alteration would have been much greater. Small dams also have a cumulative effect on flow regulation, especially at a local scale (Lehner *et al.* 2011). Although their global contribution to fragmentation has yet to be analyzed, the uncounted number of small dams that exist worldwide (there is no accurate inventory) suggests that their impacts on biota are substantial. This is confirmed by a simulation for hydropower in the Mississippi drainage, where data on dams are relatively comprehensive.

Box 5.4 *Large Dams and the Future of the Amazon*

Connectivity regulates the structure and function of freshwater eco-systems of the Amazon and hence the provisioning of services that sustain local populations. This connectivity is increasingly being disrupted: principally by the construction of dams, but also by land-use change and mining. More than 150 hydropower dams have already been built in the Amazon Basin (Castello & Macedo, 2016), which contains four of the world's 10 largest rivers (Amazon, Negro, Madeira and Japurá). Proposals for the construction of many additional dams are under consideration, amounting to a total of 428 dams, each exceeding 1 MW capacity (Latrubesse *et al.*, 2017). Castello & Macedo (2016) give a lower figure of 277 dams, while Winemiller *et al.* (2016) report that addition of 334 dams proposed for the basin to the existing stock could result in as many as 750 dams in place by 2030 or thereabouts. The accumulated environmental impacts of even a fraction of these new dams are predicted to trigger massive hydrophysical disturbances that will affect the floodplains, estuary and sediment plume of the Amazon, and which will be '... irreversible; there exists no imaginable restoration technology' (Latrubesse *et al.*, 2017: p. 363). The extent and intensity of impacts upon various elements of the biota have yet to be explored, but will likely impoverish Amazonian rivers, and will particularly impact migratory fish and fisheries (Pelicice *et al.*, 2015). And context is all important: the Amazon Basin supports 2030 fish species, almost 15 000 of which are endemic (Winemiller *et al.*, 2016).

The scale of foreseeable environmental degradation of the Amazon that would result from this rash of hydropower dams indicates the need for collective anticipatory action among the nine riparian nations at a transboundary scale. However, relevant policies are inconsistent across the basin, ignore cumulative effects and take insufficient account of hydrological connectivity (Castello *et al.*, 2013; Castello & Macedo, 2016). A basin-wide research and policy framework is needed urgently to understand and manage hydrological connectivity across multiple spatial scales and jurisdictional boundaries in order to avoid fundamental shifts in the ecology, fishery and socioeconomic components of the Amazon Basin.

Box 5.5 *The World's Largest Dam?*

The Congo (or Zaire) River, ranked second globally to the Amazon in terms of discharge, has around one sixth of the world's hydropower potential but only 40 dams, mostly on the upper course and tributaries. The government of the DR Congo, in partnership with South Africa and other nations, intends to build the world's largest hydropower plant. It will be situated on the lower course of the river at Grand Inga Falls, a complex of cataracts and rapids that include the ~100-m high Congo Falls, and its projected 40 000-MW generating capacity is close to twice that of the Three Gorges Dam on the Yangtze. The scheme envisages construction of the Grand Inga Dam in a series of phases, with the first (Inga 3, with construction initially scheduled to begin in 2017) being a successor to two smaller dams (Inga 1 and 2) completed in 1972 and 1982 but dogged by operations failures. However, suspension of World Bank funding in 2016 has delayed the project, the cost of which could eventually amount to as much as US$100 billion, although the DR Congo government continues to assert that the dam will be completed by 2022. While Inga 3 will be a run-of-the river dam, later phases will involve impoundment and reservoir formation that could harness as much as 83% of the annual discharge of the Congo, trap sediments and nutrients, block fish migrations, and surely give rise to other environmental impacts that have not been assessed adequately (Winemiller *et al.*, 2016). None of the fishes in the River Congo are threatened with extinction – at least for the present. They number around 1300 species, including more than 100 Mormyridae; remarkably, perhaps 850 of them are endemic to the river. The fish fauna remains incompletely known, but includes a component of migratory fishes that are likely to be susceptible to the impacts of a large dam.

A comparison of the impacts of 704 large dams with those of 25 857 smaller dams revealed an increase in basin-wide flow regulation from 65% to 90%; the channel fragmentation component rose from 45% to 65% (Grill *et al.*, 2015). The combined global impacts of all dams – large and small – are far more severe than those attributable to large hydropower dams only. And, while these findings cannot be directly extrapolated to impacts on biodiversity, they illustrate the pervasive threat dams and flow regulation pose to riverine plants and animals.

Are Hydropower Dams Always Detrimental to Biodiversity?

While many hydropower dams have had damaging effects on river ecology and fisheries, it may be possible to design schemes that generate electricity without excessive impacts. Pumped storage plants force water uphill to a reservoir at times when electricity is cheap – during off-peak hours at night – and release it to flow through turbines generating power at times of high demand. They act somewhat like batteries in that they 'store' power for release to the electricity grid. The turbines are actually reversible pumps that return the water to the reservoirs once the power demand has declined again. Pumped storage plants do not require a large dam and can be operated mainly as closed-loop systems that repeatedly uses the same water to generate power hence minimizing impacts on aquatic ecosystems.

Another possible alternative to the dam plus large-impoundment-model for hydropower dams is a run-of-the-river (RoR) dam where power is generated by turbines in the dam as water passes through. Proposals for RoR schemes have become more frequent in recent years, because of the perception that they are less damaging to riverine ecology than a conventional dam plus impoundment. RoR implies there is no storage of water, but typically there is some impoundment and there is no agreed definition of what qualifies a dam for the RoR label. One rule-of-thumb is that the impoundments can retain up to the volumetric equivalent of a single day's river flow, although some schemes labelled RoR have higher capacity. Because of their relatively limited storage capacity, most 'true' RoR dams tend to be built on rivers with fairly consistent year-round flow, and thus their environmental impacts are mainly confined to the barrier effect attributable to the dam. However, it is not quite that simple.

Some RoR dams are designed to meet peak-load demand, and store most of a day's river discharge only to release it over a short period when demand for electricity is greatest. In this respect they have similar impacts to the hydropeaking associated with some conventional large dams (see Box 5.6). Thus, within a day, downstream flows can range from a mere trickle, or even less, to a torrent several times the average discharge upstream. Some RoR schemes, mostly in highland regions such as the Indian Himalaya, involve diversion of a portion of the river flow through tunnels to a powerhouse downstream where there is sufficient head to generate power as the water passes through turbines and is returned to

Box 5.6 *Hydropeaking Affects Aquatic Insects and River Food Webs*

The Glen Canyon Dam has profound effects on the thermal regime of the Colorado River (see Box 5.2), but also changes the magnitude of downstream flows thereby impairing habitat conditions. More water is released during the day when electricity demand is greater, and such hydropeaking results in a fluctuating daily pattern of flows with repeated drying and wetting of the shorelines of dam tailwaters (Kennedy *et al.*, 2016). Food webs in affected reaches of the Colorado River are dominated by Diptera and lack large species of aquatic insects that are common in sections that are less influenced by hydropeaking. The reason is that most aquatic insects, especially the larger ones such as mayflies, stoneflies and caddisflies, cement their eggs just under the water surface on partially exposed boulders near the shoreline; there, falling water levels associated with hydropeaking uncover the eggs leading to desiccation and acute mortality. In contrast, many Diptera, including most blackflies and some chironomids, broadcast their eggs over the water surface. The outcome of these differences in egg-laying behaviour is that aquatic insect diversity and the biomass of emerging adults is lowest where hydropeaking is greatest, and this relationship is evident from the mainstream Colorado River as well as from a survey of insect diversity and hydropeaking intensity across dammed rivers of the western United States (Kennedy *et al.*, 2016). Non-insects also tend to dominate benthic communities immediately below hydropeaking dams. These findings have implications also for the productivity of insectivorous river fishes and the availability of food for terrestrial consumers in riparia such as spiders, bats and birds.

Although it has yet to be tested or implemented, a proscription of hydropeaking on some weekends during the times when insect emergence is highest would be beneficial, as a period of stable and low flows might reduce egg desiccation, permit population recovery by large insects and restore food-web integrity (Kennedy *et al.*, 2016). These measures could be adopted at minimal cost in terms of foregone electricity generated because energy demand at weekends is relatively low and so hydropeaking is less lucrative.

the river. If too much water is diverted into the tunnel, the channel immediately downstream may become dewatered. This becomes more likely in cases where the RoR scheme is operated by a private company that prioritizes profit over environmental protection, since they can maximize income by diverting the entire river flow. The chances of dewatering could be minimized by requiring the allocation of some proportion of the in-channel flow to downstream reaches, but implementation would depend on effective legislative oversight combined with a robust scientific assessment of the volume of water that must be left within the river. In many cases, one or other of these elements – often both – are lacking. Matters are made worse if RoR dams are built in arrays or cascades along a river since their combined impacts will accumulate downstream. Furthermore, if the schemes are operated independently, no single operator has responsibility for attempting to mitigate the cumulative impacts of the entire cascade of dams. This is yet another instance of the tragedy of the freshwater commons.

Case Study: Damming the Mekong Commons

The Lower Mekong Basin (LMB), the area of the drainage downstream of the border with China, supports the largest freshwater capture fishery in the world. It involves a large number of people with the catch contributing substantially to family welfare and food security (for details, see Box 3.6). Many species are fished, and much of the catch is made up by species that migrate within the Mekong. Some of them travel long distances along the river, traversing international boundaries. While there is evidence of overexploitation of large species and consequent fishing down the Mekong food chain (see Box 3.4), around 70% of the catch is based upon potamodromous pelagic whitefishes that migrate up- and downstream and which have not been greatly depleted (Barlow *et al.*, 2008; Dugan, 2008; Dugan *et al.*, 2010). Compared to most other large Asian rivers, levels of pollution and environmental degradation in the Mekong are low and conditions are generally good (Dudgeon, 2011). However, the river is poised to be altered irreversibly by construction of mainstream hydropower dams that will jeopardize biodiversity, fisheries and food security within the LMB (Dudgeon, 2011; Ziv *et al.*, 2012; Winemiller *et al.*, 2016). The following account, derived and updated from Dudgeon (2011), explains the sociopolitical and ecological context of these developments, and can serve as a case study of the damage caused by flow regulation, illustrating the perceived imperative to pursue river engineering despite trade-offs with fish biodiversity and food security.

China has already dammed the mainstream of the upper Mekong (known as the Lancang Jiang) in Yunnan Province, with six dams completed and two more almost finished (dams 1–8 in Plate 26), but the total along the Chinese part of the river could reach 20 or more dams within the next decade. Some of these are colossal: the Xiaowan Dam, competed in 2010, is 292 m high, and the Nuozhadu Dam (2014) is 262 m, with the other dams standing over 100 m tall. Further down-stream, there is a host of dams planned for the LMB with perhaps 71 tributary dams operating by 2030 (MRCS, 2011a), although there could be as many as 121 (Kano et al., 2016b). Eleven dams have been proposed for the mainstream in the LMB – nine in Laos and two in Cambodia (dams 9–19 in Plate 26) – and these are the primary focus here. Upon completion, they could have implications for the livelihoods of as many as 60 million people, including the inhabitants of countries with high levels of malnutrition (Cambodia) and food insecurity (Laos). Ten of the 11 dams will span the entire Mekong mainstream, with the other at Don Sahong in Laos damming the largest of several branches of the mainstream at Khone Falls immediately upstream of Cambodia. Two of the Laotian dams will be joint ventures between Laos and Thailand, with the latter being the primary recipient of the electricity generated.

Governance and Decision Making in the Lower Mekong Basin

Plans for mainstream dams in the LMB date back to the 1950s, but implementation was stalled by regional conflicts and other constraints. It was not until four decades later that they appeared to be reaching fruition with a plan to construct a cascade of 12 dams along the Lower Mekong. A range of environmental concerns, especially those relating to the effects on migratory fish and fisheries, led to suspension of the projects in 2002, largely due to qualms of the Mekong River Commission (MRC), an intergovernmental organization established by the four riparian nations of the LMB (Cambodia, Laos, Thailand and Vietnam) under the Mekong Agreement on the Cooperation for Sustainable Development of the Mekong River Basin in 1995. The MRC grew out of the Mekong River Committee, itself a modified version of an organization established in 1957 intended to coordinate water-resource development among LMB states. The 1995 agreement mandated international cooperation in all fields of sustainable development, utilization, management and conservation of the water and related resources of the Mekong River Basin (MRC, 2002). National representation from each country that provided a basis for prior consultation over water-resource developments in the LMB, allowing mutual notification about intended projects, followed by a process of review by MRC experts, offered the potential to achieve consensus on whether or not they would be beneficial for the region. On the face of it, this arrangement should have provided an excellent basis for hydropolitical cooperation and sustainable water-resource development within the LMB.

Rather than adopting a utilitarian approach to maximizing economic opportunities presented by hydropower, the MRC philosophy favoured the broader perspective of integrated management. This was reflected in an overall Basin Development Plan promulgated in 2002, intended to identify and prioritize the projects and programmes to be implemented at the basin level (MRC, 2002). In accordance with the stated objectives of the Basin Development Plan, in 2002 the MRC decided not to pursue the longstanding proposal to build a 12-dam cascade along the LMB mainstream. To further formalize the 1995 Mekong Agreement and enhance cooperation, the MRC introduced the Procedures for Notification, Prior Consultation and Agreement (PNPCA) in 2003, under which consequences or transboundary impacts arising from proposed projects can be evaluated by a MRC Joint Committee comprising representatives of the four riparian nations (MRCS, 2011a). In theory, this process had to be completed before approval of a project could be given by national regulating authorities. In addition, the MRC Council refined the Basin Development Plan and enshrined an approach based on integrated water-resources management (described in Box 5.8) as the framework for coordinated implementation of sustainable development of the LMB (MRCS, 2011b).

The current situation is far from that envisaged in the 2002 Basin Development Plan, and the MRC has been faced with proposals to build 11 mainstream dams, involving eight of the sites intended for the previous 12 dam scheme that was abandoned in 2002. Of these, the 1260-MW Xayaburi (or Sayaboury) Dam in Xayaboury Province, Laos (dam 11 in Plate 26), is now under construction, and could be completed in 2020. It will be in the third position in a six-dam cascade envisioned for northern Laos (MRCS, 2011a; dams 9–14 in Plate 26). The MRC Secretariat received the relevant consultation documents on the Xayaburi Dam from the Lao National Mekong Committee in September 2010, and the PNPCA report was published in March, 2011 (MRCS,

2011a). It was intended to inform the consultation process scheduled for completion by late April, 2011, after which a decision on dam construction was to be made by Laotian authorities. The PNPCA considered the Xayaburi Dam within the context of the planned six-dam cascade, a process that was facilitated by a strategic environmental assessment of all 11 mainstream dams undertaken by independent consultants commissioned by the MRC. The primary recommendation of the consultants (ICEM, 2010) was to defer for up to 10 years any decisions to proceed with mainstream dams, because of the scale and seriousness of potential risks of hydropower development, including livelihood impacts upon more than 2.1 million people in Laos.

After consideration of the PNPCA report, national representatives on the MRC Joint Committee were unable to reach agreement. In April 2011, they deferred any decision on the Xayaburi Dam by referring it to the ministerial level, due to the need for more thorough investigations of transboundary impacts and Cambodian concerns over knowledge gaps that needed further study. While the reprieve might have set an important precedent for the development of other mainstream dams in the LMB, the Laotian position was that the consultation process had been completed and dam construction should begin without delay. The national representatives also disagreed on the duration of the deferral; Vietnam supported the 10-year time frame advocated in the strategic environmental assessment, while Laos remained unwilling to extend consultation beyond six months. As a result, site formation for the 33-m tall Xayaburi Dam started almost immediately, and construction began in November 2012. Less than a year later, in September 2013, the Laotian government (in partnership with a Malaysian company) announced its intention to build the Don Sahong Dam, on the Mekong mainstream at Khone Falls at Siphandone, 2 km upstream from the Cambodian border. Most of the electricity would, again, be for export to Thailand, but also to Cambodia. At 32 m in height, the Don Sahong Dam would block the main channel for migrating fishes at Khone Falls and thus have potentially significant transboundary impacts. These effects were expected to be mitigated by deepening and widening two adjacent channels so that they could serve as alternative migration routes. The PNPCA process for the project, initiated by Laos in the second half of 2014, was disputed by other members of the MRC, with Cambodia, Thailand and Vietnam asking that the PNPCA period be extended beyond six months. By early 2015, when the consultation and evaluation process was due to have been completed, the PNPCA had not been endorsed. In June 2015, the MRC announced that, due to a lack of agreement, any decision on dam construction was to be deferred pending diplomatic discussions among ministers of the riparian nations. For Laos, however, the matter had been decided already: construction of the Don Sahong Dam began in January 2016.

One conclusion to draw from the resurgence of plans to build main-stream dams in the LMB, and outcome of the Xayaburi and Don Sahong PNPCA processes, is that the MRC may have little ability to influence events in the riparian nations, presumably because prior consultation provides no guarantee of cooperation when national interests conflict with (and override) international relations. This has proven to be the case with the Xayaburi and Don Sahong Dams given the Laotian preference that they proceed. Moreover, the fact that China – which has already constructed mainstream dams – is not a member of the MRC strengthens the impression that national interests trump transboundary concerns (Dudgeon, 2011). Dugan *et al.* (2010) reached a similar conclusion, finding no evidence that dam construction would not proceed as planned, their prognosis being that a large portion of Mekong fish production and its associated economic and social benefits would be lost. This accords with virtually all evaluations of the likely consequences of mainstream dam construction (*e.g.* Barlow *et al.*, 2008; Dugan, 2008; ICEM, 2010; MRCS, 2011a; MRC, 2017), as discussed in more detail in the next section, with Roberts (2001a) designating such developments as 'fluvicidal'. Although the MRC offers the best available framework to ensure conservation of migratory fishes in the Mekong, the fact remains that – thus far – it has not had sufficient influence to prevent construction of mainstream dams in the LMB.

In late 2016, the Laotian government stated its intention to proceed with plans for building a third dam, at Pak Beng, which would constitute the most upstream structure in the 11-dam LMB cascade (dam 9 in Plate 26). The MRC commenced the PNPCA process for this dam in January 2017; it was formerly concluded midyear, despite uncertainty over potential transboundary impacts, or agreement over appropriate mitigation measures. The dam, which will be 64 m tall with a 93-km-long impoundment, is to be built by a Chinese company with almost all of the electricity generated likely to be sold to Thailand. In mid-2018, the Laotian government formally notified the MRC of its initiation of the PNPCA process for a fourth mainstream dam at Pak Lay (dam 12 in Plate 26), downstream of the Xayaburi Dam, although construction is not envisaged to begin until 2022.

Potential Impacts of Mainstream Dams on Fish and Fisheries

Adequate understanding of the potential effects of dams in the LMB depends on a knowledge of patterns of fish migrations to spawning areas in the river. There are three main upstream migration systems in the LMB (Plate 26): a lower zone below the Khone Falls; a zone upstream from the Khone Falls to Vientiane; and, a zone upstream of Vientiane where a six-dam cascade is planned (Poulsen *et al.*, 2002a). A substantial number of commercially valuable whitefishes migrate longer distances, as do all five of the globally endangered Mekong megafishes (see Box 3.4). There is considerable interspecific variation in the timing of up- and downstream migration and downstream drift of larvae, but individual species appear to be cued by particular components of the annual flood cycle, such as rising water levels, with much of the upstream migration in the early wet season and least activity in the middle of the dry season (Poulsen *et al.*, 2002a). Thus, maintenance of the natural flood cycle and connectivity that allow unobstructed passage along the river is essential for fish reproduction and hence a productive fishery.

The six-dam cascade in Laos could affect more than 100 migratory fish species (MRCS, 2011a). Much of the impact would be associated with construction of the first dam at Xayaburi, and the associated transition from prevailing fast and seasonally diverse flow regimes to the limited water movement in a large reservoir. Together, the dams would convert ~40% of the mainstream riverine habitat in the LMB into a series of lacustrine water bodies, representing a loss of 90% of the upper migration system. The predicted fisheries loss to the basin-wide capture fishery due to the reduction in the area accessible to fishes migrating upstream would be approximately 66 000 t or an overall, basin-wide reduction of around 6% of the annual 2.5 million t fishery yield (MRCS, 2011a). This is significant, but not as large as might have been feared because the impacts would be largely confined to those species in the upper Mekong migration system (Plate 26). The middle and lower migration systems have much greater biomass (Poulsen *et al.*, 2002a), but the loss of 66 000 t capture capacity is equivalent to 73% of the estimated river-floodplain fishery yield for Laos, and would not be compensated by reservoir-based fisheries (Dugan, 2008) which, over the LMB as a whole, would replace only 10% of the capture fisheries' losses (Dugan, 2008; ICEM, 2010). If the impacts of the reduced capture fishery were fully absorbed by the more than two million people living in and around the dam cascade in northern Laos, the reduction in food security due to capture fish losses could be as much as 33 kg per individual per year (MRCS, 2011a). The effects on biodiversity would also be considerable: all of the IUCN Red-listed Mekong fishes (Box 3.4) are long-distance migrants, and their movement among the three migration systems along the river would be blocked by the Laotian dams. The Mekong giant catfish would likely become extinct since its only confirmed spawning site at Luang Prabang is upstream of the Laotian dam sites (MRCS, 2011a). The dams will also transform riverine habitat behind them; for instance, deep pools within the footprint of the Xayaburi Reservoir that were key dry-season habitats for megafishes and many whitefishes (Poulsen *et al.*, 2002b).

The Xayaburi Dam design provides for a fish passage, but its effectiveness for upstream migrants, and especially large species, is likely to be low (MRCS, 2011a). There are significant knowledge gaps over the ability of different migratory species to traverse a fish pass and travel upstream through the reservoir. The design of fish passages and ladders has mainly been intended for temperate-zone salmonids, but Mekong fishes such as large pangasiids and cyprinids lack their jumping ability. There is good reason to suppose that current fish-pass technology would be unable to maintain the migration in the LMB given that the range of species, body sizes and biomass of migrants is much greater than in temperate rivers, with biomass in the order of 100 times higher (Dugan, 2008; MRCS, 2011a). Downstream migration of adults (which is not typical of salmon) includes the same diversity of species and sizes, plus eggs and larvae. Downstream passage of large fishes through dam turbines would be especially problematic and likely fatal for exploited populations of large cyprinids such as *Probarbus* spp. even in the unlikely event that upstream migrations were completely unhindered (Hall & Kshatriya, 2009). Such high levels of fish-pass efficiency have rarely been achieved elsewhere (e.g. Agostinho *et al.*, 2011; see the discussion of fishways in Chapter 9). Reduced velocities of impounded water will also compromise the drift of larvae through the 100-km-long Xayaburi Reservoir (MRCS, 2011a), and may provide inadequate cues for adult migrants. Such impacts cannot be mitigated and would result in cumulative species loss or a more dramatic decline, as documented for Pak Mun Dam on the Mekong's largest tributary in Thailand where the fishery collapsed following dam completion in 1994. Scarcely one quarter of the 258 species in the Mun River could climb the fish ladder incorporated in the dam, and no gravid females of any species ascended it successfully (Roberts, 2001b). The Don Sahong Dam will also have a ladder or fishway, as well as 'fish-friendly' turbines, although their effectiveness is questionable, and a large fishway is also likely to be built into the proposed Pak Beng Dam. Despite such measures, the latest assessment by the MRC (MRC, 2017) – the so-called Council Study – conducted over six years at a cost of US$4.7 million concluded that the 11 mainstream dams and 120 tributary dams that could be completed by 2040 would seriously threaten river ecology and human food security in the LMB.

Model simulations of proposed dams in the LMB predict major reductions of migratory fish stocks by 2030 (Ziv *et al.*, 2012): the 11 mainstream dams would reduce biomass by 51% with the 78 planned tributary dams subtracting a further 19%; their combined loss could exceed 70% of migratory fish. However, construction of tributary dams alone (without any mainstream dams) would have a greater impact on migratory biomass and fish diversity than the six-dam LMB cascade alone, which is a surprising finding with important implications, given that construction of tributary dams in the LMB is a matter for national governments (and does not require any PNPCA). Obviously, projections about impacts on fisheries are subject not only to the number of dams built, but also to their type and spatial arrangement. Another simulation (Kano *et al.*, 2016b; addressed more fully in Box 7.9) that was based on a wide range of fishes, not only those that migrate or are exploited, predicts that construction of mainstream dams would be more detrimental to LMB fishes than building tributary dams, but tributary dams have smaller impacts on fish biodiversity only so long as their combined generating capacity does not exceed 15 000 MW. Moreover, any scenario involving construction of mainstream dams would have impacts on fish biodiversity, constraining the range occupancy of individual species and reducing local species richness.

Sediment Sequestration and Other Impacts

The area of the Mekong drainage basin upstream of the Manwan Dam in China (dam 3 in Plate 26) provides 45% of the total sediment load of ~160 million t annually (MRCS, 2011a). Completion of only three dams on the Lancang Jiang mainstream reduced sediment loads in the LMB by 35–40% relative to pre-dam conditions, and slowed aggradation of the Mekong Delta (Syvitski *et al.*, 2009). Trapping of nutrients bound to sediments reduced downstream fluxes by 15–35% (MRCS, 2011a). When all Lancang Jiang dams are in operation, it is estimated that they will trap more than 50% of the total sediment load of the Mekong Basin (Kummu *et al.*, 2010); when the Laotian six-dam cascade plus more than 70 planned tributary dams have been completed, at least 75% of the baseline sediment load will settle out within impoundments, leading to a 70% reduction in nutrient supply (MRCS, 2011a). A related issue is that flow alterations in the Lower Mekong due to the Lancang Jiang dams are expected to result in dry-season increases of 0.2–0.6 m in the level of Tonlé Sap floodplain lake while decreasing the extent of wet-season inundation: flood duration would be reduced by 14 days, while the floodplain area, total flood volume and amplitude would be reduced by 7–16% (Kummu & Sarkkula, 2008). Campbell *et al.* (2006) similarly envisaged a 10% decrease in the area inundated, and consequential reductions in gallery swamp forest, lake productivity and fishery yields, with implications for more than one million people who depend on ecosystem services provided by Tonlé Sap. The additional effects of the Xayaburi Dam on quantity and timing of river flows into Tonlé Sap are expected to be minor (MRCS, 2011a), but lower sediment and nutrient loads in river water could reduce productivity within the lake (Kummu *et al.*, 2008), and may have implications for agriculture in the LMB, such as the riverbank gardens cultivated during the dry season. A more recent assessment of the effects of mainstream and tributary dams in the LMB suggests that they could reduce the amounts of sediment reaching the Mekong Delta by 97% (MRC, 2017). Construction of more dams on the Lancang Jiang (such as the Wunonglong Dam in northern Yunnan, and others in the planning stage) is also likely to further alter flows, smoothing the hydrograph and increasing dry-season discharge in the lower riparian nations.

We do not know how sediment sequestration within mainstream dams and consequent nutrient reduction will interact with the effects of dams as obstacles to migration, and thereby influence fish biodiversity and fishery yields across the basin. Nor is it certain that their impacts will be exacerbated by climate change, although this seems highly likely (see Chapter 7). Other outstanding concerns relate to the impacts of dams on downstream water quality and the need to establish environmental flow allocations below dams. While the barrier effects of dams are a major focus of the PNPCA reports for mainstream LMB dams, the downstream effects of altered flow or thermal regimes below those with large impoundments should not be disregarded (MRCS, 2011a).

Dams and Food Security

Maintaining LMB food security in the face of projected fishery losses arising from dams planned to be in place by 2030 would require a substantial expansion of agricultural land to replace the fish protein with livestock products. Orr *et al.* (2012) quantified two scenarios in order to predict volume of water and area of land that would be needed to substitute alternative food supplies for lost fish catch attributable to new dams. The first assumed the construction of 11 mainstream dams in the LMB, and a projected loss of 16% of fish protein directly attributable to these dams; this is a more conservative projection than Ziv *et al.* (2012), which took account of losses of migratory species only. The second scenario envisaged construction of a further 77 proposed tributary dams (a total of 88 dams) by 2030, which could reduce fish catch by 24–38%. Under the mainstream dam scenario, there would be a 4–7% overall increase in .water use for food production across the basin, with much higher estimations for countries entirely within the LMB (Cambodia, 29–64%; Laos, 12–24%). The second scenario that took account of tributary dams predicted a 6–17% increase in water use for the basin, especially in Cambodia (42–150%) and Laos (18–56%). A more recent projection (MRC, 2017) is likewise alarming, although the numbers vary from earlier estimates, and suggests that by 2040 hydropower development will reduce the total fishery biomass in the entire LMB by 40–80%, with individual countries losing substantial percentages of current catches: Thailand 55%; Laos 50%; Cambodia 35%; and Vietnam 30%.

Projections of the additional land needed to scale up production of livestock raised and consumed in the LMB by an amount sufficient to compensate for reduced fishery yields (Orr *et al.*, 2012) were 13–27% for mainstream dams, rising to 19–63% if all 88 dams were built (even if fish yields from reservoirs were taken into account) but, again, with greater increases needed in Cambodia (up to 129%) and Laos (up to 43%). The actual amount of water and land that would be needed depends, in part, upon balance between meat and dairy products and land-use requirements of different livestock. But, irrespective of this, where would the additional land come from? Or the animal husbandry expertise? To put these percentage changes in context, Orr *et al.* (2012) estimate the additional pasture land required to replace lost fish protein with livestock would range from 4863 to 10 384 km^2 if the mainstream dams are built, and 7080 to 24 188 km^2 when the tributary dams are taken into account. Cambodia and Laos are especially vulnerable to the impacts of dams as a large portion of their populace are dependent on fishes to provide protein and micronutrients; World Bank figures (data.worldbank.org) on per capita gross domestic product of Cambodia and Laos (<US$1000

annually) are reflective of high levels of food insecurity compared to their more prosperous neighbouring states.

Lessons Learned

Although socioeconomic progress is desirable, sustainable development requires avoidance of unnecessary risks to biodiversity and ecosystem services, such as fish production. Projections of the impacts of hydropower dams in the LMB (e.g. Ziv *et al.*, 2012) can help identify those projects that are potentially most damaging, so that their construction can – or should be – avoided. The Mekong serves as a clear example of the overriding and urgent priority for scientists to convey to citizens, management organizations (such as the MRC) and governments that wise stewardship of river ecosystems will benefit humans through the provision of fishery resources, as well as the clean water upon which fishes and humans depend. Conversely, engineering works that irrevocably alter the river that provides these ecosystem services will result in a myriad of changes that could jeopardize food security for as many as 60 million inhabitants of the LMB (Orr *et al.*, 2012). Unfortunately, neither message has gained sufficient traction. The twin imperatives of economic development and livelihood improvement have led the Laotian government to prioritize economic growth over environmental protection, while failing to take due account of the fact that their decisions will impair the ecosystem services that underpin food security thereby putting any livelihood improvements at risk. It is the poorer citizens along rivers that will be most affected by hydropower dams, yet seem among the least likely to benefit from the sale of electricity generated. In addition, the Laotian disregard of the concerns of fellow riparian nations undermines the effectiveness of the MRC as a management organization, and makes it more likely that other dams will be built along the river mainstream. If upstream dams have already impoverished local river fisheries, then there is every reason for Cambodia, and then Vietnam, to build their own hydropower dams, and reap whatever economic benefits can be gained to offset fishery losses: such is the tragedy of the Mekong commons.

Environmental Flows and Water Allocations

What can be done to mitigate the impacts of the hard structures and soft practices associated with hydropower dams and other sorts of river regulation? One approach, referred to in Chapter 1, makes use of the environmental-flow concept: an allocation of '. . . the quality and quantity of water necessary to protect aquatic ecosystems and their dependent species and processes . . . in order to ensure sustainable development of water resources' (Alcamo *et al.*, 2008: p. 3). A more recent definition (Arthington *et al.*, 2018: p. 4) describes environmental flows as '. . . the quantity, timing, and quality of freshwater flows and levels necessary to sustain aquatic ecosystems which, in turn, support human cultures, economies, sustainable livelihoods, and well-being'. How can such an allocation be determined?

River flows are often summarized in terms of mean annual discharge, and it might be possible to make an allocation of, say, two thirds of mean annual flow to protecting biodiversity and ecosystem health (as an environmental reserve), with the rest as a fixed allocation for extraction to meet human needs (i.e. removal is 'capped' at one third of mean annual flow). Simplistic 'rules of thumb' about when and how much water should be allocated may work well for a single species whose response to flow has been carefully studied (e.g. a sport fish of interest), but are certain to be confounded by the range of relationships between ecosystem integrity, or individual species, and flow characteristics. And, while it is theoretically possible to derive a one-size-fits-all water allocation for all river basins globally (for example, 37% of mean annual flow: Pastor *et al.*, 2014), the amount required will vary temporally and spatially according to the climate and seasonality of individual river basins. If the flow is highly variable within or between years, removing even one third of mean annual flow can leave the river without water in low-flow periods. Furthermore, allocation of some fixed fraction of the natural flow will not allow managers to capture the ecologically important aspects of changes in flow regime, as different proportions of natural flow may be needed at different times (wet *versus* dry season).

A major challenge identified by the Global Water System Project (www.gwsp.org/; formerly based in Bonn but, after recent relocation to Australia, re-named the Sustainable Water Future Program: http://water-future.org/) is the need to develop regionally relevant hydro-ecological models that can be used to compare the environmental requirements of ecosystems with the amounts of water needed to produce goods and services for society (Alcamo *et al.*, 2008). Thus Arthington *et al.* (2010: p. 1) called for '. . . an invigorated global research programme to construct and calibrate hydro-ecological models and to quantify the ecological goods and services provided by rivers in contrasting hydro-climatic settings across the globe'. The wider intention was to contribute towards achieving the then-current water-related goals of the Millennium Ecosystem Assessment, but the aspiration is no less relevant to the Sustainable Development Goals adopted by the United Nations in 2015. The environmental-flow concept assumes the existence of optimal trade-offs that will allow water to remain in rivers, or be restored to them, in sufficient quantities to maintain ecosystem integrity along with the properties and ecological services (including water supply) most valued by humans, thereby addressing the need to protect nature and meet societal demands (Arthington *et al.*, 2010). However, striking a balance between these objectives is only possible if scientists are able to answer the question posed by managers: 'how much water does a river (or stream, or wetland) need?' A great deal of research has been devoted to addressing this matter.

There is now a general consensus among freshwater biologists that environmental water allocations must provide water levels or discharges that mimic natural hydrological variability and incorporate a range of flows, and not simply strive to maintain some minimum or threshold level (e.g. Bunn & Arthington, 2002; Arthington *et al.*, 2006, 2010). This practice has distant echoes in one credo of Leonardo Da Vinci who stated that '. . . nature should not be faced bluntly and challenged, but wisely circumvented' (translation from Charlier *et al.*, 2005), thus avoiding a wholesale transition of the natural flow regime. Mimicking that regime is important because it influences aquatic biodiversity *via* several, inter-related mechanisms that operate over different spatial and temporal scales (Bunn & Arthington, 2002; Box 5.7). Moreover, since a mimetic water allocation is not usually intended to meet the particular needs of a single species (such as, say, salmon), it has the benefit of providing the flow conditions required to sustain the whole riverine ecosystem including surface–ground water connections. Natural or semi-natural flow regimes that take account of variability and other ecologically relevant features of hydrographs permit connectedness along rivers and between rivers and their floodplains or wetlands; this is vital as it allows adaptive responses by riverine biota to the challenges of living in a warmer world. Such connectivity may also permit fishes and other animals to move among potential refugia as conditions alter.

While there are good reasons for insisting that environmental water allocations (henceforth, e-flows) mimic the natural flow regime, the principle was slow to gain acceptance, in part because it appears to result in the waste of potentially useable water during high-flow events, and because it is far simpler for consumers to remove a fixed amount of water

Box 5.7 *Why Mimic the Natural Flow Regime?*

Among the essential aspects of the natural flow regime that must incorporated in an environmental water allocation are annual or supra-annual flood events that influence channel form and govern the relationship between biodiversity and the physical nature of the aquatic habitat; that is, they shape or periodically reset the habitat template. Occasional droughts and low-flow events also play a role by limiting overall habitat availability and quality. The second element that needs to be captured in a water allocation is the periodicity, intensity and duration of particular flow events, as well as the seasonality and predictability of the overall annual flow pattern, since these influence life-history events such as the timing of reproduction or recruitment. Thirdly, sufficient water must be allocated to ensure that high-flow events will trigger longitudinal dispersal of migratory animals or transport of seeds (hydrochory) and propagules, and rising water levels will permit lateral movement and access to otherwise disconnected floodplain habitats. It is impractical to identify which particular aspects of the natural flow regime are critical for all species (or even most of them), but the indigenous biota of rivers has evolved in response to local flow conditions and the complex, variable mosaic of the habitat template that arises as a consequence. Since they are adapted to these conditions, dams, channelization, water extraction or land-use change that alter one or more aspects of the flow regime cause declines in aquatic biodiversity and create new conditions that may favour the establishment and spread of alien species. This, then, provides a fourth reason why any water allocation should emulate the timing and magnitude of natural flows. Only by embracing this mimetic application of the precautionary principle can managers expect to maintain those features of flow regimes critical to the maintenance of diversity in rivers and their associated wetlands.

from a river (determined by some cap or quota), or to release a regular volume from a hydropower dam, than to adjust the quantities according to prevailing conditions. More than 200 methods for determining water allocations had been promulgated by the end of the last century (Tharme, 2003), and additional approaches have been proposed since then (see Arthington, 2012). The main features of some of the more widely used approaches are given in Table 5.2. Determining their

Table 5.2 *Comparison of the four main types of environmental flow methods used worldwide to estimate environmental water allocations. Riverine ecosystem components, data requirements (or resource intensity), and levels of application are indicated for each method. Modified from Dudgeon et al. (2006); details of the variety of methods and their application are given by Tharme (2003) and Arthington (2012), respectively.*

Type	Riverine ecosystem components	Data requirements	Levels of application
Hydrological	Whole ecosystem, non-specific, or some components (e.g. fish)	Primarily desktop using historical flow records; some use historical ecological data	Reconnaissance level of water-resource developments, or as a tool within habitat simulation or holistic methodologies Used widely
Hydraulic rating	In-stream habitat for target biota	Desktop, limited field data; historical flow records Discharge linked to hydraulic variables – typically single river cross section. Hydraulic variables (e.g. wetted perimeter) used as surrogate for habitat-flow needs of target biota	Water resource developments where little or no negotiation is involved, or as a tool within habitat simulation or holistic methodologies Used widely
Habitat simulation	Models in-stream habitat for target biota; may include channel form, sediment transport, water quality, riparian vegetation, wildlife, recreation and aesthetics	Desktop and field data; historical flow records. Many hydraulic variables – multiple cross sections. Physical habitat suitability or preference data for target species	Water resource developments, often large-scale, involving rivers of high conservation and/or strategic importance, and/or with complex, trade-offs among users, or as method within holistic approaches Primarily used in developed countries

Table 5.2 (*cont.*)

Type	Riverine ecosystem components	Data requirements	Levels of application
Holistic	Whole river ecosystem; may also consider ground water, wetlands, floodplains, estuaries and coastal waters. Social dependence on ecosystems and related economic factors may be assessed	Desktop and field data, plus historical flow and/or rainfall records; requires multidisciplinary teams of river scientists. Many hydraulic variables assessed at multiple cross sections. Biological data on flow- and habitat-related requirements of all biota and some/all ecological components; alien species may be included in assessments of biodiversity implications. Hydro-ecological relationships and models increasingly used within holistic frameworks	Water resource developments, typically large-scale, involving rivers of high conservation and/or strategic importance, and/or with complex user trade-offs. Simpler approaches (e.g. expert panels) often used where there are limited trade-offs among users, and/or time constraints Used in developing and developed countries

relative merits is by no means straightforward, as the success of most methods in, for instance, protecting ecological integrity has not been adequately evaluated. However, any useful method for determining environmental water allocations must be based upon regionally relevant, hydro-ecological models that link changes in flow or discharge to relevant ecological responses. Thus, the success of river protection and rehabilitation/restoration will be predicated upon development of accurate models of the relationships between hydrological patterns, fluvial disturbance and ecological responses in rivers and floodplains (Poff & Zimmermann, 2010), followed by implementation of water allocations that fall within a range set by the inherent resilience of these ecosystems.

A degree of consensus has recently emerged among e-flow practitioners as to how regionally relevant hydro-ecological models and water allocations can best be determined through use of the Ecological Limits of Hydrologic Alteration (ELOHA) approach (Poff *et al.*, 2010). ELOHA represents a flexible framework for determining and implementing environmental flows at the regional scale, by drawing upon available hydrological and biological information. This e-flow method is well supported by a website in the form of a toolbox developed by The Nature Conservancy, which gives examples of ELOHA best practice, and provides instructions for its application and use (see ConservationPractices/Freshwater/EnvironmentalFlows/MethodsandTools/ELOHA/Pages/ecological-limits-hydrolo.aspx). Essentially, ELOHA consists of four steps: firstly, building a hydrological foundation of streamflow data; secondly, classifying natural river types; thirdly, determining flow–ecology relationships associated with each river type; and, finally, implementing policy to achieve river condition goals.

The ELOHA method has the advantage that it supersedes the traditional and very time-consuming approach whereby e-flow quantification was conducted at the scale of individual river reaches using one or more of a wide variety of methods to quantify environmental water needs (see Tharme, 2003). ELOHA is a systematic approach that can be applied quite rapidly, assuming that sufficient data are available, by scaling-up from site-by-site e-flow provisions to derive hydro-ecological relationships at a regional level and applying them as state, provincial or national policy. In the many parts of the world where data on explicit links between hydrological changes and ecological responses are scarce, e-flow allocations will have to be based on whatever limited data can be deployed for the ELOHA approach, supplemented by best professional judgement and risk assessment. In cases where a paucity of data on hydro-ecological relationships has effectively ruled out ELOHA applications, implementation of e-flows has required the use of more holistic approaches that attempt to strike a balance between development and resource protection (King & Brown, 2006, 2010). Under each of these circumstances, an e-flow allocation can be treated as a hypothesis-driven experiment in ecological restoration, with the outcomes monitored and evaluated, and the results can be used to refine the initial allocation. Ideally, this adaptive process would be implemented in combination with strategies for engagement of water managers, local communities and other stakeholders through a process of consultation (Arthington *et al.*, 2006; Richter *et al.*, 2006).

The ELOHA approach can be applied over a large range of scales, from huge mainstream to small tributary dams, and dam types, from large hydropower dams to various types of irrigation withdrawal. There is, however, a critical disjunction between the scales at which the patterns and processes are understood and the scale at which management occurs and policy is set (Thompson & Lake, 2010; see also Palmer *et al.*, 2010), and nor is it certain to what extent fine-scale processes (which are relatively easy to measure) operate at larger scales. In particular, the inherent logistical challenges posed by managing large rivers present many difficulties for scaling up the implementation of calculated e-flow allocations. Nonetheless, some 'proof of principle' success in adjustments of the operation of hydroelectric dams has resulted in mitigation of their downstream impacts and partial restoration of natural flow and temperature regimes (e.g. Richter & Thomas, 2007; Olden & Naiman, 2010; Renöfält *et al.*, 2013).

One impressive instance has been the implementation of e-flows downstream of the Three Gorges Dam, which had the specific intention of enhancing reproduction of major carp. Monitoring by the Yangtze River Fisheries Research Institute between 2003 and 2010 (before implementation of environmental flows) and from 2011 to 2016, during the period that e-flows were implemented, showed that the density of eggs and larvae in the reach from Yichang to Yidu was three times higher in the latter period (Cheng *et al.*, 2018). Although the complete analysis of the effects of e-flow implementation has yet to be published, changing the operation of Three Gorges Dam to provide water releases that promote carp spawning provides an important example of the scale at which regulations, stakeholder engagement and science can be combined to inform environmental management. This case study can serve as a precedent for operation of other dams in China, which has more dams than any other country in the world, and has special relevance given that Chinese companies are substantially invested in the construction of major hydropower dams in Africa, the Neotropics and elsewhere in Asia.

The e-flows paradigm that has developed over the past two decades has been judged a success (Poff, 2018), as it has reached a position of influence over local and, in some cases, national water-resources policies. In these cases, freshwater ecosystems are recognized as legitimate water users, so that decisions on water allocations take due account of the trade-off between environmental and human needs (Arthington, 2012). South Africa provides an example of one country that has promulgated legislation on an environmental water reserve, which can be viewed as a 'cap' on levels of abstraction. The National Water Act of 1998, under which water-use licences are issued, was intended to ensure that rivers and wetlands receive sufficient quantities of water of adequate quality, at the right time, in order to support ecosystem functions and the provision of goods and services. However, the 'ecological reserve' remains largely theoretical and has yet to be fully implemented, perhaps due to the complexity of the reserve determination process, or lack of political will (Still *et al.*, 2010) and, in part, because e-flow allocations to protect South African ecosystems are perceived to increase the risk of domestic and industrial water shortages (King & Pienaar, 2011). In the highly regulated lower uMngeni River (KwaZulu-Natal Province), where water releases to maintain a minimum river flow (so-called 'compensation flows, which are not intended as e-flows) have continued for over two decades, there has been an unofficial policy to allow spates for short periods each year to allow a high-value recreational canoe marathon to take place. These releases are aligned, albeit with a fraction of the necessary volume, with the e-flows that would be required if the ecological reserve had been determined, and thus they have a dual or conjunctive value, serving both environmental and recreational purposes (Still *et al.*, 2010).

In Australia, an agreement to impose a basin-wide cap on withdrawals from the Murray-Darling River was reached by the riparian states in the 1990s, and it was initially envisaged that allocation of one quarter of natural flows would be allocated to maintaining river health (Jackson *et al.*, 2001). There was certainly need for action: for instance, diversions from the Murrumbigee River, a major tributary of the Murray, which reduced water availability for floodplain wetlands, was associated with a decline in waterbird abundance during annual surveys (1983–1986 *versus* 1998–2001) of 90%, and a 21% reduction in species richness (Kingsford & Thomas, 2004). The declines in abundance were similar across all functional groups: piscivores, herbivores, ducks and waders, indicating a general impoverishment of the aquatic food sources on which these birds depended.

Although hard-won, the Murray-Darling cap arrangement was never fully realized and, despite a role in bringing about some environmental improvements (Arthington, 2012; but see Thompson *et al.*, 2018), it was replaced in 2011. The new arrangement involved a basin plan (www .mdba.gov.au/basin-plan) that included setting 'sustainable diversion limits' – i.e. maximum levels of permitted extraction – for various parts of the catchment to be implemented through a complex system of water licences. That scheme has been highly controversial, both politically and scientifically, and water continues to be over-allocated leading to potential conflicts among users (Capon & Capon, 2017), as well as significant environmental damage. Although this agreement over water allocation is far from perfect, it may be better to have an imperfect scheme in place than nothing at all. The example of the Murray-Darling exemplifies the difficulties associated with scaling up e-flow allocations, and the inevitable trade-offs to be encountered when attempting to reconcile the interests of multiple users of water within a river basin (Box 5.8).

The substantial accumulated and growing body of research on e-flows is welcome evidence of increasing awareness of threats to freshwater biodiversity and, more importantly, represents an attempt to address them. There is now good evidence that even small changes that shift the flow regimes of regulated rivers towards a more natural pattern can lead to significant improvements in river health, and can sometimes (although not always) facilitate the displacement of alien species by natives (see Box 5.9). However, a major challenge to e-flow science arises from non-stationarity in climate manifested by long-term shifts in means and ranges of temperature and precipitation (Milly *et al.*, 2008; see Chapter 7), which coupled with changes in land cover, have the effect of shifting hydrological baselines in rivers towards novel conditions. In such cases, application of the ELOHA approach, which is based on historic flow variability and restoration to reference conditions, becomes untenable (Poff, 2018). The continued proliferation of invasive species (Chapter 4), and other long-term responses to environmental change (Thompson *et al.*, 2018), including the loss of historically important fish stocks (Chapter 3) and large, charismatic species (Chapter 2), also diminish the relevance of ecological baselines as a reference for restoration endpoints. Instead, scientists will need to develop the capacity to predict under what environmental conditions e-flow interventions will be successful at protecting biodiversity in habitats undergoing continuous change; failing that, such interventions should at least give rise to outcomes that are predictable and ecologically desirable (Poff, 2018).

Box 5.8 *Integrated Water-Resources Management*

Aside from the implementation of e-flows, integrated water-resources management (IWRM) is an alternative – or, perhaps, supplementary – means of reconciling the water needs of humans and ecosystems. IWRM emphasizes the need for a holistic approach that includes allocations of water to sustain the ecological integrity of freshwater ecosystems, while attempting to manage the trade-offs that are inevitably generated as a result (Wallace *et al.*, 2003). The laudable intention of IWRM is to facilitate coordinated development of water, land and related resources to provide economic and social benefits in an equitable manner, without compromising ecosystem sustainability. As with some e-flow approaches (e.g. King & Brown, 2006, 2010), it attempts to strike a balance between development and resource protection. IWRM is intended to be applied at a wider scale than is normally considered in e-flow work, and usually involves integration of management interventions and the concerns of relevant stakeholders across an entire drainage basin or jurisdiction. Particular challenges are encountered where the drainage basin encompasses two or more countries.

While IWRM is a desirable aspiration, there is scant empirical evidence that the approach has been adopted successfully, at least in the context of biodiversity conservation. Thinking at the basin scale can highlight the need for better coordination and integration in planning and management, but is frequently hindered by inter-sectoral competition especially where water-related issues (hydropower, irrigation, navigation, flood control, etc.) fall within the bailiwick of different departments or ministries, some even belonging to different nations. Matters relating to biodiversity receive scant attention in the absence of a specific authority to represent these interests. Indeed, the experience of attempting to adopt IWRM as a basis for management of the Lower Mekong (see above) shows how local (in this case, national) interests trump collaboration and integrated management at the basin scale. The Mekong example is typical of the prevailing model of water-resource management that seems to be one of 'develop, impair-then-repair', resulting an annual expenditure of around US$0.75 trillion in water infrastructure investment directed towards limiting the threats to human water security or mitigating the environmental damage caused by inappropriate development (Vörösmarty *et al.*, 2010).

Box 5.9 *E-flows Can Help Displace Alien Fishes*

The effectiveness of changes to downstream flows as a result of changes in dam operations have had mixed effects in some rivers, for instance downstream of the Glen Canyon Dam where spring flow releases intended to favour endangered humpback chub increased the abundance of alien rainbow trout (Melis *et al.*, 2012; see Chapter 4). However, even small interventions that move a regulated flow towards a more natural regime can pay dividends. A long-term study of water releases downstream of an irrigation dam in Putah Creek, California, showed that a flow regime that combined spring flow pulses with maintenance of a minimum baseflow was sufficient to stimulate spawning and population increases of native fishes that had become constrained to habitat immediately (<1 km) below the dam. This e-flow allocation, implemented over a nine-year period, allowed natives to displace numerically dominant alien fishes from more than two thirds of the length of a 30-km section of the stream where they had flourished formerly under a flow regime that prioritized water abstraction (Kiernan *et al.*, 2012). The increase in water volume released downstream was a small proportion of the total offtake, and was allocated only during a limited, but biologically important, time of the year. Effective mitigation of the impacts of abstraction was thus achieved at a low cost and the basic objective of river regulation (in this case, irrigation) was minimally compromised. The findings validate the general notion that mimicking critical components of natural flow regimes can serve as an effective tool to manipulate and manage fish assemblages in regulated rivers

While the e-flows paradigm offers opportunities to mitigate the impacts of reductions in the quantities and shifts in the timing of water availability within river channels, it can do nothing to offset the barrier effects presented by dams nor the fragmentation of populations that result from their construction. Most concerns about these obstacles have been raised in connection with fishes – especially migratory salmon species – but other types of aquatic animals are affected also, such as amphidromous shrimps (Atyidae) and prawns (Palaemonidae). Construction of ladders or passes allowing fishes to traverse dams could help alleviate matters, but many dams do not have such facilities, or have installed structures designed to suit salmonids in countries that have a fauna lacking such fishes (as in the Lower Mekong). These fishways are seldom fit for purpose in that they tend to prevent upstream or downstream passage of migrating fishes (e.g.

Agostinho *et al.*, 2011; Brown *et al.*, 2013; Pelicice *et al.*, 2015), although aspects of their performance could be improved by better designs. The use of fishways to mitigate the effects of in-stream barriers, and restoration efforts that involve complete removal of dams, are discussed in Chapter 9.

Restoring River Flows and Enhancing Connectivity

Implementation of e-flows, along with dam removal and the installation of fishways, are – essentially – management interventions intended to restore either riverine connectivity, or some semblance of the natural flow dynamics, and sometimes both. While the science of e-flows has been successful – for instance, facilitating the restoration of native fish assemblages (see Box 5.9) – and has developed substantially over the last two decades (see the Brisbane Declaration, described in Box 9.4), e-flows are far more often planned than implemented, and very seldom is their success assessed (Thompson & Lake, 2010), although there are signs that this may be changing (Box 5.10). A notable recent advance is the use of Bayesian hierarchical modelling and decision networks to increase the inferential strength and overcome the shortcoming of weak statistical power arising from a lack of replicate sample units (rivers) in most assessments of the effectiveness of environmental flow allocations (Webb *et al.*, 2010). Such methods are especially helpful given the scarcity of cases where there has been adequate monitoring before and after implementation of an environmental flow allocation (Thompson & Lake, 2010). This is important because, as mentioned above, every opportunity should be taken to ensure that application of environmental water allocations are treated as rigorous, large-scale experiments conducted within an adaptive management framework (Richter *et al.*, 2006, 2010; King *et al.*, 2010). Bayesian decision networks have been deployed in studies that consider the relative benefits of environmental flow allocations in the all-too-common situation whereby rivers are affected by multiple stressors (Stewart-Koster *et al.*, 2010).

More generally, scrupulous analysis of outcomes is an essential part of any management intervention, be it an e-flow allocation or a stream restoration project. A meta-analysis by Palmer *et al.* (2010) shows the widely used 'if you build it, they will come' approach of restoring flows and physical habitat in rivers is unsuccessful if other stressors limiting recovery of the original community have not been alleviated. Historic disturbances (pollution, altered sediment dynamics, riparian clearance, presence of invasive species and so on) may have long-lasting legacy effects that limit ecological responses to e-flows or other restoration measures (Thompson *et al.*, 2018). In addition, the goals of freshwater restoration projects are often

Box 5.10 *Assessing Improvements Resulting from e-Flows*

River managers globally have been conducting environmental-flow experiments that mimic components of natural flow regimes in an attempt to achieve specific ecological outcomes or achieve level of habitat rehabilitation (Acreman *et al.*, 2014). For instance, controlled flood releases from dams have been employed to clear fine sediments from gravel beds and improve spawning habitat for salmonids. The first documented flow experiment in 1965 involved high-pulse releases from Glen Canyon Dam and, since then, their implementation by a variety of facilities operated for power generation and water supply has become more commonplace. While specifically targeted flow-restoration measures show promise under certain circumstances, an analysis of the outcomes of more than 113 flow-release experiments in 20 countries – mainly the United States, with Australia and South Africa a distant second and third – indicates mixed success (Olden *et al.*, 2014). Around 80% of them involved discrete flow events only, rather than changes to the overall flow regime, and while many experiments monitored both abiotic and biotic outcomes, most of the latter were confined to fishes. Significantly, even in instances where the results were scientifically valuable, many studies did not result in any change in dam operations if they failed to consider how the information gained would be used (Olden *et al.*, 2014). The increased frequency of such flow experiments in recent years, and their adoption in an increasing array of countries (Cameroon, China, New Zealand, Norway, Senegal, Switzerland and Tunisia), is evidence that scientists and water managers can, at least in some instances, collaborate to develop and implement effective measures that can benefit water-users and freshwater biodiversity. Additional research is, nonetheless, needed, as some discrete e-flow events mediated through artificial flooding sometimes fail to achieve – or can even be detrimental to – desired ecological outcomes (Melis *et al.*, 2012; Bond *et al.*, 2014a).

intended to re-establish an ecosystem service rather than a specific assemblage of species, and this may limit their value in terms of conserving biodiversity. Even where restoration projects have been targeted at individual species, success has been mixed. While the shortnose sturgeon population of the Hudson River has recovered in recent decades (see Chapter 3), other efforts directed towards restoring stocks of large or charismatic river fishes have failed (see Box 5.11).

Box 5.11 *Barriers to Restoring Salmon in the Thames*

Atlantic salmon became extinct in the Thames in 1833 due to a combination of factors: overexploitation (see Chapter 3), construction of locks and weirs from the early1800s that both obstructed passage of migrants upstream of London and degraded spawning habitat, as well as pollution from the growing city. By the mid-nineteenth century, all fishes had disappeared from the tidal reaches of the river where large sections of the Thames were anaerobic. The benefits of sewage treatment plants completed in the 1890s and thereafter were more or less cancelled out by population growth and industrialization until the 1960s, when the quality of water in the Thames began to improve steadily, and salmon were first caught again in the Thames in 1974.

Improved water quality set the scene for a decades-long effort to re-establish Atlantic salmon, involving removal of barriers to migration, and more than 30 years of stocking of hatchery-reared juveniles under the Thames Salmon Rehabilitation Scheme. Provision of fish passes on weirs began in the 1980s and was completed in 2001. Some of the stocked fish did return to the river as adults: the number peaked in 1993 but dwindled subsequently and there was no evidence of natural recruitment. Physical changes to the impounded river behind weirs had reduced the availability of gravel beds, so that salmon needed to swim 240 km upstream from the mouth of the Thames, transiting multiple fishways, before reaching a suitable spawning site (McCarthy, 2015). Stocking of juveniles into that particular section of the Thames after 2001 resulted in few returning fish and, by 2005, the number of returnees had fallen to zero, perhaps because pollution of the river had been insufficiently ameliorated (Griffiths *et al.*, 2011). Reduced flows due to overabstraction of water may also have suppressed the cues needed to kindle lengthy upstream migrations. While construction of fishways meant that it was theoretically possible for adult salmon to return to the place where they had been stocked and spawn there, they did not do so. Stocking was thus abandoned after 2011.

Stray Atlantic salmon from rivers elsewhere in southern England occasionally ascend the Thames and may eventually recolonize the river. A strategy of restore then 'protect-and-wait a bit longer' (*cf.* the shortnose sturgeon recovery described in Chapter 3) could yet pay dividends for salmon in the Thames.

Another attempt at population restoration is an ongoing scheme to remove barriers and enhance populations of migratory fishes in the River Severn – the longest river in the United Kingdom – and its tributary the Teme. The intention is to install four fish passes to allow access to historic breeding sites upstream. Migratory fishes were excluded from these tributaries when weirs were constructed during the nineteenth century. The scheme, which is the largest of its kind attempted in Europe and will cost almost 20 million pounds sterling, will enhance stocks of the only breeding population of twaite shad (*Alosa fallax*) in the United Kingdom. The more widespread allis shad, as well as Atlantic salmon, European eel and sea lamprey, are expected to benefit from the restoration of connectivity in the Severn. Apart from fishes, population recovery of endangered pearly mussels (Box 5.12) can also be facilitated by re-establishment of connectivity, because it allows free movement of the hosts needed to complete their life cycles (Box 2.2).

Box 5.12 *Fragmentation and Restoration: The Freshwater Pearl Mussel and Its Fish Hosts*

The indirect and interactive effects of flow regulation and habitat fragmentation are readily apparent from studies of the freshwater pearl mussel *Margaritifera margaritifera* (CE), which have parasitic glochidia larvae that encyst on the gills of salmonid fishes that are often migratory or highly mobile. This mussel is a long-lived lotic specialist (life span >80 years; maturing at 10 years) that is highly sensitive to flow and substratum characteristics, and especially susceptible to conditions that reduce current velocity and increase sediment infiltration (Moorkens & Killeen, 2014; Quinlan *et al.*, 2014). The historical distribution and extent of recent decline of *M. margaritifera* and other European unionids have been well documented (Lopes-Lima *et al.*, 2017). Efforts to restore pearl-mussel populations along the River Dee in Scotland involve breaking up large fishing platforms or 'croys', constructed from riverbed boulders, which altered flow patterns and substratum characteristics to the detriment of mussel habitat and salmon spawning grounds. Their removal is expected to benefit both the mussel and its host, and is one component of a larger conservation effort across multiple rivers by a coalition of UK environmental groups: the £3.5 million 'Pearls in Peril' partnership (www.pearlsinperil.org.uk/).

Elsewhere in Europe, hydropower dams along the River Ljungan in Sweden have segregated the formerly continuous habitat of brown trout, dividing these fish into a number of separate tributary-resident populations and sea-migratory strains (Martin Österling & Söderberg, 2015). This fragmentation has been associated with a decline in migratory trout, because their penetration into upstream tributaries has been severely restricted, as well as a dwindling of some isolated stream-resident populations. In parts of the Ljungan drainage, the only available hosts for pearl-mussel larvae are stream-resident trout. However, the sea-migratory strain is a better host for mussels as they are more susceptible to parasitism than stream-resident trout – perhaps because they must switch between living in the sea and in fresh water, which may dampen their immune responses. Furthermore, the seasonal timing of glochidial release by *M. margaritifera* coincides with what was the period of peak densities of sea-migrants. In this case, restoration measures that focus on creating pathways for sea-migratory trout through obstacles such as dams and weirs is likely to be more effective for the pearl mussel than stocking programmes designed to maintain isolated populations of stream-resident trout.

Attempts to enhance *M. margaritifera* populations will require the presence of the necessary hosts, but should take account of the quality of in-stream habitat and the state of the surrounding drainage basin, since efforts to restore network connectivity, flow and streambed conditions (e.g. reduction of interstitial clogging; Denic & Geist, 2015) will fail unless silt and pollutants from the catchment are reduced. One effort – termed the Life Projekt 2005-2011 (see www.margaritifera.eu), with a budget of more than €2 million – centred on the River Lutter drainage in northern Germany, involved the local government purchasing land for restoration of riparian wetlands, controlling erosion by deployment of silt traps, adding gravel to streams and removing barriers to movement of brown trout (Geist, 2010). As a result, pearl-mussel populations in the Lutter have since begun to recover, making it the only river in central Europe where *M. margaritifera* abundance has increased in recent years. But restoration success for this species has come at a substantial cost. As of

2016, there were 28 projects within the LIFE programme (the EU funding instrument for the environment) devoted to the habitat restoration of freshwater unionids. The total spend was €64 million, almost all of it (21 projects) directed towards *M. margaritifera*, including successful large-scale production of juveniles (Lopes-Lima *et al.*, 2017). Although it remains critically endangered, for this mussel species at least, there is hope for the future.

6 · *Vanishing Lakes and Threats to Lacustrine Biodiversity*

The Aral Sea is '. . . mostly desert now – albeit an entirely new kind of desert, one with ruined soil and obliterated vegetation, almost incapable of supporting life. The millions of birds that each season used the Aral Sea as a mid-migration stopping point had not alighted there for years . . . and . . . the two dozen species of fish native to the Aral Sea had been wiped out [p. 29] . . . At their peak, the Aral Sea's fishermen provided a tenth of the entire Soviet catch. . . . In Moynaq alone 10,000 fishermen plied their trade, a number more than triple its current population . . . [p. 346] . . .. For years after the sea abandoned Moynaq's shoreline, some of the town's more desperate fishermen dug numerous canals out to meet it. Each morning they patiently steered their ships down the narrow, brackish passageways. 'Chasing the sea,' they called it. In 1986, commonly regarded as the year in which the last of the Aral Sea's native fish expired, the ships were left more or less where they lay [pp. 349-350] . . . Growing from the seabed were hundreds of shrubby plants and small evil-looking trees that looked conjured up from a terrifying children's book, their graying branches covered with long, hook-tipped thorns. Revenge plants these were, the helpless counterstrike of a devastated ecosystem' [p. 347].

Tom Bissell (2003), *Chasing the Sea*

Flow regulation and water abstraction have had devastating consequences for rivers globally and are likely to affect many more of them within the next two to three decades. But what are the implications of changed water availability for the receivers of river flows? The consequences of sediment sequestration in reservoirs for deltas along the coast have been described in Box 5.3; here, attention will initially turn to the implications of reduced river flows for lakes and wetlands. Then, the focus will be broadened to give an account of some other threats to particular lakes, which have been chosen either because they are subject to especially egregious anthropogenic insults or they host iconic biodiversity (and sometimes both). The many lake ecosystems that are principally endangered by invasive species (the Laurentian Great Lakes,

Lake Victoria, among others) have been discussed in Chapter 4 but, as in most of those lakes, the main threat to their biodiversity is accompanied by other factors – most often eutrophication and land-use change – that exacerbate habitat degradation and species declines. The threats that lakes face from human-induced climate change will be described in Chapter 7.

The Aral Catastrophe

The Aral Sea in central Asia once occupied an endorheic depression between Kazakhstan, to the north and Uzbekistan to the south. The term sea is a misnomer, as it was formerly a terminal lake (hence lacking any river outflow) which, at $67\,300\,\mathrm{km}^2$, was once the fourth largest lake in the world roughly equivalent in volume to Lake Michigan. Evaporation and concentration of salts within the lake meant that it was not entirely fresh. Salinity levels were around one third those of sea water with an ionic composition that was relatively high in sulphate and low in chloride; nutrient levels tended towards oligotrophy. The Aral Sea supported a thriving fishery, supplying the then Soviet Union with annual yields of around $40\,000\,\mathrm{t}$, and ranking second in importance to the Caspian Sea, another, much larger, endorheic lake sustained by water from the Volga. A lucrative muskrat pelt industry – albeit based on an alien Nearctic mammal (see Box 4.4) that became established in the 1930s – was centred around the deltas of the two main inflowing rivers: the Syr Darya, which entered the Aral from the north, and the Amu Darya from the south.

Beginning in 1960 and continuing over the next several decades, the Soviet authorities and, subsequently, those of successor states, diverted the Amu and Syr Darya to irrigate vast expanses (over 6 million hectares) of cotton and wheat in Kazakhstan and Uzbekistan where these two rivers represent major sources of water. The irrigation systems were inefficient and leaky, leading to much waste of water as well as excessive salinization of soil. Herbicides and pesticides in runoff from the expanded cotton industry contaminated rivers and flowed into the Aral Sea. A significant portion of the flow of the Amu Darya was redirected $1300\,\mathrm{km}$ westward into the Karakum Desert via the Karakum (= Qarakum) Canal, the world's longest irrigation channel completed in 1988. As a result, annual flows into the Aral Sea had diminished from approximately 55 million km^3 during the 1950s to no more than 5 million km^3 by the mid-1980s, while salinity and chloride to sulphate ratios started to increase due to high rates of evaporation. After 1975, the flow of the Syr

Darya regularly failed to reach the Aral Sea and, after 1982, this was sometimes the case for the Amu Darya even though its annual discharge is twice that of its northern counterpart. The lake had shrunk even further by 1990, and salt concentrations continued to rise during that decade as what remained of the lake became hypersaline and polluted with agricultural chemicals.

Following the disintegration of the Soviet Union in 1991, investigations and monitoring of conditions in the Aral Sea ceased almost completely. Nonetheless, continuing shrinkage of the lake was evidence of worsening degradation. By 2000, the lake had separated into the North (Small) Aral Sea in Kazakhstan and the South (Large) Aral Sea in Uzbekistan with the latter further subdivided into western and eastern lobes. Seven years later, the remnant lake consisted of four small water bodies with a combined capacity of only a small fraction of the original volume and a salt content (in all but the North Aral Sea) of around three times that of sea water.

The ecological consequences of human expropriation of water flowing to the Aral Sea were profound and far reaching. Fisheries dwindled during the 1970s, collapsing entirely in 1983, and fishes were progressively eliminated from the Aral Sea as the salinity rose. Deltaic wetlands dried out and shrank, and the Amu Darya delta ceased to exist after river flows failed to reach the Aral consistently. Coastlines receded far from towns, and fishing vessels, harbours and even entire communities were abandoned. The former lake and its surrounds became plagued by desertification and storms of salt-laden dust contaminated by toxic agricultural residues and pesticides, causing widespread health problems for the remaining inhabitants and aggravating an already parlous situation (for more information, see Kostianoy & Kosarev, 2009). Conditions worsened further in 2014, when the eastern lobe of the South Aral dried out.

The near-obliteration of the Aral Sea represents one of the world's worst environmental disasters, representing a large-scale outcome that was both entirely man-made and completely foreseeable. The human dimensions of this tragedy have been quite well reported, although remedial action remains inadequate. The changes in Aral hydrology that took place during these decades of environmental degradation are also fairly comprehensively documented (for an encyclopaedic compendium, see Zonn et al., 2009). However, there is relatively little detail about the consequences for biodiversity in the readily accessible scientific literature; Micklin et al. (2014) is a notable exception. Most of the data that are available concern fishes: Box 6.1 gives an account of their fate.

Box 6.1 *The Collapse of Aral Sea Fisheries*

The original fish fauna of the Aral Sea – all of freshwater origin – had a rather low diversity, and far less than the Caspian Sea, to which it was somewhat hydrologically comparable. Estimates of total richness range from around 20, which seems to be the consensus figure, to as many as 33. Four sturgeons were endemic to the Aral Sea and associated rivers (see below), and among the cyprinids, which made up most of the ichyofauna, there were a number of subspecies (or, perhaps, morphotypes) endemic to the drainage: the Aral barbel (*Luciobarbus* [= *Barbus*] *brachycephalus brachycephalus*; VU), the bulatmai barbel (*L. capito conocephalus*; VU), the Aral white-eye bream (*Ballerus* [= *Abramis*] *sapa bergi* natio *aralensis*), the Aral roach (*Rutilus rutilus aralensis*), the Danube bleak (*Alburnus chalcoides aralensis*), asp (*Leuciscus* [= *Aspius*] *aspius iblioides*) and sazan (*Cyprinus carpio aralensis*) – a form of common carp. The Aral Sea stickleback (*Pungitius platygaster aralensis*: Gasterosteidae) also represents a distinct local morphotype, as did the Aral brown trout (*Salmo trutta aralensis*), which has not been recorded in recent decades and is probably extinct. Another cyprinid, the pike asp (*Aspiolucius esocinus*; VU), is endemic to the drainage where it is present in small numbers in the Amu and Syr Draya only. Differences in the tendency to include river fishes, such as the pike asp, in the pre-1960 species total for the lake probably account for the variability in estimates of Aral richness.

Other components of the former Aral ichthyofauna were wide-spread cyprinids such as common bream (*Abramis brama*), Prussian carp (*Carassius gibelio*), orfe (*Leuciscus idus*), and schel or sabre carp (*Pelecus cultratus*). Percids were represented by pike perch (*Sander lucioperca*), Eurasian perch and Eurasian ruffe, with northern pike and wels catfish constituting the apex predators. Sazan, Aral roach, bream and wels made up major components of the fishery in biomass terms. Because most Aral fishes had limited tolerance of elevated salt content (as is typical of cyprinids), by the early 1970s or thereabouts, conditions within the lake were such that the majority of species could no longer survive. However, many of them persisted in parts of the Syr or Amu Darya, or in irrigation canals and reservoirs, although populations have been affected by the same factors that led to devastation of the lake.

Until the 1960s, most Aral fishes tended to migrate into the deltas and channels of the Syr and Amu Draya for spawning, typically during the season of maximum river flow, with some species travelling far upstream. The long-distance migrants included the ship (or bastard) sturgeon (*Acipenser nudiventris*; CR) which, at up to 2 m long and 80 kg weight was the most valuable fishery species in the lake, as well as the Aral barbel that can grow to 1 m and exceed 20 kg. Both species formerly swam 1000 km along the Amu Draya, until dam construction restricted access to spawning sites and reduced recruitment. They were also threatened by overexploitation, despite regulations to control fishing effort introduced during the 1930s. The limits were brought in after a period of catastrophic mortality of ship sturgeon due to the introduction of a trematode *Nitzschia sturionis* (Capsalidae) that infested its skin and gills; the parasite arrived with stellate sturgeon (*A. stellatus*; CR) introduced from the Caspian Sea, where it is now severely threatened by overfishing. The ship sturgeon disappeared from the Aral Sea during the 1970s, although it persists in the Caspian Sea and a few localities in Russia and Europe. The Syr Darya shovelnose sturgeon (*Pseudoscaphirhynchus fedtschenkoi*; CR) is endemic to the Aral drainage but has not been seen since the 1960s (Birstein, 1997). It was vulnerable to the combined effects of dams, dewatering, eutrophication and pollution by pesticides. The dwarf sturgeon (*P. hermanni*; CR) and false shovelnose sturgeon (*P. kaufmanni*: CR) were formerly found only in the Aral Sea and Amu Darya; any remnant populations must now be confined to a fraction of their former riverine range (Mugue, 2010).

In a peculiar turn of events, the North Aral Sea supported stocks of the European flounder (*Plathichtys flesus*: Pleuronectidae) for a period around the turn of the century. This was one of a number of marine fishes that were stocked in an effort to revive the lake fishery. Attempts had also been made to introduce a variety of fishes and other species from the Caspian Sea to the Aral at various junctures during the twentieth century (for inventories, see Micklin *et al.*, 2014), but these – as well as species stocked from elsewhere in Eurasia – either failed to establish or perished as conditions in the Aral Sea became more extreme.

Changes in the Aral Sea would have had profound effects on other aquatic taxa, although these have been relatively poorly documented. Invertebrate diversity within the lake was not particularly high and, although of undoubted local ecological importance, their conservation significance was probably limited. For the majority of zooplankton, which included copepods, rotifers and Cladocera, the most that can be said is they were wiped out as conditions deteriorated. The anostrocan brine shrimp *Artemia parthenogenetica* (Artemiidae) became a temporary 'winner' after becoming established in the late 1990s. It has flourished in the hypersaline conditions, and makes up the entire zooplankton biomass in the remnant Large Aral Sea (Arashkevich *et al.*, 2009). The contemporary benthos is dominated by larvae of a salt-tolerant midge, *Baeotendipes noctivaga* (Chironomidae). Both of these taxa serve as food for migratory birds.

Waterbird abundance and richness in the Aral basin declined greatly as the lake shrank and the extent of wetland along the shores, particularly in the river deltas, has waned. Only around 10% of the original marshland area persists, and some of what remains has been transformed to salt marsh with a very different composition to the diverse assemblages of aquatic macrophytes (~30 species) that were once present. Of the 100 or so species of waterbirds in the basin, around half bred in the wetlands fringing the Aral shoreline (Kreuzberg-Mukhina, 2006) and all have experienced reductions in habitat. Among them are the pygmy cormorant (*Microcarbo pygmaeus*: Phalacrocoracidae), great white pelican, glossy ibis and the globally threatened marbled teal (*Marmaronetta angustirostris*: Anatidae; VU). The loss of natural wetlands has been partially offset by the use of irrigated land and reservoirs as wintering sites by migrants such as the Eurasian crane (*Grus grus*: Gruidae) and certain geese. These man-made wetlands have even been used for breeding by the threatened Dalmatian pelican (*Pelecanus crispus*: VU) white-headed duck and pygmy cormorant.

In 2005, completion of a 13-km dam–dyke complex funded by the World Bank effectively separated the North Aral Sea from the much larger southern portion of the lake. Water from the Syr Darya began to replenish the smaller basin; measures were also introduced to improve irrigation efficiency in Kazakhstan, so that more water was available to maintain the health of the Syr Darya. Outcomes of the Kokaral (or Kok-Aral) Dam initiative have included a substantial rise in water levels (>10 m) in the North Aral, an expansion in lake area and a reduction in salinity. Stocks of some cyprinids have begun to recover along with wels, pike and pike perch (see Box 6.1), thereby supporting a modest but growing fishery, although this has occurred at the expense of decreases in the population of European flounder. Non-native silver carp, grass carp and snakehead are also landed and were presumably deliberately stocked to increase the catch. The abundance and richness of migratory water-birds – including iconic species such as greater flamingo (*Phoenicopterus roseus*: Phoenicopteridae) – have increased also. The successful recovery of some aspects of the ecology of the North Aral Sea has spurred plans for further restoration measures, such as raising the height of the Kokaral Dam, but they have yet to be implemented. There are some grounds for optimism about the potential for ecological rehabilitation of this frag-ment of the Aral, even though it encompasses far less than 10% of the original lake area. Large parts of the former lake bed are likely to remain dry. They are the focus of attempts to plant forests of saxaul (*Haloxylon ammodendron*: Amaranthaceae), a psammophyte that can grow in near-desert conditions, in order to help stabilize the soil and reduce dust storms.

What of the future? Regional climate change, manifested in lower rain and snowfall, and glacier retreat on the Pamir Mountains where the flow of the Amu Darya originates, limits the chances of improvement of conditions in the South Aral Sea. Given continued aggressive water extraction for irrigation, the Uzbek portion of the basin appears set to complete its transition from a lake to a desert landscape. The likelihood of that outcome may well have become greater as a result of the area becoming the focus of recent exploration by international consortia for oil and gas, a sector of the economy that has been increasing in import-ance (Varis, 2014). Furthermore, unless an agreement on sharing the water of the Amu Darya is reached among the central Asian riparian nations between Uzbekistan and upstream Tajikistan or Turkmenistan for instance, there is little prospect that any flow will be available to replenish the South Aral Sea. The hydropolitics are complicated: there are ambitious plans for huge dams on a major tributary of the Amu Darya within Tajikistan. Among them, the vast 300-m Nurek Dam was com-peted in 1972 and, if sufficient investment is forthcoming, it will be joined by the Rogun Dam that, with a projected height of 335 m, would be the tallest dam in the world; it could take up to 16 years to fill. Such water-engineering schemes have raised concerns for flows in Amu Darya and water supplies in Uzbekistan and Turkmenistan. In this water-scarce region of central Asia, only Kazakhstan has shown any willingness thus far to limit the amount of water extracted for irrigation, but it has jurisdiction over a portion of the Syr Darya drainage only. In addition, there is a grandiose proposal, originating in a 2000 edict by Turkmenistan's then president, to form a huge 3500-km^2 impound-ment – the Golden Age Lake (Altyn Asyr) – within the natural Karashor

Depression in the Karakum Desert; it would be fed by water diverted by two canals running almost 3200 km from the Amu Darya (Stone, 2008). Downstream Uzbekistan relies on the river for irrigation and is hardly likely to tolerate a reduction in the share of that water, in the event that such diversions proved to be logistically feasible. Completion of the Golden Age Lake project, which has been ongoing for 15 years and included an 'opening ceremony' in 2009, would greatly worsen soil salinization problems in the region. Nonetheless, work continues at the time of writing, with the government trumpeting the potential benefits for waterbird diversity and fisheries.

The former Soviet states around the Aral Sea are among the highest per-capita users of water in the world, with each Turkmen consuming three times the water used by the average US citizen, and 13 times more than one in China. Moreover, the economic return on water in central Asia is lower than anywhere else on Earth: Turkmenistan uses almost 3 times more water than India to produce one GDP dollar (mainly through agriculture), and 14 times more than China (Varis, 2014). As the statistics demonstrate, these countries have considerable water relative to their populations, but that supply has been squandered unduly. The destruction of the Aral Sea may well represent the most egregious example of a situation where human demands for water have been permitted to compromise biodiversity, ecosystem integrity, and even the livelihoods and health of the people who depended on the lake. Nor is the Aral Sea the only such instance in central Asia (see Box 6.2) or further south.

The Hamoun Wetlands and Lake Urmia

The Hamoun (or Hamun) wetlands, located within the endorheic Sistan Basin on the border of southeastern Iran and western Afghanistan, formerly comprised a 2000-km^2 matrix of marshes and shallow (3 m or less) lakes. Like the Aral Sea, they demonstrate the potentially destructive consequences of the expansion of irrigated agriculture for freshwater ecosystems, and provide yet another example of conflicts among human users of water. Although situated within one of the driest regions in the world, the wetlands have long been important for migrating birds and other wildlife, as well as the human settlements that were the likely origin of Zoroastrianism. The wetlands received water from the Helmand River that carried snowmelt from the Hindu Kush through the Margo Desert westward into the Sistan Basin, with seasonal fluctuations

Box 6.2 *Contraction and Loss of Central Asian Lakes*

A combination of land conversion and climate change has led to the shrinkage of major lakes across central Asia: a sample of nine lakes over a 30-year period (not all endorheic, although the Aral Sea was among them) showed an average decline of 50% relative to their original area in 1975 (Bai *et al.*, 2011). Of these, the Ebinur Lake (also called Lake Aibi) is a closed inland lake located within the arid region of the Xinjiang Autonomous Region in the northwestern part of China, near the Kazakhstan border. Although this shallow lake has undergone change in size over thousands of years due to natural causes, its surface area has experienced a dramatic decrease of more than 30% from 1972 to 2013, and the dry bed is now a source of dust storms. Overextraction of water from influent rivers (fed by melting glaciers in the Tienshan Mountains) has been influential, especially from the Bortala River to the west, but much of the decrease in lake area appears to coincide with periods of rapid land reclamation (Zhang *et al.* 2015). The lake is now hypersaline and dominated by ephydrid flies and brine shrimp (probably parthenogenetic *Artemia salina*), which are harvested commercially and used as an aquaculture feed.

For now, Ebinur Lake remains an important site for migratory anatids and waders, but breeding Eurasian crane have declined; white-headed duck, relict gull (*Larus relictus*: Laridae; VU) and Dalmatian pelican are present also. The eastern part of the lake includes wetlands that have been a national nature reserve since 2007, and, in 2011, funds from the Global Environment Facility were used to launch a World Bank project intended to restore water to Lake Ebinur and protect the reserve. Thus far, there seems little evidence of improvement of the lake. The tragedy of Lake Ebinur – and the Aral Sea – are current manifestations of the consequences of overuse of water leading to increased salinity and eventual drying of lakes, particularly those in arid regions. Historically, overuse of water has likewise led to shrinkage of other water bodies in central Asia: among them Qinghai Lake and Manas Lake, as well as Aydingkol Lake and Lop Nor that are now dry.

in flow causing the inundated area to more than double during the flood season. The three main lakes began shrinking in size during the 1990s when expansion of irrigation and dam construction on the Helmand and Arghandab Rivers in upstream Afghanistan reduced the river flows into the wetlands by more than 70%. From 1998 onward, Afghanistan experienced a prolonged drought, and the sluices of the Kajaki Dam were closed by decree of the Taliban government. The lakes and much of their associated marshes had dried out by 2001, leaving mostly salt flats and dry land. All fisheries and most cultivation and livestock rearing around the remains of the Hamoun wetlands ceased, and towns have been abandoned, giving rise to much misery in an area that was already receiving large numbers of refugees fleeing armed conflict in the east. And, just as around the Aral Sea, strong winds blow fine sand and silt off the exposed lakebeds and surrounding wastelands, resulting in persistent sand storms (some lasting more than 200 days) and increased atmospheric dust and aerosol loadings, with deleterious consequences for human health and regional climate.

The impacts on biodiversity have likewise been devastating. The Hamoun wetlands no longer yield what was once an annual fish catch of 12 000 t, based on 25 species (10 of which were introduced, among them the goldfish, *Carassius auratus*). The area has been rendered largely unsuitable for waterbirds since 1999 although, in some years prior to that, it supported as many as 77 species totaling up to 0.6 million individuals. The very limited survey data that are available (Behrouzi-Rad, 2009) record a complete absence of waterbirds in some years. The wetlands once had bird species in common with the Aral Sea, but no longer offer suitable stop-over sites for winter migrants from the north, such as Dalmatian pelican, greater flamingo and various anatids including the ferruginous duck (*Aythya nyroca*; NT) and white-headed duck. The Iranian portion of the Hamoun wetlands is still listed under the Ramsar Convention, although much of the area is now mostly barren salt flats. All that remains of this once-thriving habitat is hypersaline Lake Gowd-e-Zareh, located at the southern end and the lowest level of the Sistan Basin, and a few water-storage reservoirs near the Zabol city in Iran.

Some restoration of the Hamoun wetlands could have been made possible by the release of water from the Kajaki Dam, following the fall of the Taliban government in Afghanistan, if sufficient flow had been available to permit this after irrigation needs were met. However, most water that is allowed to flow downstream is diverted by canals from the Helmand River to the Zabol and Chah-Nimeh Reservoirs. As a result, Lake Gowd-e-Zareh receives runoff only in occasional years when Helmand River flows are exceptionally high, and this allows some development of fringing vegetation in portions of the lake where salinity is temporarily reduced. Restoration of the Hamoun wetlands would depend on coordinated bilateral action on environmental water allocations, but Afghanistan seems unwilling or unable to take measures that might alleviate the situation downstream. This is notwithstanding the fact that Iran and Afghanistan have had an accord to share water since 1973, obligating some downstream release to Iran, but both countries have accused the other of breaching the agreement. Ongoing work aimed at completing the long-delayed US$100-million Kamal Khan Dam on the Helmand River is intended to allow expansion of the extent of irrigated land in Afghanistan, and is unlikely to improve the prospects for water sharing with Iran. Protracted drought conditions that may be a result of climate change impose additional constraints on an already difficult situation.

Further east, close to the border of Iran and Turkey, the surface area of Lake Urmia (also known as Lake Orumiyeh) has been gradually reduced since the 1990s to only 12% of historic values and only 5% of its former volume. This, Iran's largest lake, is endorheic and hypersaline, and thus lacks fish. Aquatic macrofauna is represented only by a brine shrimp (*Artemia urmiana*, a near-endemic species) that subsists on phytoplankton; its abundance and distribution are declining as the saline lake waters approach saturation. Sand storms and salt haze arising from the exposed dry bed have led to health concerns. The situation, caused by aggressive water extraction within the catchment, has only recently gained attention in the scientific literature, where it has been labelled as 'Aral Sea syndrome' (AghaKouchaka *et al.*, 2015). The rate of degradation seems to have been accelerating, with a 7-m reduction in water-level between 1995 and 2011. The grave consequences for humans are especially worrisome given that the population density surrounding Lake Urmia is much greater than that formerly present around the Aral Sea – 1600 km to the north east. Among possible mitigation measures that might be feasible in this arid region, given political will and funding, is a 300-km westward transfer of water from the Caspian Sea. Smaller-scale transfers of water, as has been proposed from the Zab River in Kurdistan, are likely to be inadequate and would merely shift the location of water shortages elsewhere.

Environmental deterioration in and around Lake Urmia has compromised its value as a Ramsar site (designated in 1975) and a UNESCO Biosphere Reserve (since 1976) to such an extent that the abundance of greater flamingo, which generally favour hypersaline lakes where they feed on brine shrimp, has declined greatly. Colonies of great white pelican, Dalmatian pelican and, most notably, greater flamingo no longer breed at Lake Urimia, since there is insufficient aquatic secondary production to sustain them. The lake was also formerly a globally important site for migratory waders and a variety of anatids such as gargagney (*Anas querquedula*), common teal (*A. crecca*) and marbled teal. Common and ruddy shelduck (*Tadorna tadorna* and *T. ferruginea*), white-headed duck, certain gulls, glossy ibis and Eurasian spoonbill (*Platalea leucorodia*: Threskiornithidae) once bred there; white stork and Pallas's fish eagle (*Haliaeetus leucoryphus*; EN) were present also.

Lake Chad

Tropical Africa provides another case of a lake that virtually disappeared as a result of human mismanagement of water, paralleling the instances of the Aral Sea and wetlands in Iran, although the precise causes and drivers of ecosystem degradation are particular to each water body. Lake Chad straddles the borders of Chad, Niger, Nigeria and Cameroon in the West African Sahel where it occupies part of the largest endorheic basin on Earth. The region is semiarid and prone to drought but, until relatively recently, Lake Chad was the sixth largest lake in the world (by surface area) and the third largest in Africa. Under historically 'normal' conditions, the lake had a diverse fish fauna adapted to seasonal water-level fluctuations, and supported huge congregations of waterbirds – both residents and Palaearctic migrants. It was fringed with extensive beds of sedges, reeds and other grasses, and contained dense growths of a variety of submerged macrophytes. The Chari River and its tributary – the Logone – contributes most of the inflow to Lake Chad, and its associated floodplain in the south of the lake was important for waterbirds and served as a wet-season spawning site for fishes.

All four riparian nations have large and growing human populations, and have built extensive irrigation schemes, particularly for rice, to serve the 30 million people that depend on the lake. Water is diverted from Lake Chad itself, from the Komadougou-Yobe River in Nigeria, which now only flows intermittently, and from the Chari River that enters from the southeast. As a result, Lake Chad had shrunk from approximately 25 000 km^2 in 1963 to no more than 5% of its original size little more than 30 years later. While seasonal variation in the lake depth and extent was typical of Lake Chad, in part because it is generally shallow, the remarkable shrinkage in lake size was not. Even during the wet season, the lake is now split into a northern and larger southern portion; in some years, their combined extent has been as little as 300 km^2

(Gao *et al.*, 2011). Perhaps half of this reduction in area, and an associated decline in aquifers, is attributable to regional climate change driven by overgrazing and deforestation, with a consequential decline in rainfall leading to several prolonged droughts (Coe & Foley, 2001). Extraction of water for irrigation increased severalfold as a result and became unsustainable, giving rise to widespread soil salinization and causing the lake to wither. One very marginal benefit of this outcome has been proliferation of so-called *Spirulina* cyanobacteria (actually *Arthrospira* spp.: Phormidiaceae) that thrive in hypersaline pools that form as the lake dries out, and where it can be collected or cultivated for human consumption.

Among the serious consequences of the substantial reduction in area of Lake Chad have been desertification in the northeast, crop failure, livestock deaths and increasing poverty. Shortages of water have also given rise to conflicts among farmers, herders and fishermen. Political instability in the region, largely associated with the Boko Haram insurgency and military operations to counter it, has been associated with indiscriminate violence leading to a humanitarian crisis, including *inter alia* population displacements as well as bans on artisanal fishing (van Lookeren Campagne & Begum, 2017). Conflicts among people have arisen as lake shrinkage has continued and fishers move to exploit waters claimed by other riparian nations; there are also disputes over cultivation rights to expanses of the formerly inundated lake bed. Food scarcity has led to increased hunting of wildlife, which has contributed to reductions in populations of hippo and terrestrial mammals associated with the lake shores. Spotted-necked otter (*Hydrictis maculicollis*) and African clawless otter (*Aonyx capensis*: Mustelidae) were formerly abundant. The crocodile population was substantial, and formerly assumed to consist of the widely distributed Nile crocodile, but is more likely to have been the recently discovered desert crocodile (*Crocodylus suchus*) – a species that is much rarer than the Nile crocodile and was formerly confused with it (Hekkala *et al.*, 2011). African manatee (*Trichechus senegalensis*; VU) were also present in Lake Chad during the last century, but were extirpated from the lake by around 1930 as a result of hunting for their oil and meat; they no longer occur anywhere in the basin (Keith Diagne, 2015).

Sixty-nine native fish species have been recorded from Lake Chad (although some estimates are as high as 93 species), but, unlike the East African Rift Valley lakes, none are endemic, and the lake community is relatively poor in cichlids. The entire drainage basin supports 179 fish species, more than twice the number found in Lake Chad itself, and a few of them are found nowhere else. As the lake has shrunk, there has been a shift in community composition from migratory Alestidae (for instance, the tigerfish, *Hydrocynus brevis*), cyprinids, distichodontids, mochokid catfishes and Mormyridae towards species adapted to hypoxic marshy conditions such as clariid catfishes and the lungfish *Polypterus senegalus* (Polypteridae), as well as *Oreochromis* tilapias that are tolerant of increasing salinity. Fisheries in Lake Chad are in the process of collapse, with declines in the size of landed fishes, such as schilbeid catfish (*Schilbe mystus*), alestids (particularly *Alestes baremoze*) and Nile perch, and reductions in overall catch weight from historical levels of ~140 000 t annually.

Lake Chad initially attained the status of a Transboundary Ramsar Site of International Importance in 2001 as the result of agreement between Niger and Chad, but it was not until 2010 that consensus on the matter was reached among all four riparian nations. This status is due to the large numbers of wading birds and ducks that use the lake, its associated wetlands and the southern floodplain, mainly during the Palaearctic winter. The numbers of long-distance migrants can exceed one million individuals, comprising almost 50 waterbird species. They include ruff (*Philomachus pugnax*: Scolopacidae), white-faced whistling duck (*Dendrocygna viduata*), fulvous whistling duck (*D. bicolor*), northern pintail (*Anas acuta*), gargagney, marbled teal and spur-winged goose (*Plectropterus gambensis*); black-crowned crane (*Balearica pavonina*: VU) and pink-backed pelican (*Pelecanus rufescens*) are notable also. There is no readily accessible source of recent data on how bird populations have been affected by the worsening political situation and ecological condition of Lake Chad, but the use of the area as a feeding and breeding site for wetland species is certain to have been compromised.

Designation of Lake Chad as a Ramsar site has not prevented its invasion by alien macrophytes: water hyacinth and water lettuce cover around half of the remaining water surface area, while *Typha* spp. (Typhaceae, particularly *T. 'australis'*) have also overgrown large portions of the lake bed and floodplains. A recent report on the conservation status of African freshwater fishes and molluscs (Darwall *et al.*, 2011a) mentions the deteriorating conditions in and around Lake Chad but gives no details about the effects on the aquatic fauna. However, the endemic snails, *Gabbiella neothaumaeformis* (Bythiniidae; CR) and *Biomphalaria tchadiensis* (EN), are threatened by declining lake levels and habitat change caused by invasive floating plants, as is *G. tchadiensis* (EN), which despite its name, is neither endemic to the lake nor to Chad (Kristensen & Stensgaard, 2010).

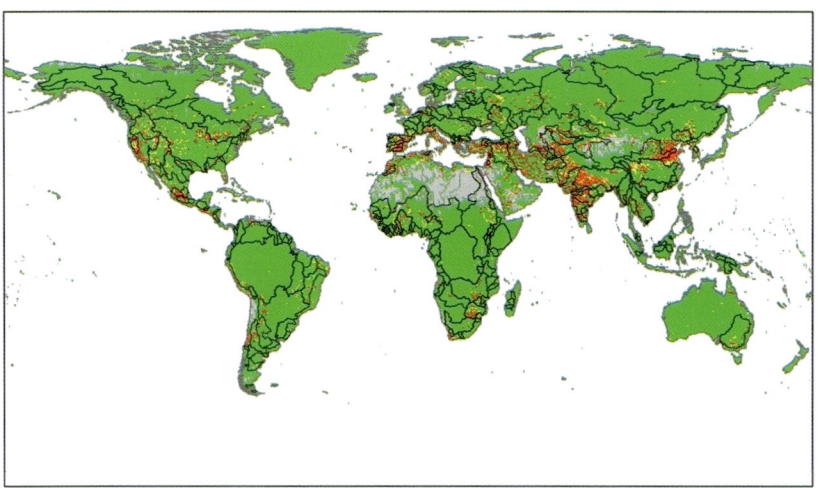

Plate 1 The distribution and current status of consumptive use of blue water (from rivers, lakes and renewable ground water), taking into account environmental water flows; i.e. basin-scale limits on the maximum rate of withdrawal from rivers in order to maintain them in a fair-to-good ecosystem state (for details see Steffen *et al.*, 2015). The green areas are within the planetary boundary for freshwater use, yellow areas are within the zone of increasing risk, and red areas are beyond the boundary and at high risk. Grey areas have very low river flow. From Steffen *et al.* (2015) with permission.

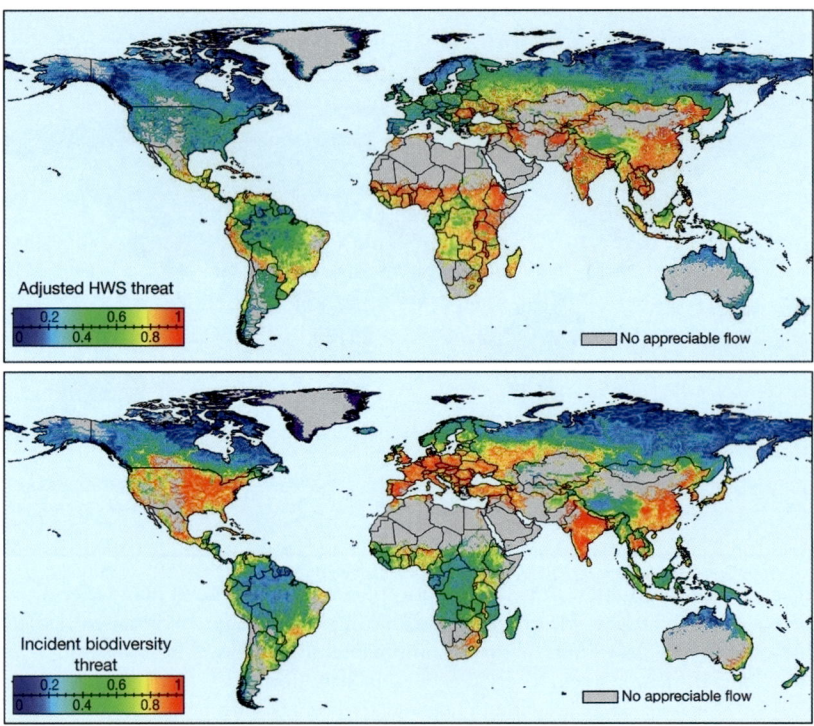

Plate 2 A global geography of river threat, showing the patterns of aggregate threat from a range of factors (see Box 1.7) to – in the upper map – human water security (HWS; adjusted to account for investments in infrastructure related to water engineering and treatment) and – in the lower map – incident threat to freshwater biodiversity. Areas shaded grey have no appreciable river flow. From Vörösmarty *et al.* (2010) with permission; see also www.riverthreat.net.

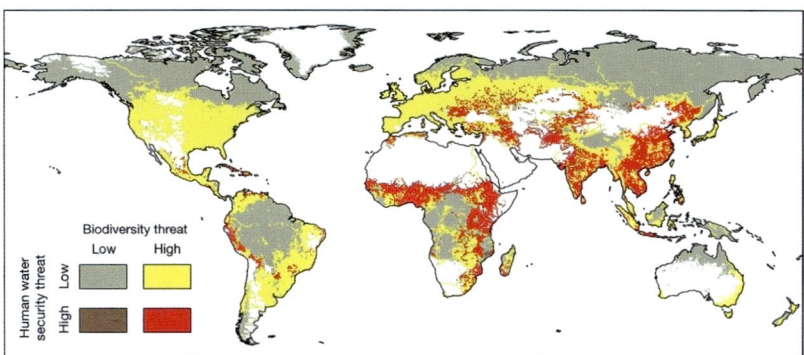

Plate 3 A global geography of river threat, showing the patterns of spatial concordance of aggregate threat from a variety of factors (see Box 1.7) to human water security and freshwater biodiversity. Especially striking is the absence of any localities with brown shading where the threat to human water security would be high while that to biodiversity would be low. Areas shaded grey have no appreciable river flow. From Vörösmarty *et al.* (2010) with permission; see also www.riverthreat.net.

Plate 4 Salmon are vulnerable to human impacts during upstream spawning migrations when they may be subject to intense fishing pressure; in many cases, stock are extirpated because dams prevent them from gaining access to spawning sites. Runs of anadromous Atlantic salmon (*Salmo salar*), pictured here, have been greatly reduced across their range, but many populations of Pacific salmon (*Oncorhynchus* spp.) have been affected also.

Plate 5 The Asian arowana or dragon fish (*Scleropages formosus*) is a symbol of good luck and prosperity in Chinese culture, and therefore much sought after as an ornamental species. Targeted fisheries exist in parts of Southeast Asia, even though commercial trade in the wild-caught individuals is prohibited under CITES Appendix I. Collection of mouthbrooding males to obtain the offspring is a highly unsustainable practice contributing to the endangered status of this species.

Plate 6 Visitors view Chinese sturgeon (*Acipenser sinensis*) in an aquarium at Ocean Park, Hong Kong. These critically endangered large fishes no longer occur in the nearby Pearl River, and wild stocks in the Yangtze River are close to extinction.

Plate 7 The Yangtze sturgeon (*Acipenser dabryanus*; CR), also known as Dabry's sturgeon, is confined to parts of the Yangtze upstream of the Three Gorges Dam and appears close to extinction. It grows to 20 kg, much smaller than the Chinese sturgeon (*A. sinensis*; CR), which occurs in the lower Yangtze and may weigh up to 450 kg and exceed 3 m in length.

Plate 8 The Chinese paddlefish (*Psephurus gladius*; CR), one of only two species of Polyodontidae on Earth, may now be extinct. Adults were reputed to grow to 7 m in length. This large and charismatic species was rapidly forgotten by local communities as soon it was no longer encountered on a regular basis, offering a striking example of rapid cultural baseline shift within a human generation.

Plate 9 The giant freshwater stingray *Himantura chaophraya* (also known as *Urogymnus polylepis*) occurs in the Mekong and other large Southeast Asian rivers, where it has been reported to weigh as much as 600 kg. It is viviparous, producing one or a very few pups at each breeding event. As with other Mekong megafishes, this endangered species is threatened by overfishing and mainstream dam construction, but, elsewhere, habitat degradation may be influential also. Photo courtesy of Zeb Hogan.

Plate 10 The Mekong giant catfish (*Pangasianodon gigas*) is a long-distance migrant endemic to the river, where it was formerly plentiful. It is now critically endangered as a result of overfishing and at risk of extinction as a result of construction of mainstream dams. This catfish (in fact, it lacks barbels) is one of the largest freshwater fishes in the world, growing to around 3 m in length and weighing more than 300 kg. Photo courtesy of Zeb Hogan.

Plate 11 The giant carp (*Catlocarpio siamensisis*) is the largest member of the Cyprinidae and occurs in the Mekong and Chao Phraya Rivers. It is critically endangered as a result of overfishing, and few – if any – individuals attain the potential maximum size of around 300 kg. Dam construction along the Mekong mainstream will also obstruct breeding migrations. Photo courtesy of Zeb Hogan.

Plate 12 Arapaima gigas is one of the biggest freshwater fishes, exceeding 2 m in length and more than 200 kg. The swim bladder is modified into an accessory breathing organ and the fish periodically gulp air at the surface. This habit makes them vulnerable to fishers, so the species has been overexploited in much of its Amazonian range (see Box 3.5).

Plate 13 Piaractus brachypomus (sometimes known as the pacu) is a frugivorous serrasalmid native to the Amazon Basin, where it plays an important role in seed dispersal (ichthyochory). That function is likely to be compromised by overexploitation of these large fish, which can grow to almost 1 m in length.

Plate 14 The North American grizzly bear (*Ursus arctos horribilis*) is a subspecies of the more widely distributed brown bear, and both bears feed on migrating salmon – in this case, a sockeye (*Oncorhynchus nerka*) – as they migrate upstream to breed. The salmon are eaten on land and represent an 'uphill' subsidy of marine-derived energy and nutrients for freshwater and riparian ecosystems.

Plate 15 The spectacled caiman (*Caiman crocodilus*) is widespread in the Amazon and is the most prevalent New World crocodilian. Following overexploitation and declines in abundance of the larger black caiman (*Melanosuchus niger*), spectacled caiman became subject to commercial exploitation, primarily for their hides, in the 1960s, and populations fell until the 1980s. Demand has since dropped and trade in caimans has been successfully regulated.

Plate 16 The pink Amazon River dolphin (*Inia geoffrensis*) or boto is the largest species of freshwater dolphin, reaching 2.5 m in length, and is widespread in the Amazon and Orinoco Rivers. Overfishing in the Amazon reduces food availability for boto and may lead to competition with fishers, increasing the chance of entanglement in nets and other gear. Boto carcasses are also used as bait to attract scavenger catfish in the piracatinga fishery (see Box 3.7).

Plate 17 The critically endangered golden coin turtle (*Cuora trifasciata*) has been subject to intense exploitation across its range in China, Laos and Vietnam because it is believed to offer a cure for cancer. Individuals can fetch several thousand dollars in illicit trade, and the few remaining wild individuals are largely confined to protected areas in Hong Kong. Photo courtesy of Yik Hei Sung.

Plate 18 The critically endangered Indian gharial (*Gavialis gangeticus*: Gavialidae) is one of the largest of all crocodilians, but subsists chiefly on fishes. It formerly ranged along rivers in southern Asia from the Irrawaddy in the east to the Indus in the west, but now occupies only a tiny fraction of that area. Major threats include degradation or loss of riverine habitat, depletion of fish stocks and entanglement in nets.

Plate 19 Beavers, such as these *Castor canadensis*, are ecological engineers and keystone species. Their dam-building activities alter riverscapes, affecting stream flows and the flux of organic matter. They are sometimes regarded as nuisance species and are subject to control in parts of their range, although Eurasian beaver (*C. fiber*) have been reintroduced to many countries in Europe where they had been exterminated by hunters.

Plate 20 The Lamington spiny crayfish (*Euastacus sulcatus*) is native to rainforest streams in the uplands of northeastern Australia. It is threatened by recreational 'fishing' and collection of the different colour morphs for the aquarium trade. Because this vulnerable species is restricted to cool upland streams, it probably has a limited thermal tolerance and will be susceptible to climate warming.

Plate 21 *Orestias cuvieri* (Cyprinodontidae: total length approximately 25 cm), an endemic Lake Titicaca killifish driven to extinction by introduced piscivores.

Plate 22 Despite its name, the Malaysian mahseer (*Tor tambroides*) is quite widely distributed in the rivers of mainland Southeast Asia. Like other members of the genus, this large species has been overexploited; it is also affected by habitat degradation. Mahseers in general are popular sport fishes, and catch-and-release angling has been used to generate funds that support conservation in Malaysia and the Indian Himalaya.

Plate 23 Various species of African tilapia (or their hybrids) have been widely introduced across the tropics for aquaculture and some, such as this *Oreochromis niloticus*, have become invasive. They are tolerant of poor water quality, and their mouthbrooding habit reduces juvenile mortality and can allow a single female carrying eggs to establish a new population.

Plate 24 The highly invasive red swamp crayfish (*Procambrus clarkii*) in defensive posture during an episode of overland travel. Although native to parts of the southern United States and Mexico, it has become established widely in Europe, Asia, Africa and South America, causing dramatic changes in native communities of freshwater macrophytes, and spreading disease (crayfish plague) to indigenous Astacidae.

Plate 25 Ponto-Caspian zebra mussels (*Dreissena polymorpha*) are highly invasive in parts of Europe and North America. They form dense accumulations on hard surfaces and even foul the shells of native pearly mussels, eventually smothering or starving them.

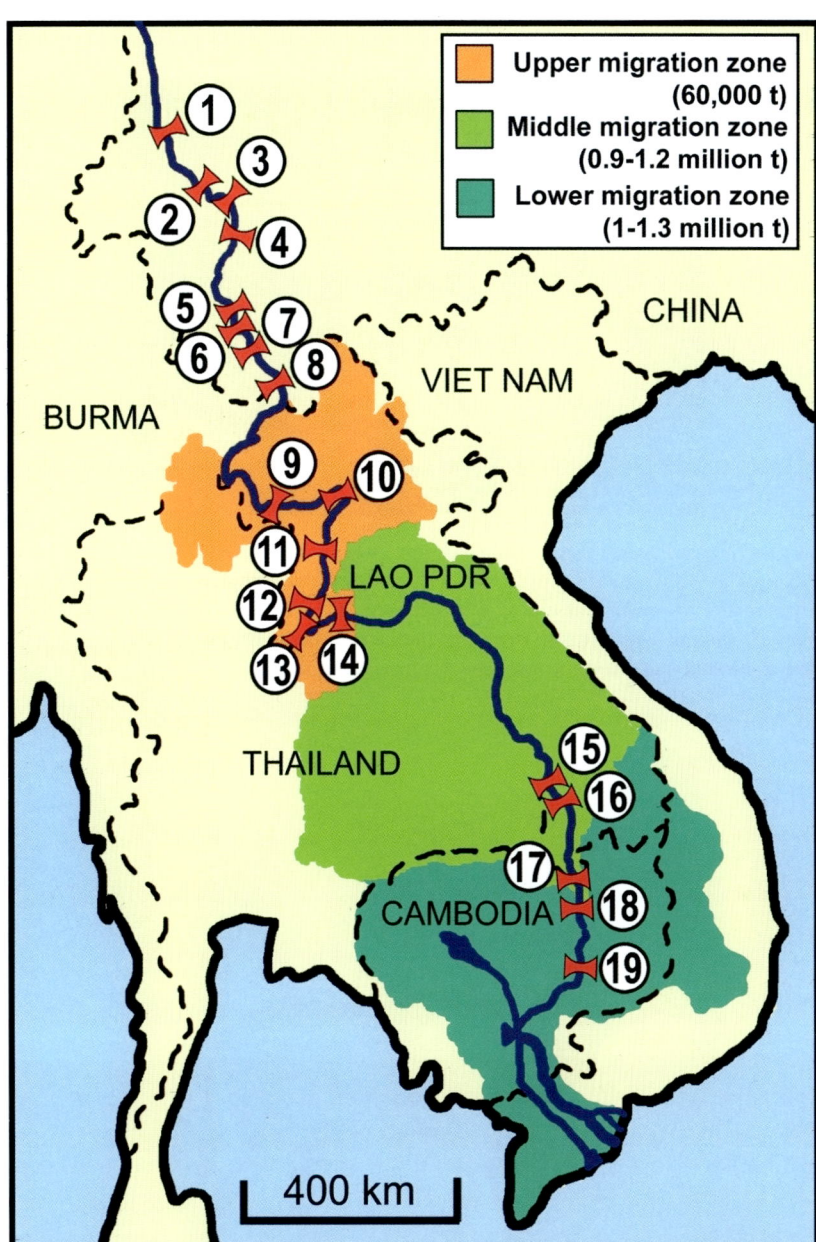

Plate 26 Map of the Mekong–Lancang Jiang showing the location of existing or planned dams. Estimates of total catch aggregated across all fish species in the upper, middle and lower migration systems of the Lower Mekong Basin are shown also, based on Halls and Kshatriya (2009). Note that other estimates of catch (e.g. Barlow *et al.*, 2008; MRCS, 2011a) differ slightly from these figures. 1 = Gongguoqiao; 2 = Xiaowan; 3 = Manwan; 4 = Dachaoshan; 5 = Nuozhadu; 6 = Jinghong; 7 = Ganlaba; 8 = Mansong; 9 = Pak Beng; 10 = Luang Prabang; 11 = Xayaburi; 12 = Pak Lay; 13 = Sanakham; 14 = Pak Chim; 15 = Ban Koum; 16 = Lat Sua; 17 = Don Sahong (at Khone Falls); 18 = Stung Treng; 19 = Sambor. Dams 1–8 are situated on the Lancang Jiang, where additional upstream dams (not shown here) have been proposed. Map based on Dudgeon (2011); used with permission.

Plate 27 Because they leap readily, salmon such as this chinook (*Oncorhynchus tshawytscha* – the largest member of the genus) are able to ascend waterfalls and man-made structures that would obstruct other fishes. They also use fish ladders that are designed to mitigate the barrier effects of dams, although these structures are of limited value to the many species that seldom jump.

Plate 28 The milu (*Elaphurus davidianus*) is sometimes referred to colloquially in Chinese by the term 'like none of the four' due to its mixture of features: the antlers of a deer, hooves of an ox, face of a horse and tail of a donkey (some accounts add the neck of a camel).

To achieve sufficiently large inflows to refill Lake Chad (after compensation for irrigation usage), inter-basin water transfer from the Oubangui River – a major tributary of the River Congo in the Central African Republic – has been mooted to supplement the flow of the Chari River. This would involve construction of a navigable 100–150 km canal, to divert 26 km^3 of water each year for a decade (Gao *et al.*, 2011). A proposed extension of this scheme, termed the Transaqua Project, envisages a massive 2400-km-long canal running through the east and north of the Congo basin, intercepting tributary waters and transferring 10 times the volume of water that would come to the Chari from the Oubangui River. Realization of either of these two inter-basin water transfers would require resolution of complicated socioeconomic and transboundary issues. The Transaqua Project lacks financial backers, and nor has it been the subject of a feasibility study in the three decades since it was first envisioned. It is, however, supported by the Lake Chad Basin Commission (LCBC), a regional organization comprised of Cameroon, Chad, the Central African Republic, Niger and Nigeria. Among other things, the LCBC is responsible for fisheries policies and coordination of regional development. It faces shortages of resources and trained manpower, although awareness of the need to improve conditions in and around Lake Chad is high. Regardless of whether the LBBC has the capacity to facilitate the audacious Transaqua Project, it could serve as a platform for integrated management and restoration of Lake Chad. As with the Aral Sea, collaboration with the World Bank may provide a potential source of funding for the measures that will be needed to bring about some recovery of the lake ecosystem, even if these are on a more local scale than the transfer of water from the River Congo. Urgent action is needed as, given the large population dependent upon Lake Chad, any further drying could give rise to a calamity that affects many more people than the Aral Sea disaster, adding to the miseries already inflicted by insurgencies in the western Sahel. Elsewhere in Africa too, other lakes are threatened by diversions or overuse of influent water (for examples, see Box 6.3).

The Drying of Lake Poopó: Bolivia's National Disaster

Lake Poopó – formerly around 2500 km^2 – ranked second in area within Bolivia to Lake Titicaca, which is South America's largest water body. Situated at approximately 3700 m elevation on the Altiplano Plateau, this shallow lake (mean depth ~3 m) underwent natural seasonal fluctuations

Box 6.3 *Smaller East African Rift Valley Lakes*

Lake Turkana, known formerly as Lake Rudolf, is a World Heritage Site in Kenya that is at risk of being degraded by the construction of the Gibe III Dam and associated irrigation projects on the Omo River in Ethiopia, which abuts the northern shore of the lake. The Omo River accounts for 90% of inflows to Lake Turkana, said to be the largest desert lake in the world, and these flows could be substantially altered by dams and reduced by irrigation of lands upstream. The reservoir behind Gibe III began filling in 2015, significantly reducing the downstream flow of the Omo River and likely leading to an increase in the salinity of Lake Turkana. This will be detrimental to biodiversity and the fishery, as well as to the livelihoods of hundreds of thousands of people that depend on the lake. Lake Turkana hosts a huge population of Nile crocodile, and around 50 fishes of which a dozen (including four cichlids and the Rudolf lates (*Lates longispinis*)) are endemic; in addition, the Turkana mud turtle (*Pelusios broadleyi*: Pelomedusidae; VU) is found nowhere else. Eighty-four waterbird species – among them wintering Palearctic migrants and passage migrants – have been recorded (Birdlife International, 2018), with globally important numbers of little stint (*Calidris minuta*: Scolopacidae). At least 23 species breed there, such as the regionally threatened African skimmer, and ardeids; greater flamingo are present also. These flamingo likewise occur in nearby Lake Naivasha, where reductions in water level are attributable to overuse of water in horticulture operations within the catchment, although other factors have also contributed to deteriorating lake conditions (see Box 4.3). Fortunately, the extent of degradation in these East African lakes does not yet match that seen in Lake Chad or the Aral Sea.

in water level driven largely by melt cycles of Andean glaciers. The low-water period had progressively lengthened until, in 2015 when El Niño was particularly intense, the basin dried out almost completely and was declared a disaster area. This occurred just over a decade after Lake Poopó had been designated a Ramsar site. Lake shrinkage took place despite a five-year restoration effort that began in 2010; an allocation of US$15 million of EU funding failed to reverse the decline. Decades of water withdrawal for mining (gold, tin) and agriculture (latterly, quinoa monoculture), as well as sedimentation, had reduced quality and quantity of inputs from the influent Desaguadero River, and heavy metals, such as lead, arsenic and cadmium, along with acid leachate from mines (regulation of mining in Bolivia is lax) had contaminated Lake Poopó (French *et al.*, 2017). Although the Desaguadero River originates as an overflow of Lake Titicaca, the release of water had been insufficient to offset the demands of mining and agriculture downstream. Climate change (reduced rainfall, warmer temperatures and glacier retreat) also contributed to the shrinkage of Lake Poopó (for details, see Satgé *et al.*, 2017), and has reduced the availability of surface fresh water across Bolivia.

A general trend towards rising salinity in Lake Poopó restricted the distribution of what was already a limited fish fauna, made up of two endemic *Orestias* spp., a trichomycterid catfish and, as in Lake Titicaca (see Chapter 4), invasive rainbow trout and pejerry. The two aliens were relatively large and constituted almost all of the fishery in Lake Poopó. Pollution combined with the effects of falling water levels led to a massive fish kill in late 2014, eradicating the fishery and precipitating a government declaration of disaster. It led to the relocation of many of the Uru people that lived on and around the lake, and coincided with the near-disappearance of waterbirds such as anatids and three species of flamingos: Andean and James's flamingoes (*Phoenicoparrus andinus* and *P. jamesi*; VU and NT) and the Chilean flamingo (*Phoenicopterus chilensis*; NT). Of these, *P. andinus* had been historically overexploited for human food.

The Salton Sea: An Accidental Lake

Although differing in origins and socioeconomic context, conditions in the Salton Sea in California are quite comparable to the Aral Sea, Lake Chad and other endorheic inland waters. The formerly dry Salton Sink basin was inundated between 1905 and 1907 as an inadvertent consequence of attempts to divert part of the flow of the Colorado River. The outcome was a lake – California's largest – extending more than almost $900\,km^2$. For the next century, the lake received some water via a canal from the Colorado, but it is nonetheless hypersaline. Pollution by agricultural runoff also makes it prone to algal blooms, some by toxic species of *Prymnesium* (Prymnesiaceae) that can cause fish kills. The fish fauna within this saline lake itself is now largely made up of omnivorous tilapia (an *Oreochromis mossambicus* hybrid), a hardy species that relies on an introduced marine polychaete (*Alitta* [= *Neanthes*] *succinea*: Nereidae) for a major part of its diet (Hurlbert *et al.*, 2007). Nationally endangered desert pupfish are also present in some less saline parts of the Salton drainage. Numerous marine species from the Gulf of California were introduced to the Salton Sea at various times during the twentieth century in attempts to establish a sustainable fishery. For a period, this resulted in the largest inland fishery in California, mainly based around introduced marine Sciaenidae (croakers). At present, the fishery remains productive, but landings comprise tilapia almost exclusively.

Changes in water apportionment strategies together with a prolonged drought in California have resulted in substantial declines in the area of the Salton Sea accompanied by a rise in salinity. This matters in terms of biodiversity because the lake – albeit man-made, but occupying part of the Salton Sink that contained water in the distant past – has become an increasingly important stop for waterbirds migrating along the Pacific Highway as degradation of wetlands elsewhere in western United States takes place. Large piscivorous species, such as the American white and brown pelicans (*Pelecanus erythrorhynchos* and *P. occidentalis carolinensis*), roseate spoonbill (*Platalea ajaja*) and double-crested cormorant (*Phalacrocorax auritus*), as well as numerous ducks, geese, grebes, gulls, terns, herons and other wading birds, occur in substantial numbers; in terms of bird diversity (424 species), the lake and its surrounds is one of the richest sites in the United States (United States Fish & Wildlife Service, 2008). It is ironic that a site of such conservation importance for waterbirds is an 'accidental lake', in which secondary production is based upon a marine polychaete and a hybrid cichlid from Africa.

Much talk has taken place in California about the need to save the shrinking Salton Sea, and so avoid the dust storms of silt and salt that would originate from the dry lake bed, but there is no agreed plan as to how this might be achieved. Given the already intense competition among users for water from the Colorado River, and a failure to establish an effective political constituency to lobby for protection of the Salton Sea, any foreseeable future for the lake is likely to involve a considerable shrinkage in area and rise in salinity. The extent to which these can occur before use of the lake by waterbirds is compromised is uncertain, but the likely first stage would be the disappearance of fish with serious consequences for piscivores (see Hurlbert *et al.*, 2007). However, potential implications for human health and economic impacts (lost recreational and tourism revenues) should serve as ample incentives for preservation of the Salton Sea. There is at least one precedent elsewhere in California, where lake shrinkage has been reversed (see Box 6.4).

Box 6.4 *Shrinkage Can Be Reversed: The Case of Mono Lake*

Endorheic and hypersaline Mono Lake was threatened by diversion of water from influent streams, particularly the Owens River, to supply Los Angeles; the scheme was extended in the 1940s causing significant falls in water levels over the following decades, a loss of marginal wetlands and a doubling of salinity levels. This could have led to conditions beyond the limits tolerated by the already simple aquatic community of algae, brine shrimp (endemic *Artemia monica*) and ephydrid flies (as in Lake Ebinur: Box 6.2), threatening the many migratory waterbirds that depend on these invertebrates during feeding stops. Huge numbers of eared (or black-necked) grebe (*Podiceps nigricollis*: Podicipedidae) are notable, as well as Wilson's and red-necked phalaropes (*Phalaropus tricolor* and *P. lobatus*: Scolopacidae), which arrive from their northern nesting grounds and feed at Mono Lake before continuing south. American avocet (*Recurvirostra americana*: Recurvirostridae) and various plovers and sandpipers are also numerous, but gulls that nested on islands became vulnerable to land-based predators as falling water levels exposed the lake bed. Reductions in waterbird abundance, which were particularly noticeable among ducks, and concern over deteriorating environmental conditions led to formation of the Mono Lake Committee in 1978. Their subsequent partnership with the National Audubon Society involved protracted court battles to restore the water supply, which were not resolved until 1994 when California State Water Resources Control Board set a target level for Mono Lake, established minimum flows and annual peak flows for influent tributary streams, and ordered development of restoration plans for waterbird habitat. After implementation, water depth increased and reductions in waterbird abundance were reversed, although lake fullness was not completely restored to pre-1940s levels (for details, see www.monolake.org/mlc/restoration), and operation of peak flows in some tributaries was not agreed upon until 2013.

The Mesopotamian Marshes: Devastation and Revival

The Tigris–Euphrates Marshes, in the south and east of Iraq and extending to Iran – collectively known as the Mesopotamian Marshes – were formerly home to as many as 500 000 Ma'dan or Marsh Arabs, a collective term referring to a rather heterogeneous group of peoples who have inhabited the marshes for more than 5000 years, perhaps since Sumerian times. They shared a common dependence on the resources of the marsh: reeds (*Phragmites australis*) were especially important for forage for domestic buffalo, for house building and for the construction of floating islands for villages. Many Marsh Arabs were displaced when the wetlands were drained and intentionally degraded over a three-decade period beginning around 1973. Initially, water was mostly diverted for irrigation, but efforts to deliberately destroy the marsh began in earnest as the region became a source of Shia insurgency, and reached a peak after uprisings in 1991 when Marsh Arabs gave sanctuary to Shia rebels fighting the regime of Saddam Hussein. Between 1973 and 2000, 9000 km^2 of wetlands became dry as a result of flow regulation in Iraq, in particular by an extensive system of canals – among them such canals as the Mother of All Battles River and the Saddam River – that diverted the flow of the Euphrates River. By the time Saddam Hussein was toppled in 2003, the Iraqi portion of the Mesopotamian Marshes had been reduced to a bare 760 km^2-remnant of the eastern Al-Hawizeh Marshes, which has an additional portion shared with Iran, while the western Al-Hammar and Central (or Al-Qurnah) Marshes had been desiccated completely. This environmental crime resulted in the destruction of almost the entirety of the largest wetland in the Middle East, with the persistence of parts of the Al-Hawizeh Marsh being attributable to the continued supply of water from the Karkheh River in Iran.

A combination of local initiatives and the commencement of a US$11 million United Nations plan (funded by Japan), led to opening of barriers along the Euphrates River in order to flood and revive habitat and thereby restore the marshes. Substantial increases in the extent of hydrophytes followed, with the area inundated growing by around ~800 km^2 annually; Al-Hammar and Al-Hawizeh Marshes were restored to about half of their former (1970s) extent, and some regrowth in the Central Marshes took place also. The recovery of marshland is, however, limited by an absolute shortage of water since both the Tigris and Euphrates are transboundary rivers, with the former receiving around 60% of its water from Turkey and 10% from Iran, while almost 90% of the flow of the Euphrates originates in Turkey, and much of the rest in Syria. Both countries, and especially Turkey, have built dams on the Euphrates since the 1990s, reducing its flow in Iraq. Iranian impoundments on the Karkheh River have reduced its discharge into Al-Hawizeh Marsh by more than half, and droughts further limit supplies in this water-scarce region. Downstream of the marshes, the Tigris and Euphrates Rivers conjoin to form the Shatt-al-Arab waterway and flow into the Persian Gulf. The low flow volume in the Shatt-al-Arab limits its potential to dilute agricultural runoff, and this pollution affects the Gulf and its inshore fisheries.

Numerous bird species were affected by declines and revival of the Mesopotamian Marshes (Birdlife International, 2015a, 2015b). Among them were the Basra reed warbler (*Acrocephalus griseldis*: Acrocephalidae; EN), Dalmatian and great white pelicans, Eurasian spoonbill, glossy ibis, various ardeids, lesser white-fronted goose (*Anser erythropus*; VU), Eurasian coot (*Fulica atra*) and numerous ducks, including species of conservation concern (e.g. marbled teal, white-headed duck and ferruginous duck), as well as an endemic race of little grebe (*Tachybaptus ruficollis iraquensis*). Al-Hawizeh is the only wetland in Iraq that holds a breeding population of African darter and African sacred ibis (*Theskiornis aethiopicus*), and was probably a refuge for wildlife, such as the Euphrates softshell turtle (*Rafetus uphraticus*: Trionychidae; EN), when the other wetlands were drained (BirdLife International, 2015c). An isolated subspecies of the smooth-coated otter (*Lutrogale perspicillata maxwelli*; VU) is endemic to the Mesopotamian Marshes and there were fears that it may have become extinct, but threats to it and the Eurasian otter are likely to have been reduced following partial revival of their habitat.

The Mesopotamian Marshes supported a fishery, formerly more than 30 000 t annually, based mostly on cyprinids. Landings included Mesopotamian bream (*Acanthobrama marmid*), *Alburnus mossulensis*, *Carasobarbus luteus*, *Carassius auratus*, common carp, *Cyprinion kais*, *Leuciscus vorax*, the 2-m long mangar and *Luciobarbus xanthopterus*, which is endemic to the Euphrates–Tigris system. Another endemic cyprinid, the binni (*Mesopotamichthys sharpeyi*; VU), was a major component of landings, and was so important that a former lake within the marshes – Umm al Binni Lake – was named after it. Hilsa shad, abu mullet (*Liza abu*: Mugilidae) and *Silurus triostegus* (Siluridae) were also exploited, and invasive *Tilapia zillii* is present (Birdlife International, 2015c).

In 2013, Iraq's Council of Ministers approved the designation of the Central Marsh as the country's first national park in the Mesopotamian Marshes of southern Iraq and agreed that a percentage of flow of the Euphrates River be set aside to maintain these wetlands. The Iraqi Ministry of Water Resources has adopted a master plan for the construction of regulators at key points to allow water to be retained in winter and released in spring, creating an artificial flood pulse. However, realization of this scheme will depend on proper water-sharing agreements for transboundary rivers in the region. A major foreseeable threat to the Mesopotamian Marshes can be expected to result from the 135-m Ilisu Dam, one of a number of impoundments under construction in Turkey, which could halve the amount of water in downstream sections of the Tigris.

Lakes of the Lower Yangtze: Diminishment and Degradation

The Yangtze floodplain is characterized by a network of more than 30 sizeable lakes and associated wetlands that formerly acted as retention reservoirs for the river, although only the two largest of them maintain their original connection with the mainstream. During the wet season (June to October), the Yangtze waters pour into these lakes raising their levels, but for much of the rest of the year more water is discharged than enters, and parts of the lake beds experience seasonal drying. Almost all of these lakes have undergone substantial reductions in area since the 1950s (Fang *et al.*, 2006), amounting to an average decline of 58% within the twentieth century (Du *et al.*, 2011), accompanied by falling fishery yields and changes in catch composition that are broadly reflective of what has occurred due to overexploitation and pollution in the Yangtze itself (see Box 3.2). Pollution has also been responsible for widespread declines in species richness and changes in the composition of lacustrine macrophyte and macroinvertebrate communities.

Shrinkage of China's largest lake, Poyang Hu, in Jiangxi Province has been such that its dry-season area has been little more than $200\,km^2$ in recent years, relative to a longer-term annual average typically more than 10 times that extent. The reduction has been variously attributed to climate change causing droughts, reduced flows in the Yangtze and influent tributaries, sand mining, and the construction of the Three Gorges Dam upstream (Feng *et al.*, 2013; Lai *et al.*, 2014; Ye *et al.*, 2014). Land reclamation (by impoldering, now proscribed by law) and sedimentation have also been historically important (Du *et al.*, 2011), but the higher rates of lake shrinkage after 2003 appear to be due to dam operations and a reduction in the forcing effect of river stage so that water flows out of the lake for longer periods each year; as a result, the inundation area of Poyang Hu has been declining by 3.3% annually (Feng *et al.*, 2013). Falling water levels have been accompanied by an increase in the coverage of wetland vegetation between 2000 and 2014, and a decrease in the extent of plant-free muddy sediments; the formerly dominant hydrophilic plants have been replaced by species adapted to drier conditions (Han *et al.*, 2018).

The macroinvertebrate benthos of Poyang Hu, which is largely made up of molluscs, has been changing too. Pearly mussels are particularly well represented (Xiong *et al.*, 2012): more than 40 species and 13 genera have been recorded, of 62 species overall in the lower Yangtze, most with highly specific habitat preferences and some apparently endangered species (such as *Solenaia carinatus*) persisting at extremely low densities (Huang *et al.*, 2013; Liu *et al.*, 2017). Poyang Hu may represent a global hotspot for these mussels (Xiong *et al.*, 2012), but recent environmental changes and long-term trends in pollution and overexploitation will have been detrimental to their diversity and abundance (see Box 2.2; see also Shu *et al.*, 2009). Although comprehensive assessments of their conservation status have yet to be undertaken, as many as 80% of unionid species in the Yangtze River could fall into threatened categories using IUCN criteria (Liu *et al.*, 2017). A comparison with historical records shows that the dominant bivalve taxa in Poyang Hu has shifted from relatively large unionids to the smaller Asian clam. This reduction in size is consistent with overexploitation (see Chapter 3), although pollution and changes in inundation patterns and water exchange with the Yangtze are also likely to have influenced unionid decline (Shu *et al.*, 2009).

Poyang Hu is a Ramsar site and an important overwintering habitat for migratory waterbirds from Siberia, among them the Oriental stork (*Ciconia boyciana*; EN), almost all of the world's population of the Siberian crane (*Grus leucogetanus*; CR), and great numbers of anatids such as the greater white-fronted goose (*Anser albifrons*) and swan goose (*A. cygnoides*; VU). Over the last two decades, geese have declined in abundance along the Yangtze to such an extent that, within China, the swan goose is now confined to Poyang Hu (Zhang *et al.*, 2015). More than 9600 dams have been constructed on the five rivers feeding into Poyang Hu, and more are planned (Birdlife International, 2016). Because it now drains more rapidly into the Yangtze during the low-water period, increasing the extent and duration of lake-bed drying, the Provincial Government has proposed measures to increase water retention that include a dam at the main outlet to Poyang Hu. The Poyang Lake Hydraulic Project is intended to ensure the maintenance of larger and more stable volumes of water during the dry season, but has received little support from scientists (Jiao, 2009): substitution of a natural wet–dry cycle with annual permanent flooding would result in substantial changes to hydrodynamics with effects on water quality (Lai *et al.*, 2016), and implications for macrophyte growth and food sources of overwintering waterbirds (Aharon-Rotman *et al.*, 2017); initial proposals for significantly increased water levels during the winter would make most or all foraging areas of Siberian crane inaccessible, triggering a population collapse (Birdlife International, 2016).

Changes in Poyang Hu have been paralleled further downstream in Hunan Province. There, Dongting Hu, China's second largest lake, remains connected to the Yangtze but has shrunk to less than half the size it was during the 1950s, with gradual reductions in mean depth and wetted area (of around 4% annually) accompanied by an increase in the duration of the low-water period – again, attributable to the Three Gorges Dam (Feng *et al.*, 2013). The lesser white-fronted goose (*Anser erythropus*; VU) has undergone considerable range contractions within China, and is now largely confined to Dongting Hu. A nature reserve and Ramsar site in the east of the lake hosts around half of the global population, and it is the only site along the Yangtze where the recessional grasslands grazed by this goose are abundant. Even here, the year-to-year availability of grassland – and hence goose body condition – depends on the vagaries of water-level fluctuations and anthropogenic influences upon them (Wang *et al.*, 2013). Variations in hydrological regime seem to affect the lesser white-fronted goose by altering food supply, making this species more sensitive to habitat change than the co-occurring bean goose (*A. fabalis*); changes in the abundance and distribution of greater white-fronted goose have occurred also but are not clearly linked to water levels (Zhang *et al.*, 2018). There have been losses of some other anatid species of national conservation from Dongting Hu, including red-breasted goose (*Branta ruficollis*; VU), as well as whooper swan (*Cygnus cygnus*) and lesser whistling duck (*Dendrocygna javanica*), while declines in unionids largely mirror those in Poyang Hu (Shu *et al.*, 2009). As in other parts of the lower Yangtze, pollution has reduced aquatic macrophyte richness in Dongting Hu, but disappearance of *Potamogeton maackianus*, a formerly dominant submerged plant, has been attributed to overstocking of phytophagous grass carp (Fang *et al.*, 2006).

In Shengjin Lake, adjacent to the Yangtze in Anhui Province, changes in inundation patterns that reduce the supply of energy-rich aquatic macrophyte tubers (principally *Vallisneria*) have been associated with declines in abundance of the hooded crane (*Grus monacha*; VU) and tundra swan (*Cygnus columbianus*); increased eutrophication may also have reduced macrophyte growth, limiting the supply of food (Fox *et al.*, 2011). Elsewhere in China, draining, impoldering and filling of Dianchi Lake and associated wetlands in Yunnan Province, plus pollution and establishment of alien species (see Chapter 4), have diminished and degraded this formerly scenic spot. Likewise, the extent of Lake Baiyangdian, which is situated in Hebei Province and is the largest lake in northern China, has been progressively reduced by land reclamation for farming, which, together with chemical and industrial pollution from nearby Baoding city, has resulted in fish kills, declines in reed beds and reduced abundance of ducks and geese. A similar situation prevails in Dianshan Lake, a major source of drinking water for Shanghai, where water quality has deteriorated and cyanobacterial blooms have occurred annually since 1985; macrophytes, which covered the whole lake in 1959, have almost disappeared (Hu *et al.*, 2014).

Pollution and other anthropogenic thumbprints are generally characteristic of Chinese lakes. While their incidence is by no means restricted to China, their intensity and prevalence is remarkable, affecting economic and recreational uses of water. The country's third largest lake, Tai Hu in Jiangsu Province, is hypereutrophic and contaminated by chemical and agricultural wastes, with frequent large-scale cyanobacterial blooms that produce toxic secondary metabolites. The contamination can be so severe as to defy treatment, leaving residents who depend on the lake – as many as 10 million people – bereft of drinking water. Macrophyte beds have diminished greatly, and marginal-zone habitats have been destroyed by land reclamation. Only 48 out of 108 fish species, and 15 of 24 families, remain in Tai Hu. The predators and migratory species are extinct, while small species such as the lake anchovy (*Coilia ectenes*: Engraulidae) and Chinese icefish (*Neosalanx tangkahkeii*: Salangidae) have become dominant (Chen & Zhu, 2008). These fishes average around 30 g in body weight, which, combined with the fact that Tai Hu has historically been overexploited, suggests that this is yet another example of fishing down the food chain (see Chapter 3). However, even these relatively tolerant fast-growing species will not be immune to the effects of worsening pollution and cyanobacterial blooms.

Endemics at Risk: Lake Baikal

Lake Baikal in Siberia is the most voluminous freshwater body on Earth, holding around one fifth of the unfrozen fresh water – considerably more than the combined volume of the Laurentian Great Lakes. It spans 4° of latitude and has a maximum depth exceeding 1.6 km. Situated in a rift valley and dating from 24 million years ago (some authorities put the age nearer 30 million), Lake Baikal is very ancient and, as a result, has more endemic species than any other lake. Estimates range from 982 to more than 1455 endemics out of around 3500 native species, but the biodiversity has not been inventoried completely, so the species totals may be considerably higher (see http://baikal.ru/en/baikal/excursion/index .html for details of some of the main taxa; Kozhov (1936, reprinted 1963) gives an encyclopedic account). This richness is well in excess of the tropical African Rift Valley lakes (Tanganyika, Victoria and Malawi), which each contain several hundred endemic species, although Lake Baikal supports far fewer fishes than its African counterparts (e.g. Box 2.6). Other Eurasian lakes, although not comparable to Lake Baikal, support remarkable endemic biodiversity, and face an array of anthropogenic threats (see Box 6.5).

The endemic fauna of Lake Baikal is distinctive, containing the world's only freshwater pinniped, the nerpa (*Pusa* [= *Phoca*] *sibirica*: Phocidae). Among other Baikalian endemics are numerous ostracods, 140 species of planarians – most notably, the 30-cm-long deep-water *Baikaloplana valida* (Dendrocoelidae) – and 160 species of oligochaetes, as well as one freshwater polychaete, *Manayunkia baicalensis*, belonging to the tube-dwelling and otherwise marine Sabellidae. There are approximately 120 endemic species of snails, among them a species flock of 25 acroloxid limpets; as with the fishes, however, the molluscan richness is lower than that of the African great lakes. Lake Baikal is especially notable for the presence of more than 300 species of Amphipoda, most of them benthic, with carnivorous *Acanthogammarus maximus* (Acanthogammaridae) reaching 7 cm in length. The majority of Baikalian amphipods occur nowhere else: they represent 45% of the species of Gammaroidea known globally and make up much of the benthos of the lake. Amphipods and other ectothermic Baikalian endemics have experienced constant temperatures for millennia, and are potentially at risk from global warming that will challenge their thermal adaptability (Jakob *et al.*, 2016).

The insects are less well represented among the endemics, but include one species of the chironomid midge genus *Sergentia*, and endemic apataniid caddisfly genera with, most conspicuously, *Baicalina bellicosa* adults emerging from the lake in vast numbers after ice-off. Lake Baikal hosts around 20 endemic sponges; the Lubomirskiidae, comprising at least 13 species (*Baicalospongia* and *Lubomirskia* spp.), is confined to the lake and contains the largest freshwater representatives of the Porifera.

Box 6.5 *Endemic Invertebrates at Risk: Lake Ohrid*

Lake Ohrid, situated in the Balkans between Albania and Macedonia, is the most species-rich lake in Europe, with remarkable endemic diversity, and the region has been a UNESCO World Heritage Site since 1979. This oligotrophic karst lake owes its richness to a combination of great age (perhaps two million years) and considerable depth (~209 m), with a well-oxygenated profundal zone spring-fed with water from underground sources. Of 1200 native species known from the lake, 586 are animals and at least 182 (around one third) of them are endemic (Albrecht & Wilke, 2008). To put this in context, Lake Baikal is almost 32 000 km^2 in extent and has at least 982 endemic species (more than any other lake), followed by the great lakes of the African Rift Valley, each with several hundred endemics and areas of ~30 000 km^2 or more. At only 358 km^2, Lake Ohrid is disproportionately rich in terms of endemics per unit area.

In contrast to African and Southeast Asian ancient lakes where vertebrates are a major component of the endemic-species compliment, most of the endemics in Lake Ohrid – as in Lake Baikal – are benthic macroinvertebrates (Albrecht & Wilke, 2008). They include a sponge genus (*Ochridaspongia*), many planarians, numerous erpobdellid leeches, ostracods and gammarid shrimps, and 56 gastropods. Some of these snails are represented by species flocks, with remarkable divergence of endemic freshwater limpets (Acroloxidae and Ancylidae), planorbids (e.g. *Carinogyraulus* spp.), valvatids (*Valvata* spp.) and hydrobiids (mainly Pyrgulinae), as well as high endemism within more prevalent gastropod genera (such as *Radix*: Lymnaeidae). Many of these endemics have ridged or complex shell armature, with a thalassoid (or 'marine-like') appearance, as occurs in molluscs of some other ancient lakes (e.g. Lake Tanganyika; see Box 2.6). There are also endemic species among the macroalgae (Charophyta) that form extensive submerged meadows within the lake. Nine out of 15 species of charophytes are endemic, but some are in decline (e.g. *Chara ohridana*: Characeae) or, in one case (*C. imperfecta*), seemingly extinct.

Lake Ohrid hosts two endemic trout. *Salmo letnica* (DD) has four distinct forms that may represent subspecies. It was formerly of considerable economic importance, but both it and the smaller *S. ohridanus* (VU) have been affected by overexploitation. Fishing of the former was banned in Macedonia from 2004 to 2014 and all fishing during the spawning period has been proscribed in Albania since 2003, but these regulations are widely flouted. Rainbow trout have also become established in Lake Ohrid and, in view of their impacts elsewhere they have been introduced (see Chapter 4), may threaten their endemic counterparts. Asian silver carp are present also. Increased urbanization and sewage discharge, as well as agricultural runoff, are increasing phosphorus levels and contributing to more eutrophic conditions that are aggravated by ongoing climatic warming (Kostoski *et al.*, 2010). These anthropogenic changes are expected to drive many endemic species in Lake Ohrid to extinction – some species may already have disappeared – and the risk is intensified because of the small size of the lake and the even smaller ranges of many of the endemics (Albrecht & Wilke, 2008). Additional challenges arise from the need for transboundary cooperation over lake management, and a paucity of conservation-related research on most taxa, although establishment of the Transboundary Biosphere Reserve Ohrid–Prespa Watershed (with EU-UNESCO support) in 2014 may help raise awareness of the need for sustainable management.

Sponges are the dominant component of the Baikalian shallow-water benthos and, historically, their filtering activity contributed to the clarity of the water. The bright-green tissue of *L. baicalensis* contains symbiotic dinoflagellates (undescribed species related to *Gymnodinium sanguineum*), and cyanobacteria have been detected in other lubomirskiids. The dinoflagellates secrete okadaic acid at low near-freezing temperatures, inducing production of polypeptides that enhance cold resistance of the sponge host (Müller *et al.*, 2007).

An estimated 60 fish species are found in Lake Baikal. At least 30 of them belong to a diverse adaptive radiation of cottoid fishes comprised of three families endemic to the lake: the Cottocomephoridae (Baikal sculpins) and Abyssocottidae (deep-water sculpins), which are mainly benthic, and the pelagic Comephoridae (golomyankas or Baikal oilfishes). The two species of golomyankas (*Comephorus baicalensis* and *C. dybowskii*) are translucent deep-water species that resemble abyssal marine fishes; together, they constitute around 95% of the pelagic fish biomass (Hampton *et al.*, 2008). The lake also contains an endemic subspecies of the more widely distributed but nonetheless endangered Siberian sturgeon: *Acipenser baeri baicalensis*.

Lake Baikal has been a UNESCO World Heritage Site since 1966. The large numbers of tourists attracted to the lake has resulted in increased pollution loads, largely from untreated sewage, and despite its huge volume, there is evidence that global warming has affected phytoplankton and their zooplankton grazers (Hampton *et al.*, 2008; see below). Plans by Mongolia to construct several dams on the Selenga River, which provides about half of the influent water to the lake, is also causing uncertainty about possible effects on water levels. The dams could also obstruct migration by breeding omul and Siberian sturgeon. The former, mentioned in Chapter 3, is the main commercial species in the lake, and is heavily overexploited; fishing of both species is banned.

Recent signs of environmental degradation of Lake Baikal are evident from proliferation of the invasive macrophyte *Elodea canadensis*, which forms huge beds along sandy shorelines where it displaces native counterparts (Kobanova *et al.*, 2016). Lake Baikal has also experienced blooms of filamentous algae such as *Stigeoclonium tenue* (Chaetophoraceae) and *Spirogyra* spp. (Zygnemataceae), and mats of these algae smother sessile macrobenthos, such as the photosymbiotic sponges. Non-native *S. fluviatilis* has blanketed extensive areas of inshore habitat, leading to episodes of hypoxia, and algal mats are washed up to rot along the shore (Timoshkin *et al.*, 2015; Kobanova *et al.*, 2016). Similar accumulations occur in Lake Michigan also (see Chapter 4), although the underlying drivers are different. The cause in Lake Baikal appears to be increased nutrients (especially phosphorus from untreated sewage) originating in settlements adjacent to the lake – a situation is aggravated by seasonal influxes of tourists. There have also been increases of planktonic Cryptophyta (mainly *Cryptomonas* spp.) and heterotrophic flagellates, as well as blooms of cyanobacteria (*Anabaena* spp.), in coastal embayments. These, together with the increase of *Spirogyra*, are signs of ongoing environmental degradation, but widespread pelagic eutrophication has not yet been reported (Izmest'eva *et al.*, 2016; Kobanova *et al.*, 2016).

Lakes in general are sentinel ecosystems of climate change (Adrian *et al.*, 2009), with changes in epilimnetic temperature and ice phenology driving shifts in distribution and abundance of planktonic and benthic organisms. Climate change effects in Lake Baikal (for details, see Box 7.2) will likely combine with eutrophication to exacerbate cyanobacterial blooms (Wiedner *et al.*, 2007), and contribute to higher loadings of industrial pollutants such as polychlorinated biphenyls and dioxins into Lake Baikal because soil warming and thawing of permafrost within the lake catchment will release stored chemicals. The Irkutsk region southwest of Lake Baikal is heavily industrialized and a likely source of such pollutants (Moore *et al.*, 2009). It seems possible that Lake Baikal is on the brink of ecological change, which may be irreversible. However, when compared to some of the other examples of lake degradation given above, there is reason to be hopeful about the persistence of Baikalian freshwater biodiversity.

Lake Nicaragua: Distinctive Biodiversity in a Threatened Ecosystem

At $8264\,km^2$ in extent, Lake Nicaragua is the largest topical lake outside Africa. It is remarkable for the occurrence of elasmobranchs that were once sufficiently abundant to support a commercial fishery. The lake was formerly well populated with bull shark. This euryhaline elasmobranch frequently swims up rivers to spend considerable time in fresh water, and, like many large river fishes that are apex predators, it was overexploited (see Chapter 3) – a consequence of being targeted for its fins. A second elasmobranch breeds in the lake: the largetooth sawfish was also the subject of a dedicated commercial fishery until it too was overexploited. It is now critically endangered throughout its global range. These two examples clearly demonstrate the effects of overexploitation of species with low reproductive rates (both are viviparous), which drives the great shrinking in body size of freshwater fishes described in Chapter 3. It also serves to reiterate the tendency for elasmobranchs in fresh water to be more vulnerable to human exploitation than their marine counterparts.

Apart from remnant populations of bull shark and largetooth sawfish, Lake Nicaragua supports at least 16 native cichlids (mainly *Cichlasoma* spp.), 11 of which are endemic to San Juan Province where the lake is situated. The cichlids are an important but overexploited fishery resource, formerly constituting more than half of landings from the lake. Their decline has been partially offset by floating cage culture of *Oreochromis* tilapias. Three species of tilapias (*O. aureus*, *O. mossambicus* and *O. niloticus*) were also stocked in the lake in the early 1980s. Their growing contribution to fishery landings has been correlated with further declines in native cichlids, attributed to introduced diseases and competition, as the tilapias spread through the lake (McKaye *et al.*, 1995). The combined effects of overfishing and invasive species on native cichlids have been exacerbated by environmental degradation: waste feed from tilapia aquaculture operations is a significant source of pollution of Lake

Nicaragua, adding to contamination of the waters by agricultural runoff and urban wastewater.

The ecology and fisheries of Lake Nicaragua are at further risk from an ambitious scheme to build a giant canal allowing ship passage between the Atlantic and Pacific coasts of Nicaragua. It was first mooted over a century ago (McKaye *et al.*, 1995), and is currently under active consideration by the Nicaraguan government in partnership with a Chinese enterprise: the Hong Kong Nicaragua Canal Development Investment Company. The canal would traverse 105 km of Lake Nicaragua, and the entire waterway would be more than three times the length of the Panama Canal. The canal would need to be 30 m deep in order to allow the passage of bulk carriers, but the average lake depth is only around half this, and the removal of vast quantities of bed sediment would be required. The consequences of this mega-project for Lake Nicaragua would thus be far reaching.

What might those consequences be? An environmental and social impact assessment of the proposed Nicaragua Canal was completed in mid-2015 and was approved by the country's Grand Canal Development Commission six months later. The report asserts that the canal construction will be safe and feasible. Scant attention is paid to impacts on the lake or Rio San Juan, nor concerns over cichlids of conservation concern or, indeed, the social impacts of the project. For example, the excavation of wet sediment to create the channel traversing the lake would involve five times more material than was moved during construction of the offshore Chek Lap Kok airport in Hong Kong, which is the largest such wet-excavation project ever conducted. Some of the dredged sediment would be used to construct three islands in Lake Nicaragua, with the rest dumped on the lake bed to form berms along the channel margins. Likely impacts of this organic-rich material during the construction phase are increased turbidity, depleted oxygen and resuspension of silt and nutrients that could degrade water quality and cause algal blooms; fisheries yields would likely be reduced, and use of the lake as a drinking-water source would be compromised. Continual resuspension of lake-bed silt by bulk carriers during the operations phase constitutes a further concern, as is the routine dredging needed to maintain the channel depth. Impacts to benthic communities and overall food-web structure (among other things) cannot be predicted with certainty, because of limited understanding of the ecology and biodiversity of Lake Nicaragua. The national government seems determined to press ahead with the canal and has issued a 50-year renewable concession for the channel construction and accompanying land expropriation. However, for the time being at least, the project is stalled because the developer is experiencing funding difficulties.

Lake Nicaragua faces perhaps the most unusual threat of any lake discussed within these pages, but it is notable that its effects are likely to play out against the broader context of ecosystem degradation due to eutrophication, land-use change, overexploitation and invasive species. And, as the examples given in this chapter show, as well as others not considered herein, such threats are frequently accompanied by reductions in influent water and declining lake levels. Furthermore, lakes are among the first ecosystems to manifest the effects of climate change (Adrian *et al.*, 2009) – a subject that is addressed more fully in Chapter 7 – and so, in many instances, lakes and the biodiversity they contain face a 'perfect storm' of anthropogenic threats. The consequences of such threats are most severe when they affect ancient lakes that support globally important concentrations of biodiversity found nowhere else. Nevertheless, degradation of almost any lake has importance for human water security and culture, irrespective of any implications for biodiversity.

7 · *How Will Climate Change Affect Freshwater Biodiversity?*

We stand now where two roads diverge. But unlike roads in Robert Frost's familiar poem, they are not equally fair. The road we have long been traveling is deceptively easy, a smooth superhighway on which we progress with great speed, but at its end lies disaster. The other fork of the road — the one 'less traveled by' — offers our last, our only chance to reach a destination that assures the preservation of our earth.

Rachel Carson (1962: p. 277), *Silent Spring*

Climate change is no longer some far off problem; it is happening here, it is happening now.

Barack Obama speaking at the World Economic Forum in Davos, November 2015.

Predicting the General Effects of Temperature Rise

The global hydrological cycle is driven by warmth from solar radiation; most freshwater animals are ectothermic or 'cold-blooded'. These two fundamental facts lead to the inescapable conclusion that freshwater biodiversity and ecosystems will be affected profoundly by global climate change, since the first of them influences the availability of surface water (when and where it will be) and hence the extent and duration of habitat for aquatic life. The second relates to the quality of that habitat: will it be too warm? Will species be able to adjust their distributions (by means of altitudinal or latitudinal shifts) so that they can remain within their evolved physiological tolerances? Or, will adaptation be sufficiently rapid to cope with rising temperatures? The possible questions are legion, because the range of outcomes under climate change are wide (Table 7.1 gives a list of examples) and it is possible that each species may respond differently. A disparity in the timing of (say) spring phytoplankton blooms and the peak abundance of Cladocera in a warming temperate lake could decouple the grazing interaction, and disrupt pelagic food

Table 7.1 *Examples of the range of potential effects of global climate change on fresh waters*

Impacts will arise from warming temperatures, shorter ice cover in boreal lakes and rivers, and long-term shifts in rainfall patterns, as well as medium-term effects such as glacial melt and increased frequency of extreme climatic events

— Higher temperatures will mean greater water use by plants (crops, pasture and natural vegetation) and thus more water abstraction for irrigation, and less water for rivers, lakes and wetlands; some perennial waters may become intermittent or seasonal, and some seasonal waters may become ephemeral or dry

— Warming will change the thermal budget of lakes, shortening periods of ice cover, and affect stratification, turnover and nutrient availability (with consequences for food webs and bentho-pelagic linkages)

— Conditions in water bodies may no longer be favourable for species that evolved there; opportunities for dispersal to suitable habitat (e.g. cooler waters at higher altitude) may be limited

— Human adaptation to a more uncertain climate is likely to encourage dam construction for water storage, flood control and hydropower, thereby magnifying impacts of flow regulation on biodiversity

— Warmer temperatures, altered flows and increased floods and droughts will interact with other threat factors (e.g. warmer temperatures may increase the toxicity of pollutants, or affect responses of phytoplankton to eutrophication), leading to great uncertainty about the severity of their combined impacts

— Increased fragmentation of river networks by dams, as well as the establishment of alien species, will limit the ability of native ectotherms to shift their ranges (e.g. upward or northward) in order to remain within their thermal envelopes

— Warmer temperatures will affect phenology and life-cycle events with temporal shifts in calling and breeding by temperate amphibians, shifts in timing of phytoplankton blooms, and so on; changes in timing of biological events could decouple interspecific relationships and disrupt food chains

— Atmospheric warming can deplete ozone, increasing ultraviolet exposure of lake plankton, especially in stratified surface waters; impacts will depend on the presence of dissolved organic carbon (DOC) which, at high levels, limits UV penetration; DOC concentrations are reduced by lake acidification, but increase when runoff (rainfall) is higher

chains as well trophic interactions and energy flow within the lake more generally.

Warmer atmospheric temperatures will ramp up the global hydrological cycle leading to periods of more intense precipitation but, in some places, blue-water availability will be reduced because terrestrial plants will need to take up and transpire more soil water under warmer conditions. Predictions of future climatic conditions include substantial increases in temperature and river discharge in far north-temperate latitudes, but declines in arid regions already suffering water stress, such as the dry tropics (e.g. Lake Chad: see Chapter 6), and parts of the temperate zone where some major European rivers are situated (Van Vliet *et al.*, 2013). The frequency of floods and droughts will increase in many parts of the world, with consequences that include increased unpredictability of hydrographs, and the duration and extent of inundation, as well as patterns of seasonality. That this is already occurring is evident from the observation that median global temperatures rose by 1.8°C between 1955 and 2011; over the same period, the proportion of the Earth's surface affected by extreme events (those exceeding three standard deviations from mean conditions) increased from 1% to 15% (Isaak *et al.*, 2012). Greater unpredictability and higher frequency of extreme events (see Ledger & Milner, 2015) could mean that formerly perennial water bodies will become intermittent or retain water for part of the year only. Such circumstances would be highly detrimental to the many freshwater species that cannot tolerate a period of drying. Vernal pools and other temporary water bodies may become too ephemeral to suit even those species with life cycles adapted to short-lived aquatic habitats. Potential sources of physical disturbance and stress on riverine species include increased scouring and washout associated with snow melt and other flood events, as well as saline intrusion caused by sea-level rise in coastal areas.

Because of the dominance of ectothermic vertebrates (and invertebrates) in fresh waters, and the physical phenomenon that dissolved oxygen concentrations decline as temperature rises, the effects of warming on the freshwater biota will surely be ubiquitous. This could be especially challenging since the metabolic rates of – for instance – fishes scale directly with water temperature, thus they will require more oxygen under warmer conditions where it is less available, with broader implications for overall metabolism (Sheridan & Bickford, 2011). Conversely, growth and production by those ectotherms that are currently temperature limited may increase under moderate warming. There is also good evidence that sensitivity to pollutants is temperature related, as chemical toxicity is exacerbated when conditions are warmer than the optimal temperature tolerated by particular freshwater ectotherms (Wang *et al.*, 2019). However, in many fresh waters, any impacts rising from changed flow and inundation patterns may be inseparable from those of temperature. For that reason, it has been argued that understanding of the responses of freshwater ecosystems to climate change is too incomplete to permit broad-scale generalizations about potential impacts on biodiversity (e.g. Wilby *et al.*, 2010).

Vulnerability to climate change will be increased because many species living in isolated inland-water bodies (or rivers fragmented by dams) have limited abilities to disperse as the environment changes (Woodward *et al.*, 2010); again, fishes are an obvious example, and they are far more constrained than most terrestrial vertebrates. Warming of streams in the United States has raised concerns over reductions in thermally suitable habitat for salmonids, and may compromise efforts to rehabilitate populations that are already under stress (Isaak *et al.*, 2012). Nonetheless, in certain circumstances, freshwater animals can adapt to warming. Directional selection in anadromous pink salmon (*Oncorhynchus gorbuscha*) has resulted in upstream migrations in Alaska beginning two weeks earlier than such movements 40 years previously. The mechanism appears to involve loss of the gene associated with later migration, and has taken place within 17 generations (Kovach *et al.*, 2012).

A second example of adaptation is increased tolerance of high temperatures by European *Daphnia magna* – occurring, once more, over a 40-year period (Geerts *et al.*, 2015). *Daphnia* reproduce asexually under good conditions, but undergo sexual reproduction under conditions of stress (e.g. high temperatures), producing dormant embryos enclosed within ephippia that can remain buried in benthic sediments for decades, constituting an egg bank. Ephippia raised from core samples taken in a shallow English lake were used to compare *D. magna* raised from recent (1995–2005) eggs and those hatched from eggs that dated back 40 years (1955–1965), allowing reconstruction of the evolutionary changes that might have occurred over that period, when lake temperatures had risen 1.15°C and the frequency of heatwaves had increased. Individuals reared from recent samples of the egg bank were more tolerant of heat stress, maintaining activity at a higher maximum temperature than those hatched from the older eggs. Not only had these cladocerans adapted to ongoing climate warming, but they may also have the genetic capacity to evolve even greater tolerance to future warming (see Geerts *et al.*, 2015).

A similar approach extending over a longer time span, covering centuries rather than decades, involved hatching ephippia of *Daphnia pulicaria* recovered in cores taken from the bed of a Minnesota lake, allowing resurrection of natural subpopulations from past environments. These 'physiological fossils' were then used in experiments that tested their response to environmental variables (Yousey *et al.*, 2018). Temperatures experienced by the oldest genotypes dating back to the 14th century were considerably cooler (by perhaps as much as 10°C) than those to which their modern descendants had been exposed. Thermal-shock experiments revealed that more recent genotypes exhibited higher thermal tolerance than older subpopulations, suggesting the role of temperature as a selective agent. The two *Daphnia* examples, and that of the pink salmon, show that provided there is sufficient genetic variability, some temperate species can adapt sufficiently quickly to cope with rising freshwater temperatures that occur at decadal scales as well as those that take place more gradually.

Winners and Losers

There has been insufficient research on the implications of climate change for freshwater biodiversity, especially in the tropics, and the potential for adaptation to warmer temperatures is unknown for the vast majority of species. The best we can do is make extrapolations from the studies of temperate species, especially 'cold-blooded' ectothermic animals as well as plants and algae that all have temperature-sensitive metabolic rates. That might allow identification of 'winners' – species that may thrive under the changed conditions, particularly those that have metabolic rates that are currently limited by temperature – and 'losers' – those that fail to adjust and perish. Such extrapolation could, however, prove misleading for tropical ectotherms since they may already be close to their upper tolerance limits (Deutsch *et al.*, 2008; Kellermann *et al.*, 2012; Sunday *et al.*, 2014), so the impacts of any further warming could be considerable. The inverse relationship between temperature during growth and body size in amphibians and many aquatic invertebrates will lead to smaller size at metamorphosis, plus decreased body mass due to increased metabolism and hence reduced adult fitness (e.g. Bickford *et al.*, 2010); the same relationship applies to fishes (Daufresne *et al.*, 2009; Sheridan & Bickford, 2011). Further demographic consequences are possible in aquatic reptiles that have environmentally determined sex ratios (reviewed by Pezaro *et al.*, 2016; see below). One prediction that seems likely to be robust is that the species most vulnerable to climate change will be those that are highly specialized, with complex life histories, restricted ranges or limited distributions, or highly specific habitat requirements. The pearly mussels discussed in Box 2.2 have most, if not all, of these attributes, and are sure to be placed at further risk by climate change. Niche specialization for particular fish hosts by these mussels in North America is already driving their decline as changes in fish distribution decouple host–affiliate relationships, and this process will continue as climate change proceeds (Spooner *et al.*, 2011). Inevitably, pearly mussels – which are known to be highly temperature sensitive (Lopez-Lima *et al.*, 2017) – will be climate-change losers.

Climate-change winners will be species that are generalist in their habits and habitat requirements, and have short generation times that will increase the possibility of rapid adaptation to changed conditions. But there may be other options allowing persistence. If, for the purposes of simplification, we assume that climate change only affects median water temperature of rivers, one option for species that lack the evolutionary capacity to adapt to rising temperatures (or cannot do so quickly enough) is to shift their distribution. For instance, animals in rivers could, conceivably, compensate for rising water temperatures by moving upstream to higher – and cooler – elevations or latitudes. This could be especially important for species in the tropics that are already close to their upper thermal tolerances and might be feasible for (say) fishes in north-to-south flowing rivers, such as occur in the United States, although such movements would be subject to limitations imposed by river topography, the presence of dams or other in-stream barriers, availability of suitable habitats upstream, or some combination of these. Compensatory movements north or south by fully aquatic animals are not possible where drainage basins are oriented east–west as in, for example, China. And even flying adults of aquatic insects or amphibians that can travel over land might find their dispersal opportunities limited in human-dominated environments. While some range adjustment may be possible, the extent of movements needed to compensate for the upper bounds of the range of temperature rises predicted for the next century seem insurmountable for most freshwater species (for example, Bickford *et al.*, 2010), and any shifts in distribution will be hostage to vagaries of local geography such as the presence of sea barriers, absence of nearby highlands or drainages that extend across latitudes.

Climate-Change Interactions with Other Threats to Freshwater Biodiversity

Whatever the direct effects of climate change on freshwater ecosystems, human responses to the unpredictability inherent in such change could give rise to indirect impacts on biodiversity that will be as strong as or even greater than the direct effects. One such indirect consequence is likely to become pervasive as more water is used for irrigation to offset reduced precipitation. Thus, by the 2050s, up to 80% of African freshwater fish biodiversity will experience hydrological conditions substantially different from the present (Thieme *et al.*, 2010). Many systems are already exposed to numerous anthropogenic stressors, and climatic drivers will interact with land-use changes and nutrient loading in future (e.g. Woodward *et al.*, 2010; Moss *et al.*, 2011; Mantyka-Pringle et al., 2014). There is experimental evidence (from amphibians) that even moderate climate change can cause mortality and population declines under scenarios that involve multiple stressors (Rohr & Palmer, 2013). Warming waters and rising acidity combine to threaten populations of cold-water stenotherms such as brook trout (*Salvelinus* spp.) by reducing the extent of suitable habitat. Their combined effects progressively narrow fish ranges because warming typically acts at lower elevations while acidity is a major problem higher in drainages where waters are typically less well buffered. Projections from bioclimatic models suggest that distributional shifts by brook trout and sculpins (*Cottus* spp.: Cottidae) to colder, higher-elevation streams in order to compensate for warming will be constrained by acidification of headwater streams (McDonnell *et al.*, 2015).

By and large, the impacts of climate change on fishes will likely be most apparent in water bodies already subject to eutrophication, poor water quality or overexploitation (e.g. Jeppesen *et al.*, 2012). Warming magnifies the toxicity effects of some contaminants on a wide range of freshwater animals (e.g. Dinh Van *et al.*, 2013; Moe *et al.*, 2013; Wang *et al.*, 2019), although the direction of this interaction can be inconsistent and, occasionally, ameliorative (Rohr *et al.*, 2011). Further hazard arises from climatic shifts that allow the establishment of alien species and the proliferation of parasites (e.g. MacNab & Barber, 2012). Furthermore, clearing riparian forest amplifies the effects of warming in streams, creating conditions that suit invasive piscivores to the detriment of stenothermic salmonids (Kuehne *et al.*, 2012; Lawrence *et al.*, 2014). The likely facilitating effect of climate change on the spread and impact of aliens will be considered in more detail later in this chapter (see also Box 7.7). Finally, it must be reiterated that anthropogenic alteration of freshwater wetlands tends to enhance the rates at which they emit carbon (Nahlik & Fennessy, 2016), as exemplified by the peatswamp forest of Southeast Asia (see Box 2.7), transforming them from sinks into sources of the greenhouse gases responsible for global warming.

Since climate change will create or exacerbate water-supply shortages and cause floods that threaten human life and property, it will surely encourage hard-path engineering solutions to mitigate these problems (Palmer *et al.*, 2008), including new dams, dredging, levees and water diversions to enhance water security for people and agriculture, so altering flow and inundation patterns in ways that will not augur well for biodiversity. To this must be added the increasing impetus to install new hydropower facilities along rivers to reduce dependence on fossil fuels and meet growing global energy needs (Zarfl *et al.*, 2014; Winemiller *et al.*, 2016). These engineering responses magnify the direct impacts of climate change because they limit the natural resilience of ecosystems: for instance, by restricting the ability of animals to make compensatory movements to cooler conditions. A related problem is that hard-path solutions initiated in response to disasters, such as severe floods caused by extreme rainfall, may be allowed to circumvent environmental reviews and regulations because of the urgent need for project implementation. The effects of dams already built on the global water system has been profound: dams retain more than $10\,000\,\text{km}^3$ of water, the equivalent of five times the volume of the Earth's rivers; the associated reservoirs trap around one third of the total sediment load that formerly reached the oceans (Chapter 1). Offsetting some of the effects of dams will require that their operation be adjusted to ensure allocation of sufficient water – e-flows – to sustain ecosystems and biodiversity downstream (see Chapter 5). For instance, hypolimnetic releases of cool water from impoundments could create thermal refuges for cold-water stenothermic fishes (Matthews *et al.*, 2015), and although their implementation may be limited to a few specific circumstances, there are other options for 'climate sustainable water resource management' (*sensu* Matthews *et al.*, 2011). For instance, restoration of riparian habitats can reduce nutrient loads, slow terrestrial runoff and provide shade, thereby mitigating the combined impacts of climate change and land-use transformation on the freshwater biota (Mantyka-Pringle *et al.*, 2014).

Making precise – or even rather general – predictions of how ecosystems and communities will respond to climate change in the context of multiple stressors is inherently difficult given that they will occur in the context of a 'no-analogue' future (*sensu* Strayer, 2010) where novel combinations of species will be interacting under conditions far beyond those experienced during their evolutionary histories. In many instances, the physical architecture of the water bodies within which interactions take place will be far from those typical of the pre-Anthropocene world. One possibility is that fish assemblages will become more homogeneous, because disturbance favours communities with a combination of common traits (Buisson *et al.*, 2013) that will tend to include widely distributed alien species.

In this chapter, I outline the physical shifts that are happening in fresh waters in response to climate change, describe the types of biological responses to these shifts and consider their significance for biodiversity. This account is necessarily selective since the published literature on climate change and its effects (especially those that can be modelled or predicted) has expanded exponentially in recent years. Even for fresh waters, it has become too unwieldy to parse adequately, as the density of citations herein reflects. The primary focus will be upon shifts or responses that have been measured and documented, including local extinctions (for instance, see Box 7.1), rather than those that can be

---

Box 7.1 *Is There a Global Footprint of Climate Change?*

While it is possible to envisage doomsday scenarios of fresh waters under climate change, what is really happening now? What effects on biodiversity have been unequivocally demonstrated and can be supported by published work in primary journals? An early literature review of ecological and evolutionary responses to recent climate change (Parmesan, 2006) had little to say about fresh waters, but drew attention to the loss of thermal habitat of some salmonids and mountaintop frogs. A decade later, the 'broad footprint' of climate change (Scheffers *et al.*, 2016) was clearly evident in fresh waters, with effects on almost three quarters (23 out of 31) of ecological processes investigated at different levels of biological organization. These included changes manifest at the levels of the organism (genetics, physiology, morphology), population (phenology, dynamics), species (distribution) and community (productivity, composition, interspecific relationships). Each of the 23 processes where an effect was seen had been documented in a least one case study. Surprisingly, given the expectation that freshwater species would be relatively susceptible to climate impacts, the footprint in marine (81% of processes affected) and terrestrial realms (91%) was heavier. However, the review only considered field-based case studies that reported changes in the processes through time, and if an effect was missing for some processes, it would be more likely to reflect data deficiencies than the absence of any response to climate. Nevertheless, a pervasive influence of climate change throughout the biological hierarchy had occurred with just 1°C of average atmospheric warming since preindustrial times (Scheffers et al., 2016).

Another approach is to use data on range shifts to calculate the frequency of local extinctions related to recent climate change. These have occurred already in more than 450 species, or 47% of 976 species surveyed (Wiens, 2016). The frequency of local extinctions was significantly higher in fresh waters relative to terrestrial and marine habitats (74% versus 46% and 51%), and more common in animals (50% versus 39% in plants) and among tropical than temperate species (55% versus 39%). While we might thereby extrapolate that tropical freshwater animals are especially vulnerable to climate change, the freshwater species for which range-shift data were available were temperate, and none of the fishes was tropical. For fishes alone, local extinctions were significantly more common in fresh water. Marine fishes have a greater capacity to adjust the temperatures that they experience by vertical or horizontal movement within the water column and are not vulnerable to the same landscape barriers that constrain range shifts by fishes limited to inland waters. Many will be unable to undergo range adjustments quickly enough to prevent local extinction from climate change. This is a matter of particular concern in the tropics, where ectotherms seems to have small thermal safety margins (Deutsch et al., 2008), and are unlikely to be able to adapt to warming (Kellermann et al., 2012).

modelled and predicted, since the latter are highly sensitive to assumptions about future rates of carbon emissions and are subject to a host of biological uncertainties. Nevertheless, some of these projections can be informative, especially where they have been backed up by experimental studies.

Lakes As Sentinel Ecosystems

There is no doubt that fresh waters are warming, and warming rapidly. Lakes, in particular, are sentinel ecosystems of climate change, because they are sensitive to temperature, respond rapidly and integrate information about changes in the catchment (Adrian et al., 2009). Climate-related signals, such as water-level fluctuation, and shifts in the timing of ice cover, are visible and easily measured even in the largest lakes, but other changes can be detected through palaeolimnological examination of sediment cores. In a worldwide synthesis of *in situ* and satellite-derived data from 235 lakes, O'Reilly *et al.* (2015) found that summer temperatures of surface waters rose rapidly between 1985 and 2009 (global mean = 0.34°C per decade). The signal was especially conspicuous at higher latitudes, suggesting that biophysical changes in global freshwater resources are already under way. For individual lakes, air and lake temperature trends often diverged, reflecting the variety of factors controlling lake heat budgets, and water temperatures did not simply track air temperatures. The most rapidly warming lakes were geographically widespread, with warming dependent on combinations of climate and local characteristics instead of, as might be expected, lake location. However, overall, ice-free lakes were warming more slowly, frequently at rates similar to air temperature, than lakes with seasonal ice cover. Surface waters of large deep lakes warmed more rapidly than air, while small, shallow lakes tracked rising air temperature more closely. Lakes with high rates of surface temperature change may be susceptible to major ecosystem alteration, because greater warming of the epilimnion will increase thermal stratification. Nonetheless, even lakes with lower rates of warming may be at risk of major ecological change, particularly if the initial water temperatures are already near the physiological maxima of key taxa (O'Reilly *et al.*, 2015).

Lake Baikal is an outstanding example of a climate sentinel, with progressive and significant increases in mean surface temperature and the ice-free period, and associated reductions in ice thickness over the past few decades, which have been linked to changes in phytoplankton and zooplankton (Hampton *et al.*, 2008; see Box 7.2**)**. Baikalian summer phytoplankton abundance is positively correlated with the long-term warming trend, whereas in Lake Tanganyika – another large-volume, ancient lake – warmer water temperatures over much of the last century enhanced water column stability and reduced vertical mixing of nutrients, thereby limiting primary production and algal biomass (O'Reilly *et al.*, 2003; Verburg *et al.*, 2003). The consequences of this warming, which include a decline in lake fisheries, are set out in Box 7.3. Epilimnetic warming of Lake Baikal is increasing temperature differences between surface and deeper waters, which may reduce mixing and restrict nutrient availability for phytoplankton. There are no signs of such limitation yet, because a substantial temperature difference between surface and deep water would be required to impede mixing in Lake

Box 7.2 *Lake Baikal on the Brink?*

There are clear signals of climate warming in Lake Baikal. These are of particular significance since, due to its tremendous volume, the lake might have been expected to be relatively resistant to climate change. During the last several decades, average water temperature has risen (1.21°C over a 60-year period since 1946), and biomass of phytoplankton (300% since 1979) as well as populations of grazing Cladocera (335% since 1946) have increased. Statistical modelling of these data shows that Cladocera have responded strongly to temperature, which has risen in Lake Baikal at a rate well in excess of the global average, but their abundance was not correlated with increased phytoplankton (Hampton *et al.*, 2008). Warming surface temperatures during the period of summer stratification of Lake Baikal (between 1955 and 2000) led to local strengthening of the thermal gradient and reduced mixing within the top 50 m of the water column, with the result that diatoms were found deeper in the epilimnion whereas most zooplankton shifted to shallower depths, perhaps to feed on picoplankton (Hampton *et al.*, 2014). Reduced mixing and changes in the vertical distribution of plankton seem certain to have implications for pelagic food webs and nutrient cycling in Lake Baikal. A shorter time series (1977–2003) showed that, lake-wide, surface waters warmed 2.0°C, and were associated with increased abundance of Cladocera and the cyclopoid copepod, *Cyclops kolensis* (Izmest'eva *et al.*, 2016). The endemic calanoid copepod, *Epischura baikalensis*, a cold-water stenotherm that typically comprises 90% of the zooplankton biomass in the lake and is a putative keystone species, has yet to show a numerical response to warming (Izmest'eva *et al.*, 2016), although its depth distribution has shifted upward (Hampton *et al.*, 2014).

As Lake Baikal has warmed, the annual ice-free period of around six months has lengthened: earlier spring ice-off in Lake Baikal threatens recruitment of nerpa, both through declines in adult fertility and reduced rearing success of seal pups (Moore *et al.*, 2009). The shorter duration of ice cover will not only affect seals. Ice is an important positive driver of productivity in Baikal. The spring algal bloom typical of most temperate lakes occurs under the ice, which is kept free of snow by strong winds, and there is sufficient light penetration to sustain the bloom. Large endemic diatoms (e.g. *Aulacoseira baicalensis*) that occupy interstitial spaces within the ice are an important component of the bloom, forming long (>10 cm) filaments that hang from the ice into the water beneath. After ice-off, these diatoms sink to the lake bed where they are eaten by benthic amphipods and gastropods (Hampton *et al.*, 2008; Moore *et al.*, 2009).

Summer warming of the surface waters of Lake Baikal has allowed temperatures in deeper waters (below 25 m) to increase significantly during the fall and winter. Such long-term warming may jeopardize endemic, cold-water stenotherms unable to adapt physiologically or behaviourally (Hampton *et al.*, 2008; Jakob *et al.*, 2016). Among them are the two species of deep-water golomyanka fishes (see Chapter 6) and the pelagic amphipod *Macrohectopus branickii* (Macrohectopidae), which has a filiform appearance and is ecologically similar to a mysid shrimp (see Chapter 4). These zooplanktivores migrate vertically at night, ascending from the depths below 150 m into the top 50 m to feed on copepods. If they avoid the warmer upper waters, their food intake will be reduced. Nerpa that feed on golomyankas will need to dive deeper for prey, with consequences for seal energetics and growth.

Climate change and warmer water in Lake Baikal will most likely favour small cosmopolitan phytoplankton and tiny picoplankton, as well as cyanobacteria (see Wiedner *et al.*, 2007), over the larger cold-stenothermic endemic diatoms. Such a transition is already evident from phytoplankton observations between 1951 and 2000 (Izmest'eva *et al.*, 2011, 2016) and, as it progresses, will have profound repercussions on pelagic and benthic food webs (see also Hampton *et al.*, 2014), although the ramifications for consumers at higher trophic levels remain uncertain. Endemic Baikalian ectotherms will be challenged by climate warming because, in the case of *Eulimnogammarus* amphipods for example, oxygen supply becomes limiting at higher temperatures, and this correlates with observed shifts into deeper water and reductions in the abundance of *E. verrucosus* along the shallow marginal zone (Jakob *et al.*, 2016). The retreat of native gammarids could provide an opportunity for establishment of *Gammarus lacustris*; this ubiquitous Holarctic species inhabits various shallow-water bodies in proximity to Lake Baikal, and its invasion would be facilitated by a combination of warming and ongoing coastal eutrophication.

Box 7.3 *Conspicuous Effects of Climate Change on Fisheries in Lake Tanganyika*

Lake Tanganyika, the second deepest (1470 m) lake in the world and (excluding the Caspian Sea) the second largest by volume, is hyperdiverse in terms of its fishes as well as other fauna (see Box 2.6). The lake historically supported a highly productive pelagic fishery (O'Reilly *et al.*, 2003); it currently provides 25–40% of the animal protein supply for the populations of the four riparian nations. A rise in surface-water temperature of around 1.3°C attributable to climatic warming since the beginning of the twentieth century has increased the stability of the water column (Verberg & Hecky, 2009). A regional reduction in wind velocity over the same period has been associated with reduced mixing of lake waters, decreasing nutrient upwelling from deeper waters into the euphotic zone, and limiting growth of planktonic diatoms. Less phytoplankton means fewer zooplankton, and a reduction in food supply for the sprat or dagga (*Stolothrissa tanganyicae*: Clupeidae) that is the basis of a major fishery in Lake Tanganyika; the confamilial 'sardine' (*Limnothrissa miodon*) is also an important component of landings. Carbon isotope records in sediment cores suggest that primary productivity may have decreased by about 20% over the last century, implying a roughly 30% decrease in fish yields (O'Reilly *et al.*, 2003). The regional effects of global climate change seem to be having a major influence on provisioning ecosystem services provided by Lake Tanganyika, and may exceed any impacts attributable to local anthropogenic activity (pollution, sedimentation) or overfishing. The importance of this finding must be put in context: a human population of 10 million (growing at ~2.5% annually) live in the catchment of the lake, and diminishing fish catches due to climate change will result in protein shortages and reduced food security.

Others dispute the conclusion that climatic signal is so conspicuous in view of the multiple effects that people may have on the lake. Rates of land-use change in the catchment are rapid, and fishers from all four riparian nations exploit Lake Tanganyika so that the total catch effort is not known, contributing to overexploitation of large, long-lived endemic *Lates* species. However, a detailed analysis of lake temperature and stratification over a 1500-year period revealed that warmer surface temperatures were associated with increased lake stratification, preventing nutrient recharge from deep water and limiting primary productivity (Tierney *et al.*, 2010). Both remote-sensing estimates and modelling studies indicate that photosynthetic production in Lake Tanganyika has declined as warming has increased in recent decades (Bergamino *et al.*, 2010), and confirm that high temperature and low wind stress reduce phytoplankton biomass (Naithani *et al.*, 2011). Lake transparency has increased also, because reduced vertical mixing results in lower phytoplankton densities and greater water clarity; the same phenomenon has been seen in Lake Malawi, another deep tropical lake affected by warming (Verburg & Hecky, 2009). Warming of Lake Tanganyika in the past few decades has exceeded previous natural variability, and the corresponding decrease in primary productivity (~15%) is likely to have knock-on effects on the fishery (Tierny *et al.*, 2010). The physical changes of warming and mixing, rather than increases in fishing pressure, may well be responsible for lower fish biomass and observed declines in catch per unit effort. Changes in algal production are rapidly reflected in the abundance of dagga, because it is short-lived (typically less than one year) with massive annual recruitment, so stocks closely track food availability. These life-history attributes would also tend to increase the resilience of dagga populations to intensive exploitation.

Using palaeoecological records of fish bones and scales, gastropods, and ostracods from the same 1500-year period as the earlier study of temperature change in Lake Tanganyika, Cohen AS *et al.* (2016) showed that declines in commercially important fishes (mainly dagga plus *Lates* spp.) and endemic molluscs have accompanied lake warming over the last five centuries, as demonstrated by a negative correlation between lake temperature and fish and snail fossils. Ongoing declines in fishery species began well before the advent of commercial fishing in the mid-twentieth century, and sustained warming during the last ~150 years has affected the biota by strengthening stratification of the water column and reducing the depth of the epilimnion. Much of the lake volume (and the hypolimnion) has long been free of oxygen at depths below 100–200 m, with the anoxic zone tending to be deeper in the south. Reductions in lake mixing due to warming have not only depressed phytoplankton production but have reduced the extent of the oxygenated benthic habitat – by almost 40% in some parts of the lake. As a result, the abundance of deep-water snails such as *Tiphobia horei* and

Tomichia gulleimei (Paludomidae) has decreased substantially (Cohen AS *et al.*, 2016). Continued warming will exacerbate stratification and can be expected to increase the extent of benthic habitat loss, particularly affecting deep-water specialist fishes and invertebrates, and could ultimately impact the many species that typically inhabit lake margins (Vadeboncoeur *et al.*, 2011).

Despite a general negative correlation with lake temperature, the abundance of dagga and their predators have undergone a series of natural boom-and-bust cycles; these took place well before the advent of commercial fishing in the mid-twentieth century. The highest abundance of fishes occurred in the seventeenth and early eighteenth century, when Lake Tanganyika temperatures were low and phytoplankton production was high (Cohen AS *et al.*, 2016). Whatever overfishing has occurred (as has affected predatory *Lates* spp.), this pressure is operating in the broader context of warming-induced shifts in ecosystem production that limit the overall biomass of pelagic fishes. Successful management of this fishery will need to account for the diminished food base of dagga and co-occurring species, as it makes them highly susceptible to overexploitation. Climate-change effects on the productivity of Lake Tanganyika, and a loss of deep-water habitat, is likely to have consequences for species persistence – at least for profundal specialists – and changes to the pelagic food web is a matter for concern, particularly because of its implications for fishery yields.

Baikal. Water column stability increases less sharply with temperature in cooler waters than in the warm (>20°C) waters typical of tropical lakes such as Tanganyika (Hampton *et al.*, 2008). The different responses of these two ancient lakes to climatic forcing is a good example of the dependence of response variables to the local or geographical context of the sentinel lake.

What Physical Shifts Have Taken Place in Streams and Rivers?

Signs of climate change in running waters include a consistent trend in stream warming across the United States of 0.2°C per decade since 1980

(Isaak *et al.*, 2012), which is about 60% of that reported for lakes at the global scale over almost the same period (0.34°C; O'Reilly *et al.*, 2015). Higher air temperature was the dominant factor accounting for both long-term trends and inter-annual variability in stream temperature, but summer discharge also had an influence. Many streams in the United States are exhibiting a coherent response to climate forcing, despite some serious deficiencies in the stream temperature monitoring record (Isaak *et al.*, 2012), and half of 40 major rivers experienced significant warming trends over the 50–100 years prior to 2010 (Kaushal *et al.*, 2010). The effects of such warming are exacerbated by land-use change and riparian clearance (Lawrence *et al.*, 2014). At the global scale, a direct carbon dioxide signal has been detected in continental runoff, which increased overall throughout the twentieth century despite intensifying human water consumption (Gedney *et al.*, 2006). This trend was consistent with a suppression of plant transpiration due stomatal closure induced by elevated atmospheric carbon dioxide concentrations. A subsequent spatial analysis deployed a model of global vegetation and hydrology to quantify the contributions of shifts in climate, carbon dioxide levels and land use to changes in river discharge (Q) during the twentieth century (Gerten *et al.*, 2008). Q declined in some regions, such as central and southern Asia, but increased in others, especially in parts of North America and western Asia.

On the global scale, however, Q rose almost 8% between 1901 and 2002, which is broadly consistent with the findings of Gedney *et al.* (2006), although the cause was attributed primarily to increasing precipitation. Elevated carbon dioxide concentrations and land-use changes (mainly deforestation) also contributed to higher Q, while global warming (which could have increased evapotranspiration) and irrigation were associated with discernible reductions in discharge, although the effect of the latter was minor (Gerten *et al.*, 2008). The contrasting effects of carbon dioxide concentrations and warming are due to the fact that evapotranspiration is reduced as carbon dioxide becomes more available (because plants can keep their stomata open for shorter periods), whereas plants use more water as temperatures warm. Despite rising carbon dioxide, however, higher evapotranspiration due to global warming could lead to a 6% reduction in Q by 2100 (Gerten *et al.*, 2008). While there is both spatial and temporal variability within the original data, and in projections of the relative influences of temperature, carbon dioxide and land use on Q, the anthropogenic influence on the global climate and water cycle is clear, and it will grow in future. Whether flows of particular rivers will increase or decrease will depend upon the climate of their catchments. However, streams will be more susceptible than larger rivers to low flows and spates associated with changes in rainfall patterns and earlier snowmelt, simply because of local affects arising from the smaller size of their drainage basins. In addition, small streams will more closely track air temperatures than large rivers, making their biota vulnerable to short-term extremes such as heat waves.

The extent and pervasiveness of changes in Q and in the frequency of extreme events have led Milly *et al.* (2008) to declaim that, with respect to hydrology and water management, stationarity is dead. Floods that formerly had a return time of one hundred years can be expected to occur more frequently than once in a century, and extreme flow events that might have taken place, on average, once in a decade may recur twice or more within a 10-year period. This non-stationarity exposes the riverine biota to hydrological regimes that differ for those to which they were adapted (or, more strictly, those to which they were *ab*apted from), and the imperative to find engineering solutions to increase predictability and restore stationarity for humans (see Palmer *et al.*, 2008) will, as mentioned above, create novel flow environments with conditions that vary greatly from those in the rivers of the pre-Anthropocene world.

Impacts on Phenology, Life Histories and Population Dynamics

Global warming has affected lake phytoplankton populations through changes in the onset and timing of thermal stability or stratification, and earlier ice-break up, with advances in the start or shifts in the duration of spring-time algal blooms attributed to elevated temperatures (e.g. Moss *et al.*, 2011; Vadadi-Fülöp *et al.*, 2012; de Senerpont Domis *et al.*, 2013; Rühland *et al.*, 2015). Such temperature-driven increases in diatom growth can lead to an earlier onset of silica limitation in temperate lakes (Meis *et al.*, 2009), and climate change has been implicated as the main cause of the proliferation of small planktonic diatoms that has been reported in many relatively oligotrophic lakes (Rühland *et al.*, 2015). However, the effects of warming on phytoplankton phenology and dynamics will, to a large extent, be latitude- or system-specific (de Senerpont Domis *et al.*, 2013), as the examples of Lakes Baikal and Tanganyika show clearly (Boxes 7.2 and 7.3). Variations in lake physical properties that influence phytoplankton populations will, in turn, have a substantial influence on the phenology and population dynamics of zooplankton (Vadadi–Fülöp *et al.*, 2012). Warming will tend to amplify the consequences of nutrient loading for phytoplankton and their grazers, and this indirect influence could be more influential than the direct effect of temperature on plankton phenology or growth (Moss *et al.*, 2011; de Senerpont Domis *et al.*, 2013; Jeppesen *et al.*, 2014), with implications for water quality, humans and biodiversity (see Box 7.4).

Turning to another, relatively large-bodied, component of the freshwater biota, a 30-year study of the timing of amphibian arrival at ponds in South Carolina revealed that all 10 species altered their reproductive period in response to the combined effects of temperature and rainfall, with some autumn-breeding species tending to arrive later and winter species earlier; shifts in phenology ranged from six to 37 days per decade (Todd *et al.*, 2010). Likewise, the timing of peak calling periods in a Canadian anuran assemblage shifted in response to warming over a 14-year period, and began earlier among spring-breeding species as temperatures increased; late-breeding species showed no change (Walpole *et al.*, 2012). Earlier spring calling associated with rising temperatures has also been reported for the European common frog, *Rana temporaria*, in the United Kingdom (Phillimore *et al.*, 2010). In both this species and in the Canadian study, earlier calling in spring has been predicted to increase the duration of the breeding period by around one third by 2100. Spawning dates of the European common frog have been reported to shift as an adaptation to local temperature differences across the United Kingdom, and also show plasticity among years, such that they breed earlier in warmer years. Despite these observations, the current capacity for temporal flexibility will be insufficient to adjust breeding dates to compensate for projected temperature increases, and a substantial micro-evolutionary challenge will need to be overcome before the common frog can advance the onset of breeding sufficiently to adapt to warming conditions (Phillimore *et al.*, 2010).

Higher temperatures in marginal habitats have the potential to benefit montane frogs, as shown by a nine-year demographic study of the Columbia spotted frog (*Rana luteiventris*) in Montana: reductions in winter severity correlated with increased breeding probability and greater survival, thereby promoting population viability (McCaffery & Maxell, 2010). These and other alpine or boreal ectotherms at or near their thermal lower limits are expected to benefit from the milder winters provided by a warming climate so long as habitats remain intact (Al-Chokhachy *et al.*, 2013). For instance, growth and production of

Box 7.4 *Interactive Effects of Warming and Nutrients on Lake Phytoplankton*

Empirical space-for-time studies (e.g. Meerhoff *et al.*, 2012, Jeppesen *et al.*, 2014) indicate that the combined effects of increased nutrients and warmer temperatures will have impacts on lake water quality, as warming amplifies the indirect consequences of nutrient loading for plankton. For example, higher temperatures enhance rates of phosphorus release from lake-bed sediments which, together with changed evapotranspiration and runoff, exacerbate eutrophication by increasing phytoplankton abundance and growth rates, as well as the relative proportions of cyanobacteria, especially nitrogen-fixing forms (Moss *et al.*, 2011; Kosten *et al.*, 2012; de Senerpont Domis *et al.*, 2013). The resultant blooms can constrain water use if they involve toxin-producing cyanobacteria (Trolle *et al.*, 2015). Intensification of blooms, which arise from the ability of cyanobacteria to make use of nutrient pools that are usually unavailable to other phytoplankton, has been widely reported as a symptom of climate change (e.g. Wiedner *et al.*, 2007; Kosten *et al.*, 2012; Cottingham *et al.*, 2015). Furthermore, the dominance of cyanobacteria is not due to their higher growth rates under warmer conditions but, instead, to their ability to migrate vertically in strongly stratified waters and their resistance to grazing, especially when elevated temperatures reduce zooplankton body size (Daufresne *et al.*, 2009; Lürling *et al.*, 2013).

As global warming continues, nutrient concentrations in lakes may have to be brought down substantially from present values if cyanobacterial dominance is to be prevented (Kosten *et al.*, 2012; Olrik *et al.*, 2013; Trolle *et al.*, 2015), and critical nutrient loadings for good ecological state in lakes will need to be reduced in order to secure water supplies (Moss *et al.*, 2011). For large, deep tropical lakes that are dominated by internal nutrient cycling (as in Lake Tanganyika; Box 7.3), then the outcomes of warming are different, involving reductions in aquatic primary productivity that reduce fish production and affect human food security.

the least cisco (*Coregonus sardinella*) in Arctic Alaskan lakes is projected to increase as warming proceeds (Carey & Zimmerman, 2014), and the same is likely to apply to other fishes and ectotherms with feeding rates that are correlated with temperature (e.g. Bickford *et al.*, 2010; Beer &

Anderson, 2011; Pease & Paukert, 2014). Such animals will become winners that benefit from faster growth rates, larger size or greater reproductive output under moderate warming scenarios.

A variety of lines of evidence suggest that reduced body size is a frequent response to climate warming, with increases in the proportion of small species, a decrease in size-at-age and a higher proportion of younger age classes in freshwater fishes, zooplankton and phytoplankton (Daufresne et al., 2009; Sheridan & Bickford, 2011;Vadadi-Fülöp et al., 2012). Although the metabolic rates of freshwater ectotherms are highly sensitive to temperature, which has implications for growth, body size and reproductive effort, data on the thermal tolerance of most of the freshwater fauna (even many fishes) within their natural habitats are scant (Li et al., 2013). This knowledge gap is especially apparent outside the northern hemisphere (Matthews et al., 2015), and little of the vast literature on the temperature physiology of freshwater fishes is useful for predicting the effects of global warming (Morgan et al., 2001). Nonetheless, it seems probable that warmer temperatures will be detrimental to fishes living towards their upper thermal tolerance limits either seasonally (in summer) or in low (tropical) latitudes. Conversely, higher temperatures may benefit species that face physiological constraints at higher latitudes or altitudes, or during winter.

Although smaller body size and reduced fitness is a common response of ectotherms to rising temperature (Daufresne et al., 2009), actual outcomes are not always as binary as the winner and loser dichotomy might suggest. For example, projected temperature increases are expected to enhance growth of some salmonids during the spring, but have the opposite effect in summer (Hardiman & Mesa, 2014), and there is experimental evidence of significant sublethal costs of warming in terms of salmonid growth, metabolism, development and disease resistance (Kuehne et al., 2012); elevated temperatures also tend to favour fish parasites (McNab & Barber, 2012). Climate change has the potential to disrupt gametogenesis in the European bullhead (Cottus gobio), causing complete reproductive failure at the highest temperature (8°C) these fish were exposed to in the laboratory (Dorts et al., 2012). These are not particularly warm conditions and might imply that reproduction by cool-adapted fishes in small streams, which track air temperatures closely, are more susceptible to warming than their counterparts breeding in larger water bodies.

Modelling studies based on amounts of warming that has already taken place (as opposed to extrapolations of future temperatures) suggest that salmonids could be eliminated from some nuptial streams in North America due to loss of thermal habitat (Lawrence *et al.*, 2014; MacDonald *et al.*, 2014). Higher mean temperatures (and reduced snowfall) will enhance growth of juvenile salmonids in streams that currently experience cool spring conditions, but truncate the optimal-growth period in summer-warm streams (Beer & Anderson, 2011). Some Rocky Mountain streams will warm during summer only, but colder thermal regimes may prevail during freshets caused by earlier snowmelt; outcomes could include earlier hatching of some trout species, and delayed hatching by others, depending on their breeding season (MacDonald *et al.*, 2014). Changes in timing or volume of peak flows, rather than temperature *per se*, can also affect recruitment by scouring streams and the gravel beds that some salmonids use for spawning (Mantua *et al.*, 2010; Wenger *et al.*, 2011). Increases in precipitation due to climate change has also been correlated with long-term decline of the spring salamander (*Gyrinophilus porphyriticus*: Plethodontidae) because higher flood frequency over a 12-year period reduced survival during metamorphosis (Lowe, 2012).

Overall, the predicted relative impacts of climate change on salmonids and other fishes will be context specific, varying among species and across ecoregions – especially for widely distributed species such as brown trout (e.g. Filipe *et al.*, 2015). Moreover, local conditions, such as the moderating effects of cool ground-water inputs, offer thermal refuges that could modify the outcome of coarse-scale predictions about the extent of range shrinkage in cold-water fishes (Snyder *et al.*, 2015). It is not only the variability of local conditions that limit our ability to generalize about the influence of climate across the entire geographical range of a species of interest, but also the manifold effects of warming on life histories. Yellow perch in Lake Erie grow more rapidly during the short winters resulting from climate change, but eggs produced in long-winter conditions are up to 40% larger and their hatching success is two- to four-fold greater than the smaller eggs produced during warmer short winters. The larvae produced in short winters are therefore smaller and more vulnerable to predation with low survival. Successive recruitment failures associated with warmer winters have been invoked to account for the fact that yellow perch abundance in Lake Erie is currently only around half that in the 1960s and 1970s (Farmer *et al.*, 2015).

The overall association between freshwater warming and smaller body size or shifts in phenology of ectotherms has been observed also in some aquatic insects. One of the first was reported for Odonata in the United

Kingdom where, between 1960 and 2004, there had been a significant, consistent advance of adult flight period of 1.51 days per decade or 3.1 days per degree rise in temperature (Hassall *et al.*, 2007). The emergence of the aerial adult stage of many aquatic insects is stimulated by water temperatures and/or flow (see Heino *et al.*, 2009), and warming plus changes in peak discharge caused by snow melt can result in earlier emergence of mayfly females that are smaller in size and hence produce fewer eggs (e.g. Harper & Peckarsky, 2006). An experimental study using outdoor mesocosms also showed that stream warming induced earlier emergence of adult insects, mainly mayflies (Greig *et al.*, 2012). Another such investigation reported an increase in the average size of emergent aquatic insects with a 3°C temperature rise, reflecting a decline in abundance of adults of smaller taxa, mainly chironomids; total emergent biomass was unchanged, due to increased proportions of caddisflies and mayflies (Jonsson *et al.*, 2015). This finding is in direct contradiction to the generalization (Daufresne *et al.*, 2009) that climate warming favours the small in freshwater ecosystems.

Climate warming poses a particular threat to freshwater herpetofauna, especially in taxa where sex is determined by the incubation temperature of the eggs (Pezaro *et al.*, 2016). Warmer conditions can result in female-skewed sex ratios in turtles (Schwanz *et al.*, 2010; Refsnider *et al.*, 2013) and crocodilians (Simoncini *et al.*, 2014; López-Luna *et al.*, 2015), as well as certain amphibians (Bickford *et al.*, 2010). It may be reasonable to conclude that female-biased populations (provided some males are still present) will be at lower risk of extinction than those dominated by males, and thus an increase in the production of daughters at higher temperatures is preferable to an increase in the proportion of sons. The directional effects of temperature on turtle sex ratio can, to some extent, be moderated by thermal fluctuations that reduce tendency towards biased sex ratios at high (or low) temperatures (Neuwald & Valenzuela, 2011). Furthermore, there is a possibility that variability in nest-site selection by some aquatic reptiles could ameliorate the effects of temperature on sex ratio in some species (Bickford *et al.*, 2010; Simoncini *et al.*, 2014; Refsnider *et al.*, 2014), but not all (Refsnider *et al.*, 2013). As well as herpetofauna, environmental sex determination has been reported in a small proportion of freshwater fishes (mainly cichlids and poeciliids): elevated temperatures skew sex ratios towards males, potentially increasing extinction risks for isolated populations of susceptible species (Ospina-Álvarez & Piferrer, 2008; Brown *et al.*, 2015).

Effects on Distribution

The limited availability of data on historical species distributions hinders our ability to investigate range shifts by many freshwater animals (see review by Heino *et al.*, 2009), although information for fishes are far more comprehensive than for most invertebrates. In view of the importance and economic value of fishes to humans, there is a host of studies modelling changes in fish distributions under climate change, with recent examples ranging from a single widely distributed taxon (brown trout in Europe: Filipe *et al.*, 2015) to 40 species within French rivers (Conti *et al.*, 2015). There are far fewer investigations on observed changes in distribution as a result of climate change that has already taken place, and they are strongly biased towards Holarctic cold-water fishes (Comte *et al.*, 2013). That tendency could affect perceptions of the influence and geographical importance of climate change on fish and fisheries but, regardless, a focus on documented changes in the distribution of fishes should offer a more robust basis for assessment of outcomes than projections modelled for an uncertain climatic future. Predictions of distributional change for non-fishes may, however, have potential value where the data sets are large or take account of many taxa.

A global meta-analysis (Comte *et al.*, 2013) reveals that freshwater fish distributions have been affected by contemporary climate change in ways consistent with anticipated responses under future climate-change scenarios: the range of cold-water stenotherms has shrunk or shifted to higher altitudes or latitudes, whereas that of cool- and warm-water species has either expanded or contracted, with most warm-water species showing expansions. Observed changes could be attributed solely to trends in climate in 55% of articles, but interacted with other anthropogenic drivers in the remainder, producing distributional changes that were much larger than expected from climate alone. However, most evidence about impacts of climate change in the meta-analysis was derived from the many studies devoted to salmonids, and most of the rest involved cold-water species (Comte *et al.*, 2013). Some of these stenotherms are expected to undergo substantial range shrinkage in future: one study forecast a mean 47% decline in total suitable habitat for all trout species in the western United States by 2080, the extent of the reduction ranging from 35% to 77%, depending on the temperature and flow requirements of individual species (Wenger *et al.*, 2011). Such projections could be alarming to recreational anglers (see also Box 7.5). Conversely, species that have been living close to their lower thermal limits, may undergo

Box 7.5 *Distributional Changes of Sport Fishes*

As conditions warm, the northern range boundaries of all warm- and cool-water sport fishes – including introduced species – in Ontario lakes have been shifting poleward, but northern range boundaries of most small bait fishes have contracted southward, perhaps because of interactions with piscivorous sport fishes (Alofs *et al.*, 2014). The range retreat by smaller fishes is consistent with observations that warming temperatures have facilitated range expansion of smallmouth bass, allowing these highly invasive native piscivores access to habitats where they harry or prey upon juvenile salmonids (Kuehne *et al.*, 2012; Lawrence *et al.*, 2014). Such shifts could have economic impacts: annual expenditure in recreational freshwater fishing in the United States, involving an estimated 27 million people in 2011, totalled more than US$25.7 billion; reductions in trout habitat, and more limited opportunities for ice fishing (!), could result in annual economic losses of US$6.4 billion by 2100 (Jones *et al.*, 2013).

range expansion under warming. Between 1979 and 2006, duration of available preferred thermal habitat in Lake Superior increased at a mean rate of five to six days per decade for three fish species, as reflected in growing spatial extent of their preferred habitat (Cline *et al.*, 2013). A fourth species lost three days per decade and underwent range reduction within the lake. The rate of loss of thermal habitat amounts to less than half a day per year over almost three decades, suggesting a very limited capacity to adapt to warming conditions.

Although distributional limits of fishes have generally shifted as a result of climatic warming, a comprehensive study of longitudinal occurrence along French rivers appears to demonstrate that these shifts are insufficient to compensate for measured temperature rises, lagging especially at range centres (Comte & Grenouillet, 2013). Comparing an initial period (1980–1992) with a more contemporary one (2003–2009), expansion in upper altitudinal range (at a rate of $61.5\,\text{m decade}^{-1}$), was accompanied by substantial contractions at the lower limit ($6.3\,\text{km decade}^{-1}$), but mean shifts in range centres ($13.7\,\text{m}$ in elevation and $0.6\,\text{km}$ upstream decade^{-1}) of river fishes failed to keep pace with recent warming, raising the question of whether they will be able to cope with future changes in climate. Shifts in fish distributions may well involve both colonizations and extirpations playing counterbalancing roles in the reshuffling of communities, with one estimate predicting species turnover amounting to ∼60% of the current composition of French river fishes (Conti *et al.*, 2015). Such turnover could reduce the functional diversity of fish assemblages, because each will be dominated by species with shared disturbance-tolerance traits (Buisson *et al.*, 2013).

A major study that made projections about future distributions of 413 western-hemisphere amphibians, with temperature and precipitation as drivers, predicted large reductions in range size, with net losses in habitable area expected for 85% of species assuming unlimited dispersal ability, and for 95% of species when dispersal was assumed to be limited to a 50-km radius. As many as 13% of the former group would lose their entire range, and this proportion would rise to 15% assuming a 50-km limitation on dispersal (Lawler *et al.*, 2010). Large changes in local faunal composition were anticipated, and models of 1099 restricted-range amphibians suggested species turnover rates in some neotropical locations might exceed 60% (Lawler *et al.*, 2010). Interestingly, this is the same proportionate estimate for turnover of fishes attributable to climate warming in French rivers (see above).

Projections for uphill movement of amphibians under climatic warming on a tropical mountain in Colombia had the majority of species (32 out of 46) shifting part of their ranges to unoccupied high-elevation sites, but this would have the effect of reducing the range of many species by from 30% (11 species) to more than 70% (7 species; Forero-Medina *et al.*, 2011). Despite the potential for altitudinal adjustments to cope with warming temperature, evidence is lacking for poleward shifts in amphibian distributions (Li *et al.*, 2013). In at least one aquatic reptile, however, there have been changes in the latitude of the distribution centre and the northern boundary of the Hainan water skink (*Tropidophorus hainanus*: Scincidae) in China over the last 50 years, which is consistent with warmer temperatures increasing the habitable range of this amphibious lizard (Wu, 2015).

Among freshwater macroinvertebrates, Odonata show phenological and distributional responses to climate warming. They are considerably more responsive in this regard than other aquatic insects so that direct observations can be made, often by citizen scientists, of shifts in range and emergence phenology (Bush *et al.*, 2013; Hassall, 2015). However a growing body of published literature confirms that climate change now affects macroinvertebrates in streams in Europe, North America, Australia and New Zealand (Durance *et al.*, 2009). Unsurprisingly, these effects are clearest at sites where other stressors, such as poor water quality, are absent. They indicate that some taxa (e.g. certain mayflies) have more flexible life histories than others (e.g. stoneflies), implying contrasting, and potentially facultative, responsiveness to climate change. For streams at different altitudes in the United Kingdom with multiple years of contemporaneous data and no confounding influence of water quality, macroinvertebrate composition and abundance at upland sites tracked temperature, with abundance declining in response to warming, while functional composition of lowland chalk-stream assemblages shifted in response to discharge, with grazers and filter feeders increasing in years with higher flows (for details, see Durance *et al.*, 2009).

Other information on the likely impacts of climate change on macro-invertebrates can be inferred from modelling studies. Among these, Domisch *et al.* (2011) projected the effects of warming on the distribution of 38 species (from nine orders) in central European rivers. An average altitudinal shift of 83 to 122 m was predicted, depending on the warming scenario, but a contraction in ranges, with associated declines in popula-tion size and genetic variability, was anticipated for species that currently occur in streams with low mean annual air temperatures. Conversely, range sizes of macroinvertebrates that inhabit rivers where air tempera-tures are currently relatively high were expected to become larger, and this mirrors the observed dichotomy of range shrinkage and expansion observed for cold- versus warm-water fishes. At least one 25-year study in Wales provides circumstantial evidence of this: local extinction of a cool-adapted planarian (*Crenobia alpina*: Planariidae) appeared to be a consequence of warming temperatures in upland streams (Durance & Ormerod, 2010).

The biodiversity of unique cold-water-dependent alpine macroinver-tebrates – worldwide – is likely to be threatened under a warming climate, as glacial retreat leads to a loss of their habitat, and permits an upward extension of the ranges of more generalist species (Finn *et al.*, 2010; Jacobsen *et al.*, 2012, 2014). For example, the western glacier stonefly (*Zapada glacier*: Nemouridae) is found only in streams within Glacier National Park, Montana, where it was first identified in 1963. The glaciers within the park are predicted to disappear by 2030, and, as the extent of permanent snow and ice shrinks, the western glacier stonefly is responding by retreating upstream in search of higher, cooler habitat directly downstream of retreating glaciers. Sampling in 2011–2012 detected the stonefly in only one (of six) previously occupied stream and in two new locations at higher elevations (Giersch *et al.*,

2015). Similar upward shifts in the distribution of other montane stone-flies have also been recorded (over a 30-year period: ~25 m per decade) in an apparent response to stream warming (Sheldon, 2012). This is in agreement with the general conclusion that stoneflies are particularly sensitive to warming, especially those species occupying higher-altitude sites (Durance *et al.*, 2009).

For a broader insight into the effects of climate change on freshwater macroinvertebrates, we must draw upon the results of studies that attempt to predict shifts in distribution or range size during the next century. A trait-based approach combined with field observations was used to predict the effects of climate change on freshwater insects in Swedish lakes and rivers by 2100 (Sandin *et al.*, 2014). Taxon-specific trait information – for instance, feeding specialism, preference for high altitude and sensitivity to temperature – was combined and compared with possible climate futures for communities at more than 1300 sites in different parts of Sweden. Predicted temperature change increased from south to north, and combined trait scores indicated the highest potential impact on aquatic insects in the arctic-alpine and northern boreal ecor-egions, so that there was congruence between exposure and sensitivity to climate change. While these results are projections only, they could be of value when planning and implementing management and conservation strategies, such as the use of riparian buffer strips to mitigate stream warming (Sandin *et al.*, 2014), assuming that the climate warmed suffi-ciently to support the growth of trees in alpine or boreal ecoregions. A similar trait-based study, but on a larger scale, modelled likely shifts in the distribution of European aquatic insects (Conti *et al.*, 2014), conclud-ing that species in southern ecoregions (mainly in the Iberian Peninsula, Italy and Greece) and in alpine areas of central Europe, which typically have narrow distributions, had the highest potential vulnerability to climate change.

In the most taxonomically comprehensive of recent modelling studies, Markovic *et al.* (2014) combined catchment-scale species data and pro-jections from multiple climate models to assess the climate-change impacts on the distributions of 1648 European freshwater plants, fishes, molluscs, odonates, amphibians, crayfishes and turtles, including both common and rare species, by the 2050s. Six per cent of common and 77% of rare species were predicted to lose >90% of their range, with eight fishes and nine mollusc species expected to experience 100% range loss. Molluscs, with the most rare species, were the most affected taxon (60% of species would lose >70% of their range). An overlay of freshwater protected areas revealed that around half of fish and mollusc species would have no protected-area coverage within their projected ranges. Although Markovic *et al.* (2014) did not demonstrate impacts, but instead predicted them, their work is notable for the range and number of taxa encompassed, and for including some assumptions about differences in taxon dispersal ability (e.g. odonates *versus* molluscs). Even if these climate-related projections are robust, they may turn out to be conservative, as they did not take into account any additional reductions in range size that could be brought about by interactions with invasive species, nor anthropogenic alterations in habitat suitability due to flow regulation, fragmentation or reduced water quality. An additional point, which needs emphasis here, is that none of the studies mentioned above deals with tropical freshwater species.

Community-Level Effects

Due to the overarching effect of temperature on metabolism, activity, feeding interactions, growth and so on, as well as the shifts in distribution of species that have occurred and will continue to take place, differences in the fortunes of individual species under warming will scale up to affect community composition (e.g. Jeppesen *et al.*, 2012). These direct consequences of climate change will be accompanied by indirect effects on interspecific interactions and changes in competitive dominance that will influence food webs and ecosystem functioning (Woodward *et al.*, 2010). Furthermore, streams and the adjacent terrestrial environment are characterized by permeable boundaries that are crossed by reciprocal resource subsidies between land and water (e.g. allochthonous litter inputs, emergence of adult aquatic insects), and we can anticipate that (*inter alia*) higher temperatures and altered hydrology will shift the temporal and spatial synchrony of resource availability and result in trophic mismatches that scale from populations to communities and to ecosystems (Larsen *et al.*, 2016). However, subsidy dynamics will be affected also by land-use change, increasing water use, invasive species and so on, such that the climate signal may be difficult to discern within this broader threat array.

There are, nonetheless, clear signs that community-level effects attributable to climate change have already taken place in freshwater ecosystems (see Box 7.1). For instance, there have been well-documented inter-year shifts in the functional composition of chalk-stream macroinvertebrate communities in the United Kingdom

(Durance *et al.*, 2009), and large-scale changes in production and energy flow in Lakes Baikal and Tanganyika (see Boxes 7.2 and 7.3). Given the huge volumes and rich and endemic biotas of these two lakes, there can be no doubt about the existential threat climate change poses to global freshwater biodiversity. But the impacts of climate change in Lakes Baikal and Tanganyika are apparent precisely because these water bodies have been relatively unaffected by other stressors. Elsewhere, climate change is simply one of a panoply of threats to fresh waters, and its influence *may* appear slight relative to eutrophication, pollution, invasive species and so on. But that is to ignore the evidence that warming amplifies the effects of eutrophication (see Box 7.4), and the likelihood that it facilitates range spread of non-native species, as discussed below (in particular, Box 7.7).

The gradual and pervasive effects of climate change may well be overlooked or disregarded – at least by policy makers, much of the public and some water-resource managers – in the face of immediate concerns about water quality, fish kills, algal blooms and the proliferation of nuisance species that represent some of the more egregious outcomes of human mismanagement of fresh waters. These are certainly more obvious than the trophic uncoupling that seems to be occurring within Lakes Baikal and Tanganyika, and the impacts on production that are consequences of food-web restructuring resulting from shifts in the spatial and temporal dynamics of producers and consumers. The outcomes may be capricious because trophic uncoupling can involve more than one life stage of any particular consumer species, and several components of the food web (see Box 7.6).

Community-level effects of climate change will, by definition, involve multiple species, each of which may be affected by primary (or direct), as well as secondary (or indirect) responses of other species, with consequences that may cascade through several trophic levels giving rise to tertiary (where one species reacts to the secondary response of another species) or even quaternary responses. Almost every instance of the many shifts in phenology and distribution of individual species mentioned earlier in this chapter could have community-level effects. In the face of such complexity, it is hardly surprising that our ability to document, analyze and predict the extent of community modifications lags far behind the rate at which they are occurring in nature, and probably means that the ubiquity of community-level shifts has been underestimated. But they become more apparent when the arrival and subsequent spread of a non-native species is involved.

Box 7.6 *Trophic Uncoupling and Double Jeopardy*

The great Arctic charr (*Salvelinus umbla*) is an apex predator in Lake Vättern, Sweden – the sixth largest lake in Europe. There, warm winters that have occurred periodically over 25 years are correlated with poor commercial catches of adult charr five to six years later (Jonsson & Setzer, 2015). Warming could have an effect via one or more of three potentially critical life stages of charr – the fry, small juveniles and large juveniles – and there is evidence of trophic mismatches for both the zooplanktivorous fry and large piscivorous juveniles. One of then acts directly on charr fry, because warmer winters speed up egg development (spawning occurs in October) leading to earlier hatching. Thus, fry become free-swimming before the summer peak in zooplankton abundance, which is largely controlled by spring temperatures, and therefore suffer food shortages. A second mismatch indirectly affects large juveniles, via another species, the zooplanktivorous vendace (or European cisco). Prior to 1990, the abundances of vendace and Lake Vättern charr followed a closely linked predator–prey cycle. This diminished and disappeared during the subsequent 25 years, indicative of a second tropic mismatch caused by poor recruitment of vendace after warmer winters that (as with charr) result in fry hatching before the spring zooplankton peak (Jonsson & Selzer, 2015). Charr in Lake Vättern are thus subject to double jeopardy arising from phenological changes that cause trophic uncoupling affecting two life stages. Risks arise because charr fry share resources with vendace, which is also major prey of large juvenile charr, and a similar trophic mismatch is affecting both species during the fry stage.

It remains to be seen whether adaptation, which might involve delaying the spawning date, could aid population recovery of charr in Lake Vättern. Whatever plasticity is inherent in this trait has been insufficient to narrow the timing mismatch gap between charr fry and zooplankton over a 25-year period, implying that genetic change will be needed. Its likelihood will depend on the extent of genetic variability, which may be small within an isolated lacustrine population.

Climate Change and Alien Species

Changes in community structure combined with warmer temperatures can provide new opportunities for the establishment of invasive species (Rahel & Olden, 2008; Strayer, 2010) and may accelerate the 'great mixing' of biotas that is a characteristic signature of the Anthropocene (see Chapter 4). For instance, climatic warming has facilitated the spread of an invasive tropical cyanobacterium *Cylindrospermopsis raciborskii* in temperate zones, where it has become established in Europe, the Laurentian Great Lakes, and as far afield as New Zealand; '... no other cyanobacteria or freshwater phytoplankton has undergone such an incredibly rapid and successful spread from topical to temperate regions all over the world' (Wiedner *et al.*, 2007: p. 274). Warming likewise underlies the expanding global distributions of other cyanobacteria (e.g. Sukenik *et al.*, 2012). Similarly, increasing temperatures in Lake Tahoe during the second half of the twentieth century allowed the establishment and subsequent proliferation of the Asian clam, which rapidly became the dominant taxon in the lake (for details, see Chapter 4). Both this clam, and the invasive golden mussel (see Box 4.8), which also originates in Asia, are thought to be limited by cold winters and could substantially expand their invaded range as climate moderates (Wittmann *et al.*, 2013), forcing greater economic investment in their control (Nakano & Strayer, 2014). Conversely, warmer temperatures could restrict the range of heat-sensitive invasive biofoulers such as the zebra mussel, providing scope for some cost savings. Climate change does not necessarily benefit all invaders (Strayer, 2012), although by increasing disturbance intensity and the frequency of extreme events, climate change will tend to make fresh waters more susceptible to invasion, as will human responses to water shortages and flooding so further transforming freshwater ecosystems and increasing their susceptibility to invasion. Climate change may also increase the intensity of the impact of alien species, as discussed in Box 7.7.

How will the presence of aliens, and their probable increased prevalence, affect the resilience of freshwater ecosystems to global warming and climate disturbances? They could influence resilience (defined here as the persistence of the major taxa defining the ecosystem) by altering resistance to change or the rates of return to the initial state (i.e. recovery), or both. Much will depend on other features of the particular water body under consideration (as in coastal environments; see O'Leary *et al.*, 2017), especially connectivity (and opportunities for recruitment), management of local-scale stressors and the extent of remaining natural habitat. In fresh water, it is very likely that the presence of invasive alien species – perhaps best regarded as a chronic biotic disturbance – has a strong influence on resilience and works against maintenance of native communities.

Box 7.7 *Will Climate Change Increase the Impacts of Alien Species?*

Alien species in freshwater ecosystems show an array of transformative effects (see Chapter 4), by means of direct (predation, grazing, competition) and indirect (changing habitat conditions) interactions, which are, in part, attributable to the strong trophic linkages that characterize aquatic ecosystems (Gallardo *et al.*, 2015). These interactions are all likely to be temperature dependent and so would be amplified by global warming. As warming will also remove range limitations on species currently constrained by temperature, freshwater ecosystems will experience growing pressure from invasive species, although expectations about the extent of this increase vary, with some (e.g. Havel *et al.*, 2015) being more alarmist than others (e.g. Strayer, 2010). Regardless of which prognosis is correct, it would be prudent to step up efforts to limit or prevent invasions in an attempt to minimize their ecological and economic consequences.

In general, freshwater aliens can be expected to flourish in a warmer world because they are typically species tolerant of a wide range of environmental conditions (Karatayev *et al.*, 2009; see also Chapter 4). The size of the native geographical range is positively related to invasion success, presumably because aliens that are broadly distributed are more likely to be ecological generalists. Furthermore, successful aliens have broader geographical ranges and greater heat tolerance than related non-invasive species (Bates *et al.*, 2013), making it likely that the incidence of aliens supplanting native species will increase as the world warms. Changes in weather patterns could also modify species' interactions in favour of non-native species that are already established because more extreme events may be more detrimental to natives (Havel *et al.*, 2015). In other words, as the climate shifts away from that to which the natives are adapted, then coexistence with relatively tolerant invaders becomes less likely.

Global warming will modify the ecological impacts of some pathogenic invasive species, as in the case of *Myxobolus cerebralis*, which causes whirling disease in fishes, and becomes more virulent with increasing temperature (Rahel & Olden, 2008). Climate-related changes could also affect the dispersal of pathogenic aliens (e.g. through flooding), giving rise to conditions that favour particular pathogens and their hosts – a similar process seems to underlie the increasing incidence of invasive bloom-forming cyanobacteria that contaminate lakes and reservoirs (Wiedner *et al.*, 2007; Havel *et al.*, 2015). The greater demand for water storage and conveyance structures in a warmer world will spur construction of more reservoirs and a greater extent of irrigation canals which, in the tropics, create conditions that are favourable to the snail hosts of *Schistosoma* (Schistosomatidae) and other parasitic flukes, increasing the incidence of schistosomiasis and other water-borne diseases (Rahel & Olden, 2008; Sokolow *et al.*, 2015). Either the snail host or the parasite, or both, may be invasive. And, as mentioned in Chapter 4, invasive species are far more likely to occur in impoundments and reservoirs than in natural lakes and rivers (Johnson *et al.*, 2008). As the number of these man-made water bodies grows, they will serve as 'stepping stones' that facilitate the spread of aliens.

What Can Be Done?

Recent and continuing climate change has been driven by anthropogenic carbon emissions, and any attempt to halt or reverse such change will require substantial and rapid emissions reductions, requiring principally – but not only – a global transition to non-carbon-based energy sources. Many efforts towards that end are ongoing, but that important work lies beyond the scope of this book. Suffice to say, it is essential that – sooner or later – the transition is completed, since the habitability of the planet depends upon it. In the meantime, what needs to be done to ameliorate the effects of climate change on freshwater biodiversity? Here the focus is only on the direct consequences of such change, principally warming, and not the effects of their interaction with other stressors (exacerbation of eutrophication, facilitation of alien species and so on).

One consequence of the insular nature of freshwater bodies is a constraint upon adjustment to rising temperatures by way of compensatory movements into cooler habitats further from the equator or at higher altitudes. Shifts in distribution may not be possible, and populations may face isolation in thermally unsuitable habitats. For instance, the carmine shiner (*Notropis percobromus*) is listed as nationally (in Canada) and regionally (in the United States) threatened due to its restricted range and sensitivity to water quality and temperature. Projections of habitat change by 2050 suggest that the spatial extent of the current distribution of carmine shiner would shift north as the southern portions became unsuitable (Pandit *et al.*, 2017). However, much of the potential northern habitat gains will be inaccessible due to dispersal limitations, suggesting that the carmine shiner could experience an extinction debt within the next half century.

A conservation initiative that could help address this problem of population isolation is translocation or, more specifically, assisted migration or managed relocation of species at risk from warming to more suitable water bodies within their thermal range (Olden *et al.*, 2011; Thomas, 2011). Such actions would be costly, requiring detailed information about life history and physiological tolerances, currently available for only a tiny fraction of freshwater species that might be imperilled by climate change. They would also be controversial (Ricciardi & Simberloff, 2009; Schwartz *et al.*, 2009), not least because of the potential risk of ecological outcomes of the type associated with the spread of species beyond their natural geographical distribution (see Chapter 4). The *ex situ* practice of translocation also conflicts with established conservation paradigms that favour maintaining the *status quo* of species ranges, and *in situ* management (see review by Hewitt *et al.*, 2011). Nonetheless, the argument that we should not move animals such as fishes to new environments so as to avoid causing unanticipated harm deserves close scrutiny. It cannot be equated with adopting the precautionary principle, because climatic shifts as the world warms may leave freshwater organisms stranded within water bodies where temperatures exceed those to which they are adapted or to which they can adjust, ultimately dooming them to extinction. Under these circumstances, doing nothing could result in more harm than would arise from managed relocation (Thomas, 2011; but see Schwartz *et al.*, 2009). Implementation of such measures would need to be predicated on robust risk assessments. But if freshwater animals are relocated to isolated water bodies, which they could not reach by themselves and from which they cannot spread, then the ecological risk may be quite small, assuming appropriate quarantine procedures have been followed and the species moved is not a rapacious predator or an ecosystem engineer. Translocations will be revisited in Chapter 9, where the broader issue of the use of reintroductions in conservation is considered.

Predictions about the availability of thermal habitat for species that may be threatened by climatic warning, such as the carmine shiner, or more general projections about distribution shifts in the context of such change (e.g. Lawler *et al.*, 2010; Domisch *et al.*, 2011), are generally based upon species-distribution models that use present-day relationships between occurrence and temperature to predict future ranges based on particular climate-change scenarios. Model outcomes often project a loss of thermal habitat within current range boundaries but can also produce valuable information on locations that could become habitable in future. Inclusion of predicted range shifts within conservation planning can inform the spatial prioritization of habitats (or potential habitats) and has potential to increase the likely benefits for biodiversity under alternative realized climates (e.g. for fishes: Bond *et al.*, 2014). A similar approach has been advocated for forecasting the location of future cold-water refugia for montane salmonids so that these waters can be designated for protection (Isaak *et al.*, 2015). Sites that may be thermally suitable for translocation of freshwater species at risk could likewise be identified using species-distribution models.

How Much Does Climate Change *Really* Matter for Freshwater Biodiversity?

In view of the extensive evidence of a climatic fingerprint on freshwater ecosystems, this question may well be unexpected. In addition, freshwater animals appear to be highly vulnerable to climate change because many of them appear incapable of making range adjustments quickly enough to compensate for temperature rise (e.g. Comte & Grenouillet, 2013), which could eventually result in local extinction. Tropical ectotherms that are already close to their upper thermal limits may be particularly vulnerable (Kellermann *et al.*, 2012). However, grim projections of the frequency of local extinctions in fresh waters due to recent climate change (e.g. Wiens, 2016) depend mainly on the use of data on range shifts among the set of species surveyed to calculate extinction frequency. These are neither confirmed *local* extinction rates nor *global* extinctions, which, in any case, are notoriously difficult to verify (Harrison & Stiassny, 1999). While extrapolations about climate-driven extinctions should not be dismissed entirely, studies from the terrestrial realm show that, in addition to poleward and upward range shifts, many reports document other types of range shifts – in east–west directions or even towards tropical latitudes and lower elevations – that may be a response to complex local climate changes (reviewed by Lenoir & Svenning, 2015). A more multidimensional approach seems mandated for the development of improved predictive models of climate-related range shifts and species extinction risks; such models would need to be designed to account for the hierarchical structure of drainage networks, and the dispersal challenges faced by freshwater species.

It would be obtuse to disregard the evidence that climate change is, to a greater or lesser extent, affecting population dynamics, changing communities and altering food webs in many of the world's fresh waters. And, it appears certain that these effects will become greater as the world warms – something that appears 'locked in' for at least the first part of this century. However, it is arguable whether the climate poses a more imminent threat to freshwater biodiversity than other elements of global change, such as catchment alteration, flow regulation, invasive species, overexploitation, pollution and eutrophication. These factors, on some occasions acting in isolation and, on others, in consort, have been responsible for the 'great thinning' and the 'great shrinking' of freshwater biodiversity, and have facilitated the 'great mixing', so giving rise to patterns of endangerment that are now globally epidemic. Among them, land use is subject to rapid alteration with immediate impacts on fresh water. A comparison of scenarios that modelled macroinvertebrate distribution in a Chinese river basin between 2021 and 2050 under climate change only, land-use transformation, and the two combined showed that land use alone reduced local species richness by 20%, with climate change having a secondary effect only (Kuemmerlen *et al.*, 2015). In combination, both drivers reduced biodiversity, although some species increased their range occupancy and became winners us a result of climate change. For streams and rivers, land-use transformation, pollution, dams and the like represent immediate existential threats that may well be greater than the longer-term and more subtle – albeit pervasive – consequences of climate change. This could even be the case in highly biodiverse regions of the world (Box 7.8) where the prevalence of anthropogenic influences presently is far less than, say, northwest Europe (Vörösmarty *et al.*, 2010). Lakes will experience many of the same stressors as rivers, but climate change will also affect their internal dynamics (stratification, bentho-pelagic linkages) in ways that have no equivalent in running waters. Whether such effects will lead to extinctions has yet to be seen, but the likely consequences already evident in biodiverse ancient lakes such as Tanganyika (see Box 7.3) portend substantial changes in lake ecology as warming proceeds.

One of the first attempts to model the impacts of the effect of climate change on riverine biodiversity at a global scale produced an alarming projection (Xenopoulos *et al.*, 2005): reductions in discharge due to the

Box 7.8 *Fish Futures in the Amazon Basin*

Threats to fish communities through rapid expansion of human infrastructure and economic activities in the Amazon Basin could be far greater than those anticipated from future climate change (Oberdorff *et al.*, 2015). There are already more than 100 hydropower dams in the Amazon Basin, and proposals for the construction of many more could lead to an eventual total of 428 dams each exceeding 1 MW capacity. The accumulated hydrophysical impacts of even a fraction of these will result in irreversible changes to the floodplains, estuary and sediment plume of the Amazon (Latrubesse *et al.*, 2017), and will have far more immediate and profound impacts on freshwater biodiversity than any conceivable effects arising from climate change. Thus, the priority for this region should be a reduction in the impacts from planned or ongoing dam development, rather than the less tangible effects of future climate change (Oberdorff *et al.*, 2015; Latrubesse *et al.*, 2017). Moreover, there is already strong evidence of overexploitation and fishing down of Amazonian waters (see Box 3.5), that will likely become more evident as dams and hydrophysical alterations limit breeding migrations and recruitment. Focusing on present-day threats is not to assert that global warming is unimportant or should not be addressed by way of robust international action. Rather the point is that the imperilment of biodiversity will increase far more as a result of other anthropogenic threats, and that halting climate change will do little or nothing to staunch projected species losses that could occur in the near future. In addition, many of the measures that would limit non-climate-related threats can be implemented at national or, sometimes, regional levels, which, in either case, is far less challenging than achieving an international consensus on climate action.

combined effects of climate change and increased water withdrawal by 2070 would result in rivers with reduced discharge, where up to 75% of local fish biodiversity would be threatened with extinction. While 25% of rivers would lose fewer than 4% of their fish species, another 25% of rivers were forecast to lose more than 22%. That study was based on a total of 325 rivers. The findings can now be reassessed in the light of a subsequent analysis (Tedesco *et al.*, 2013) using a more comprehensive database of more than 1000 drainages to model fish extinction rates under climate-change projections that resulted in habitat loss (i.e. reduction in the extent of perennial flow). The number of predicted extinctions by 2090 was rather low: most basins experiencing no losses at all, and only 20 (1.9%) were expected to lose fish species as a result of reduced water availability. However, data from 20 well-sampled central and North American river basins show that contemporary anthropogenic insults (especially dams and pollution) are currently causing fish extinctions estimated to be ~150 times higher than natural rates (Tedesco *et al.*, 2013). These findings reinforce the view that it would be more beneficial for conservation action to prioritize amelioration of immediate anthropogenic threats, rather than be diverted towards measures intended to mitigate the impacts of future climate change.

The Earth is currently in the midst of a hydropower boom (Grill *et al.*, 2015; Zarfl *et al.*, 2015; Winemiller *et al.*, 2016; see Chapter 5) and an expansion of huge, long-distance, inter-basin water transfers (Shumilova *et al.*, 2018), with grand water-engineering schemes projected for the world's greatest and most biodiverse rivers. Among these, the Mekong is of particular concern given its role as a source of fish protein for millions of people (see Box 3.6): 11 dams have been proposed for the river mainstream in the Lower Mekong Basin (LMB), along with perhaps as many as 121 tributary dams (Kano *et al.*, 2016b). The Mekong is a relatively pristine river in terms of pollution and contamination, but the scale of dam construction in the LMB threatens fish biodiversity, especially the many migratory species. Fish distributions can also be expected to shift in response to climatic warming, but the barrier effects of dams will constrain range adjustments by fishes that might compensate for warming temperatures. To what extent will habitat transformation by dam construction affect fishes, and how will it compare with the consequences of climate change? To answer that question, Kano *et al.* (2016b) projected how the distribution of 363 fish species found in the Indo-Burma global biodiversity hotspot would change under the separate and joint impacts of global warming and dam construction along the Lower Mekong and its tributaries. The modelled outcomes, which are described in Box 7.9, were complex because they depend on which of the planned dams are actually built, where they are situated (e.g. tributaries versus mainstream) and assumptions about the magnitude of warming derived from likely carbon emissions. Overall, however, they underscore the greater incipient threat of dam construction for biodiversity, with some projections even suggesting an increase in fish species richness under climate change, and evidence of synergistic effects of dams and climate

Box 7.9 *Fish Futures in the Lower Mekong Basin*

The results of projections of fish diversity associated with 81 scenarios for dam construction within the LMB were as expected, with progressive reductions in overall species richness and mean habitable area (roughly equivalent to range size) for fishes as hydropower-generating capacity increased (Kano *et al.*, 2016b). Changes in the proportion of threatened species – defined as those predicted to lose more than 30% of their habitat compared to 'pre-dam' conditions – were modelled also. This threshold was chosen because the IUCN Red List classifies any species that has experienced a decline of this extent as 'vulnerable' or, in more extreme cases, as 'endangered' (or 'critically endangered'). The proportion of threatened fishes increased to around 15% at maximum generating capacity and, unsurprisingly, mainstream dams had a larger impact n than dams on tributaries for all three biodiversity response variables.

Projections from 126 global-warming scenarios (for 2050 and 2070) included a rise in species richness, a reduction in habitable area and an increase in the proportion of threatened species. However, there was substantial variation in the extent of these responses among various warming projections, and they were smaller and less consistent than changes attributable to dam construction. Projections from scenarios that included both the effects of dams and global warming combined them in a synergistic manner that took account of the likelihood that habitat shifts under global warming would be constrained by river fragmentation due to dams. Reductions of overall species richness under the synergistic projections were similar to those for dam-building scenarios, but declines in habitable area were up to 20% greater under the synergistic scenarios and were exacerbated as generating capacity increased – particularly if carbon emissions remained high. The proportion of threatened species under synergistic scenarios was negatively correlated with total generating capacity, increasing between 2050 and 2070, with values (7–44%) exceeding those projected for dam construction or global warming alone. Overall, these predictions indicate plainly that hydropower dams will have greater and more imminent impacts than global warming on fish biodiversity within the LMB.

change that will reduce habitat occupancy by fishes and increase the proportion of threatened species.

Although Kano *et al.* (2016b) did not attempt to make projections about fishes of commercial importance, other modelling studies have concluded that, by themselves, climate-change impacts on the LMB fisheries' yield would not be detectable because of the high levels of natural flow variation that would overshadow climate change effects (Welcomme *et al.*, 2016). As with river fishes on a global scale (Tedesco *et al.*, 2013), the LMB findings suggest that conservation would be best served by addressing immediate anthropogenic threats, including a cessation of dam building and water-engineering works, and assigning climate-change mitigation as a secondary priority. This arrangement does not mean that efforts to limit climate change should cease. Instead, it recognizes the relative imminence of impending threats, and acknowledges that slowing warming, or stopping it entirely, will require sustained commitment on the part of many governments. In contrast, conservation interventions to protect freshwater diversity within particular basins involve only one or a few riparian states or regulatory agencies, and can produce results relatively rapidly. And the most effective of these is – unsurprisingly – to abandon plans for large-scale water-engineering schemes, so avoiding the potential devastation they wreak upon freshwater ecosystems. Where they are unavoidable, and enhance human water security and livelihoods, such developments must be designed and operated in ways that minimize their ecological impacts. Crucially, local or regional action to limit further endangerment of freshwater biodiversity in the Amazon and Mekong, among others, needs to take place in parallel with broader international engagement to limit climate change. It is not a matter of choosing to do one thing or the other, but doing both. If, eventually, climate change is brought under control, we must have taken steps in the interim to ensure endangered biodiversity has been able to persist in the face of other, more pressing, threats.

8 · Ecosystem Services and Incentivizing Conservation of Freshwater Biodiversity

Some conservationists '. . . disdain moral persuasion and . . . purchase what they can afford or argue that the market should be used to preserve biodiversity . . . defending endangered species and rain forests on economic grounds. Instead of seeing modern economics as the problem, they see it as the solution. . . . The new economic conservationists think they are being rational; I think they treat Mother Nature like a whorehouse'.
Jack S. Turner (1996: pp. 57–58), *The Abstract Wild*

Linking Human Needs to Protection of Fresh Waters

Inland waters and the biodiversity that they support are a valuable natural resource. They are a source of clean fresh water or other hydrological services (e.g. water filtering or purification by wetlands), and provide habitat for animals and plants that may be eaten or used (e.g. fishes, reeds and so on). Their conservation and management must be seen as critical to the interests of all nations and governments. Emphatically, the importance of freshwater biodiversity to society must be communicated successfully to all. Although these threats are becoming increasingly known among scientists (e.g. Darwall *et al.*, 2018), they are insufficiently incorporated within water development. Immediate conservation action is needed in some instances where opportunities exist to set aside relatively pristine lake and river systems in large protected areas (see Chapter 9). Nonetheless, it must be recognized that there are significant portions of the Earth's surface where it would be impractical to dedicate any freshwater body to the single purpose of biodiversity conservation (so-called fortress conservation), if that meant humans were denied access or their use of the water was substantially restricted. There may thus be inevitable trade-offs between conservation of biodiversity versus human use of the global freshwater commons, because even well-protected conservation areas can become focal points for tourism and recreational activities that may reduce habitat quality and biodiversity.

In the near future, clean fresh water could become the Earth's scarcest and most sought-after natural resource. This is not only a reflection of the absolute limitations on the quantities of fresh water available for human use but also because, in many countries, water-supply improvements outpace progress towards sanitation, and because sanitation systems in the developing world seldom include adequate sewage treatment (Lodge, 2010). The important objective of water supply is therefore being achieved at the expense of environmental goals and biodiversity conservation. The failure to provide sanitation, especially sewage treatment for burgeoning urban populations, has direct consequences for human health and sustainability. It is a major cause of pollution that compromises biodiversity and benefits to be gained for healthy ecosystems, such as fisheries, and principally affects those humans who already have unacceptably low living standards, increasing their exposure to water-borne diseases and environmental degradation.

Pollution and widespread reductions in water quality are not the only global-scale challenge to humans and biodiversity. Water regimes shaping the evolution of freshwater biodiversity and the life-history adaptations of individual species during the Holocene and earlier will be different in the Anthropocene when major shifts in water regimes will be (and are) taking place (see, for example, Chapter 1) with only a rudimentary understanding of the organisms and ecosystems being affected, or the larger-scale consequences of those changes. To reverse things, and improve the condition of fresh waters globally, it will be imperative to link the conservation or restoration of freshwater ecosystems and biodiversity conservation to human well-being and the benefits accruing to society. The concept of ecosystem services, which are the benefits that humans derive from ecosystems, came to prominence after publication of the Millennium Ecosystem Assessment in 2005 (MEA, 2005; see www.maweb.org), which offers a means to reinforce this link or, at least, make it explicit. It may also help to resolve the conflict between human use of fresh water and protection of biodiversity.

Freshwater Ecosystem Services

Apart from water, which is essential for life, fish and fishing are among the most conspicuous services that humans can derive from freshwater ecosystems. Fishes represent a major source of animal protein that has long been essential for the livelihoods of people in many countries (see Chapter 3). For instance, the Lower Mekong Basin supports the world's largest inland fishery and has been central to Cambodian culture since ancient times, sustaining the Khmer civilization that gave rise to the temple complex at Angkor Wat (~800 AD). However, food is just one of the benefits that can be derived from intact freshwater ecosystems, and the MEA (2005) provides a framework for the characterization. Four categories of services can be recognized. First, provisioning services are goods that can be derived from ecosystems, such as water, animals for food, and plants for food, fuel and medicines. Second, regulating services are the benefits obtained from water purification, and regulation of floods and extreme events, local climate, and parasites or diseases. Third, cultural services are the non-material benefits derived from ecosystems, such as recreational, spiritual or educational benefits. Finally, supporting services such as soil formation, recharging ground water, nutrient cycling, primary production, carbon sequestration and so on are necessary for maintaining the three other service categories. Provisioning services can be thought of as providing a 'direct-use' value for humans, whereas the third category of cultural services represents a 'non-use' value; the second and fourth are 'indirect-use' values that nonetheless support livelihoods.

It has been estimated that more than 60% of ecosystem services have deteriorated or been overused (MEA, 2005), and matters have surely worsened of late as the cases of overexploitation of provisioning services in Chapter 3 attest. This corrosion has taken place, at least in part, because the contribution of intact ecosystems to human well-being is little recognized and undervalued; the outcomes have been species loss, habitat transformation and degradation of the resource base on which people rely. The fact that 53% of surface-derived drinking water in the United States flows from forests should provide a compelling rationale for the maintenance of the intact land cover within catchments (Brown *et al.*, 2008), but can also be taken as a caution about the potential fragility of ecosystem-service provision in a rapidly changing world.

To what extent is maintenance of the four categories of ecosystem services dependent on biodiversity? It seems obvious that a more speciose fish community is necessary to maintain a productive fishery (as, indeed, seems to be the case: Brooks, 2016), but, in other instances, biodiversity plays a variety of roles in relation to providing ecosystem services (Mace *et al.*, 2012): as a regulator of underpinning ecosystem processes, as a final ecosystem service and as a good that is subject to valuation. Some confusion can arise from this multilayered relationship, as described in Box 8.1. Nonetheless, whatever services are to be gained from ecosystems require the presence of a range of interacting taxa transferring

Box 8.1 *Biodiversity and the Ecosystem-Services Concept*

If we are to use the provision of ecosystem services as a proxy for biodiversity on the assumption that maintaining the former will serve to protect the latter, then a robust understanding of the relationship between the two is needed. One widely held notion about this relationship is that maintenance of ecosystem services depends upon activities carried out by the associated biodiversity, and thus they are inextricably linked. Continued provision of ecosystem services can thus be secured by protecting all biodiversity, or as much as is feasible. The literature that focuses explicitly on the relationship between biodiversity and provisioning or regulating services suggests that it is frequently positive. This is the case in the repeatable and consistent effects of species diversity on primary productivity, which is frequently driven by complementarity (Tilman *et al.*, 2014). However, the relationship between biodiversity and ecosystem services can be both variable and complex (reviewed by Cardinale *et al.*, 2012). Complexity abounds because, in some circumstances, the terms 'biodiversity' and 'ecosystem services' have been treated almost synonymously (Mace *et al.*, 2012); in extreme instances, biodiversity itself has been regarded as an ecosystem service, with environmental management directed only towards the desired biodiversity components.

An additional complication concerns potential misunderstanding of the relationship between ecosystem functioning and ecosystem-service provision. These are sometimes treated as synonymous: for instance, a well-cited review (Cardinale *et al.*, 2012) treats organic-matter dynamics and nutrient remineralization as regulating ecosystem services. In contrast, Lele *et al.* (2013) advocate thinking of ecosystem services as only those stock or flow variables that are socially valuable, treating the rest (supporting or regulating services) as ecosystem functioning or processes that have no intrinsic value, and result from interactions among organisms and their environment. This strict interpretation of the ecosystem-services concept excludes any processes that have no direct value or explicit link with human well-being, referring to them as ecosystem functions (Lele *et al.*, 2013). Mace *et al.* (2012) use the term 'final ecosystem services' to encompass the same idea of explicit benefits to humans, separating them from the underpinning ecological and environmental processes within ecosystems, but these authors nonetheless retain the same general ecosystem-services framework used in the Millennium Ecosystem

Assessment. However, reference to final ecosystem services may leave the unfortunate impression that the value of ecosystems depends on the extent to which they directly benefit humans.

The strict interpretation of the ecosystem-services concept does not provide any rationale for protection of all diversity for its own sake, or its functional role in the ecosystem, but only if biodiversity benefits humans as a good (e.g. fishery yields) or directly underpins some service (such as water purification) that is amenable to valuation. Any intrinsic value of biodiversity is subsumed under cultural services; i.e. the religious, recreational and aesthetic values provided by an ecosystem. Managing ecosystems solely to maximize service provision for humans is unlikely to result in comprehensive protection of biodiversity and will benefit only those species that are known to contribute to socially valuable processes.

energy through food webs and recycling nutrients. It thus seems reasonable to suppose that maintaining services is positively correlated with species richness and the degree of ecosystem intactness, and might even be maximized in environments where natural, highly diverse communities are retained. This supposition forms the basis for the search for 'win–win' conservation solutions that simultaneously protect biodiversity and provide valued ecosystem services.

What Has Freshwater Biodiversity Ever Done for Us Anyway?

Of the various ecosystem services offered by freshwater biodiversity, at least one is apt to be overlooked, but the need to maintain it can offer an opportunity to protect certain aquatic species. Anthropogenic activities that degrade habitats, alter food webs and significantly reduce freshwater biodiversity tend to increase the abundance of parasite vectors and facilitate pathways of disease transmission. The effects on human health include rising disease exposure due to increases in vector populations unchecked by predation, and this 'disease control' is an important regulating service provided by freshwater biodiversity (Naiman & Dudgeon, 2011). The example of schistosomiasis, caused by *Schistosoma* blood flukes, is informative in this regard. Schistosomiasis affects people in more than 70 countries, and ranks second only to malaria as a cause of human morbidity by a metazoan parasite. The fluke alternates between human and snail hosts, favouring planorbid snails (*Biomphalaria* and *Bulinus*) that proliferate in irrigation canals, reservoirs and man-made water bodies. There, snail hosts establish high densities because flushing flows are artificially stabilized, producing favourable substrate conditions, and natural dry periods are eliminated by year-round provision of irrigation water. As a result, high parasite burdens are typical of residents in the vicinity of dams and new irrigation schemes (Sokolow *et al.*, 2015), especially in places where sanitation is poor. Aquatic predators of parasite vectors are often the first to be impacted by habitat degradation and other anthropogenic threats. For instance, overexploitation of molluscivorous fishes and declines in water quality have been associated with increased incidence of schistosomiasis among people living along the shores of Lake Malawi (Stauffer *et al.*, 2006). In addition, dams that block reproductive migrations of river shrimps that eat the snail intermediate hosts of schistosomiasis have been implicated as an indirect cause of higher disease burdens (Box 8.2). Increased incidences of parasites or diseases will offset some of the benefits of food and water security that accrue from water-engineering schemes, and they are seldom or, at best, incompletely, taken into account when assessing the net gains or environmental impacts of such projects (Naiman & Dudgeon, 2011). Certainly, if diverse communities in general can be shown to provide net benefits to human well-being through 'disease control' or other mechanisms, this could provide a powerful motivation for preserving the Earth's remaining biodiversity (Kilpatrick *et al.*, 2017).

Payment for Ecosystem Services

Intact, fully-functioning freshwater ecosystems provide a number of valuable services for humans. At least two billion people depend directly upon rivers for provision of ecosystem services in the form of 'food', including the benefits to be derived from fisheries, flood-recession agriculture and dry-season grazing (Richter *et al.*, 2010). One estimate (Costanza *et al.* 1997) puts the global value of ecosystem services provided by 'fresh' waters (i.e. wetlands, lakes and rivers, including brackish waters) at US\$9.0 trillion annually (1997 dollar values), 20% of the value of all ecosystems combined, despite their very small relative extent. An update of this estimate (using 2011 dollar values), yielded a value of US\$28.9 trillion, 23% of the global total, despite a reduction in wetland area of around 40% during the intervening 14-year period (Costanza

Box 8.2 *River Shrimps As Ecosystem-Service Providers*

Dams have many ecological and socioeconomic effects, including a general increase in burdens of human diseases such as schistosomiasis. One contributing factor is that dams block reproductive migrations of amphidromous river shrimps (*Macrobrachium* spp.; larger species are often referred to as 'prawns') that prey upon the snail intermediate hosts of schistosomes. This relationship became evident when shrimp populations declined and schistosomiasis caused by *Schistosoma hematobium* increased after completion of the Diama Dam, situated immediately upstream of the estuary of the Senegal River, in 1986. Sokolow *et al.* (2015) found that restoring shrimp (through stocking of *M. vollenhoveni* collected elsewhere) to reaches upstream of the dam reversed the effect via a reduction in snail (*Bulinus* spp.) densities; a decline in reinfection rates of people treated with an anthelminthic drug was reported also. Because the Diama Dam prevented shrimp recruitment, their decline and subsequent extirpation contributed to a steep increase of schistosomiasis after dam completion. To test whether such a cascade effect (shrimps-to-snails-to-schistosomiasis) was widespread, Sokolow *et al.* (2017) collected data from 14 large dams in sub-Saharan Africa. Dam construction was followed by significant increases in schistosomiasis incidence within upstream catchments that had previously supported amphidromous shrimps; the effect was not seen in catchments where shrimps were lacking historically. This finding points to shrimp restoration as an ecological solution for reducing human disease. An estimated 277 to 385 million people live within schistosomiasis-endemic regions where migratory *Macrobrachium* spp. are (or were) present, representing around 40% of the total of ~800 million people who are at risk from this disease. As many as one third of potential sufferers could benefit from the presence or restoration of snail-eating shrimps and the ecosystem service that they provide (Sokolow *et al.*, 2017).

et al., 2014). While valuation estimates such as these are subject to controversy (a matter that will not be pursued here), the general message that fresh waters have immense economic importance seems self-evident. Moreover, their value is bound to increase in future as fresh waters are further degraded by anthropogenic activities and climate change that will compromise continued service provision.

The fact that nature provides ecosystem services or, at least, 'final ecosystem services' (see Box 8.1) that benefit humans can, in certain cases, be deployed in service of conservation to arrive at outcomes that incorporate 'wins' for people and biodiversity. This depends on the acknowledgement that such services have potential monetary value, and therefore market-based mechanisms can be called upon to fund conservation of drainage basins, and their associated water bodies and biodiversity, through a system of payment for ecosystems services (PES) by governments, communities or individuals (e.g. Ormerod, 2014). Implementation of PES is based on two essential ingredients: quantifying the economic value of ecosystem services, which generally go unmeasured, and persuading someone to pay for them. With regard to the latter, payment may be enforced by legislation (requiring compliance), or by government action (where, for instance, public funds are used to pay for appropriate management of drainage basins), or may come about voluntarily for ethical reasons, or can occur in cases where corporations include such payments in their business model in anticipation of future regulation.

New York State offers an early example of a PES model in which land purchase and payments to landowners (totaling more than US$1 billion in 1996) were made to maintain or restore tree cover in the Catskill Highlands. The objective was to enhance the ecosystem functions of water filtration and purification and ensure a sustainable supply of clean stream water to city-dwellers downstream. The costs were offset against an estimated US$8 billion needed to build a new water-treatment plant, plus a further annual operating cost of US$300 million that would have been required to provide the same quality of water in the absence of the 'free' ecosystem services (Chichilnisky & Heal, 1998). Elsewhere, a water fund based on New York's model has been in operation in Quito (Ecuador), based on a partnership between private-sector companies (e.g. the regional bottler of Coca-Cola) and The Nature Conservancy (TNC: a US-based non-governmental organization) who pay around US$1 million annually to Andean communities in order to protect the drainage basins they inhabit; this compensation is designed to secure a sustainable supply of potable water for Quito residents. A parallel initiative was established in Kenya in 2015, when TNC launched the pilot phase of the Upper Tana–Nairobi Water Fund, intended to raise US$15 million, using the same business model as in Ecuador in partnership with businesses (again including Coca-Cola), utilities (water, sewerage and electricity), conservation groups and farmers. In this instance, the fund invests in activities to manage and protect land cover in the upper catchment of the

Tana River that supplies virtually all the water used by residents of Nairobi and generates hydropower that meets around half the city's needs. These two examples, based on the principle of paying farmers or landowners to maintain a certain amount of natural vegetation on hillsides and encouraging practices such as terracing and drip-feed irrigation, can increase agricultural productivity and help provide a reliable supply of clean water downstream. Similar schemes have been adopted in almost 30 countries, such as Costa Rica, Mexico, the Philippines and Vietnam, and are likely to spread within sub-Saharan Africa.

This category of PES, often termed payment for watershed services (PWS), provides water to downstream users and has ecological benefits that include reduced sedimentation, lowered incidence and intensity of flooding (representing a regulating ecosystem service), and enhanced conservation of terrestrial and aquatic biodiversity. Such programmes are particularly well-developed in China and the United States, which, together, have more than half of all PWS initiatives. One estimate of their global extent in 2011 was 117 million hectares – equivalent, approximately, to the area of South Africa – with payments valued at US\$8.17 billion (Bennett et al., 2013). Most were made in China, where the government has forcefully promoted PWS. Provincial authorities pay suppliers of ecosystem services for lost income or land-use rights through 'eco-compensation' that is a component of national environmental protection policies. Not only is a substantial land area protected by PWS schemes globally, but they kept approximately 44 t phosphorus and 1.55 t nitrogen out of fresh waters in 2011 (Bennett et al., 2013).

Can PES Provide a Basis for Freshwater Biodiversity Conservation?

While PES schemes offer the hope that it is possible to reconcile preservation of biodiversity with maintenance of ecosystem services that sustain livelihoods, such 'win–win' outcomes are possible only when the value of the ecosystem service is maximized by high biodiversity, and is compromised by reductions in species richness. Unfortunately, the converse is frequently true. A focus on maintaining the processes that underpin and deliver desired ecosystem services can have detrimental effects on total biodiversity or species of conservation concern. For example, evaluation of stream restoration projects in the US state of Maryland, where channels were re-engineered to provide the ecosystem services of storm-water management, nutrient storage and retention of suspended sediments, revealed that service maximization had impacts on biodiversity, with stream biota replaced by taxa more characteristic of wetlands, and a loss of riparian trees (Palmer *et al.*, 2014). Moreover, the services were provided irrespective of the recovery or re-establishment of the original stream biota. Thus, restoration or management aimed solely at provision of services (which, in this case, might be thought of as equivalent to ecological engineering) may transform ecosystems into something quite different from their existing or historical state. In extreme cases, streams can be converted to little more than storm-water conveyance systems.

It is seldom possible to predict how much biodiversity can be lost from an ecosystem before it ceases to provide the services valued by humans. Furthermore, the rate at which societal benefits associated with an ecosystem service diminish can be expected to be specific to the particular service and the relative strength of the biotic and abiotic processes that influence it. As mentioned above, fishery yields are sensitive to changes in biodiversity and, in the case of recreational angling (see Box 8.3), may even depend on the presence of a particular species, while reductions in the service of biofiltration have been directly linked to losses of pearly mussels (Vaughn, 2018; see Box 3.10). Conversely, flood-control capability is determined mainly by the physical form of stream channels and the type of land use in the drainage basin. The capacity for water purification is likely to depend upon the interaction of both biological and physical processes, but not on the persistence of a particular species of conservation concern, such as charismatic vertebrates, nor even a particularly broad range of taxa. In advocating the ecosystem-services paradigm for river conservation, Ormerod (2014) makes the important point that it should be used as an adjunct to existing approaches rather than as a replacement for them.

Disbenefits to biodiversity can arise where humans attempt to manage freshwater ecosystems to maximize service provision. Damming a river can generate hydropower, provide water for irrigation and could also lead to the development of a fishery in the impoundment, as well as offering recreational opportunities and facilitation of navigation in the regulated river channel. But such benefits would come at a loss to native biodiversity as flow, sediment and temperature regimes downstream and immediately upstream of the dam were altered. Within the impoundment, an assemblage of indigenous lotic fishes would be supplanted by species adapted to lacustrine conditions, some of them likely to be species introduced from elsewhere to boost fishery yields. This is one ecosystem service that can be maximized by the substitution of indigenous species

Box 8.3 *PES and the Preservation of Iconic Fishes*

As explained in Chapter 3, large-bodied river fishes are particularly vulnerable to human impacts arising from overexploitation because many of them are slow growing and/or late maturing and migratory, and are apt to encounter a variety of threats or stressors at different times and locations during their lives. Economic value can be a contributing factor to their exploitation and subsequent decline, but can also provide an opportunity for species protection through the adoption of a PES model. For example, a rural community in the northern Indian Himalaya derives benefits from protection of an iconic, flagship fish species in the Western Ramganga River (Everard & Kataria, 2011). The endangered golden mahseer, which may exceed 2.5 m in length and weigh 50 kg, is a favoured species for recreational angling. Along with associated cultural and wildlife tourism, angling generates income that creates incentives for protection of the river by local inhabitants. They benefit economically from sustainable exploitation of mahseer through a catch-and-release fish sport fishery, leading to establishment of a PES market involving local people, tour operators and visiting anglers. This scheme enjoys support among stakeholders (Gupta *et al.*, 2014), and is sustainable so long as people profit economically to a greater extent from the benefits generated by recreational angling than they would through killing golden mahseer for sale and consumption.

Use of PES to create incentives derived from angling and tourism may offer an effective means of preventing overexploitation of some large fishes, especially where such species also have symbolic or cultural value. However, sharing the benefits of recreational angling markets is essential to promote self-interested resource stewardship of the type practiced along the Western Ramaganga River. If profits accrue to a few business operators only, without distribution of the revenues from tourism, local people are unlikely to have any incentive to protect iconic freshwater fishes.

by aliens: for instance, reservoir fisheries based on tilapias in Sri Lanka (Fernando, 2000; see Chapter 4). The service provided by a novel ecosystem, in this case, has societal benefit even if it is to the detriment of indigenous biodiversity. Although assertions that we should protect intact ecosystems in order to conserve biodiversity can be bolstered by assertions about the valuable services they provide, attempts to enhance or maximize service provision through ecosystem management will seldom, if ever, preserve native biodiversity. Highly modified or novel freshwater ecosystems, even (or especially) if they contain non-native species, often provide more readily valued commodities than intact ecosystems. Lake Victoria (Box 8.4) offers a salutary example of the detrimental consequences for endemic biodiversity of an attempt to increase provisioning services.

Another potential shortcoming of PES schemes is that some ecosystem services are amenable to valuation (fishes or other exploited aquatic animals; see Box 8.3) but others (supporting services in general) are less so, and changing market prices may incentivize overexploitation of rare species (Box 8.5). Furthermore, the beneficiaries of an ecosystem service

---

Box 8.4 *Ecosystem Services and Disbenefits to Biodiversity*

Lake Victoria provides a good example of how maximizing ecosystem service provision can be detrimental to biodiversity. As described in Chapter 4, the initial impetus for introducing Nile perch to Lake Victoria was to boost fishery yields and economic returns from the lake. Once established, however, this piscivore devastated the endemic assemblage of haplochromine cichlids with catastrophic species losses. Thus, the benefit of the ready availability of Nile-perch fillets from Tanzania in overseas' supermarkets can be unequivocally related to substantial reductions in biodiversity. Furthermore, the changes in fishery yields from Lake Victoria have led to trade-offs between the benefits from ecosystem services enjoyed by different stakeholders. Artisanal fishers had gear suitable for catching haplochromines that could be readily processed by drying in the sun; Nile perch are caught by large, motorized vessels and need to be preserved by refrigeration, necessitating the development of commercial fish processing and an associated export operation. These larger-scale commercial fishers derive economic benefit from Nile perch at the expense of subsistence fishers of haplochromines who must seek supplemental means of making a living. That might include increased land clearance for farming, which aggravates existing sedimentation and eutrophication problems in Lake Victoria. A naïve economic valuation of ecosystem services from Lake Victoria would likely highlight the worth of Nile-perch exports, while externalizing the livelihood benefits received from native haplochromines.

Box 8.5 *Valuing Species: Monetization versus Intrinsic Worth*

The concept of ecosystem services is based around the idea that biodiversity and ecosystems will benefit if the costs of actions that exploit or degrade them are made explicit. One or more of three outcomes will result: either nature will be preserved because its true value is recognized; or the true cost of ecological damage will be reflected by higher prices and hence reduced market demand; or there will be an incentive to restore any damage. However, these outcomes all reflect a tendency to reduce the overall worth of nature to those components that can be readily monetarized, and ecological or environmental gains therefore become secondary to financial ones, with the result that public investment in conservation comes under market control. This monetization of ecosystem services devalues nature: firstly, because it cannot accommodate the possibility that nature has non-market worth (i.e. that it is a public good) and, secondly, because it makes biodiversity vulnerable to the vagaries of changeable markets. For instance, the value of an exploited species can increase as populations decline (Courchamp *et al.*, 2006; see Box 3.9); the incentive to exploit rises as the species becomes rarer and its monetary value grows ever larger.

PES differs from what might be termed 'biodiversity banking', which hinges on how much one might pay to protect a species, in and of itself, because the PES model puts value on a species according to its role in providing final ecosystem services. In either case, the fundamental point is that biodiversity and ecosystem services are no longer regarded as external to economic systems, in which case their loss would not be factored into monetary models. Instead these externalities are assigned a non-zero economic value. An evident complication is the decision on what that value might be, as it could be set by government, or decided between those who use the service (e.g. water consumers) and those who provide it (e.g. by managing upstream catchments), or some combination thereof. There is also the more general matter of whether putting a price on species is even desirable, as it implies that anyone with sufficient funds could make a payment that would compensate for them destroying the remaining habitat or individuals of that particular species. Also, those lacking such means might be encouraged to kill a species or degrade its habitat before their existence came to the notice of others. Nevertheless, placing a non-zero value on biodiversity or habitats at least imposes some cost on those seeking to alter or degrade the environment, forcing proponents to factor in the impacts of any proposed development into the total cost of their project. And, if the valuation is high enough, the project may no longer be economically viable.

Conservation biologists often argue that non-human species and ecosystems (or entities) should be attributed intrinsic value, thereby securing an ethical basis for conserving them. This value is said to hold irrespective of instrumental or utilitarian value, and can be taken as a normative principle (Soulé, 1985). While, at first glance, this seems defensible and appears to have the desirable attribute of transcending economic or market-based measures of value (e.g. through use or exchange), much depends on what is actually meant by 'intrinsically valuable' – apart from its converse: i.e. lacking instrumental value. This is not mere semantics: it is possible to argue for conservation without invoking that entities have intrinsic value – provided that instrumental value is not confused with market value (see Justus *et al.*, 2009). The former can include values arising for aesthetic, religious or spiritual reasons, as well as monetary gains derived from recreation, tourism, bio-prospecting for pharmaceuticals and so on. A dragonfly may have aesthetic value to someone who could not ascribe a market-based value to that insect. Instead of appealing to intrinsic value, more clarity and consistency can result from recognizing that entities can have non-monetary instrumental value – because their continued existence is valued – regardless of whether that is due to aesthetic, cultural or other reasons. Solving the problem of assessing and measuring such worth may not be easy given that 'valuing nature' has often meant monetizing it. However, adopting an instrumental-value framework when making decisions about conservation allows different values to be compared without any one form (typically, that based on markets) having precedence. In addition, most people who value nature respond to it aesthetically or emotionally, implying that conservation has an instrumental basis and that the urge to preserve particular entities does not depend solely on their intrinsic worth.

Monetization of nature was initially intended as a means of connecting conservation of biodiversity with policy-making, but it fails to capture the aesthetic, moral or spiritual values that people attach to species and ecosystems. It results ultimately in financialization of nature (*sensu* Silvertown, 2015) in which derivatives of ecosystem services (biodiversity offsets, carbon permits) become tradeable assets. While there may well be instances where monetization of ecosystem services can generate win–win outcomes for people and biodiversity, these are insufficient and probably occur insufficiently often (see main text) to substitute adequately for the non-monetary worth of nature that many feel justifies action to conserve it (e.g. Dudgeon, 2014). Nor does monetization take adequate account of the value of biodiversity to future generations (its intergenerational value), which, if embraced fully, would warrant widespread application of the precautionary principle ('do no harm') in environmental management. Ideally, freshwater biodiversity might be seen as a public good providing a service that has value to society without generating a direct financial return. However, as political philosopher Michael Sandel (2012: p. 16) notes, '. . . market values crowd out non-market values worth caring about'. At a time where markets govern decision making as never before, only the most optimistic would expect arguments for conservation couched in non-monetary terms to gain widespread traction.

may choose not to pay for the service provided. If the vegetation covering the upper part of a catchment under a PWS scheme is protected by those foregoing the short-term economic benefits of clearing forest, a predictable supply of clean water is provided to those dwelling downstream. But, in this situation, the former group (the sellers) have no assurance that the latter (the buyers) will pay for the ecosystem service, and hence no incentive to continue to maintain it. Nor can they cease providing the service – by turning off the tap – until such payments are forthcoming. In this situation, intervention and initial payments by government or non-government organizations (such as The Nature Conservancy, see above) may be needed to fund establishment of the PWS (or another type of PES) scheme, and to develop education programmes that increase awareness among buyers of potential reductions in water supply that would accompany upstream deforestation (for a Bolivian case study, see Asquith *et al.*, 2008).

Unfortunately, win–win outcomes that achieve conservation objectives, or even good environmental management, while simultaneously meeting the needs of humans appear to be rare, with trade-offs between biodiversity conservation and human well-being appearing to be far more common (McShane *et al.*, 2011). This is well illustrated by a meta-analysis of 231 studies of ecosystem-services trade-offs and synergies (Howe *et al.*, 2014); in this case, ecosystem services were defined broadly to encompass 'species diversity and abundance traits'. Trade-offs were recorded almost three times as often as synergies, and there was no generalizable context for win–win outcomes. There is, however, evidence that taking account of why trade-offs occur (for instance, when one of the parties has a private interest in a service), and using that information to design a PES scheme, is more likely to lead to a win–win synergy than planning on the assumption that such an outcome is inevitable. A pragmatic approach will probably require acceptance that ascribing value to ecosystem services within a PES framework will have beneficial outcomes for biodiversity conservation in some circumstances, but not in others. Much depends on whether stakeholders value final ecosystem services, scenic wild places, charismatic megafauna or a subset of particularly appealing taxa deemed to have high existence value, or have the goal of maximizing biodiversity. If a utilitarian focus on maintaining a particular service is the only aim, any intervention should be considered as environmental management rather than conservation, since the benefits to non-human biodiversity would be merely incidental.

Monetization (or financialization) may or may not be beneficial for biodiversity conservation, but three general points can be made (from Silvertown, 2015). First, there is a scant evidence that monetization of ecosystem services has resulted in benefits for nature that would not have accrued otherwise, nor has it led to widespread incorporation of the value of nature into decision making. Second, win–win actions that protect ecosystem services and benefit humans are the exception rather than the rule, and trade-offs where gains in human welfare are made at the expense of ecosystem services are common, especially in situations that involve markets or private interests. Third, there is insufficient evidence to determine whether PES schemes operate to produce intended outcomes, although they do show a strong tendency to benefit property owners (see also Howe *et al.*, 2014) and thus enlarge wealth inequalities. Placing biodiversity in the context of ecosystem services can provide additional justifications or opportunities for conservation, but monetization and PES cannot be relied upon as a silver bullet that will bring about the preservation of nature. However, PWS schemes should certainly be increased in both scale and number globally, since they enhance human water security and – in certain cases – can be beneficial for biodiversity (Box 8.6).

Box 8.6 *Does PWS Offer a Basis for Biodiversity Protection?*

Does the world need more PWS schemes? In order to answer this question, we need to know how much contribution expanses of less-disturbed landscape, such as protected areas, make to the provision of fresh water. Harrison *et al.* (2016) measured the quantity of water that is provided by protected areas on a global scale and assessed how threatened these areas were in terms of their water provision. The threats were multiple drivers under three main categories of human activities: pollution, flow modification and land-use change (as in Vörösmarty *et al.*, 2010; see Box 1.7). The results were compared with water provision from unprotected drainages. Protected areas – which occupy 15% of the Earth's land mass – delivered 21% of the global annual total of continental runoff ($\sim$40 000 km^3). Only 10% of this total provision was regarded as under high threat from human activities, although 63% was moderately threatened. Remarkably, no continent has fewer than 59% of users receiving water from upstream protected areas under high threat, while some regions with the greatest downstream flow from protected areas (as in the Amazon Basin) are used by relatively few people.

These findings indicate that reducing threats to protected areas, and enhancing their existing protection, coupled with designation and management of new reserves, will boost the water security of people downstream. Globally, almost 70% of reaches in the upper parts of river catchments are unprotected (Harrison *et al.*, 2016); designating them as protected areas would be beneficial to biodiversity as well as water provision. In addition, the flows, habitat conditions and riparia of low-order streams are generally less modified by humans than reaches downstream, offering practical, achievable opportunities for conservation. This is not to suggest that high-order downstream sites are unimportant for conservation (they are) but reflects the reality that PWS schemes tend to protect the upper portion of catchments, which are more likely to have intact vegetation cover than arable lowlands.

To identify global priorities for new protected areas, it will be necessary to compare their potential water-provisioning role with spatially equivalent global data for freshwater species distributions and conservation status. At present such data are insufficiently comprehensive to make this comparison at a global scale. However, the spatial distribution of threats to human water security and freshwater biodiversity are similar in many parts of the world (Plates 2 and 3; Vörösmarty *et al.*, 2010), implying that establishment of new protected areas to increase water provision for people should also benefit freshwater biodiversity.

Valuing Freshwater Biodiversity through Mitigation Banking

Another attempt to incorporate freshwater biodiversity and ecosystems in markets is by way of mitigation banking. Legislation in place in the United States requires anyone wishing to drain or fill a wetland to obtain a permit under which, assuming the work cannot be avoided, the proponent must minimize the impact, and mitigate, compensate or offset any damage to the habitat in an amount that is at least equal to, and preferably greater than, that which has been lost. The offset may involve wetland creation, restoration or enhanced protection elsewhere (i.e. offsite). This obligation has inspired the business of wetland mitigation banking, where natural wetlands owned by companies – or expanses of wetland habitat they have created or restored – are sold to project proponents so that they can meet the requirement of compensation for other wetlands that have been destroyed. In this instance, introduction of government legislation has had the effect that wetlands are no longer regarded as worthless externalities in the United States' economic system and, where their loss cannot be avoided, it is compensated by protection or creation of similar habitat. There are, inevitably, problems ensuring that the compensation includes wetland 'credits' of similar value to the sites that have been destroyed; that they support a comparable array of species (especially those of conservation interest); that there is an appropriate ratio of area compensated to area lost; that the mitigation wetlands provide the same type and magnitude of ecosystem services; and that they are managed appropriately and sustainably over the long term. However, with appropriate regulatory oversight by government (or its proxies), combined with financial assurances (such as establishment of trust funds) by project proponents, wetland mitigation banking has the potential to contribute to the conservation of freshwater biodiversity. It certainly offers a better option than the practice of treating wetlands as worthless.

The same principle could be used for biodiversity protection if, for example, an individual who owns a lake- or riverfront property that supports a species of interest could obtain credits for protecting existing habitat, or creating new habitat in compensation for some damage caused. Another approach would be to reward owners for wetlands or riparian zones set aside and remaining under native vegetation cover; where these incentives are in the form of tax deductions or credits (as in parts of the United States), the practice constitutes a conservation easement. However, the scope for mitigation banking or easements as a conservation tool for riverine or lacustrine species and habitats is probably more limited than for land-based initiatives for terrestrial biota, and protection of the upper parts of catchments, simply because relatively few individuals own lakes or rivers. Quite another approach to valuing or monetizing fresh water that might have benefits for biodiversity is described in Box 8.7.

Box 8.7 *Privatization of Water?*

Although it is a controversial idea, when a significant proportion of the Earth's population still lack access to clean drinking water, privatization in developed economies might give rise to a system of realistic pricing that reflects the scarcity of water as well as the ecological and environmental benefits associated with leaving water within rivers. This would ensure that the largest consumers have an economic incentive to increase efficiency of use and introduce measures to conserve water. For example, the agricultural sector often receives water for free or enjoys a subsidized allocation that takes priority over other competing uses; consequently, water costs are not reflected in the market price paid for crops and other produce. Privatization could lead to increases in investment in water infrastructure and efficiency of service provision, since these things would minimize wasteful losses along the supply chain and maximize customer base (i.e. number of those paying for water), thereby allowing the water company to make a profit (see elaboration by Lymas, 2011). However, it would be necessary to ensure that all potential users had access to water, irrespective of their ability to pay. While no panacea, privatization could well be a better option than leaving the task of supplying water in the hands of the public sector, given a context where governments typically invest minimal amounts in maintenance and upgrades of ageing infrastructure, with the result that much water is wasted and only a fraction of blue-water withdrawals arrive at the taps of consumers. Putting a realistic price on water can be seen as analogous to taxing carbon emissions; if citizens pay nothing for the environmental effects of burning carbon and producing carbon dioxide, then there will be no incentive for them to stop doing it. Privatization would need to be implemented in such a way that the per-unit cost of water increased as quantities consumed rose, thus discouraging profligacy, while minimizing costs for small-scale individual users.

There may be other benefits associated with privatization of water. Those actors concerned with providing water for irrigation and supplying potable water to urbanites tend to be preoccupied with building large dams and hard-engineering solutions aimed at generating economic growth in large cities, downgrading sewage treatment and the need to allocate water to support aquatic biodiversity and ecosystem services. They become matters of secondary concern, in part because the costs associated with their neglect are more difficult to quantify or monetize and become apparent only after a lag (Lodge, 2010). Projected trajectories of human population growth and diets that include more meat, which is water-intensive to produce (Vanham et al., 2018), combined with rising global temperatures and climatic unpredictability, will surely exacerbate human demands for fresh water. It would be unwise to dismiss the potential for market incentives to improve water-use efficiency and so play a role in protection of freshwater ecosystems and biodiversity.

Putting Things Together and Creating Incentives

What would need to be included within a unified management structure that reconciled the trade-offs between ecological sustainability, public health and infrastructural initiatives? Lodge (2010) suggests that it should incorporate at least five components: (i) a decision framework that incorporates benefits and costs to all sectors; (ii) consideration of the impact of management on both water quantity and quality; (iii) a sufficiently broad spatial scale to integrate humans and ecosystems within entire drainage basins; (iv) a temporal scale long enough to include the effect of lag-times on society and ecosystem services; and (v) creative government and market incentives that both encourage and implement such management. Taking greater account of provision of ecosystem goods and services in development decisions could increase the net, long-term benefits of water projects and, perhaps, facilitate win–win outcomes. An excellent example is the decision to spend money improving forest management in upstream drainages of New York City instead of constructing a multibillion-dollar water treatment facility (as described above under Payment for Ecosystem Services), and would likely be a transferrable model given how much surface-derived drinking water flows from vegetated catchments (Brown *et al.*, 2008).

As part of a unified management structure, protection of freshwater ecosystems and biodiversity requires education – so that people are aware of the threats and the need to abate them. Legislation, plus the necessary enforcement, to protect habitats and prevent their degradation is needed also: for example, by implementation of water-quality standards via the 1972 US Clean Water Act or the 2000 EU Water Framework Directive. Under such legislation, poor stewardship of inland waters can be punished by fines or other penalties. Education and legislation are necessary for protection of fresh waters, but they are not sufficient. They need to be supplemented by incentivization or inducements that encourage superior performance by stake holders so that they 'do the right thing' and make the necessary connections between knowledge, awareness and environmental behaviour (Sweeney & Blaine, 2016). For instance, landowners could be encouraged, through credits, easements or payments, to protect water quality in drainage basins by maintaining riparian buffers. Both the extent of stream length protected and the width of the buffer could be subject to incentives. The latter is important because buffers ≥ 30 m wide are needed protect against changes in thermal regime, to remove the fine sediments that degrade stream habitat, and maintain the biological integrity of streams (Sweeney & Newbold, 2014). However, narrower buffers also bring some benefits (Box 8.8). Incentivization might involve giving landowners proportionately more money or higher reimbursements for maintaining a wider buffer; or scaling the level of incentives to buffer width; or creating disincentives to creating inadequate buffers by offering greatly reduced credit for narrow buffers. The goal would be to develop ways to scale incentives to the adoption of better practices for drainage-basin management and thereby protect the receiving waters. This is but a short step from PWS initiatives or other PES schemes that reward landowners for management practices that enhance ecosystem services.

A paradigmatic example of an initiative to enhance ecosystem health and guide investment in catchment protection and rehabilitation is the

Box 8.8 *Protection of Tropical Streams by Riparian Buffers*

The use of riparian buffers to protect streams in temperate regions is well established (Richardson *et al.*, 2010; Sweeney & Newbold, 2014), although the precise width needed may depend upon local circumstances (e.g. Cole & Newton, 2013; Adkins *et al.*, 2016), but their usefulness has been less studied in the tropics. Research on oil-palm (*Elaeis guineensis*: Arecaceae) plantations in central Kalimantan (Indonesia) suggests that buffers can be beneficial for fishes even if the buffers are as narrow as 10 m (Giam *et al.*, 2015). The benefits of buffers increased up until widths of 200 m. Streams that lacked riparian buffers had around 40% fewer species than those that retained them, and buffered streams tended to maintain coarse-grained bed sediments and higher stocks of allochthonous litter. This finding is particularly interesting as Giam *et al.* (2015) asserted that theirs was the only study that had documented the benefits for aquatic taxa of retaining tropical forest patches within plantations. More recent studies of aquatic insects in streams draining oil-palm plantations in Brazil also indicate that community structure is more likely to be intact at sites where riparian vegetation has been retained (Luiza-Andrade *et al.*, 2017).

Research in northern Kalimantan (Sabah, Malaysia) demonstrates that the presence of riparian buffers in oil-palm plantations offered some protection for forest-associated dragonflies, and streams with wider riparian buffers supported adult assemblages more similar to – although less speciose – than those found in forest (Luke *et al.*, 2017). However, the beneficial effect of narrow (<20 m) buffers for dragonflies was negligible, resulting in assemblages similar to those of non-buffered oil-palm plantation streams. Another study in Sabah reported that buffers reduced impacts of plantations on macroinvertebrate communities, but less protection was offered by narrow buffers (Chellaiah & Yule, 2018). Evidently, the width of buffers (20 m) required by law in Sabah for streams more than 3 m wide do little to maintain forest-stream-dependent dragonfly communities (Luke *et al.*, 2017). Legislation in Indonesia mandates the retention of riparia along streams within oil-palm plantations, with widths of 50 to 100 m – according to river size – prescribed by law. Unfortunately, enforcement is weak, and clear definitions of river size and the required composition of vegetation within buffers are lacking (Giam *et al.*, 2015). Elsewhere, however, there is evidence that legislation on buffer-zone widths does provide adequate protection for tropical headwater stream communities (e.g. fishes and macroinvertebrates in Costa Rica: Lorion & Kennedy, 2009a, 2009b).

Healthy Waterways Programme in southeastern Queensland, Australia (Bunn *et al.*, 2010). Multiple stressors, acting at different spatial and temporal scales, interact to affect river water quality, biodiversity and ecosystem processes in this region. The Programme monitors 16 variables twice a year at 250 sites, and these data form the basis of an annual report card that is presented in a public ceremony to local politicians and the broader community. Identification of the primary causes of degradation and the appropriate spatial scale for rehabilitation or protection allow targeted management investments in catchment protection and rehabilitation; greater public confidence that limited funds are being well spent; and better outcomes for ecosystem health.

The Healthy Waterways Programme represents an instance where scientific approaches to managing freshwaters have been leveraged by adding a social and political dimension to a research and monitoring effort so as to build consensus over what can and should be done to protect fresh waters. A similar partnership in the northeastern United States, involving researchers, conservation organizations, citizens, business and different levels of government, has allowed development of a long-term collaboration resulting in new land-use and management practices to achieve the joint aims of economic development and vernal pool protection (Hart & Calhoun, 2010). Both of these initiatives represent instances of the application of technical tools and knowledge to areas that have traditionally been seen as beyond the ambit of researchers, a practice that has been identified by Ehrlich and Pringle (2008) as key to the success of initiatives intended to prevent biodiversity loss and protect ecosystems (see also Chapter 9). The incentivization of sustainable management of catchments in order to increase the public good is one example of a practice that could generate conservation gains, with benefits for terrestrial and aquatic biota as well as people.

The incentivization approach to drainage-basin management could be extended if the rewards were scaled non-linearly to the area involved. Or, payments could be on a sliding scale that reward demonstrated environmental improvements, based on evidence from monitoring that environmental targets have been achieved or exceeded. Incentive programmes that provide additional payments for measurable mitigation of target substances, such as nutrients and sediment, or desirable changes in streamflow may also be feasible (Sweeny & Blaine, 2016). The monitoring equipment needed to demonstrate such improvements could be subsidized by downstream users paying into a clean-water fund, with the money used to pay for solar-powered data loggers, or sensor units, as well as the necessary drainage-basin management practices. Precedents for water funds are well established by, for instance, The Nature Conservancy (https://waterfundstoolbox.org/), although the operational details vary on a case-by-case basis. Through the use of incentives for appropriate drainage-basin management practices that reward private behaviour and increase the public good by enhancing the freshwater commons, Sweeney and Blaine (2016: p. 760) '... envision a system that landowners will embrace, rather than one that merely penalizes them, which, if history is any guide, is a strategy doomed to fail'. Given that headwater streams comprise over two thirds of total channel length in a typical river drainage and directly connect the upland and riparian landscape to the rest of the ecosystem (Freeman *et al.*, 2007), protection of catchment land cover or, at least, maintenance of adequate buffers of intact riparia, would make a significant contribution to conserving freshwater biodiversity (see also Box 8.6). It could be both achievable and, through provision of downstream ecosystem services, help meet the needs of growing human populations.

9 · *Conservation of Freshwater Biodiversity*
Opportunities and Initiatives

O star-eyed science, hast thou wander'd there, to waft us home the message of despair?
  Thomas Campbell (1829: p. 39), *The Poetical Works of Thomas Campbell*

To be truly radical is to make hope possible, rather than despair convincing.
  Raymond Williams (1989: p. 118), *Resources of Hope: Culture, Democracy, Socialism*

We cannot win this battle to save species and environments without forging an emotional bond between ourselves and nature as well – for we will not fight to save what we do not love ... We really must make room for nature in our hearts.
  Stephen Jay Gould (1991: p. 14), Unenchanted evening. *Natural History*, September 1991

Meeting the Grand Challenges: Will Freshwater Protected Areas Help?

Freshwater biodiversity is subject to three 'great' forces: thinning, shrinking and mixing. They will be augmented by changes in the Earth's climate, and many disruptive effects of warming are already readily apparent (see Chapter 7). Avoiding further change would require a dramatic shift in the way that humans use fresh water and the value that they place upon it – something that is unlikely to occur in the near future, although the PES and PWS schemes described in the previous chapter might offer some basis for optimism – so long as they can be scaled to encompass the middle and lower parts of catchments. What other initiatives or interventions might provide opportunities for conserving freshwater biodiversity?

In cases where the main threat to a species or assemblage of conservation concern is overexploitation (see Chapter 3), then fisheries management or regulation of capture and international trade (as with turtles and frogs, for example) should be legislated, implemented and enforced. Assisted translocations offer a potential – albeit controversial – means by which species at risk from excessive warming in their native ranges could be spared (see Chapter 7), or could be relocated from highly degraded habitat or sites where they are threatened by an aggressive alien species (see Chapter 4). Thus far, such interventions are rare (but see Box 9.1). Immediate conservation action is needed in some instances where opportunities exist to set aside relatively unspoiled lake and river systems in large protected areas but, as noted in Chapter 8, there are few places on Earth where a freshwater body could be dedicated to the exclusive purpose of biodiversity conservation, with humans denied access or largely prevented from using the water resource. While this 'fortress conservation' approach can be applied to areas of land with high-quality habitat that can be bounded and protected, it is likely to fail for river segments or lakes embedded in unprotected drainage basins unless the boundaries are drawn at a catchment scale, which is virtually never the case. For instance, protection of a particular constituent of the riverine biota requires control over the upstream drainage network, the surrounding land, the riparian zone and – in the case of migrating aquatic fauna – downstream reaches. Conservation action at a large scale, involving interconnected landscape units, is needed also for certain terrestrial taxa that undertake seasonal migrations, but the shortcomings inherent in fortress conservation are particularly acute for freshwater biodiversity, not least because of the strong directional connectivity of river ecosystems.

Box 9.1 *Translocations to Support Porpoise Conservation*

The Yangtze finless porpoise (*Neophocaena asiaeorientalis asiaeorientalis*: Phocoenidae; CR) is an isolated subspecies of a widespread East Asian cetacean and the only freshwater porpoise in the world. It is endemic to the Yangtze River, where it is confined to the floodplain and, since 2014, has been categorized as a National First Grade Key Protected Wild Animal. Steep declines in the abundance and distribution of this porpoise (also known as 'river pig') are attributable to the same array of factors that eliminated the baiji (Box 2.1) – entanglement with fishing gear, pollutants, collision with vessels, sand mining and dams limiting access to floodplain lakes – raising understandable concerns that it may soon become extinct. One estimate of the remaining population was ~1800 individuals, with an annual rate of decline of at least 5% (Zhao *et al.*, 2008); four years later, a comprehensive survey put the total number in the Yangtze mainstream and floodplain lakes at only 1040 (Mei *et al.*, 2104). These accelerating rates of decline, combined with population fragmentation, are predicted to lead to a high probability of extinction (almost 90%) within the next 100 years (Mei *et al.*, 2012).

In 1992, Chinese authorities established a national reserve at Tian-'e-Zhou in Shishou County (Hubei Province), incorporating a 21-km-long oxbow lake cut off from the Yangtze. It had been hoped that the oxbow could support a breeding population of the now-extinct baiji (see Box 2.1), providing a basis for its *ex situ* conservation as part of the 'Baiji and Yangtze Finless Porpoise Protection Act' approved by the Chinese Ministry of Agriculture in 2001. Collaboration between the Ministry and the Institute of Hydrobiology of the Chinese Academy of Sciences led to the capture and translocation of porpoises to the oxbow on several occasions. They bred successfully, and the 'semi-natural' population has grown to around 90 individuals. Following experience gained in Tian-'e-Zhou, a second translocation site was identified at the He-Wang-Miao (or Ji-Cheng-Yuan) oxbow, which spans Hubei and Hunan Provinces (and has different names in each). It was incorporated into a translocation site network established by the Ministry of Agriculture in 2015, and a third site – Xijiang oxbow within Anqing Nature Reserve, Anhui Province – was added in 2016. Translocations among the oxbows and fresh captures from the Yangtze have been used to boost genetic diversity of the three porpoise populations.

Six *in situ* reserves have also been established along the Yangtze in localities where the porpoise was formerly abundant, and attempts have been made to limit boat traffic and harmful fishing practices within them. The Saving Yangtze Finless Porpoise Alliance was launched in 2017 and is made up of government institutions, research centres, nature reserves and non-government organizations, coordinated by the Ministry of Agriculture and the Yangtze Fishery Administration Office. It intends – among other things – to establish new *in situ* porpoise reserves, patrolled by former fishermen. However, it seems unlikely that management efforts alone can significantly reduce the impacts of anthropogenic activities within the Yangtze, limiting the effectiveness of *in situ* conservation measures for the porpoise

(Wang *et al.*, 2013). Although it could be argued that the three oxbow sites were historic porpoise habitat, and hence constitute part of an *in situ* conservation initiative, establishment of the breeding populations depended on translocations, and thus they are more analogous to *ex situ* efforts. Effective management of these three populations will likely be an essential component of attempts to conserve the Yangtze finless porpoise.

In spite of the difficulties of establishing appropriate boundaries, it should be possible to identify priority areas for the conservation of freshwater biodiversity, so long as they are informed by planning efforts focused on aquatic species of interest rather than better known terrestrial taxa (typically birds and mammals) that are sometimes assumed to be effective surrogates. Furthermore, criteria and thresholds employed during conservation planning for terrestrial habitats are frequently inappropriate for freshwater ecosystems (Dunn, 2003). One method for identifying global priorities for freshwater protected areas involves the detection of key biodiversity sites based on the presence of threatened and endemic species or ecologically unique assemblages (Holland *et al.*, 2012). They are then mapped using HydroBASINS (Lehner & Grill, 2013), the best global-scale digital tool for mapping hydrology and connectivity of river networks and lakes within catchments that is currently available. The application of this approach at a continental scale in Africa and parts of Asia (Darwall *et al.*, 2011b; Allen *et al.*, 2012; Holland *et al.*, 2012) has shown that many key freshwater biodiversity areas do not coincide with the existing terrestrial protected-area network in terms of either spatial coverage or management focus. Similarly, most of the protected areas in the Brazilian Amazon established for terrestrial vertebrates – some of which encompass large areas – do not correspond to places of high conservation value for fishes, leaving a high portion of the regional vertebrate fauna inadequately protected (Guimarães Frederico *et al.*, 2018). This confirms the results of much smaller-scale studies of inland waters in Michigan and the Murray–Darling Basin, showing that terrestrial reserves do not adequately represent or protect freshwater biodiversity and ecosystems (Herbert *et al.*, 2010; Chessman, 2013), in part due to the high levels of natural variability and spatiotemporal connectivity that characterize inland waters (Adams *et al.*, 2015). The mismatch is surely one of the reasons for the relatively high levels of endangerment of the freshwater fauna. Complementary planning and management for both freshwater and terrestrial conservation targets is needed to rectify this discrepancy.

It is essential that decisions about prioritizing areas for protection in river and stream drainages must take account of connectivity, which has a very strong directional component, if they are to lead to efficient and targeted conservation action (Adams *et al.*, 2015). High levels of connectivity must be maintained to ensure free movement of organisms, but also have implications for the transfer of materials (sediment, detritus) and the potential for threat propagation within drainages (for instance, through the downstream flow of pollution). Attempts to identify sites for inclusion in expanded national protected-area systems that improve the representation of freshwater biodiversity typically combine a geographical information system with conservation planning algorithms, so that the network of potential sites is mapped in a spatially efficient manner (e.g. Nel *et al.*, 2009; Lira-Noriega *et al.*, 2015). If riverine connectivity is ignored or not included appropriately within decision-support tools and computational approaches for identifying priority areas, the sites selected tend to be highly fragmented. The solution is to model directional connectivity as part of the site-selection process, in which case high-priority conservation areas tend to consist of entire river basins or headwater sub-catchments (Moilanen *et al.*, 2008).

While there is an urgent need to identify sites where freshwater protected areas should be established, it is just as important to decide upon proper management plans for these locations. An integrated approach would consist of three stages: first, identification of focal sites or habitats that are important for species or communities of concern; second, definition of critical management zones that would support the integrity of these areas; and third, embedding these zones within a wider catchment-management scheme that integrates multiple-user needs (Abell *et al.*, 2007). The focal sites and critical management zones would form components of the management plans for key freshwater biodiversity areas, which would scale up to incorporate protection of the wider catchment (Linke *et al.*, 2011). Unfortunately, biodiversity objectives and management actions are rarely captured across freshwater and terrestrial realms, despite awareness that such integration is needed for managing landscape-level threats and the functional connectivity between land and water (Linke *et al.*, 2011), probably because topical but separate analyses are more expedient (Leonard *et al.*, 2017). In cases where the management focus is primarily on the aquatic environment, attention will need to be paid to particular interventions, such as greater control of alien species and adequate e-flow allocations, if freshwater protected areas are to realize their full conservation potential (Chessman, 2013).

One exception to the generalization that terrestrial reserves offer insufficient protection for freshwater species can be seen in hyperdiverse Lake Tanganyika (see Box 2.6), where the limited extent of shoreline included within terrestrial protected areas increased the taxonomic and functional diversity of near-shore assemblages of endemic cichlids (Britton *et al.*, 2017). Despite not being designed for fresh waters, the protected areas improve conditions in Lake Tanganyika through local reduction of sedimentation and pollution. Elsewhere, a study of 175 lakes across Ontario evaluated whether terrestrial protected areas benefitted lake-fish assemblages but failed to detect any influence on overall abundance or diversity. There was, however, a tendency towards greater abundance of small-bodied species in lakes where protection was lacking, and this trend was accompanied by higher dissolved-solid loadings and greater angling pressure (Chu *et al.*, 2018).

Most modern systematic planning approaches to identifying areas in need of protection are based on the CARE principles: comprehensiveness, adequacy, representativeness and efficiency (Linke *et al.*, 2011, and references therein). Efficiency is usually provided by a complementarity-based strategy, aiming to select new areas in the light of previously protected features (see also Nel *et al.*, 2009). As explained above, these strategies require supplementation to account for the connectivity of rivers because setting adequacy targets – the most challenging aspect in planning – needs to be evaluated in a freshwater-specific context that takes account of the distribution of diversity in dendritic drainages. Further complexity is added to the conservation planning process by the fragmentation of these drainage networks by barriers, such as dams, which compromise connectivity. New tools are becoming available to model the location of priority areas within fragmented drainages, with the aim of reducing the number of disrupted connections at no additional cost in terms of the extent of protected area needed (Hermoso *et al.*, 2018).

The Strategic Plan for Biodiversity (2011–2020), adopted by all parties to the Convention on Biological Diversity, set 20 Aichi Biodiversity Targets to address biodiversity loss and ensure its sustainable use by 2020. Target 11 describes what an improved global conservation network (including inland-water ecosystems) should look like, but there has been no comprehensive assessment of what needs to be achieved under Target 11 for conservation of freshwater biodiversity. Juffe-Bignoli *et al.* (2016)

identify the need for concerted action on several fronts. Designation of new protected areas or expansion of existing ones to encompass areas of importance for biodiversity and ecosystem services, including a representative sample of biodiversity, should take place at national and global levels, combined with the application of other area-based conservation measures in places where designating a protected area is not feasible or appropriate. Promotion and implementation of better management strategies for freshwater protected areas is required also, and must take account of connectivity, contextual vulnerability, and the necessary technical and human capacity. To these can be added the need for additional resources devoted to freshwater conservation management; the necessity for better understanding of complex management problems beyond the boundaries of the protected area; and the implementation of monitoring programmes to assess the effectiveness of protected areas for freshwater biodiversity (Hermoso *et al.*, 2016, and references therein). This last point is key, since it is not known whether the subset of freshwater species that are included within existing protected areas are actually secure.

The imperative to find solutions that generate co-benefits for humans as part of efforts to conserve freshwater biodiversity, perhaps by ensuring the delivery of ecosystem services (as discussed in Chapter 8), may present opportunities for the creation of new freshwater protected areas (or improved management of existing ones) in the context of wider - catchment-scale initiatives, such as PWS schemes (see Box 8.6). Nonetheless, given the inadequate safeguards that terrestrial protected areas provide for freshwater biodiversity, and the inherent difficulties in establishing new and effective freshwater protected areas, there seems little chance of meeting Aichi Target 11 for freshwater biodiversity by 2020. The greatest uncertainty may be just how far we will fall short of what is needed.

Habitat Restoration and Rehabilitation

In the many cases where habitat destruction or degradation are primary threats to freshwater biodiversity, restoration measures may ameliorate their worst effects, allowing species persistence or even recovery. Restoration or rehabilitation of degraded streams, rivers and other water bodies is a topic that has growing importance and increased in complexity. With regard to improving water quality, the application of biomanipulation to lakes is quite well established in north-temperate latitudes (e.g. Mehner *et al.*, 2002; Bernes *et al.*, 2015), and has begun to receive attention in the southern hemisphere also (e.g. Sierp *et al.*, 2008; Burns *et al.*, 2014). As a result, biomanipulative techniques have been incorporated into integrated lake-specific programmes to enhance water quality in many locations. The practice is particularly successful in shallow lakes where submerged macrophytes can thrive, and in water bodies where control of nutrient inputs and water quality can be combined with fisheries management that includes stocking of piscivores. The literature on biomanipulation is large, as might be imagined given the global extent of research on lake eutrophication; space precludes further consideration of the topic here.

River and stream restoration can encompass both the aquatic habitat and associated riparian zones, and has grown in prevalence worldwide during the past three decades. Restoration can be defined broadly to encompass any activity that assists the recovery of reaches that have been degraded, damaged or destroyed; habitat enhancement for biodiversity is just one of the many reasons for undertaking such work. Formerly, restoration was seldom planned and executed with inputs from ecological theory (Lake *et al.*, 2007), but there are now some generally agreed principles of good practice, as well as recognition of the need to establish clearly defined goals and outcomes (Palmer *et al.*, 2005; Jansson *et al.*, 2005). Evaluations of river-restoration projects that use channel reconfiguration as a methodology for improving stream ecosystem structure and function have found little evidence of measurable ecological improvement (e.g. Bernhardt & Palmer, 2011), and restoring physical processes is more straightforward than the rehabilitation of community structure and function (Palmer *et al.*, 2010, 2014; Englund & Wilkes, 2018). There appear to be few convincing examples of restoration success in multiple-stressor situations at scales sufficient to have ecosystem-level benefits (Collier, 2017; see also Box 5.11). One exception involves a large-scale restoration project in lowland reaches of the Rhône River, which involved making changes to discharge and the morphology of floodplains and channels. There was good evidence that diversity of fish and invertebrate communities increased (especially lotic species), and social benefits were derived from renewal of the relationships among the stakeholders linked to the river (Lamouroux *et al.*, 2015). As mentioned in Chapter 5, a failure of the 'if you build it, they will come' approach to restoring flows and physical habitat (for instance, by increasing heterogeneity) can be attributed to lack of alleviation of stressors or legacy effects limiting recovery of the original community (Palmer *et al.*, 2010; Thompson, 2018). However, to date, the outcomes of river restoration have seldom been compared across projects, which makes it difficult to identify factors that influence their success or failure (Weber *et al.*, 2018).

Even if restoration interventions can increase biodiversity, the evidence for positive effects on ecosystem processes or functioning, such as organic-matter transformation, is equivocal (Lake *et al.*, 2007). And, returning once more to the matter of the relationship between biodiversity and ecosystem-service provision (see Chapter 8), restoration that refashions stream channels in an attempt to maximize ecosystem services (flood control, nutrient storage and sediment retention) can transform community composition into something quite different from the historical state (Palmer *et al.*, 2014; see Chapter 8). In such cases, the terms 'rehabilitation' or 'ecological engineering' seem more appropriate than 'restoration', and such a stream or river may well provide ecosystem services in the absence of conservation gains; it may also be possible that interventions intended to maximize biodiversity compromise ecosystem-service benefits.

The general topic of river restoration, which has only been touched upon here, deserves a book of its own. Examples of such include Darby and Seer (2008) and Speed *et al.* (2016), with the latter being notable for incorporating case studies from outside Europe and North America. Online information on good practice can be obtained from a number of sources: for instance, the UK River Restoration Centre (www.therrc .co.uk/) provides an online Manual of River Restoration Techniques that aims to help potential users identify and implement a range of techniques for use in sustainable river management. The topic will not be addressed in further detail here, although some aspects of river restoration have already been considered in relation to e-flows in Chapter 5. Targeted initiatives to restore riverine connectivity will be discussed below.

Re-establishing Connectivity: Fishways and Dam Removal

Fragmentation of watercourses arising from impoundment, flow regulation and water abstraction poses a major threat to freshwater biodiversity worldwide, especially fishes (reviewed by Reidy Liermann *et al.*, 2012; see also Chapter 5) but also amphidromous *Macrobrachium* shrimp (Olivier *et al.*, 2013) that can support important 'fisheries' in tropical rivers (see Chapter 3). Some of these obstacles can be made 'permeable' or, to be more accurate, 'semipermeable' by the provision of suitable passageways within the original structure, or by retrofitting to mitigate observed barrier effects. Fishways (including fish passages, fish ladders) – defined here as any structure intended to facilitate safe and timely fish movement past an obstacle – date back at least several centuries; nonetheless, the performance of these structures remains low in many regions (Silva *et al.*, 2018). Furthermore, many dams lack fishways, or have installed structures designed to suit salmonids (Plate 27) in countries that have a fauna lacking such fishes; consequently, up- or downstream passage of migrating fishes may be prevented (e.g. Bunt *et al.*, 2012; Pelicie *et al.*, 2015). Other reasons for the failure of fishways are a lack of relevant biological knowledge – with designs formerly dominated by an engineering-focused approach – and flaws in fishway construction and/or operation (Kemp, 2016). Nonetheless, there remains an erroneous view in some of the technical literature and communities of practice that fish passages are largely a proven technology (reviewed by Silva *et al.*, 2018).

Installation of fishways at dams that previously lacked them are, essentially, restoration projects with the goal of re-establishing longitudinal connectivity. Some partial success in re-establishing up- and downstream migrations by anadromous brown trout has been reported, although turbine-induced losses during downstream movement remained high (Calles & Greenberg, 2009). Successful passage in both directions by Atlantic salmon took place after replacement of a steep fishway associated with a Norwegian hydropower dam by one with a more natural, lower-gradient design (Nyqvist *et al.*, 2017), demonstrating the importance of matching fishway features to the target species. Fishways installed at many previously disconnected sections of the Murray–Darling River have had mixed success depending on the height of the dam and the target species, even though extensive research on the biology of native fishes was undertaken to inform fishway design and operation (Baumgartner *et al.*, 2014). However, construction of a rock ramp fishway in one Murray-Darling tributary did allow Macquarie perch (*Macquaria australasica*) – a nationally endangered species – to gain passage from a reservoir to areas of previously inaccessible spawning habitat (Broadhurst *et al.*, 2013).

Assessment of the effectiveness of different designs and types of fishway for representative species of migratory fish is needed urgently, especially in the tropics, and such targeted research would pay conservation dividends as the results could be applied readily. A range of stream types should be assessed to identify the most effective design for multispecies fishways (Steffensen *et al.*, 2013; Yoon *et al.*, 2015), but one obvious generalization is that – irrespective of design details and ecological context – fishway effectiveness is inversely proportionate to dam height. Improved international collaboration, information sharing and method standardization are required, accompanied by development of regional expertise in South America, Asia and Africa where hydropower dams are being planned and constructed at an unprecedented rate (see Chapter 5).

Whatever measures are put in place within river drainages to rehabilitate or restore them, or maintain network connectivity, anadromous fishes must pass through estuaries which, in the Anthropocene world, are often the location of cities, with seawalls and harbour facilities that provide an unfavourable environment for fishes. However, the effects of shoreline armouring can be mitigated by restoring shallow waters and substrate complexity (by enhancing texture and relief), and minimizing shading underneath overwater structures (reviewed by Munsch et al., 2017). For instance, adding steps (protruding 'shelves') to sea walls in Seattle had the effect of creating complex near-shore habitat, providing refuges and enhancing the abundance of invertebrate food. In consequence, juvenile chinook, pink and chum salmon could feed, grow and avoid predators as they travelled along the engineered shoreline through Puget Sound to the ocean.

At best, fishways are no more than a partial solution to the obstacles presented by dams, because the associated reservoirs are also a barrier to migration – especially in a downstream direction (Agostinho et al., 2011; Pelicice et al., 2015). The scant hydraulic cues in lentic water bodies discourage downstream migration by rheophilic adult fishes, while the lack of flow is detrimental to the downstream drift of eggs and larvae that would carry them to nursery habitats; a similar requirement for transport of larvae is also necessary for river shrimp (Novak et al., 2017). The magnitude of barrier effects is positively correlated with reservoir size (especially length) and water residence time, and thus, in some cases, fishways that permit upstream migration have little conservation value (Pompeu et al., 2012). The only known technical solution to mitigate the impacts of reservoirs on migratory river fishes is to open or completely remove dams.

Restoring riverine connectivity by opening dams or completely removing them has been undertaken in a few countries only. More than 1200 dams have been removed in the United States during the last 40 years: the decadal rate of removal is increasing exponentially (O'Connor et al., 2015; Bellmore et al., 2017). Such events are often spurred by the hazards posed by ageing infrastructure but could give rise to conservation gains if migration routes of fishes are re-established. The governments of California and Oregon have announced plans to remove four hydropower dams on the Klamath River, as part of an effort to restore salmon fisheries. Larger dams are becoming subject to attention, as evinced by removal of the 64-m-tall Glines Canyon Dam from the Elwah River in Washington during 2014.

Rivers respond quickly to dam removal (Tullos *et al.*, 2016), eroding and redistributing sediment, re-establishing connectivity and returning to pre-impoundment conditions within years, rather than decades. Most scientific evaluations of removal (involving fewer than 10% of total removals) have focused on these rather rapid geomorphic changes, whereas ecological recovery is slower but nonetheless fairly rapid: salmonids and other migratory fishes readily colonize newly available habitat upstream (Hitt *et al.*, 2012) exemplifying a more general replacement of lentic fishes by lotic species (O'Connor *et al.*, 2015). Macro-invertebrate richness tends to increase after dam removal, typically reaching a maximum within 5–20 months (Sullivan & Manning, 2017; Carlson *et al.*, 2018), but responses are sometimes more equivocal with suppression of richness (Renöfält *et al.*, 2013), reflecting short-term declines that might result from increased sediment export (Carlson *et al.*, 2018). Unexpectedly, removal of low-head dams has rather small ecological effects, leading only to relatively nuanced shifts in water-to-land trophic dynamics, and minor consequences for terrestrial consumers (birds, spiders) feeding on volant aquatic insects, despite major changes to the physical structure of river channels (Sullivan *et al.*, 2018).

While rates of construction of dams – especially large ones – far outpace the number of removals, the practice has momentum: if current trends continue, the United States can expect between 4000 and 36 000 total removals by 2050 (Grabowski *et al.*, 2018). The characteristics and ages of those dams removed are not, however, representative of the current inventory. The majority are privately owned hydropower and water-supply dams, mostly <8 m tall (Bellmore *et al.*, 2017). Climate change will likely lead to a further increase in the rate of dam construction to satisfy increasing demands for low-carbon energy sources and growing consumptive use of water in a warmer world (Chapter 7). In addition, the numbers and types of dams being removed is hardly likely to be sufficient to offset the anticipated transformation of freshwater ecosystems associated with plans for large-scale water-engineering schemes (e.g. Shumilova *et al.*, 2018). Nonetheless, 'river reclamation' by dam removal is a conservation intervention with the potential to re-establish connectivity, enhancing overall ecosystem resilience, and benefits migratory species.

Reintroduction, Translocation and *Ex Situ* Conservation

In cases where restoration of the physicochemical features of a formerly degraded habitat has been accomplished, there is scope for considering reintroduction of species of conservation concern – either to boost a depleted population or to establish a new population in localities where that species flourished in the past. The approach is applicable also in cases where a species has been eliminated from parts of its geographical range as a result of overexploitation. It depends on the existence of historic records of what species were formerly present, as was the case for Eurasian beaver (see Box 9.2). However, such information is not invariably available.

Box 9.2 *Beaver Reintroduction to the United Kingdom*

Historically, the Eurasian beaver was hunted to near-extinction over much of its range and, by 1900, had been reduced to a few small relict populations (see Chapter 3). Numbers began to recover during the twentiethcentury in places where this beaver became legally protected from hunting, and, since the 1920s (beginning in Sweden), there have been separate reintroductions to 15 European countries. However, the United Kingdom, where Eurasian beavers were probably eliminated in the sixteenth century, has been an exception to this general trend. There were no officially endorsed attempts to reintroduce Eurasian beavers until relatively recently, although some individuals had been held in fenced semi-natural enclosures within reserves. The first experimental release of a handful of animals in Scotland did not take place until 2009 (Law *et al.*, 2014). There had been some earlier informal introductions of beaver to sites on private land (Law *et al.*, 2016), and some escapees from captivity had become established in the River Tay (western Scotland), from where they began to spread to the River Earn and beyond. A handful of individuals of unknown origin were present in the River Otter (southern England) in 2010, where they have bred, and there has been a government-sanctioned trial release and monitoring of two individuals in the same part of the country.

The apparent reluctance to undertake beaver reintroductions in the United Kingdom is surprising given that they were formerly part of the native fauna. Beavers are known to act as keystone species elsewhere (see Chapter 3), and reintroductions in some parts of Europe were not undertaken to improve aquatic habitat and fulfil a range of management objectives (Rossel *et al.*, 2005; Kemp *et al.*, 2011). Dam building by Eurasian beaver released into degraded streams draining private pastureland in eastern Scotland had environmental benefits, including increased retention of organic matter and higher biomass of aquatic plants (Law *et al.*, 2016). Nonetheless, there is a persistent view that beavers are nuisance animals (and they are treated as such in some American states), responsible for increased risk and extent of flooding, damage to trees and reduced timber yields, as well as impacts on fisheries. All have been invoked as reasons not to reintroduce Eurasian beaver to the United Kingdom. In 2016, however, the Scottish government granted beaver protected-species status, and its range expansion has continued.

Shifting baseline syndrome (described in Chapter 3) has important implications for restoration efforts, because it is accompanied by 'ecosocial anomie' where there is a breakdown in expectation of what species should be present in particular water bodies, and thus what needs to be restored. If people cannot remember what has been lost, or what conditions were formerly like, then it becomes difficult to manage degraded fresh waters in ways that will allow the recovery of rare or threatened species – in part because the target conditions for restoration have been forgotten (Turvey *et al.*, 2010). Furthermore, the accelerating and effectively irreversible environmental change of the Anthropocene has put any historical baseline out of reach (Corlett, 2016), and thus restoration to a historical reference state may no longer be feasible, or would be prohibitively expensive and, hence, not societally acceptable. Restoration or rehabilitation to some desired state (e.g. enhanced populations of native species, increased ecosystem-service provision), or to a counterfactual (relative to what would have happened without intervention), or simply improvement of conditions relative to present baselines, may represent a sounder, more achievable approach than efforts based on appeals to some imagined Edenic past. It is only in a few serendipitous instances – as in the case of the milu (Box 9.3) – that opportunities remain for conservation of near-extinct species that have long been forgotten by local communities. In this instance, however, centuries–old historic records do attest to the presence of Père David's deer along the Yangtze.

Box 9.3 *Back from the Brink*

Père David's deer or milu (*Elaphurus davidianus*: Cervidae) is – or was – an inhabitant of swampy river floodplains in central and southern China (Plate 28). These deer are amphibious and strong swimmers, spending considerable time in the water as well as on grasslands and in reed beds where they graze a mixture of semi-terrestrial and aquatic plants. Their hooves resemble those of cows and are adapted to soft ground. Because of the productivity of the floodplain habitat, milu can reach 200 kg and are larger than the majority of terrestrial deer; the males have large and many-branched antlers. The colloquial Chinese name of milu translates as 'the four unlikes', although the accounts vary as to what the constituents might be: a horse's face, a cow's hooves (and nose), a deer's antlers, a donkey's tail and a camel's neck. Although formerly abundant and widespread in China, milu populations had already been greatly depleted 1500 years ago – and perhaps much earlier – by habitat loss, due to conversion of floodplain to rice paddy, and hunting (Jiang & Harris, 2016).

Milu became known to western science in the 1860s through the observations of missionary Père Armand David who encountered the animals in Beijing (then Peking). Numbers elsewhere in China had been greatly reduced around 500 years before, and almost all of the surviving milu were part of a herd in the royal hunting garden where they had been maintained – by a succession of emperors – since the thirteenth century. A few milu were subsequently transported to Europe, which proved fortunate since series of accidents and political upheavals in the late nineteenth and early twentieth century depleted the last herd, and the survivors were slaughtered during the Boxer Rebellion. Subsequent survival of milu was a result of maintenance and captive breeding of descendants of the animals exported from China at Woburn Abbey in England.

Although currently classified by the IUCN as 'extinct in the wild' (Jiang & Harris, 2016), captive specimens were first returned to China (as zoo animals) in 1956. Then, in 1985, milu from the English herd were sent to China for reintroduction, and two semi-wild populations were established: the first at Dafeng Reserve (Jiangsu Province) in 1986 and, later, in 1993 at Tian-'e-Zhou Reserve on the Lower Yangtze floodplain. An oxbow lake in the latter is an important site for *ex situ* conservation of the Yangtze finless porpoise (see Box 9.1). Both deer populations have expanded considerably, leading to incursions from reserves into surrounding farmland, but the nuisance caused is an indication of how far milu have been brought from the brink of extinction. Memories of this deer have long-since disappeared as a result of centuries of cultural baseline shift, and any restoration measures along the Yangtze would have been unlikely to include re-establishment of milu if they had relied upon local recollections of what the floodplain was once like.

Reintroductions, such as those involving the Eurasian beaver and milu, have seldom been undertaken in fresh waters, but usually include an *ex situ* component where the species concerned is removed (or rescued; see Box 9.1) from its original habitat to be maintained and bred in captivity until such a time that it can be reintroduced to a place at or near the location of the source population. For instance, the ovoviviparous Kihansi spray toad (*Nectophrynoides asperginis*: Bufonidae) was displaced from its sole habitat at the Kihansi River Falls in Tanzania by construction of a hydropower dam completed in 1999. The toad was declared extinct in the wild in 2004 (IUCN SSC Amphibian Specialist Group, 2015), but individuals transported to the United States in 2000 were the basis of breeding populations established at the Bronx and Toledo zoos. In 2010, captive-bred individuals were sent to a propagation centre in Dar es Salaam, and toad reintroductions to the Kihansi River began in 2012; additional releases occurred in subsequent years, with at least some of those individuals surviving in the wild. It is currently uncertain if the reintroduced animals have established a self-sustaining population.

In amphibians and other threatened taxa, the selection of species for reintroduction will depend on the feasibility of mitigating or reversing the causes of population decline within the original habitat (Griffiths & Pavajeau, 2008). Species that are doomed to extinction in the wild as a result of irreversible threats to their habitat must either be relocated to novel sites or maintained in perpetuity as captive 'heritage' species. As an alternative, where it is not possible to reintroduce a species to its original habitat, it may be relocated to a similar site within – or in proximity to – its natural geographical range. This represents a translocation or relocation, rather than a reintroduction, and – as described in Chapter 7 – could be a means to conserve species confined to an unfavourable thermal environment as a result of climate change. In view of the many uncertainties about the ecological ramifications of relocating any species to a new environment where it does not occur naturally (discussed in Chapter 7), and the multifarious impacts of introduced alien species (see Chapter 4), managed relocation has seldom been used as a conservation intervention – most instances involve terrestrial plants (Hewitt *et al.*, 2011). One exception involves the desert pupfish in the southwestern United States, which was translocated to new habitats (mostly constructed refugia) in Arizona after its native haunts had been rendered uninhabitable after invasion by mosquito fish (Chapter 4). This small cyprinodont is an ideal candidate for relocation: it is inoffensive, inhabits small isolated drainages within a semiarid environment, and has limited dispersal abilities so that its survival is to a large degree reliant on human intervention.

In addition to the desert pupfish, proof-in-principle that translocation can be successful and lead to establishment of breeding populations has been demonstrated for the fringed darter (*Etheostoma crossopterum*: Percidae) in Illinois (Poly, 2003). Furthermore, in one of the first studies of its kind, four species of small cyprinids in Sri Lanka translocated in 1981 were represented by flourishing populations at new stream sites four years later, and are present to this day (Wikramanayake, 1990, pers. comm.). Apache trout and flannelmouth sucker have undergone translocations in the Colorado River (see Chapter 4), although more extensive schemes involving more species were not pursued (Minckley, 1995). Captive breeding and subsequent translocation has been used successfully as a conservation strategy for an endemic Hong Kong frog (*Liuixalus romeri*: Rhacophoridae; EN), with animals translocated in 1992 and 1993 maintaining breeding populations at seven out of eight sites more than a decade later (Banks *et al.*, 2008). While there may be scepticism over the risks associated with translocations, there is evidence that such conservation interventions do work, at least in the short to medium term (see also Box 9.4).

Translocations have also been used to supplement populations of threatened pearly mussels, sometimes combined with artificial propagation and culture (Lopez-Lima *et al.*, 2017); restoring river connectivity has also been proven to be beneficial for these bivalves and their fish hosts (see Box 5.12). Research on six species of unionids from the Great Lakes region of Canada (Galbraith *et al.*, 2015) demonstrated that translocation was a useful conservation tool with minimal genetic risk so long as it was carried out within the same drainage basin. Re-introducing mussels into locations where they were historically found was also viewed as appropriate, provided the donor stock originated from within the same drainage. Despite drastic reductions in freshwater mussel abundance, remarkably little evidence of reduced population genetic variability within drainages has been found, perhaps because of the extremely long lifespan of these molluscs. Whether these findings are more widely transferrable for management of unionids beyond the Great Lakes remains to be seen. But pearly mussels are expected to be climate-change losers, and they could well be the subject of future translocations carried out to keep pace with global warming (see Chapter 7). Unless a high degree of genetic similarity between translocated individuals and residents has been confirmed, it is preferable that introductions take place into water bodies that lack conspecifics. Mixing populations among drainages should also be avoided given the extent of interpopulation genetic variability at this scale in fully aquatic freshwater animals (Galbraith *et al.*, 2015), unless such mixing is part of a deliberate strategy to increase heterogeneity and population viability (see Box 9.1).

Given the extent of endangerment of freshwater fishes, and the sheer variety of species maintained and bred by aquarists, there should be scope for captive breeding and *ex situ* conservation of the most threatened species. Some that are close to extinction in the wild, such as the red-tailed shark (*Epalzeorhynchos* [= *Labeo*] *bicolor*: Cyprinidae; CR), dwarf chain loach (*Ambastaia* [= *Botia*] *sidthimunki*: Botiidae; EN) and White Cloud Mountain minnow (*Tanichthys albonubes*: Cyprinidae; CR), are popular in trade (see also Box 3.1). The matter has received scant attention in the primary scientific literature – Reid (1990) remains a key reference (see also Koldewey *et al.*, 2013; Cochran-Biederman *et al.*, 2014) – but, in 2004, a group of North American hobbyists launched the CARES Fish Preservation Programme (https://caresforfish.org/) to enable hobbyists to get involved in species preservation. CARES has since expanded beyond North America, and this voluntary network, which has grown to involve thousands of participants, maintains a priority list of hundreds of at-risk species. Aquarists keeping any of these species can register them on a central database and exchange information on their husbandry. The primary focus of the CARES programme is maintaining species in captivity, including around two dozen species that are extinct in the wild; reintroduction is a secondary concern. The *ex situ* populations could provide the stock for fish reintroductions although, thus far, there are plans for only one such species maintained by CARES to be returned: the Potosi pupfish (*Cyprinodon alvarezi*), which became extinct in its Mexican home range after the introduction of predatory largemouth bass; control of this alien could render the former habitat of pupfish suitable for its return.

CARES has recently partnered with another aquarist group, African-cichlids.net (www.africancichlids.net/), to raise awareness of the need for captive breeding of Lake Victoria haplochromines (as advocated by Reid, 1990), and this model could be extended to encompass *ex situ* conservation of a wide range of threatened African cichlids. A second international organization of hobbyists established in 2009 – the Goodeid Working Group (www.goodeidworkinggroup.com/home) – has similar objectives to CARES, although the focus is narrower and entirely devoted to the Goodeidae. Comparable initiatives also involve other freshwater taxa: the Turtle Survival Alliance (www.turtlesurvival.org/) and the Amphibian Conservation Action Plan (www.amphibians.org/ acap/) coordinated by the IUCN have, among other things, created a network of zoos, aquaria and private breeders to maintain viable populations of turtles and amphibians, with a long-term view of undertaking reintroductions to the wild.

The various groupings that advocate captive breeding provide an opportunity for citizens to become involved in freshwater biodiversity conservation and, at the very least, can contribute to public education on the serious topic of actual and prospective extinctions of freshwater species. There is no prospect that the captive breeding can alone halt population declines or prevent the loss of taxa in nature. Nevertheless, captive breeding provides conservation options for individual heritage species, and can generate conservation gains in cases where it leads to translocations or stocking and reintroductions in the wild, as has already taken place among some amphibians (Box 9.4), and would perhaps be feasible for the axolotl (Box 2.5). Examples of reintroductions of freshwater fishes are given by Cochran-Biederman *et al.* (2014), who report that inadequately addressing the initial cause of decline was the best predictor of failure, while availability of high-quality habitat was an important determinant of successful reintroductions. Variables associated with stocking (such as genetic diversity) were of subsidiary importance to habitat conditions.

Box 9.4 *Captive Breeding and Conservation of Amphibians*

Initial efforts (1996–2006) at captive breeding and reintroduction or translocation programmes for amphibians focused on threatened species from north-temperate countries with relatively low amphibian diversity (Griffiths & Pavajeau, 2008). Of 110 species involved, 39 were in captive-breeding programmes with the objective of reintroduction or translocations of wild animals; a further 19 species were in programmes involving only relocations of wild animals. There were no plans for reintroduction of any of the other 52 species. Eighteen out of 58 reintroduced or trans-located species bred successfully in the wild, with 13 of them establishing self-sustaining populations. In the period after 2007, when the Amphibian Conservation Action Plan was initiated, there were relatively few new reintroductions; most programmes focused on conservation-related research and securing species in captivity as a precaution against extinctions in the wild (Harding *et al.*, 2016). There was a welcome increase in the number of conservation programmes undertaken in Central and South America and the Caribbean, where amphibian biodiversity is high, as well as a shift towards a broader representation of taxa. However, as of 2014, zoos around the world held only 6.2% of threatened amphibian species, a much smaller figure than for most other vertebrate groups, and one that falls considerably short of the number of species for which *ex situ* management is likely to be needed (Dawson *et al.*, 2016).

Scientific Programmes and Organizations Promoting Freshwater Biodiversity Conservation

A number of international science programmes have the objective of better understanding aspects of the global water system, and integrating conservation and sustainable management of fresh waters in general with attempts to meet water-related Millennium Development Goals. Some are supported by intergovernmental organizations (e.g. the Division of Water Sciences at UNESCO, unesco.org/water), others by individual countries or non-profit organizations having an international perspective: for instance, the Freshwater Sustainability

Project of The Nature Conservancy, the Stockholm World Water Week, the World Water Forum and the Global Water System Project. They have coalesced scientific, policy and user communities, enabling them to agree on – and articulate – guidelines for improved water management.

In addition to large international programmes and non-government organizations with a conservation focus (e.g. Conservation International, The Nature Conservancy, WWF), or a particular concentration on inland waters (e.g. International Rivers, formerly the International Rivers Network), there have been numerous efforts by smaller – usually science-based – teams to quantify the global scope of issues related to freshwater quality and quantity, identify basic principles that could guide sustainable water resource development, and devise practical approaches for balancing water uses. They often have origins at meetings of scientific societies (e.g. the Society for Freshwater Science, the Freshwater Biological Association, the International Society for Limnology), when discussions centre on topical issues relating to conservation and management of freshwater ecosystems. Table 9.1 provides one example of the outputs from such a team who formulated a general set of eight guidelines (Bernhardt *et al.*, 2006) for balancing water-resource development and environmental well-being that can be used to underpin sustainable planning decisions.

One specialized but nonetheless significant grouping of water scientists was responsible for the 2007 Brisbane Declaration, which was promulgated at the annual meeting managed by the International RiverFoundation based in Australia. The Declaration was intended to draw attention to the scientific consensus regarding the importance of e-flow allocations for humans and nature (http://riverfoundation.org.au/wp-content/uploads/2017/02/THE-BRISBANE-DECLARATION.pdf; see also Arthington, 2012). It was revisited a decade later during the 20th International River*symposium* and Environmental Flows Conference (again held in Brisbane), and has now been revised and updated as The Brisbane Declaration and Global Action Agenda on Environmental Flows 2018 (http://riversymposium.com/about/brisbane-declaration/; see Box 9.5). It sets out an urgent call for action on e-flows to protect and restore riverine ecosystems for their biodiversity, intrinsic value and ecosystem services (Arthington *et al.*, 2018), as a central element of integrated water-resources management and a foundation for achieving water-related Sustainable Development Goals (SDGs).

Table 9.1 *Eight general guidelines for balancing water-resource development and management of freshwater ecosystems (considerably modified from Bernhardt et al., 2006)*

Understand the aggregative effects of human activities within a catchment
Determine the optimal spatial configuration of development, protection and
restoration that minimizes constraints on water resources yet maintains provision
of ecosystem goods and services

Embrace environmental uncertainty
Recognize that freshwater ecosystems are inherently dynamic and variable; find ways
to accommodate or adjust to the natural range variability rather than using hard-
path engineering approaches in an attempt to achieve stability and predictability

Accept that rivers and estuaries need environmental flows
Ensure that rivers, lakes and wetlands are provided with water of the right quality at
the right times in sufficient amounts to sustain biodiversity and ecosystem
functioning, and hence the provision of the goods and services for humans

Manage connectivity of freshwater systems
Transport of materials (nutrients, sediments) and movement of animals, plants or their
propagules within catchments, along rivers and into associated wetlands, and
longitudinal connections to coasts and estuaries must be maintained or enhanced;
measures should be taken to prevent or limit transfer of contaminants and invasive
species

Understand the consequences of biodiversity loss for ecosystem services
Component species within ecosystems are involved in an array of complementary
processes (e.g. primary and secondary production, decomposition, nutrient cycling)
but the contribution of individual species is often unclear; protect all elements of
freshwater biodiversity, including species that may seem redundant or unimportant,
as 'insurance' to maintain ecosystem functioning in the context of environmental
change, and minimize the likelihood of unexpected tipping points or shifts in state

**Contribute to development and evaluation of new technologies to manage
water in heavily modified catchments**
Make better use of remote-sensing technologies and computer modelling tools to
improve the efficiency of water use and reuse with consequent reductions in
economic costs, degradation of ecosystems and threats to freshwater biodiversity

Undertake research that will facilitate the recovery of degraded ecosystems
Establish innovative collaborations to gain a better mechanistic understanding of how
rivers provide ecosystem services for humans; develop effective measures for restoration
and rehabilitation of fresh waters so as to maximize the biodiversity they can support

Improve the ecological understanding of the global water system
More must be learned about the functioning and integration of components of the global
water system, particularly the impacts of human activities (e.g. the consequences of
virtual water trading), in order to improve the prospects of remaining within planetary
limits for blue-water sustainability in a warmer, more crowded world

Box 9.5 *E-flows: From 2017 to 2018*

While the 2018 Brisbane Declaration recognizes that there has been substantial progress in environmental flows science and water management since 2007, major challenges in protecting and restoring the integrity of freshwater ecosystems remain, and these compromise ecological services that sustain human livelihoods. E-flow requirements have yet to be adequately assessed for most aquatic ecosystems (see Chapter 5) and have been implemented in even fewer. There is still no comprehensive global record of environmental flow implementations, nor a good understanding of why some projects have succeeded, while other initiatives have failed to materialize. Major obstacles to e-flow implementation lie largely outside the realm of ecology: a lack of political will and public support; constraints on resources, knowledge and local capacity; and institutional barriers and conflicts of interest (Arthington *et al.*, 2018). They are matters of particular concern due to the number of large hydropower dams under construction (Zarfl *et al.*, 2015; Winemiller *et al.*, 2016). Furthermore, demands for water will continue to grow, especially in arid regions and parts of the world that are likely to experience shortages as a result of climate change. The inevitable outcome will be increased flow alteration and less water for the environment in coming decades. The need for action to bring about effective e-flow implementation has never been more urgent.

The 2018 Declaration sets out 35 actionable recommendations on e-flows under the three general headings of 'leadership and governance', 'management' and 'research'. The first of these urges all levels of government '. . . to take part in the development of legislation, policies, regulations and funding mechanisms to institutionalize, promote, and support e-flow science and management within the broader context of jurisdictional natural resource management' (Arthington *et al.*, 2018: p. 6). 'Management' refers broadly to planning, assessing, implementing, monitoring and then adaptive management of the e-flow allocation – essentially, learning by doing. 'Research' is added to emphasize the need for better mechanistic understanding but, understandably, focuses on monitoring and evaluating e-flow outcomes and adjusting the implementation plans accordingly – this is adaptive management in another guise. While the 2018 Declaration is a worthy and comprehensive document, it is not apparent how the lack of progress in e-flow implementation between 2007 and 2017 will change post-2018, unless the necessary political will can be brought to bear. E-flow science has become quite well developed, with many skilled practitioners, but unless the socio-economic context can be transformed significantly, allocation of water to the environment will likely remain a low public and administrative priority. Well-intended exhortations such as the Brisbane Declaration are unlikely to bring about change in the absence of a broader shift in societal priorities.

The Freshwater Animal Diversity Assessment (FADA; Balian *et al.*, 2008a, 2008b) is another instance of scientific collaboration, in this case among 163 experts, to provide the first global overview of genus- and species-level diversity among the major groups of freshwater animals and macrophytes in different biogeographical regions. Although far from complete, especially in tropical latitudes, the FADA provided a far more detailed overview of freshwater biodiversity than had been available previously, with the raw data underlying the assessment made publically available (http://fada.biodiversity.be/). The FADA also helped draw attention to the remarkable richness of freshwater ecosystems. The BioFresh project (http://project.freshwaterbiodiversity.eu/) was built upon the FADA and focused directly on the plight of biodiversity. Funded by the EU for five years (2009–2014), BioFresh constructed a global information platform to host databases on the geographical distribution, trends and status of freshwater species. A major outcome was establishment of a clearing house allowing sharing of resources and data that were previously scattered and difficult or impossible to access. It included a repository of information (the BioFresh BioMatrix) on the contemporary distributions of more than 2300 freshwater species of (predominately European) animals and plants mapped to the latest catchment layers of HydroBASINS (www.iucn.org/sites/dev/files/import/downloads/deliverable_d4_4__biomatrix_manual.pdf). After funding for BioFresh ended, four partner institutes committed to extend the clearing house and widened its scope from biodiversity issues to offer online resources dealing with all aspects of the ecology and management of fresh waters. This expanded online resource was rebranded as the

Freshwater Information Platform (www.freshwaterplatform.eu/; see Turak *et al.*, 2017a).

The last decade has seen a rise in the number of global networks aiming to develop Earth observation systems, most notably the Group on Earth Observations (GEO: www.earthobservations.org/index.php) and its Biodiversity Observation Network (GEO BON) formed in 2008 (Scholes *et al.*, 2008). GEO BON aims to build on the evolution of remote sensing systems, and developments in the fields of macroecology, environmental DNA (see Box 9.6), metagenomics and ecoinformatics to produce global maps to infer the state of biodiversity. The maps are generated by combining *in situ* observation with abiotic remote-sensing data, supplemented by modelling. This mapping work has engendered the Essential Biodiversity Variables (EBVs) framework (Pereira *et al.*, 2013), which comprises six broad components of biodiversity: genetic composition, species populations, species traits, community composition, ecosystem structure and ecosystem function. The EBVs are significant because meeting the 2020 Aichi Biodiversity Targets, as well as the UN SDGs proposed for 2030, will require improved capacity to determine how and where biodiversity is changing across the world. Recent advances in freshwater monitoring may soon increase the feasibility of measuring global changes in biodiversity. The freshwater working group of GEO BON have identified priorities for EBV assessment by 2020 under three of the EBV categories (species populations, community composition and ecosystem structure) – most notably, a complete assessment of the state of global freshwater biodiversity using IUCN Red List criteria (Turak *et al.*, 2017b). A list of requirements for assessing variables in other EBV categories by 2030 was drawn up also. They include refinement of environmental DNA methods (see Box 9.6); the development of species-traits databases; globally consistent estimation of wetland extent; and the use of crowdsourcing and citizen science to expand the coverage of data collection and validate remote-sensed information (Table 9.2). While some of the short-term priorities may not be achieved by 2020, it is nevertheless useful to separate the more immediate objectives that require urgent attention from others that will take longer to realize. There may even be sufficient tools to provide the potential for making freshwater biodiversity observations in 'real-time' or close to it by 2020, when periodic evaluations across large regions could become a distinct possibility (Turak *et al.*, 2017a).

Box 9.6 *Use of Environmental DNA Techniques in Freshwater Biodiversity Conservation*

Organisms inhabiting lakes and rivers release DNA to the water column in the form of secretions, cells, tissues, faeces or gametes that can be transported through drainage networks. Fragments of this environmental DNA (eDNA) can be isolated from organic matter in water samples, sequenced and (at least in theory) assigned a species using metabarcoding (Elbrecht & Leese, 2017). The technique has stimulated a great deal of research, some of it drawing attention to the potential conservation applications of eDNA techniques in fresh waters (Goldberg *et al.*, 2015), for instance, as an aid to detection of rare and endangered species whose presence cannot be confirmed easily by more conventional means (Jerde *et al.*, 2011; Laramie *et al.*, 2015; Sigsgaard *et al.*, 2015; Eva *et al.*, 2016), and for monitoring colonization of new habitat by potentially invasive species (Goldberg *et al.*, 2013; Jerde *et al.*, 2013; Rees *et al.*, 2014). Targeted or 'active' surveillance directed towards detection of eDNA for a single species of conservation interest can be contrasted with 'passive' surveillance, using high-throughput sequencing approaches, whereby sampled eDNA is used to assess community composition and happens to reveal the presence of a particular species (Simmons *et al.*, 2016) The latter approach could have applications for biomonitoring or bioassessment, since the eDNA signal of a community of macroinvertebrates could be used to estimate diversity with less investment of time and effort than the benthic sampling methods that are widely used currently (Rees *et al.*, 2014; Elbrecht & Leese, 2017). When combined with next-generation sequencing methods, and markers that target many species, metabarcoding enables the concurrent detection of multiple taxa without any prior knowledge of community composition (Valentini *et al.*, 2016). In addition, analysis of eDNA transported in river networks offers a spatially integrated way to assess the species richness (both aquatic and terrestrial) of entire drainage basins (Deiner *et al.*, 2016), and could well transform biodiversity data acquisition (Turak *et al.*, 2017b).

An important global collaboration was initiated during the International Year of Biodiversity in 2010 at a UNEP-facilitated meeting at Busan, South Korea, where agreement was reached among 85 nations to

Table 9.2 *Short- and medium-term priorities for measuring global change in Essential Biodiversity Variables (EBVs) in fresh waters (modified from Turak et al., 2017b)*

EBV category	Short-term priorities	Medium-term priorities
Genetic composition	- Expand reference DNA sequence databases - Improve methods for characterizing eDNA samples - Integrate eDNA into freshwater biomonitoring	- Complete barcoding library of freshwater species - Standardize global protocols for eDNA and monitoring intraspecific genetic change - Establish a global network of genetic monitoring sites
Species populations	- Expand the geographical scope to complete a global baseline for Red List assessments of fresh waters - Repeat assessments of taxa that already have baselines in place to produce index of change - Broaden geographical and taxonomic coverage in the freshwater Living Planet Index - Refine species distribution modelling and validation - Improve tools and opportunities for public participation	- Repeat Red List assessments at regular intervals to monitor progress towards targets - Integrate new sources of remotely sensed data with *in situ* biodiversity measurements, and use to improve distribution modelling - Global networks of citizen scientists and taxonomic experts collectively recording and verifying species occurrences
Species traits	- Collect trait data for more species, especially in the tropics - Build a global, open-access species-trait database - Begin to relate species traits to climate-change vulnerability	- Complete trait descriptions of all species in the expanded Red List - Develop models that use traits to predict species susceptibility to climate change and other stressors
Community composition	- Map subcatchments for species richness and turnover to assess global state of biodiversity	- Establish reporting structure and repeat global assessments with updated threat data at

Table 9.2 (*cont.*)

EBV category	Short-term priorities	Medium-term priorities
	- Develop guidelines for globally consistent monitoring of fish and invertebrate assemblages	regular intervals, with additional sampling to improve spatial resolution in data-poor regions - Build an open-access database for reporting monitoring results
Ecosystem structure	- Develop global standards and programmes for citizen scientists to record wetland extent - Complete globally consistent high-resolution mapping of wetland extent - Establish network of wetlands for monitoring biodiversity change, by combining remote-sensing technologies and *in situ* observations, as well as crowdsourced data	- Automate high-resolution mapping of wetlands calibrated by crowdsourced *in situ* observations - Connect habitat structure and species models to develop a hierarchical system of ecosystem bio-regionalization - Enhance public access to increasingly complex data sources
Ecosystem function	- Identify functional attributes that can be readily quantified and linked to changes in species composition or populations - Develop methods for assessing temporal trends in fishery yields (an important ecosystem service)	- Establish monitoring programmes for ecosystem functioning, and link these to the underlying mechanisms, including changes in species diversity - Undertake global scale assessments of temporal trends in fishery yields

establish an Intergovernmental Science-policy Platform on Biodiversity and Ecosystem Services (IPBES; see http://ipbes.net). It was intended to mirror the Intergovernmental Panel on Climate Change and work to integrate data on ecosystem degradation – specifically, declines in bio-diversity and provision of goods and services – with the government action required to reverse them (Larigauderie & Mooney, 2010; Diaz *et al.*, 2015). Formal establishment of the IPBES by the UN General Assembly, and endorsement by a global ministerial meeting of UNEP, took place in 2012, and a work programme was drawn up in 2013. The IPBES coordinates global-scale peer reviews of research on the status and trends of biodiversity and ecosystem services, and is intended to serve as a conduit to deliver 'gold standard' reports and policy recommendations to governments. Unfortunately, there is no topic relating to freshwater biodiversity among the assessment reports included in the first IPBES work programme (Darwall *et al.*, 2018), and this should be a priority for the immediate future.

Another ambitious collaborative effort comprises an overarching consortium of international non-governmental scientific organizations, such as the GWSP, the conservation group DIVERSITAS, the International Geosphere–Biosphere Programme (IGBP) and the International Human Dimensions Programme on Global Environmental Change (IHDP), as well as higher-order groups such as UNESCO and UNEP. At the UN Conference on Sustainable Development (Rio+20) in June 2012, they agreed to come together as the Future Earth research initiative, to be coordinated by the International Council for Science. Future Earth became fully operational, with a permanent secretariat, at the end of 2015 (see www.futureearth.org), and thousands of scientists are involved in the consortium's initial 10-year programme. Its mission is to provide global-scale solutions to ensure that human activities remain with planetary boundaries for water use, biodiversity conservation and other aspects of global environmental changes in the Anthropocene. Achieving these goals will likely require wholesale societal changes extending far beyond concerns related to fresh water (Box 9.7).

Box 9.7 *Implications of the Future Earth Agenda*

The overarching goal of Future Earth is to improve lives and livelihoods by promoting sustainable access to food, water and energy while protecting biodiversity and ecosystem services (Griggs *et al.*, 2013), and there are a number of areas where Future Earth research could be expected to make a major contribution to sustainability challenges relating to fresh water. The term sustainable, in this context, means '... development that meets the needs of the present while safeguarding the Earth's life-support system, on which the welfare of current and future generations depends' (Griggs *et al.*, 2013: p. 306). Clearly this is not possible without a shift in economic models and governance structures upon which development is predicated (Biermann *et al.*, 2012). Such a change would have to encompass a new global vision of the relationships between the three fundamental components of sustainable development (i.e. economy, society and environment) in which the economy is nested within society that is contained by the Earth's life-support system (Griggs *et al.*, 2013), necessitating a move away from treating ecosystems as economic externalities and leading to a widespread adoption of PES and related models (see Chapter 8). Nesting economic development within the planetary life-support system also has the implication that the costs of doing business as usual must change, because private-sector profits are generally made at the expense of environmental degradation and species declines, and these revenues would no longer be sustainable. Instead, it might be reasonable to expect businesses to contribute to funds or trusts that support habitat restoration and species-recovery plans.

The newest of the various global science initiatives, and probably the one most relevant to this book, is the Alliance for Freshwater Life (https://allianceforfreshwaterlife.org/), launched at the World Water Week, Stockholm, in August 2018. It aims to unite specialists in five core areas of work: research, data synthesis, conservation, education and outreach, and policy-making. A key driver of the Alliance is recognition that although there are policy initiatives intended to protect freshwater biodiversity, they are seldom implemented with sufficient conviction and enforcement (Darwall *et al.*, 2018). The mission of the Alliance is to halt and reverse the global decline of freshwater biodiversity, and ensure that it is better understood, valued and safeguarded. To achieve this mission, the expert network plans to undertake a united effort to provide the critical mass required for the effective representation of freshwater biodiversity at policy meetings; to develop solutions balancing the needs of development and conservation; and to better convey the important role freshwater ecosystems play in human well-being (Darwall *et al.*, 2018). Among actions as cited examples for making positive change are augmenting the online Freshwater Information Platform to collate, harmonize and visualize data on biodiversity; broadening the representation of freshwater species in the WWF Living Planet Index; promoting a specific freshwater biodiversity target in the post-2020 Aichi targets; and establishing a freshwater biodiversity topic in the second (2020–2030) work programme of IPBES (see Table 9.3).

Table 9.3 *Short- and medium- (to long-) term priorities for action under the five core areas of work under the Alliance for Freshwater Life (modified from Darwall et al., 2018)*

Core work area	Short-term priorities	Medium-term priorities
Research	– Identify (using modelling) and survey locations where biodiversity is predicted to be rich or understudied – Support development of tools to collect data of freshwater biodiversity (e.g. using EBVs) – Support development of a global classification of freshwater ecosystems	– Implement biodiversity assessments in poorly known locations – Support completion of global assessments of the conservation status of representative taxa – Enhance understanding of drivers of biodiversity decline, and compile evidence about workable solutions to prevent or reverse declines
Data and synthesis	– Build on the Freshwater Information Platform to collate, harmonize and visualize biodiversity data for dissemination – Review major initiatives on freshwater biodiversity to identify gaps and partners	– Broaden representation of freshwater species in the Living Planet Index – Identify key freshwater biodiversity areas as global priorities for conservation action – Enhance representation of freshwater species within global and regional biodiversity analyses
Conservation	– Support development of a funding platform for site based freshwater conservation action – Complete a gap analysis of freshwater biodiversity within the global network of protected areas	– Support efforts to expand the global protected-areas network to more fully encompass freshwater biodiversity – Develop conservation action plans for the most threatened freshwater species

Table 9.3 (*cont.*)

Core work area	Short-term priorities	Medium-term priorities
Education and outreach	– Develop and publicize a website for the Alliance – Engage with private sector and citizens using social and multimedia	– Establish programmes on freshwater ecosystems in educational curricula, and develop massive open online courses – Promote general interest in freshwater biodiversity through documentaries, mass media articles, and other events
Policy	– Promote a specific post-2020 freshwater biodiversity target – Develop a network of freshwater biodiversity experts to participate in international policy forums – Establish a freshwater biodiversity topic in IPBES	– Make recommendations to international conventions for development of freshwater biodiversity resolutions – Support a review of the SDGs and the role of freshwater ecosystems – Ensure strong representation of freshwater biodiversity in relevant policies

Will the SDGs Adequately Protect Freshwater Biodiversity?

A global effort is needed to address and reverse global trends in the degradation of freshwater ecosystems, which is to the detriment of humans and nature, and it is plain from the foregoing that many programmes and organizations have a role to play in this undertaking. How well are the 17 SDGs, and their 169 constituent targets (see https://sustainabledevelopment.un.org/sdgs), promulgated in 2015, likely to facilitate conservation of freshwater biodiversity? That will depend, in no small part, on how well the topic is represented. Goal 6, termed 'Clean Water and Sanitation', is intended '. . . to ensure availability and sustainable management of water and sanitation for all'. Among its seven listed targets (and subtargets), 6.6 states: 'By 2020, protect and restore water-related ecosystems, including mountains, forests, wetlands, rivers, aquifers and lakes'. Under the broader context of meeting human needs for water, there is specific acknowledgement of the need to protect fresh waters. Goal 14 is rather misleadingly titled 'Life below Water' and is concerned exclusively with the oceans; the constituent targets are not relevant to inland waters. Instead, freshwater biodiversity falls under Goal 15 'Life on Land' envisioned to 'Protect, restore and promote sustainable use of terrestrial ecosystems, sustainably manage forests, combat desertification, and halt and reverse land degradation and halt biodiversity loss'. Although not mentioned within the overarching objectives of Goal 15, fresh water is included in target 15.1: 'By 2020, ensure the conservation, restoration and sustainable use of terrestrial and inland freshwater ecosystems and their services, in particular forests, wetlands, mountains and drylands, in line with obligations under international agreements'. (In fact, there are few binding obligations to protect freshwater ecosystems under international agreements.) Target 15.5 is generic, but must surely also be intended to encompass fresh waters: 'Take urgent and significant action to reduce the degradation of natural habitats, halt the loss of biodiversity and, by 2020, protect and prevent the extinction of threatened species'. Target 15.8 makes specific mention of fresh waters: 'By 2020, introduce measures to prevent the introduction and significantly reduce the impact of invasive alien species on land and water ecosystems and control or eradicate the priority species'. It is the only one of two targets (the other being 15.1) of the 12 listed under Goal 15 that specifies the aquatic realm. Although the SDGs do make reference to the need to protect freshwater biodiversity, as was the case with the 2005 Millennium Development Goals (and as the trend line for the freshwater Living Planet Index in Fig. 2.1 makes clear), aspirational statements about environmental protection do not necessarily lead to changes in relevant policy or practice.

An ecologically based set of SDGs should include protection of freshwater ecosystems as fundamental for achieving the other goals, since food security and poverty alleviation can hardly be achieved without the fresh water that supports irrigated agriculture and inland fisheries yields, nor can human health or social and economic development be sustained in the absence of access to clean fresh water. Such an understanding is, for example, embedded within the concept of the global water system (see Box 1.1). Furthermore, 'Life on Land' (Goal 15) is not possible without fresh water, and maintenance of healthy ecosystems in all realms would seem an essential prerequisite for achievement of the other SDGs. It is a pity that, as formulated, the UN SDGs fail to adequately reflect the fact that protection of freshwater ecosystems and their biodiversity is a precondition for ensuring social and economic well-being in the Anthropocene world.

Prior to promulgation of the UN SDGs, Griggs *et al.* (2013) proposed six universal SDGs that cut across economic, social and environmental domains and drew upon the science relating to planetary limits. Of these, Goal 3, to 'Secure Sustainable Water' proposed a 2030 target to '. . . limit volumes withdrawn from river basins to no more than 50–80% of mean annual flow' and to restrict consumptive use of runoff resources to <4000 km^3 annually (Griggs *et al.*, 2013: p. 307). A second target envisaged that humans would consume no more than one third of the accessible global runoff of 12 000 km^3/y but, since (under the first target) some rivers would retain only 20% of their mean annual flow, as much as 80% of water could be extracted at the scale of individual rivers (see also Box 1.2). And, since not all extracted water is consumed, water withdrawal would evidently exceed one third of accessible global runoff. Nevertheless, these 'alternative' SDGs forcefully put the case for protecting freshwater ecosystems to safeguard areas crucial for biodiversity and ecosystem services. They have the added value of incorporating some numerical targets. Among those relevant to fresh waters were halving phosphorus runoff into lakes and rivers by 2030; keeping extinctions to within 10 times natural background rates; and retaining at least 70% of species in any ecosystem (Griggs *et al.*, 2013).

Whether these 'alternative' SDGs could maintain life-support systems in rivers that may contain as little as 20% of their annual flow, have much less water to dilute phosphorus inputs, and have lost as many as 30% of their constituent species, is far from certain. Nor can we be sure that national or international action predicated on the UN SDGs will be beneficial for freshwater biodiversity. Nonetheless, the proposed sustainability goals represent initial targets that could be refined by Future Earth researchers and others. Furthermore, it could be argued that such protection is warranted irrespective of the utilitarian or economic value of the services provided by intact ecosystems. Adopting an ethical stance that asserts all species have an inherent right to exist would imply that extinctions should be avoided on principle rather than being hostage to (say) economic rationalization. One possible outcome would be to protect species by assigning their habitats – and nature in general – 'rights' that must be recognized under law (see Box 9.8). While bestowal of rights or personhood on rivers, lakes and other ecosystems may be viewed as fanciful by some, and less effective than specific legislation directed towards environmental protection, it nonetheless represents a significant shift away from treating nature as an externality or something lacking economic value.

Box 9.8 *Rights for Rivers and Nature?*

In an unprecedented 2017 decision, the courts in New Zealand recognized the Whanganui River – the third largest in the country – as a legal entity with a voice, rights and interests equivalent to that of a company or a person under the law. The recognition gives the representatives of the iwi (indigenous people) of the Whanganui River responsibility for protecting and managing the river, by treating it as a living entity rather than viewing it from the perspective of ownership. This means abusing or degrading the river would, in law, be seen as little different from harming the iwi themselves. Although the details of a whole-river management strategy involving collaboration among iwi, government, commercial and recreational stakeholders have yet to be worked out, bestowal of personhood rights on a river, making it a protected entity, would be one way to ensure conservation of a freshwater habitat.

The Whanganui is the first river to achieve such status, but there is at least one similar precedent. In 2008, the constitution of Ecuador was modified by referendum to ensure that its lakes, rivers and forests would have similar rights to humans with regard to protection from harmful exploitation under a Rights for Nature constitutional article (http://therightsofnature.org/wp-content/uploads/pdfs/Rights-for-Nature-Articles-in-Ecuadors-Constitution.pdf). It states that nature '... has the right to exist, persist, maintain and regenerate its vital cycles, structure, functions and its processes in evolution'. The extent to which the article will be implemented is not yet known, because the constitution allows the Ecuadorian government to override legal protection of the environment if it is in the national interest. Bolivia passed a bill giving similar legal rights to nature under The Law of Mother Earth (2011), requiring all other existing legislation to be adjusted to conform to ecological limits. Specific clauses of the law refer to the right to clean water; the right to continue processes free from human alteration; and the right to not be affected by infrastructure and development projects that affect the 'balance' of ecosystems.

Challenge and Change

The development and implementation of e-flows and PES initiatives offer some hope of improved sustainability for people and freshwater ecosystems (see also Ormerod, 2014). Nonetheless, much more needs to be done to reduce the anthropogenic threats responsible for the trends described in Chapter 2, as well as the great thinning, shrinking and mixing of freshwater biodiversity. Current conservation practices alone appear inadequate to staunch ongoing declines and species loss (Strayer & Dudgeon, 2010), and policies that focus on the development and management of fresh waters as a resource for people almost universally neglect the biodiversity that they contain (Darwall *et al.*, 2018). In fact, governments worldwide do not adequately protect their inland waters and therefore place freshwater functions and attendant ecosystem services at risk (Creed *et al.*, 2017). It remains a major and urgent challenge to find satisfactory ways to manage inland waters for multiple uses; to restore or rehabilitate already damaged systems; and, where possible, set restoration targets that take account of the need to compensate for shifting baselines. Climate change leads to even greater uncertainty over the likely success of conservation measures in a world where stationarity is dead (Chapter 7).

Because of the large risks attending a business-as-usual model that will surely fail to meet the anthropogenic challenges to fresh waters, it has been argued that these ecosystems should rank second to none in terms of priority for research funding (Ormerod *et al.*, 2010). Moreover, the tasks that must be completed to provide a robust underpinning for actions to preserve freshwater biodiversity have been outlined by the Alliance for Freshwater Life (see Table 9.3). However, financial constraints may not be the greatest impediment to conservation. The total costs of funding *all* conservation globally are substantial, but not unduly burdensome (see Box 9.9). Irrespective of whether the necessary monies will be made available sufficiently quickly to substantially alleviate the threats facing freshwater biodiversity, this is not the time to indulge in speculative studies that may fail to bring measurable conservation benefits. Time is short. Research, management and conservation must proceed quickly and in tandem with public engagement, which – among other things – will be an essential prerequisite for continued funding support.

Box 9.9 *How Much Would Conservation Cost?*

McCarthy *et al.* (2012) calculated that an annual investment of US$4 billion, used wisely, could improve the status of all globally threatened species on the IUCN Red List, and virtually halt human-driven extinctions. A further US$76 billion could effectively protect and manage all known sites of global conservation significance. The total annual sum of US$80 billion, which would need to be adjusted for inflation, is insignificant in comparison with both the size of the global economy (comprising ~0.1% of around US$70 trillion annually) and an estimate of the total value of ecosystem services provided by nature each year (US$22–74 trillion; Costanza *et al.*, 2014). Making a sum of this magnitude available would require conservation funding to increase by at least an order of magnitude, but this would not be unduly burdensome on the global economy, especially in view of benefits to be gained. For comparison, the total amount needed represents a mere 5% of global military spending, and only 20% of what is spent on soft drinks each year. The estimated investment is intended to successfully conserve threatened species in all realms, and is so far more than the sum required to protect all imperilled fresh-water species, but it gives an idea of the scale of financial commitment that would be required.

It is difficult to exaggerate what may be at stake. Ever-increasing demands for water (see Chapter 1), coupled with concerns arising from deteriorating water quality, are inextricably linked to ensuring food security for growing human populations. Appropriation of sufficient water resources, of the required quality, will be essential for sustaining human well-being. Large-scale engineering solutions to store, redistrib-ute and treat water resources have been initiated globally, and there are now 33 existing water-transfer megaprojects with a further 76 planned or under construction (Shumilova *et al.*, 2018). These schemes transfer water between formerly separate donor and recipient river basins, with little attention paid to the environmental consequences. Of the future projects, 43 are intended to benefit agriculture development, 14 transfer water for hydropower, and 10 combine both purposes. Seventeen pro-jects will transfer water over a distance that exceeds the length of the River Rhine and, if all planned schemes are realized, they will move $1923 \, \text{km}^3$ of water each year over a total distance of more than twice the circumference of the Earth! Given that global freshwater withdrawal is currently around $4000 \, \text{km}^3$, the amount of water transferred would amount to almost half of this, or around 5% of river discharge to the oceans. This will represent a profound alteration of the global water system, at least as it pertains to organisms living in surface fresh waters. The economic costs of water-transfer projects are very substantial

(Shumilova *et al.*, 2018), yet internationally agreed standards to evaluate their impacts are lacking and their cumulative consequences are unknown. Although some transfers might have positive environmental impacts – if implemented, the Transaqua Project could begin to refill Lake Chad by relocating water from the River Congo (see Chapter 6) – most would transform both donor and recipient waters, and further advance the great mixing of freshwater biotas.

The United States provides another example of likely future changes to fresh waters, arising in this case from the outcome of an ongoing debate over which water bodies are protected under federal law by the Clean Water Act. Although there is scientific evidence that headwater streams and isolated wetlands provide services for humans (e.g. Freeman *et al.*, 2007; Cohen MJ *et al.*, 2016), convincing politicians of this is quite another matter (Creed *et al.*, 2017). The Clean Water Rule (also called the Waters of the United States Rule), which was codified in 2015 to clarify the jurisdictional scope for federally protected waters, was almost immediately subject to litigation, resulting in a stay by the US Court of Appeals for the Sixth Circuit; the subsequently elected administration placed the rule (and hence the definition of what type of waters could be protected) under review. At the time of writing, the Environmental Protection Agency seemed likely to rescind legal protection of river tributaries and wetlands that do not have year-round flows and are not considered to be navigable waters. This is despite research demonstrating the ecological importance of intermittent streams and the fact that around half of the global river network comprise channels that periodically cease to flow (Datry *et al.*, 2016; Leigh *et al.*, 2016). Creed *et al.* (2017) provide three policy options that would allow the US government to enhance protection of these 'vulnerable' fresh waters (i.e. intermittent streams and ephemeral wetlands): a strategy of protecting all vulnerable waters; a strategy that protects those vulnerable waters with quantifiable effects downstream; and a strategy that protects a portfolio of hydrological, chemical and biological functions that vulnerable waters provide downstream. There are trade-offs inherent in all three options, and the opportunity costs of protecting all vulnerable waters under option one are particularly high. Nonetheless, a 'do nothing' approach, representing a fourth option, is wholly inconsistent with the scientific evidence of the importance of these vulnerable waters, and would put ecosystem services at risk.

Society appears to have accepted a trade-off between rapacious exploitation of fresh waters and a sustainable future (Moss, 2010). If scientists wish to rectify this situation, a behavioural adjustment will be needed: specifically, a greater willingness to extend our influence well beyond the expected ambit of researchers (see also Ehrlich & Pringle, 2008). A more enlightened attitude on the part of policy-makers towards maintenance of biodiversity and ecosystem processes will be essential also. One way forward would be a shift in emphasis from doing research that may turn out (eventually) to be useful to society and, instead, focus upon science that will actually be used (Rogers, 2008). Thus, research must be salient to the concerns of managers and stakeholders, the information provided must be that which is needed, and the source of the information must be credible. While this approach could create problems of ensuring objectivity and independence of scientists (Hart & Calhoun, 2010), it is essential that all parties collaborate upon decisions about the multiple uses of water, and agree upon those that have sustainable outcomes. This would involve researchers in a three-step process: firstly, proactive engagement with stakeholders and managers; secondly, recognition that hypothesis testing through poorly targeted research (however complex it may be) is no supplement for problem-focused research that directly informs decision-making; and thirdly, presentation of research outputs in ways intended to be useful and solve problems, instead of simply cataloguing on-going environmental degradation (Knight, 2013).

Can We Conserve Freshwater Biodiversity in a Rapidly Changing World?

Only if we treat this endeavour with the seriousness that it deserves. While biodiversity constitutes the underpinning of ecosystem services, even where such a link is abundantly clear (as in the case of provisioning services), its promotion has not gained sufficient traction. Ongoing construction of mainstream dams in the Lower Mekong Basin (see Chapter 5), where the world's largest freshwater capture fishery is located (Box 3.6), is a clear instance where one group of humans have chosen to degrade a river that supplies the essential needs of other people.

Perhaps the underlying reason for decisions that damage the resource base upon which humans depend is a lack of interest or awareness of the importance of freshwater biodiversity that prevails among some sectors of water users or managers and, more generally, among citizens of many countries. Such an information gap leads to 'inadvertence', whereby the impacts of anthropogenic activities on biodiversity are overlooked. But, increasingly, environmental groups and human-rights advocates draw attention to the potential impacts of unsustainable development, such that ignorance or inadvertence can seldom serve as an excuse for self-interested decision making. Convenience may dictate that the likelihood of potential impacts is ignored due to economic, political or technical expediency that favour development. Alternatively, consideration of potential impacts may be set aside on the assumption that they can be addressed later. This is particularly likely if proposed projects can address pressing water-resource needs, or where hydropower dams can be expected to yield economic benefits from selling electricity (as along the Mekong). In such cases, development is often allowed to proceed without due accounting of the long-term environmental costs. Governments are unable or unwilling to invest in funding monitoring or surveys that will yield information on the likelihood of impacts or environmental degradation, and a shortage of trained personnel may hinder such investigations. Even where data on potential impacts are readily available, the consequences may not be readily understood by decision makers, or not perceived as relevant to local circumstances because it is too site-specific or sectoral. This is a particular problem for the integrated management of inland waters, which usually requires incorporation and integration of different types of data gathered at various scales from a wide range of sources, including information on hydrology, water quality, vegetation cover, land-use, ecology, socioeconomics and so on.

Additional complications arise because conservation of biodiversity takes place in a context where some parties can be characterized as dishonest, self-interested and hostile to nature; their motivations may be neither open nor fair, and may be driven by corporate considerations or, more simply, profit. As Schaller (1993: p. xv) sets out: 'A conservation project is always divided between politics and science . . .' yet accounts of attempts to save species '. . . of public concern tend to shy away from disclosing the true conservation conflicts, the basic issues of human greed and indifference'. The enthusiasm and goodwill of conservationists '. . . count for little when the enemy is a vast bureaucracy of local officials who myopically use obstruction, evasion, outdated concepts, activity without insight and other tragic traits to . . . create ecological mismanagement on a dismaying scale' (Schaller, 1993: p. 234). Vested interests tend to be set against the disinterested presentation of evidence favoured by scientists. At worst, selective presentation of data by project proponents may be used to spread doubt and confusion despite the existence of a scientific consensus (as in the case of anthropogenic global warming), and thereby attempt to undermine evidence or obfuscate issues that are not in dispute. Oreskes & Conway (2010) present compelling instances where those with vested interests act to keep controversies alive by exaggerating environmental uncertainties and expressing contrarian views that often receive extensive media coverage.

One way to push back against this behaviour, and the threat it poses to biodiversity, is to assert clearly that anthropogenic species loss is morally repugnant. Without such an affirmation, we cannot expect that biodiversity will be regarded as having sufficient importance to warrant consideration when decisions are made about human welfare and water needs on a warmer and more crowded planet. However, ethical or moral arguments about the need to protect all species because they have an inherent right to exist, or because they have intergenerational value (see Dudgeon, 2014), or because biodiversity represents a public good or service (see Box 8.5), can fail for a number of reasons. They include an unwillingness to pay to protect nature or to refrain from damaging actions that are profitable in the short term. Any affirmation that human-driven extinctions are morally repugnant is, on its own, unlikely to be persuasive when there are economic reasons for people to behave in ways that threaten species. A powerful rational argument for conservation will be essential to change things. That said, ethical or moral arguments for protecting biodiversity are not weakened by drawing attention to the economic benefits that some species provide (despite the reservations some may have about the monetization of nature). Utilitarian justifications relating to maintenance of ecosystem functioning and provision of goods and services should be made wherever this is possible and supported by science, although, as the case of the Mekong shows, there is no guarantee of success. Furthermore, emotional pleas for conservation action may fail to convince simply because much of society does not share the same concerns, and cares little about nature, biodiversity or ecosystems. For instance, Google search patterns suggest declining interest in all aspects of the environment, except climate change (Mccallum & Bury, 2013; but see Ficetola, 2013).

It is plausible that the virtual world of the internet makes it more likely that people will become more distant from – or care less about – nature. It is certainly true that nature has to compete for attention with many other activities, distractions, advertisements and messages conveyed by far-from-disinterested parties who make highly effective use of various media. (Conservation biologists may need to learn from that.) Another reason conservation messages fail to connect is that citizens may feel their actions can do little to change things, asking 'what difference could

I make?' It is essential to identify something that the target audience can do to make things better. In that connection, the term sustainability may well be demotivating to many individuals. Efforts to bring about sustainability may be interpreted merely as harm reduction, and insufficient to meet the challenges of damage already done to the environment. Hes & Du Plessis (2014) have argued for the need to go beyond harm reduction, advocating the requirement to make regenerative sustainability – rather than sustainability – a normative ethical principle. Regardless, it is essential that conservation scientists remain hopeful about the prospects of preserving a significant fraction of the Earth's species, even though it remains essential to monitor and report rates of decline and habitat loss. To quote Swaisgood & Sheppard (2010: p. 629), '. . . we do not believe that the environmental crisis can be averted . . .' but '. . . even in a dramatically altered world we can find meaning and a place for nature' (see also Knight, 2013). They are not optimistic, but hopeful. This view has echoes in other disciplines: 'Like all conservationists, a cultural conservationist in today's America can only act in hope while living with amply justified fear' (Jacoby, 2008: p. 317).

It is a challenge to remain optimistic about the future of freshwater biodiversity. The increasingly uneven distribution of fresh water, in space, time and quality, will engender greater reliance on large-scale engineering solutions (such as water transfers) and necessitate a hybrid management approach that – somehow – recognizes the value and benefits of freshwater ecosystems for people and nature alike (Shumilova et al., 2018). Recent history suggests that, even with the best of motivations and appropriate legislation, maintaining or improving the condition of freshwater ecosystems is hard to accomplish: for example, the EU Water Framework Directive has not delivered on its main objective of achievement of good ecological status for European lakes and rivers (Gilbert, 2015; Vouvoulis et al., 2017). However, presentiments of doom are unlikely to improve matters, and may have the opposite effect if they discourage those who might work to bring about positive change. New approaches to freshwater (regenerative) sustainability – implemented through scientifically informed adaptive management – are going to be needed to protect freshwater ecosystems through periods of changing societal needs (Creed et al., 2017).

Scientists must continue to insist that actions taken now to improve conditions will pay off in the form of future conservation gains. But, to reiterate, a transformation in the nature and shift in the direction of research, plus greater collaboration with non-scientists, will be needed to ensure that scientific knowledge is included effectively in societal decisions that affect freshwater biodiversity. If scientists hope to alleviate the tragedy of the freshwater commons, we must become facilitators of change by establishing partnerships that ensure whatever the scale of the challenge, our response is at least equivalent. We must look '. . . for some wise principle of coexistence between man and nature, even if it has to be a modified kind of man and a modified kind of nature' (Elton, 1958: p. 145).

Afterword

You must unite behind the science. You must take action. You must do the impossible. Because giving up can never ever be an option.

Greta Thunberg speaking before the US Congress in Washington, DC, September 2019

Conservation of freshwater biodiversity is not merely a matter of preservation of the products of evolution for their own sake, or for whatever intrinsic worth they may be perceived to have, but because of the services they provide. For instance, freshwater capture fisheries have relevance for human livelihoods, food security and sustainable development, as described in Chapter 3. This is made clear in a recent assessment of the importance of such livelihood contributions in countries with limited marine fisheries or aquaculture (Funge-Smith & Bennett, 2019). Freshwater fisheries are crucial for many socially, economically and nutritionally vulnerable groups around the world, but the challenges in monitoring these fisheries preclude a complete understanding of the magnitude of their contributions and limit the capacity to manage them sustainably. Reviewing global catch statistics (including recreational fisheries), Funge-Smith & Bennett (2019) confirm the widely held view that freshwater catches have been underestimated (see also Chapter 3). They also report some declines in yields from significant fisheries in parts of the Mekong and Amazon basins that may be indicative of overexploitation. The economic (first-sale) value of the global freshwater catch, estimated at US\$24 000 million, is almost one quarter that of marine fisheries – a larger fraction than might have been expected, apparently reflecting relatively high first-landing prices of freshwater fishes.

Given their bearing on human livelihoods, effective management of freshwater capture fisheries will be essential for sustainable development, which will require conservation of biodiversity and maintenance of ecosystem services in the face of multiple threats to inland waters. Protected areas offer one means of accomplishing these ends (see Chapter 9). Models of the benefits of no-take areas in riverine floodplains, where fisheries are typically indiscriminate, have been parameterized for Tonlé Sap by Hannah *et al.* (2019). They predict that protected areas can be associated with substantial increases in catch, especially as coverage increases up to around 50%, because the current extent of protected areas within Tonlé Sap is low. A reduction in the area fished would reduce catches initially, so implementation of no-take zones would be problematic for subsistence fishers in Cambodia (or elsewhere) who could not reduce capture effort now to gain more fish in future. Bridging funds or government subsidies would be needed to tide them over. Fisher communities could contribute by enforcing no-take zones and controlling poachers, if they had ownership or responsibility for the protected area (Hannah *et al.*, 2019). Regrettably, fisheries management and conservation at the local scale can be rendered irrelevant by interventions that change the 'rules of existence' for biodiversity within water bodies or drainages. Thus construction of dams on the Mekong mainstream in Laos (see Chapter 5) – as at Don Sahong, immediately adjacent to the border with Cambodia – will be detrimental to the ecology and fisheries of Tonlé Sap, irrespective of management efforts and protected areas within the lake.

Whatever prospects the future holds for conservation of freshwater biodiversity, much will depend on the efforts of scientists and non-government organizations who focus primarily on inland waters, as well as others – such as WWF and TNC – with a wider remit that encompasses both terrestrial and aquatic species. A recent initiative by a group of European scientists focused on the Freshwater Information Platform (FIP; see Chapter 9), which was launched in 2015 in the context of BioFresh with the aim of facilitating public access to primary data, metadata and reports on freshwater biodiversity derived mainly from EU-funded research projects. The near-term plan is for the FIP to be hosted by the Global Biodiversity Information Facility (GBIF; www.gbif .org/) – an international network and research infrastructure, funded by participating governments, intended to provide open-access data about life on Earth. GBIF currently lacks extensive repositories of data on freshwater species, and incorporation of the FIP could help remedy that situation. Management responsibility for biodiversity data from different realms would be taken up by the GBIF Secretariat in Copenhagen (for details, see Schmidt-Kloiber *et al.*, 2019). The proposal is fully supported by groups such as the Alliance for Freshwater Life (see Table 9.3) and Freshwater GEO BON (described in Chapter 9). Ultimately, the FIP will evolve to incorporate data from a variety of sources beyond the EU, thereby offering an integrated and comprehensive global information portal hosting tools and resources for freshwater biodiversity science.

Initiatives such as the FIP will help raise awareness about the need to conserve freshwater biodiversity and provide a knowledge base for action. But they will not be sufficient to address the scale, extent and variety of impending and emergent anthropogenic threats (Reid *et al.*, 2019), some of which are far from realm-specific. Climate change is the most palpable of these. In September 2019, a high-level synthesis describing the current state of the global climate was published. *United in Science* (World Meteorological Organization (WMO), 2019) was prepared for the United Nation's Secretary-General's Climate Action Summit. It was coordinated by the WMO, with contributions from the IPCC, Future Earth and UNEP. Notable conclusions were that the signs and impacts of climate change – such as sea level rise, ice loss and extreme weather – increased during 2015–2019, which is projected to be the warmest five-year period on record. Greenhouse gas concentrations in the atmosphere have risen to their highest levels in three million years, with CO_2 concentrations increasing at rates nearly 20% faster than during the previous five years; they could reach 410 ppm by the end of 2019 (WMO, 2019). The global average temperature has warmed by 1.1°C since the preindustrial period, and by 0.2°C compared to 2011–2015. Mass loss of reference glaciers for 2015–2019 has been the highest for any five-year period since 1950. The WMO concludes that there have been anthropogenic influences on the incidence of extreme events, including almost every major heatwave since 2015, and on the risk of torrential rainfall and droughts. Climate changes will affect freshwater ecosystems and their biota profoundly, as described in Chapter 7, and the speed of warming and its consequences are alarming. They will make it harder to deal with other exigent threats to biodiversity, such as dams and flow regulation, and place limits on attempts to mitigate them – by e-flows, for example (see Chapter 5). Warmer temperatures will also facilitate the establishment or spread of invasive species (Chapter 4), and intensify contaminant toxicity (Wang *et al.*, 2019). New approaches to adaptive management of fresh waters will be needed to gain better mechanistic understanding of these ecosystems, strengthen their resilience, and improve forecasting tools to limit risk (Tonkin *et al.*, 2019).

A 2018 IPCC special report on *Global Warming of 1.5°C*, which was alluded to in the Foreword, sets out just how high the stakes have become. It states that limiting warming to 1.5°C, as envisaged under the 2015 Paris climate agreement, is not physically impossible but would require unprecedented transitions in all aspects of society. There will be clear benefits from limiting warming to 1.5°C compared to 2°C, and every additional increment of warming matters a great deal. Impacts on biodiversity and ecosystems (not only inland waters) are projected to be lower at 1.5°C compared to 2°C of warming, and will allow freshwater ecosystems to retain more of their services to humans. Of 105 000 species (in all realms) assessed, 18% are projected to lose over half of their climatically determined geographical range with 1.5°C of warming, compared to 42% with 2°C of warming (IPCC, 2018). One thing is certain: the rapid rise in greenhouse gas levels during the past five years indicates that emissions reductions to date have been insufficient; governments or supranational institutions must ramp up commitments to reduce carbon emissions and implement effective strategies for climate mitigation.

Limiting warming to 1.5°C should not be incompatible with realizing the global Sustainable Development Goals (SDGs) and eradicating poverty, but will necessitate substantial and immediate cuts to carbon emissions. It will also require the widespread adoption of new and possibly disruptive technologies and practices, and enhancement of climate-resilient development pathways (IPCC, 2018). The necessary transitions will have profound implications for global and regional land use, and hence for biodiversity and ecosystems. Moreover, establishment of effective governance at a scale appropriate to ensure that the required changes can be made without damaging trade-offs remains a distant prospect. This wider planetary context places real constraints on the likelihood of achieving improved conservation outcomes for freshwater biodiversity.

It is not only the IPCC and WMO reports on the Earth's climate that have bleak messages to convey. The IPBES Global Assessment on Biodiversity and Ecosystem Services, presented to the IPBES Plenary in May 2019, expresses grave concern over the condition of the global life-support system. Fourteen of 18 ecosystem services – most of them regulating services and non-material contributions – have declined over the past four decades (IPBES, 2019). Global degradation of ecosystems by multiple drivers is epidemic, and, as the 'great thinning' proceeds, more species than ever before are threatened with extinction (an average of ~25% in assessed groups across all realms). Freshwater ecosystems show among the highest rates of species decline (see Fig. 2.1), and only 13% of the wetland area present in 1700 remained by 2000; contemporary losses have been especially rapid (0.8% annually since 1970). Around 75% of the land surface has been significantly transformed, altering runoff and sedimentation regimes in fresh waters. The rate of introduction of invasive alien species seems higher than ever before and shows no signs of slowing (IPBES, 2019). As a manifestation of this 'great mixing', biological communities are becoming more similar to each other within and across regions. And, as has been described in this book, freshwater ecosystems face '... a series of combined threats that include land-use change, water extraction, exploitation, pollution, climate change and invasive species...' (IPBES, 2019: IPBES/7/10/Add.1, para B1).

Sustaining freshwater ecosystems in the context of climate change, growing human demands for water, widespread pollution and contamination will require cross-sectoral and sector-specific interventions to improve water-use efficiency, reduce sources of pollution, minimize habitat degradation and restore natural flow regimes (IPBES, 2019). Integrated water-resource management practices have a role to play where they can foster collaborative management and equity between water users. Other actions needed (some of which are mentioned in this book) include expansion of protected areas for freshwater biodiversity; reduction and reversal of vegetation clearance within catchments; cessation of unsustainable agriculture; and widespread adoption of practices that reduce erosion, sedimentation and polluted runoff. Maintaining connectivity within drainages can help minimize the negative effects of dams and flow regulation, and may conceivably be brought about through blending natural (green) and built (hard or grey) infrastructure; design improvements of the latter are needed to produce better outcomes for biodiversity. Sector-specific interventions could include improved water-use efficiency in agriculture, application of locally developed water conservation techniques, and water pricing and incentive programmes (such as PWS schemes; see Chapter 8). Promotion of investment in water projects with clear sustainability criteria is advocated also (IPBES, 2019).

As with the *United in Science* synthesis, the IPBES assessment concludes that the SDGs cannot be met merely by continuing on current trajectories: transformative changes to the economic and political underpinnings of society will be needed. Failing that, ecosystem degradation and declines in biodiversity will undermine progress towards 80% of SDG targets (35 out of 44; IPBES, 2019). There is a critical need for policy targets, indicators and datasets that explicitly account for, and effectively track, the consequences of the direct drivers of nature degradation for human well-being and attainment of the SDGs. In addition, several steps must be taken to alleviate the indirect drivers: (1) capacity-building and inducements to promote environmental responsibility and eliminate perverse incentives; (2) cooperation to improve integration across sectors and jurisdictions; (3) pre-emptive and precautionary actions by regulatory and management institutions to avoid or remedy degradation of nature; (4) enhanced decision making that manages for greater ecological resilience in the face of complexity and uncertainty; and (5) strengthening environmental laws and their implementation. These transformations will require the commitment of more sectors and stakeholders than are currently engaged in the conservation of freshwater biodiversity or, indeed, biodiversity in any realm.

Many of the threats that imperil freshwater biodiversity are components of a wider syndrome of global environmental degradation affecting humans and many other species. The impacts on humans are felt mainly through impairment of livelihoods and constraints upon opportunities or prospects, while the effects on biodiversity can result in local or global extinction. None of these consequences are acceptable, nor are they sustainable. Unless global-scale drivers of degradation are halted through resolute action and urgent step-change – including, but not limited to, establishment of binding targets for greatly reduced carbon emissions – local efforts to protect biodiversity in particular, lakes or rivers represent little more than sticking plasters on a patient who suffers from a terminal infection. While every effort, however small, should be made to keep the patient alive, real progress can be made only at the whole-body scale. Failing that, it is difficult to be optimistic about the eventual fate of many freshwater species and ecosystems. But perhaps I am wrong. Or something may change. Irrespective, we can at least hope that it will be possible to preserve the vestiges when (and if) humanity finds a means to exist within ecological limits and planetary boundaries. In the interim, we should strive to make the best of all opportunities for conservation. Because we can, and because we must.

References

Abell, R., Allan, J.D. & Lehner, B. (2007). Unlocking the potential of protected areas for freshwaters. *Biological Conservation* **134**: 48–63.

Abell, R., Thieme, M.L., Revenga, C., Bryer, M., Kottelat, M., Bogutskaya, N., Coad, B., Mandrak, N., Contreras Balderas, S. Bussing, W., Stiassny, M.L.J., Skelton, P., Allen, G.R., Unmack, P., Naseka, A., Ng, R., Sindorf, N., Robertson, J., Armijo, E., Higgins, J.V., Heibel, T.J., Wikramanake, E., Olson, D., Lopez, H.L., Reis, R.E., Lundberg, J.G., Sabaj Perez, M.H. & Petry, P. (2008). Freshwater ecoregions of the world: a new map of biogeographic units for freshwater biodiversity conservation. *BioScience* **58**: 403–414.

Acreman, M., Arthington, A.H., Colloff, M.J., Couch, C., Crossman, N.D., Dyer, F., Overton, I., Pollino, C.A., Stewardson, M.J. & Young, W. (2014). Environmental flows for natural, hybrid, and novel riverine ecosystems in a changing world. *Frontiers in Ecology and the Environment* **12**: 466–473.

Adams, V.M., Setterfield, S.A., Douglas, M.M., Kennard, M.J. & Ferdinands, K. (2015). Measuring benefits of protected area management: trends across realms and research gaps for freshwater systems. *Philosophical Transactions of the Royal Society B* **370**: 20140274. https://doi.org/10.1098/rstb.2014.0274.

Adkins, J., Barton, C., Grubbs, S., Stringer, J. & Kolka, R. (2016). Assessment of streamside management zones for conserving benthic macroinvertebrate communities following timber harvest in eastern Kentucky headwater catchments *Water* **8**: 261. https://doi.org/10.3390/w8060261.

Adrian, R., O'Reilly, C.M., Zagarese, H., Baines, S.B., Hessen, D.O., Keller, W.K., Livingstone, D.M., Sommaruga, R., Straile, D., Van Donk, E., Weyhenmeyer, G.A. & Winder, M. (2009). Lakes as sentinels of climate change. *Limnology & Oceanography* **54**: 2283–2297.

AghaKouchaka, A., Norouzib, H., Madanic, K., Mirchid, A., Azarderakhshe, M., Nazemif, A., Nasrollahia, N., Farahmanda, A., Mehrana, A. & Hasanzadeh, E. (2015). Aral Sea syndrome desiccates Lake Urmia: call for action. *Journal of Great Lakes Research* **41**: 307–311.

Agostinho, C.S., Pelicice, F.M., Marques, E.E., Soares, A.B. & Almeida, D.A. (2011). All that goes up must come down? Absence of downstream passage through a fish ladder in a large Amazonian river. *Hydrobiologia* **675**: 1–12.

Aharon-Rotman, Y., McEvoy, J., Zheng, Z., Yu, H., Wang, X., Si, Y., Xu, Z., Yuan, Z., Jeong, W., Cao, L. & Fox, A.D. (2017). Water level affects availability of optimal feeding habitats for threatened migratory waterbirds. *Ecology and Evolution* **7**: 10440–10450.

Albrecht, C. & Wilke, T. (2008). Lake Ohrid: biodiversity and evolution. *Hydrobiologia* **615**: 103–140.

Albright, T.P., Moorhouse, T.G. & McNabb, T.J. (2004). The rise and fall of water hyacinth in Lake Victoria and the Kagera River basin, 1989–2001. *Journal of Aquatic Plant Management* **42**: 73–84.

Alcamo, J.M., Vörösmarty, C.J., Naiman, R.J., Lettenmaier, D.P. & Pahl-Wostl, C. (2008). A grand challenge for freshwater research: understanding the global water system. *Environmental Research Letters* **3**: 010202. http://iopscience.iop .org/1748-9326/3/1/010202.

Alcaraz, C. & Garcia-Berthou, E. (2007). Life history variation of invasive mosquito fish (*Gambusia holbrooki*) along a salinity gradient. *Biological Conservation* **139**: 83–92.

Al-Chokhachy, A., Alder, A., Hostetler, S., Gresswell, R. & Shepard, B. (2013). Thermal controls of Yellowstone cutthroat trout and invasive fishes under climate change. *Global Change Biology* **19**: 3069–3081.

Allan, J.D. & Castillo, M.M. (2007). *Stream Ecology: Structure and Function of Running Waters*, 2nd ed. Springer, Dordrecht.

Allan, J.D., Abell, R., Hogan, Z., Revenga, C., Taylor, B.W., Welcomme, R.L. & Winemiller, K. (2005). Overfishing of inland waters. *BioScience* **55**: 1041–1051.

Allen, D.J., Smith, K.G. & Darwall, W.R.T. (2012). *The Status and Distribution of Freshwater Biodiversity in Indo-Burma*. IUCN, Cambridge and Gland.

Allison, E.H., Perry, A.L., Badjeck, M.C., Adger, W.N., Brown, K., Conway, D., Halls, A.S., Pilling, G.M., Reynolds, J.D., Andrew, N.L. & Dulvy, N.K. (2009). Vulnerability of national economies to the impacts of climate change on fisheries. *Fish and Fisheries* **10**: 173–196.

Alofs, K.M., Jackson, D.A. & Lester, N.P. (2014). Ontario freshwater fishes demonstrate differing range-boundary shifts in a warming climate. *Diversity and Distributions* **20**: 123–136.

Alroy, J. (2015). Current extinction rates of reptiles and amphibians. *Proceedings of the National Academy of Sciences of the United States of America* **112**: 13003–13008.

Anderson, C.B. & Rosemond, A.D. (2007). Ecosystem engineering by invasive exotic beavers reduces in-stream diversity and enhances ecosystem function in Cape Horn, Chile. *Oecologia* **154**: 141–153.

Anderson, J.T., Saldaña Rojas, J. & Flecker, A.S. (2009). High-quality seed dispersal by fruit-eating fishes in Amazonian floodplain habitats. *Oecologia* **161**: 279–290.

Andreou, D., Arkush, K.D., Guégan, J.F. & Gozlan, R.E. (2012). Introduced pathogens and native freshwater biodiversity: a case study of *Sphaerothecum destruens*. *PLoS One* **7**: e36998. https://doi.org/10.1371/journal.pone.0036998.

Angulo, A. (2008). Conservation needs of *Batrachophrynus* and *Telmatobius* frogs of the Andes of Peru. *Conservation & Society* **6**: 328–333.

Antunes, A.P., Fewster, R.M., Venticinque, E.M., Peres, C.A., Levi, T., Rohe1, F. & Shepard, G.H. (2016). Empty forest or empty rivers? A century of commercial hunting in Amazonia. *Science Advances* **2**: e1600936. http://advances .sciencemag.org/cgi/content/full/2/10/e1600936/DC1.

Arantes, M.L. & Freitas, C.E.C. (2016). Effects of fisheries zoning and environmental characteristics on population parameters of the tambaqui (*Colossoma macropomum*) in managed floodplain lakes in the Central Amazon. *Fisheries Management and Ecology* **23**: 133–143.

Arashkevich, E.G., Sapozhnikov, P.V., Soloviov, K.A., Kudyshkin, T.V. & Zavialov, P.O. (2009). *Artemia parthenogenetica* (Branchiopoda: Anostraca) from the Large Aral Sea: abundance, distribution, population structure and cyst production. *Journal of Marine Systems* **76**: 359–366.

Arthington, A.H. (2012). *Environmental Flows: Saving Rivers in the Third Millennium.* University of California Press, Oakland.

Arthington, A.H., Bunn, S.E., Poff, N.L. & Naiman, R.J. (2006). The challenge of providing environmental flow rules to sustain river ecosystems. *Ecological Applications* **16**: 1311–1318.

Arthington A.H., Naiman, R.J., McClain, M.E. & Nilsson, C. (2010). Preserving the biodiversity and ecological services of rivers: new challenges and research opportunities. *Freshwater Biology* **55**: 1–16.

Arthington, A.H., Bhaduri, A., Bunn, S.E., Jackson, S.E., Tharme, R.E., Tickner, D.,Young, B., Acreman, M., Baker, N., Capon, S., Horne, A.C., Kendy, E., McClain, M.E., Poff, N.L., Richter, B.D. & Ward, S. (2018). The Brisbane Declaration and Global Action Agenda on Environmental Flows (2018). *Frontiers in Environmental Science* **6**: 45. https://doi.org/10.3389/fenvs.2018.00045.

Asquith, N.M., Vargas, M.T. & Wunder, S. (2008). Selling two environmental services: in-kind payments for bird habitat and watershed protection in Los Negros, Bolivia. *Ecological Economics* **65**: 675–684.

Audzijonyte, A., Kuparinen, A. & Fulton, E. A. (2013). How fast is fisheries-induced evolution? Quantitative analysis of modelling and empirical studies. *Evolutionary Applications* **6**: 585–595.

Bai, J., Chen, X., Li, J., Yang, L. & Fang, H. (2011). Changes in the area of inland lakes in arid regions of central Asia during the past 30 years. *Environmental Monitoring and Assessment* **178**: 247–256.

Bain, M.B., Haley, N., Peterson, D.L., Arend, K.K., Mills K.E. & Sullivan, P.J. (2007). Recovery of a US endangered fish. *PLoS ONE* **2**: e168. https://doi .org/10.1371/journal.pone.0000168

Baird, I.G. (2006a). *Probarbus jullieni* and *Probarbus labeamajor*: the management and conservation of two of the largest fish species in the Mekong River in southern Laos. *Aquatic Conservation: Marine and Freshwater Ecosystems* **16**: 517–532.

Baird, I.G. (2006b). Strength in diversity: fish sanctuaries and deep-water pools in Lao PDR. *Fisheries Management and Ecology* **13**: 1–8.

Baird, I.G. (2013). *Boesemania microlepis*. The IUCN Red List of Threatened Species 2013: e.T181232A7664209. http://dx.doi.org/10.2305/IUCN.UK.2011-1.RLTS.T181232A7664209.en

Baird, I.G., Phylavanh, B., Vongsenesouk, B. & Xaiyamanivong, K. (2001). The ecology and conservation of the smallscale croaker *Boesemania microlepis* (Bleeker 1858-59) in the mainstream Mekong River, southern Laos. *Natural History Bulletin of the Siam Society* **49**: 161–176.

Baldwin, A.K., Corsi, S.R. & Mason, S.A. (2016). Plastic debris in 29 Great Lakes tributaries: relations to watershed attributes and hydrology. *Environmental Science & Technology* **50**: 10377–10385.

Balian, E.V., Lévêque, C., Segers, H. & Martens, K. (2008a). *Freshwater Animal Biodiversity Assessment*. Springer, Berlin.

Balian, E.V., Segers, H., Lévêque, C. & Martens, K. (2008b). The freshwater animal diversity assessment: an overview of the results. *Hydrobiologia* **595**: 627–637.

Barlow, C., Baran, E., Halls, A. & Kshatriya, M. (2008). How much of the Mekong fish catch is at risk from upstream dam development? *Catch and Culture* **14**: 16–21.

Barnhart, M.C. (2008). Unio Gallery. http://unionid.missouristate.edu

Banks, C.B., Lau, M.Y.N. & Dudgeon, D. (2008). Captive management and breeding of Romer's tree frog *Chirixalus romeri*. *International Zoo Yearbook* **42**: 99–108.

Barnosky, A.D., Matzke, N., Tomiya, S., Wogan, G.O.U., Swartz, B., Quental, T.B., Marshall, C., McGuire, J.L., Lindsey, E.L., Maguire, K.C., Mersey, B. & Ferrer, E.A. (2011). Has the Earth's sixth mass extinction already arrived? *Nature* **471**: 51–57.

Barrett, J.H., Locker, A.M. & Roberts, C.M. (2004). The origins of intensive marine fishing in medieval Europe: the English evidence. *Proceedings of the Royal Society of London B* **271**: 2417–2421.

Barthem, R.B., Goulding, M., Leite, R.G., Cañas, C., Forsberg, B., Venticinque, E., Petry, P., Ribeiro, M.L.B., Chuctaya, J. & Mercado, A. (2017). Goliath catfish spawning in the far western Amazon confirmed by the distribution of mature adults, drifting larvae and migrating juveniles. *Scientific Reports* **7**: 41784. https://doi.org/10.1038/srep41784

Bates, A.E., McKelvie, C.M., Sorte, C.J.B., Morley, S.A., Jones, N.A.R., Mondon, J., Bird, T.J. & Quinn, G. (2013). Geographical range, heat tolerance and invasion success in aquatic species. *Proceedings of the Royal Society B: Biological Sciences* **280**: 20131958.

Baumgartner, L., Zampatti, B., Jones, M., Stuart, I. & Mallen-Cooper, M. (2014). Fish passage in the Murray-Darling Basin, Australia: not just an upstream battle. *Ecological Management and Restoration* **15**: 28–39.

Baxter, C.V., Fausch, K.D., Murakami, M. & Chapman, P.L. (2004). Fish invasion restructures stream and forest food webs by interrupting reciprocal prey subsidies. *Ecology* **85**: 2565–2663.

Beard, T.D., Jr., Arlinghaus, R., Cooke, S.J., McIntyre, P.B., De Silva, S., Bartley, D. & Cowx, I.G. (2011). Ecosystem approach to inland fisheries: research needs and implementation strategies. *Biology Letters* **7**: 481–483.

Beatty, S.J., Morgan, D.L. & Lymbery, A.J. (2014). Implications of climate change for potamodromous fishes. *Global Change Biology* **20**: 1794–1807.

Becker, L.A., Pascual, M.A. & Basso, N.G. (2007). Colonization of the southern Patagonia ocean by exotic chinook salmon. *Conservation Biology* **21**: 1347–1352.

Beer, W.N. & Anderson, J.J. (2011). Sensitivity of juvenile salmonid growth to future climate trends. *River Research and Applications* **27**: 663–669.

Beggel, S., Brandner, J., Cerwenka, A.F. & Geist, J. (2016). Synergistic impacts by an invasive amphipod and an invasive fish explain native gammarid extinction. *BMC Ecology* **16**: 32. https://doi.org/10.1186/s12898–016-0088-6

Behrouzi-Rad, B. (2009). Waterbird populations during dry and wet years in the Hamoun Wetlands Complex, Iran/Afghanistan border. *Podoces* **4**: 88–99.

Bellard, C., Cassey, P. & Blackburn, T.M. (2016). Alien species as a driver of recent extinctions. *Biology Letters* **12**: 20150623.

Bellmore, J.R., Duda, J.J., Craig, L.S., Greene, S.L., Torgersen, C.W., Collins, M.J. & Vittum, K. (2017). Status and trends of dam removal research in the United States. *WIREs: Water* **4**: e1164. https://doi.org/10.1002/wat2.1164

Bennett, G., Carroll, N. & Hamilton, K. (2013). *Charting New Waters: State of Watershed Payments 2012*. Forest Trends, Washington, DC. www.ecosystemmarketplace.com/reports/sowp2012

Benson, A.J., Raikow, D., Larson, J. & Fusaro, A. (2014a). *Dreissena polymorpha*. USGS Nonindigenous Aquatic Species Database, Gainesville, FL. http://nas.er.usgs.gov/queries/FactSheet.aspx?speciesID=5

Benson, A.J., Richerson, M.M., Maynard, E., Larson, J. & Fusaro, A. (2014b). *Dreissena rostriformis bugensis*. USGS Nonindigenous Aquatic Species Database, Gainesville, FL. http://nas.er.usgs.gov/queries/factsheet.aspx?speciesid=95

Bergamino, N., Horion, S, Stenuitec, S., Cornet, Y., Loiselle, S., Plisnier, P. & Descy, J. (2010). Spatio-temporal dynamics of phytoplankton and primary production in Lake Tanganyika using a MODIS based bio-optical time series. *Remote Sensing of Environment* **114**: 772–780.

Bernes, C., Carpenter, S.R., Gårdmark, A., Larsson, P., Persson, L., Skov, C., Speed, J.D.M. & Van Donk, E. (2015). What is the influence of a reduction of planktivorous and benthivorous fish on water quality in temperate eutrophic lakes? A systematic review. *Environmental Evidence* **4**: 7. https://doi.org/10.1186/s13750–015-0032-9

Bernhardt, E.S. & Palmer, M.A. (2011). River restoration: the fuzzy logic of repairing reaches to reverse catchment scale degradation. *Ecological Applications* **21**: 1926–1931.

Bernhardt, E.S., Bunn, S.E., Hart, D.D., Malmqvist, B., Muotka, T., Naiman, R.J., Pringle, C., Reuss, M. & van Wilgen, B. (2006). Perspective: the challenge of ecologically sustainable water management. *Water Policy* **8**: 475–479.

Bianchi, T.S., Davis, G.M. & Strayer, D.S. (1994). An apparent hybrid zone between freshwater gastropod species *Elimia livescens* and *E. virginica* (Gastropoda: Pleuroceridae). *American Malacological Bulletin* **11**: 73–78.

Bickford, D., Howard, S.D., Ng, D.J.J. & Sheridan, J.A. (2010). Impacts of climate change on the amphibians and reptiles of Southeast Asia. *Biodiversity and Conservation* **19**: 1043–1062.

Biermann, F., Abbott, K., Andresen, S., Bäckstrand, K., Bernstein, S., Betsill, M.M., Bulkeley, H., Cashore, B., Clapp, J., Folke, C., Gupta, A., Gupta, J., Haas, P.M., Jordan, A., Kanie, N., Kluvánková-Oravská, T., Lebel, L., Liverman, D., Meadowcroft, J., Mitchell, R.B., Newell, P., Oberthür, S., Olsson, L., Pattberg, P., Sánchez-Rodríguez, R., Schroeder, H., Underdal, A., Vieira, S.C., Vogel, C., Young, O.R., Brock, A. & Zondervan, R. (2012). Navigating the Anthropocene: improving Earth system governance. *Science* **335**: 1306–1307.

BirdLife International (2014). *IUCN Red List for Birds*. www.birdlife.org

BirdLife International (2015a). *Important Bird Areas Factsheet: Haur Al Hammar*. www.birdlife.org

BirdLife International (2015b). *Important Bird Areas factsheet: Central Marshes.* www.birdlife.org

BirdLife International (2015c). *Important Bird Areas Factsheet: Hawizeh.* www.birdlife.org

BirdLife International (2016). *Leucogeranus leucogeranus. The IUCN Red List of Threatened Species 2016*: e.T22692053A98336905. http://dx.doi.org/10.2305/IUCN.UK.2016-3.RLTS.T22692053A98336905.en

BirdLife International (2018). *Important Bird Areas Factsheet: Lake Turkana.* www.birdlife.org

Birstein, V.J. (1997). Threatened fishes of the world: *Pseudoscaphirhynchus* spp. (Acipenseridae). *Environmental Biology of Fishes* **48**: 381–383.

Blaber, S.J., Milton, D.A., Brewer, D.T. & Salini, J.P. (2003). Biology, fisheries, and status of tropical shads *Tenualosa* spp. in South and Southeast Asia. *American Fisheries Society Symposium* **35**: 49–58.

Blinn, D.M. & Poff, N.L. (2005). Colorado River Basin. In *Rivers of North America* (A.C. Benke & C.E. Cushing, eds), Elsevier Academic Press, Amsterdam: pp. 483–538.

Boethius, A. (2016). Something rotten in Scandinavia: the world's earliest evidence of fermentation. *Journal of Archaeological Science* **66**: 169–180.

Bogardi, J.J., Dudgeon. D., Lawford, R., Flinkerbusch, E., Meyn, A., Pahl-Wostl, C., Vielhauer, K. & Vörösmarty, C. (2012). Water Security for a planet under pressure: interconnected challenges of a changing world call for sustainable solutions. *Current Opinion in Environmental Sustainability* **4**: 35–43.

Böhm, M., Collen, B., Baillie, J.E.M., Bowles, P., Chanson, J., Cox, N., Hammerson, G., Hoffmann, M., Livingstone, S.R., Ram, M. *et al.* (2013). The conservation status of the world's reptiles. *Biological Conservation* **157**: 372–385.

Boltovskoy, D. & Correa, N. (2015). Ecosystem impacts of the invasive bivalve *Limnoperna fortunei* (golden mussel) in South America. *Hydrobiologia* **746**: 81–95.

Boltovskoy, D., Karatayev, A., Burlakova, L., Cataldo, D., Karatayev, V., Sylvester, F. & Marinelarena, A. (2009). Significant ecosystem-wide effects of the swiftly spreading invasive freshwater bivalve *Limnoperna fortunei*. *Hydrobiologia* **636**: 271–284.

Bond, N., Costelloe, J., King, A., Warfe, D., Reich, P. & Balcombe, S (2014a). Ecological risks and opportunities from engineered artificial flooding as a means of achieving environmental flow objectives. *Frontiers in Ecology and the Environment* **12**: 386–394.

Bond, N.R, Thomson, J.R. & Reich, P. (2014b). Incorporating climate change in conservation planning for freshwater fishes. *Diversity and Distributions* **20**: 931–942.

Braulik, G.T., Arshad, M., Noureen, U. & Northridge, S.P. (2014). Habitat fragmentation and species extirpation in freshwater ecosystems; causes of range decline of the Indus River dolphin (*Platanista gangetica minor*). *PLoS ONE* **9**: e101657. https://doi.org/10.1371/journal.pone.0101657

Britton, A.W., Day, J.J., Doble, C.J. Ngatunga, B.P., Kemp, K.M. Carbone, C. & Murrell, D.J. (2017). Terrestrial-focused protected areas are effective for conservation of freshwater fish diversity in Lake Tanganyika. *Biological Conservation* **212**: 120–129.

Britton, J.R., Boar, R.R., Gray, J., Foster, J., Lugonso, J. & Harper, D.M. (2007). From introduction to fishery dominance: the initial impacts of the invasive carp *Cyprinus carpio* in Lake Naivasha, Kenya, 1999 to 2006. *Journal of Fish Biology* **71** (Suppl. D): 239–257.

Broadhurst, B.T., Ebner, B.C., Lintermans, M., Thiem, J.D. & Clear, R.C. (2013). Jailbreak: a fishway releases the endangered Macquarie perch from confinement below an anthropogenic barrier. *Marine and Freshwater Research* **64**: 900–908.

Brooks, E.G.E, Holland, R.A., Darwall, W.R.T. & Eigenbrod, F. (2016). Global evidence of positive impacts of freshwater biodiversity on fishery yields. *Global Ecology and Biogeography* **25**: 553–562.

Brooks, S.E., Reynolds, J.D. & Allison, E.H. (2008). Sustained by snakes? Seasonal livelihood strategies and resource conservation by Tonlé Sap fishers in Cambodia. *Human Ecology* **36**: 835–851.

Brooks, S.E., Allison, E.H., Gill, J.A. & Reynolds, J.D. (2010). Snake prices and crocodile appetites: aquatic wildlife supply and demand on Tonle Sap Lake, Cambodia. *Biological Conservation* **143**: 2127–2135.

Brooks T.M., Mittermeier R.A., da Fonseca G.A.B., Gerlach J., Hoffmann M., Lamoreux J.F., Mittermeier C.G., Pilgrim J.D. & Rodrigues A.S.L. (2006). Global biodiversity conservation priorities. *Science* **313**: 58–61.

Brosse, S., Beauchard, O., Blanchet, S., Dürr, H.H., Grenouillet, G., Hugueny, B., Lauzeral, C., Leprieur, F., Tedesco, P.A., Villéger, S. & Oberdorff, T. (2013). SPRICH: a database of freshwater fish species richness across the World. *Hydrobiologia* **700**: 343–349.

Brown, A.R., Owen, S.F., Peters, J., Zhang, Y., Soffker, M., Paull, G.C., Hosken, D.J., Abdul Wahab, M. & Tyler, C.R. (2015). Climate change and pollution speed declines in zebrafish populations. *Proceedings of the National Academy of Sciences of the United States of America* **112**: E1237–E1246.

Brown, J.J., Limburg, K.E., Waldman, J.R., Stephenson, K., Glenn, E.P. & Juanes, F. (2013). Fish and hydropower on the U.S. Atlantic coast: failed fisheries policies from half-way technologies. *Conservation Letters* **6**: 280–286.

Brown, K.J., Rüber, L., Bills R. & Day, J.J. (2010). Mastacembelid eels support Lake Tanganyika as an evolutionary hotspot of diversification. *BMC Evolutionary Biology* **10**:188. https://doi.org/10.1186/1471-2148-10-188

Brown, T.C., Hobbins, M.T. & Ramirez, J.A. (2008). Spatial distribution of water supply in the conterminous United States. *Journal of the American Water Resources Association* **44**: 1474–1487.

Buisson, L., Grenouillet, G., Villéger, S., Canal, J. & Laffaille, P. (2013). Toward a loss of functional diversity in stream fish assemblages under climate change. *Global Change Biology* **19**: 387–400.

Bunn, S.E. & Arthington, A.H. (2002). Basic principles and ecological consequences of altered flow regimes for aquatic biodiversity. *Environmental Management* **30**: 492–507.

Bunn, S.E., Abal, E.G., Smith, M.J., Choy, S.C., Fellows, C.S., Harch, B.D., Kennard, M.J. & Sheldon, F. (2010). Integration of science and monitoring of river ecosystem health to guide investments in catchment protection and rehabilitation. *Freshwater Biology* **55** (Suppl. 1): 223–240.

Bunnell, D.B., Barbiero, R.P., Ludsin, S.A., Madenjian, C.P., Warren, G.J., Dolan, D.M., Brenden, T.O., Briland, R.,Gorman, O.T., He, J.X., Johengen, T.H., Lantry, B.F., Nalepa, T.F., Riley, S.C., Riseng, C.M., Treska, T.J., Tsehaye, I., Walsh, M.G., Warner, D.M. & Weidel, B.C. (2014). Changing ecosystem dynamics in the Laurentian Great Lakes: bottom-up and top-down regulation. *BioScience* **64**: 26–39.

Bunt, C.M., Castro-Santos, T. & Haro, A. (2012). Performance of fish passage structures at upstream barriers to migration. *River Research and Applications* **28**: 457–478.

Burney, D.A. & Flannery, T.F. (2005). Fifty millennia of catastrophic extinctions after human contact. *Trends in Ecology & Evolution* **20**: 395–401.

Burns, C.W., Schallenberg, M. & Verburg, P. (2014). Potential use of classical biomanipulation to improve water quality in New Zealand lakes: a re-evaluation. *New Zealand Journal of Marine and Freshwater Research* **48**: 127–138.

Butchart, S.H.M., Walpole, M., Collen, B., van Strien A., Scharlemann, J.P.W., Almond, R.A.E., Baillie, J.E.M., Bomhard, B., Brown, C., Bruno, J., Carpenter, K.E., Carr, G.M., Chanson, J., Chenery, A.M., Csirke, J., Davidson, N.C., Dentener, F., Foster, M., Galli, A., Galloway, J.N., Genovesi, P., Gregory, R.D., Hockings, M., Kapos, V., Lamarque, J.F., Leverington, F., Loh, J., McGeoch, M.A., McRae, L., Minasyan, A., Hernández Morcillo, M., Oldfield, T.E., Pauly, D., Quader, S., Revenga, C., Sauer, J.R., Skolnik, B., Spear, D., Stanwell-Smith, D., Stuart, S.N., Symes, A., Tierney, M., Tyrrell, T.D., Vié, J.C., & Watson, R. (2010). Global biodiversity: indicators of recent declines. *Science* **328**: 1164–1168.

Bush, A., Theischinger, G., Nipperess, D., Turak, E. & Hughes, L. (2013). Dragonflies: climate canaries for river management. *Diversity and Distributions* **19**: 86–97.

Cael, B.B., Heathcote, A.J. & Seekell, D.A. (2017). The volume and mean depth of Earth's lakes. *Geophysical Research Letters* **44**: 209–218.

Cairns, A. & Yan, N.D. (2009). A review of the influence of low ambient calcium concentrations on freshwater daphniids, gammarids, and crayfish. *Environmental Reviews* **17**: 67–79.

Calles, O. & Greenberg, L. (2009). Connectivity is a two-way street – the need for a holistic approach to fish passage problems in regulated rivers. *River Research and Applications* **25**: 1268–1286.

Campbell, I., Poole, C., Giesen, W. & Valbo-Jorgensen, J. (2006). Species diversity and ecology of Tonle Sap Great Lake, Cambodia. *Aquatic Sciences* **68**: 355–373.

Capon, S.J. & Capon, T.R. (2017). An impossible prescription: why science cannot determine environmental water requirements for a healthy Murray-Darling Basin. *Water Economics and Policy* **3**: https://doi.org/10.1142/S2382624X16500375

Capps, K.A. & Flecker, A.S. (2013a). Invasive aquarium fish transform ecosystem nutrient dynamics. *Proceedings of the Royal Society of Biology: Series B* **280**: 20132418.

Capps, K.A. & Flecker, A.S. (2013b). Invasive fishes generate biogeochemical hot-spots in a nutrient-limited system. *PLoS ONE* **8**: e54093. https://doi.org/10.1371/journal.pone.0054093

Capuli, E. & Froese, R. (1999). Status of the freshwater fishes of the Philippines. In *Proceedings of the 5th Indo-Pacific Fish Conference* (B. Seret & J. Y. Sire, eds), Societe Francaise d'Ichthyology, Paris: pp. 381–384.

Cardinale, B.J., Duffy, J.E., Gonzalez, A., Hooper, D.U., Perrings, C., Venail, P., Narwani, A., Mace, G.M., Tilman, D., Wardle, D.A., Kinzig, A.P., Daily, G.C., Loreau, M., Grace, J.B., Larigauderie, A., Srivastava. D.S. & Naeem, S. (2012). Biodiversity loss and its impact on humanity. *Nature* **486**: 59–67.

Carey, M.P. & Zimmerman, C.E. (2014). Physiological and ecological effects of increasing temperature on fish production in lakes of Arctic Alaska. *Ecology and Evolution* **4**: 1981–1993.

Carlisle, D.M., Wolock, D.M. & Meador, M.R. (2011). Alteration of streamflow magnitudes and potential ecological consequences: a multiregional assessment. *Frontiers in Ecology and the Environment* **9**: 264–70.

Carlson, P.E., Donadi, S. & Sandin, L. (2018). Responses of macroinvertebrate communities to small dam removals: implications for bioassessment and restoration. *Journal of Applied Ecology* **55**: 1896–1907.

Carlsson, N.O.L., Brönmark, C. & Hansson, L.A. (2004). Invading herbivory: the golden apple snail alters ecosystem functioning in Asian wetlands. *Ecology* **85**: 1575–1580.

Carrete, G. & Wiens, J.J. (2012). Why are there so few fish in the sea? *Proceedings of the Royal Society B: Biological Sciences* **279:** 2323–2329.

Carroll, M.J., Heinemeyer, A., Pearce-Higgins, J.W., Dennis, P., West, C., Holden, J., Wallage, Z.E. & Thomas, C.D. (2015). Hydrologically-driven ecosystem processes determine the distribution and survival of ecosystem-specialist predators under climate change. *Nature Communications* **6**: 7851. https://doi.org/10.1038/ncomms8851

Castello, L. & Macedo, M.N. (2016). Large-scale degradation of Amazonian freshwater ecosystems. *Global Change Biology* **22**: 990–1007.

Castello, L., McGrath, D.G., Hess, L.L., Coe, M.T., Lefebvre, P.A., Petry, P., Macedo, M.N., Renó, V.F. & Arantes, C.C. (2013). The vulnerability of Amazon freshwater ecosystems. *Conservation Letters* **6**: 217–229.

Castello, L., Arantes, C.C., McGrath, D. G., Stewart, D.J. & De Sousa, F.S. (2015a). Understanding fishing-induced extinctions in the Amazon. *Aquatic Conservation: Marine and Freshwater Ecosystems* **25**: 587–598.

Castello, L., Isaac, V.I.N. & Thapa, R. (2015b). Flood pulse effects on multispecies fishery yields in the Lower Amazon. *Royal Society Open Science* **2**: 150299.

Cataldo, D., O'Farrell, I., Paolucci, E., Sylvester, F. & Boltovskoy, D. (2012). Impact of the invasive golden mussel (*Limnoperna fortunei*) on phytoplankton and nutrient cycling. *Biological Invasions* **7**: 91–100.

Ceballos, G., Ehrlich, P.R. & Dirzo, R. (2017). Biological annihilation via the ongoing sixth mass extinction signaled by vertebrate population losses and declines. *Proceedings of the National Academy of Sciences of the United States of America* **114**: E6089–E6096.

Charlier, R.H., Chaineux, M.C.P. & Morcos, S. (2005). Panorama of the history of coastal protection. *Journal of Coastal Research* **21**: 79–111.

Chellaiah, D. & Yule, C. (2018). Riparian buffers mitigate impacts of oil palm plantations on aquatic macroinvertebrate community structure in tropical streams of Borneo. *Ecological Indicators* **95**: 53–62.

Chen, D., Duan, X., Liu, S. & Shi, W. (2004). Status and management of the fisheries resources of the Yangtze River. In *Proceedings of the Second International Symposium on the Management of Large Rivers for Fisheries, Vol. 1* (R. Welcomme, R. & T. Petr, eds), FAO Regional Office for Asia and the Pacific, Bangkok: pp. 173–182.

Chen, D., Xiong, F., Wang, K. & Chang, Y. (2009). Status of research on Yangtze fish biology and fisheries. *Environmental Biology of Fishes* **85**: 337–357.

Chen, Y. & Zhu, S. (2008). Change of fish fauna and long-term dynamics of the harvest of aquatic product in a large shallow lake (Lake Taihu, China). *Journal of Fisheries and Aquatic Science* **3**: 72–76.

Cheng, L., Opperman, J.J., Tickner, D., Speed, R., Guo, Q. & Chen, D. (2018). Managing the Three Gorges Dam to implement environmental flows in the Yangtze River. *Frontiers in Environmental Science* **6**: 64. www.frontiersin.org/article/10.3389/fenvs.2018.00064

Chessman, B.C. (2013). Do protected areas benefit freshwater species? A broadscale assessment of fish in Australia's Murray-Darling Basin. *Journal of Applied Ecology* **50**: 969–976.

Cheung, S.M. & Dudgeon, D. (2006). Quantifying the Asian turtle crisis: market surveys in southern China 2000-2003. *Aquatic Conservation: Marine and Freshwater Ecosystems* **16**: 751–770.

Chichilnisky, G. & Heal, G. (1998). Economic returns from the biosphere. *Nature* **391**: 629–630.

Chowdhury, G.W., Zieritz, A. & Aldridge, D.C. (2016). Ecosystem engineering by mussels supports biodiversity and water clarity in a heavily polluted lake in Dhaka, Bangladesh. *Freshwater Science* **35**: 188–199.

Chu, C., Ellis, L. & Kerckhove, D.T. (2018). Effectiveness of terrestrial protected areas for conservation of lake fish communities. *Conservation Biology* **32**: 607–618.

Chucholl, C. (2013). Invaders for sale: trade and determinants of introduction of ornamental freshwater crayfish. *Biological Invasions* **15**: 125–141.

Ciechanowski, M., Kubic, W., Rynkiewicz, A. & Zwolicki, A. (2011). Reintroduction of beavers *Castor fiber* may improve habitat quality for vespertilionid bats foraging in small river valleys. *European Journal of Wildlife Research* **57**: 737–747.

Cinner, J.E., Daw, T. & McClanahan, T.R. (2009). Socioeconomic factors that affect artisanal fishers' readiness to exit a declining fishery. *Conservation Biology* **23**: 124–130.

Clarke, A., Mac Nally, R., Bond, N. & Lake, P.S. (2008). Macroinvertebrate diversity in headwater streams: a review. *Freshwater Biology* **53**: 1707–1721.

Clausnitzer, V., Kalkman, V.J., Ram, M., Collen, B., Baillie, J.E.M., Bedjanič, M., Darwall, W.R.T., Dijkstra K.-D.B., Dow, R., Hawking, J., Karube, H., Malikova, E., Paulson, D., Schütte, K., Suhling, F., Villanueva, R.J., von Ellenrieder, N. & Wilson, K. (2009). Odonata enter the biodiversity crisis debate: the first global assessment of an insect group. *Biological Conservation* **142**: 1864–1869.

Clavero, M. & García-Berthou, E. (2005). Invasive species are a leading cause of animal extinctions. *Trends in Ecology & Evolution* **20**: 110.

Cline, T.J., Bennington, V. & Kitchell, J.F. (2013). Climate change expands the spatial extent and duration of preferred thermal habitat for Lake Superior fishes. *PLoS ONE* **8**: e62279. https://doi.org/10.1371/journal.pone.0062279

Cochrane, K., De Young, C., Soto, D. & Bahri, T. (2009). *Climate Change Implications for Fisheries and Aquaculture: Overview of Current Scientific Knowledge*. FAO Fisheries and Aquaculture Technical Paper. No. 530. Food and Agriculture Organization of the United Nations, Rome.

Cochran-Biederman, J., Wyman, K., French, W. & Loppnow, G. (2014). Identifying correlates of success and failure of native freshwater fish reintroductions. *Conservation Biology* 29: 175–186.

Cochran-Biederman, J.L., Wyman, K.E., French, W.E. & Loppnow, G.L. (2015). Identifying correlates of success and failure of native freshwater fish reintroductions. *Conservation Biology* 29: 175–186.

Coe, M.T. & Foley, J.A. (2001). Human and natural impacts on the water resources of the Lake Chad basin. *Journal of Geophysical Research* **106**: 3349–3356.

Cohen, A.S., Gergurich, E.L., Kraemer, B.M., McGlue, M.M., McIntyre, P.B., Russell, J.M., Simmons, J.D. & Swarzenski, P.W. (2016). Climate and fishery declines in Lake Tanganyika. *Proceedings of the National Academy of Sciences of the United States of America* **113**: 9563–9568.

Cohen, M.J., Creed, I.F., Alexander, L., Basu, N.B., Calhoun, A.J.K., Craft, C., D'Amico, E., DeKeyser, E., Fowler, L., Golden, H.E., Jawitz, J.W., Kalla, P., Kirkman, L.K., Lane, C.R., Lang, M., Leibowitz, S.G., Lewis, D.B., Marton, J., McLaughlin, D.L., Mushet, D.M., Raanan-Kiperwas, H., Rains, M.C., Smith, L.C. & Walls, S.C. (2016). Do geographically isolated wetlands influence landscape functions? *Proceedings of the National Academy of Sciences of the United States of America* **113**: 1978–1986.

Cole, E. & Newton, M. (2013). Influence of streamside buffers on stream temperature response following clear-cut harvesting in western Oregon. *Canadian Journal of Forest Research* **43**: 993–1005.

Collen, B., Loh, J., Whitmee, S., McRae, L., Amin, R. & Baillie, J.E.M. (2009). Monitoring change in vertebrate abundance: the Living Planet Index. *Conservation Biology* **23**: 317–327.

Collen, B., Whitton, F., Dyer, E.E., Baillie, J.E.M., Cumberlidge, N., Darwall, W.R.T., Pollock, C., Richman, N.I., Soulsby, A. & Böhm, M. (2014). Global patterns of freshwater species diversity, threat and endemism. *Global Ecology and Biogeography* **23**: 40–51.

Collier, K.J. (2017). Measuring river restoration success: are we missing the boat? *Aquatic Conservation: Marine and Freshwater Ecosystems* **27**: 572–577.

Compagno, L.J.V. & Cook, S.F. (1995). The exploitation and conservation of freshwater elasmobranchs: status of taxa and prospects for the future. *Journal of Aquariculture & Aquatic Sciences* **7**: 62–90.

Comte, L. & Grenouillet, G. (2013). Do stream fish track climate change? Assessing distribution shifts in recent decades. *Ecography* **36**: 1236–1246.

Comte, L., Bouisson, L., Daufresne, M. & Grenouillet, G. (2013). Climate-induced changes in the distribution of freshwater fish: observed and predicted trends. *Freshwater Biology* **58**: 625–639.

Convention on Biological Diversity (2016). *Inland Waters Biodiversity*. Convention on Biological Diversity [online]. www.cbd.int/waters

Conti, L., Comte, L., Hugueny, B. & Grenouillet, G. (2015). Drivers of freshwater fish colonisations and extirpations under climate change. *Ecography* **38**: 510–519.

Conti, L., Schmidt-Kloiber, A., Grenouillet, G. & Graf, W. (2014). A trait-based approach to assess the vulnerability of European aquatic insects to climate change. *Hydrobiologia* **721**: 297–315.

Copp, G.H., Bianco, P.G., Bogutskaya, N.G., Eros, T., Falka, I., Ferreira, M.T., Fox, M.G., Freyhof, J., Gozlan, R.E., Grabowska, J., Kovac, V., Moreno-Amich, R., Naseka, A.M., Penaz, M., Povz, M., Przybylski, M., Robillard, M., Russell, I.C., Stakenas, S., Sumer, S., Vila-Gispert, A. & Wiesner, C. (2005). To be, or not to be, a non-native freshwater fish? *Journal of Applied Ichthyology* **21**: 242–262.

Cordell, J.R. (2012). Invasive freshwater copepods of North America. In *A Handbook of Global Freshwater Invasive Species* (R.A. Francis, ed.), Earthscan, Oxford: pp. 161–172.

Corlett, R.T. (2016). Restoration, reintroduction, and rewilding in a changing world. *Trends in Ecology & Evolution* **31**: 453–462.

Correa, S.B., Costa-Pereira, R., Fleming, T., Goulding, M. & Anderson, J.T. (2015). Neotropical fish-fruit interactions: eco-evolutionary dynamics and conservation. *Biological Reviews* **90**: 1263–1278.

Costanza, R., d'Arge, R., de Groot, R., Farber, S., Grasso, M., Hannon, B., Limburg, K., Naeem, S., O'Neill, R.V., Paruelo, J., Raskin, R.G., Sutton, P. & van den Belt, M. (1997). The value of the world's ecosystem services and natural capital. *Nature* **387**: 253–260.

Costanza, R., de Groot, R., Sutton, P., van der Ploeg, S., Anderson, S.J., Kubiszewski, I., Farber, S. & Turner, R.K. (2014). Changes in the global value of ecosystem services. *Global Environmental Change* **26**: 152–158.

Cottingham, K.L., Ewing, H.A., Greer, M.L., Carey, C.C. & Weathers, K.C. (2015). Cyanobacteria as biological drivers of lake nitrogen and phosphorus cycling. *Ecosphere* **6**: 1–19.

Courchamp, F., Angulo, E., Rivalan, P., Hall, R.J., Signoret, L., Bull, L. & Meinard, Y. (2006). Rarity value and species extinction: the anthropogenic Allee effect. *PLoS Biol* **4**: e415. https://doi.org/10.1371/journal.pbio.0040415

Cowx, I.G., Arlinghaus, R. & Cooke, S.J. (2010). Harmonizing recreational fisheries and conservation objectives for aquatic biodiversity in inland waters. *Journal of Fish Biology* **76**: 2194–2215.

Craig, J.F., Halls, A.S., Barr, J.J.F. & Bean, C.W. (2004). The Bangladesh floodplain fisheries. *Fisheries Research* **66**: 271–286.

Creed, I.F., Lane, C.R., Serran, J.N., Alexander, L.C., Basu, N.B., Calhoun, A.J.K., Christensen, J.R., Cohen, M.J., Craft, C., D'Amico, E., DeKeyser, E., Fowler, L., Golden, H.E., Jawitz, J.W., Kalla, P., Kirkman, L.K., Lang, M., Leibowitz, S.G., Lewis, D.B., Marton, J., McLaughlin, D.L., Raanan-Kiperwas, H., Rains, M.C., Rains, K.C. & Smith, L. (2017). Enhancing protection for vulnerable waters. *Nature Geoscience* **10**: 809–815.

Crutzen, P.J. (2002). Geology of mankind. *Nature* **415**: 23.

Cucherousset, J., Boulêtreau, S., Azémar, F., Compin, A., Guillaume, M. & Santoul, F. (2012). 'Freshwater killer whales': beaching behavior of an alien fish to hunt land birds. *PLoS ONE* **7**: e50840. https://doi.org/10.1371/journal.pone.0050840

Cumberlidge, N., Ng, P.K.L., Yeo, D.C.J., Magalhaes, C., Campos, M.R., Alvarez, F., Naruse, T., Daniels, S.R., Esser, L.J., Attipoe, F.Y.K., Clotilde-Ba, F.L.,

Darwall, W., McIvor, A., Ram, M. & Collen, B. (2009). Freshwater crabs and the biodiversity crisis: importance, threats, status, and conservation challenges. *Biological Conservation* **142**: 1665–1673.

Dahanukar, N. & Raghavan, R. (2011). *Hypselobarbus mussullah. The IUCN Red List of Threatened Species* 2011: e.T172446A6893728. http://dx.doi.org/10.2305/IUCN.UK.2011-1.RLTS.T172446A6893728.en.

Dalin, C., Wada, Y., Kastner, T. & Puma, M.J. (2017). Groundwater depletion embedded in international food trade. *Nature* **543**: 700–704.

Danell, K. (1996). Introduction of aquatic rodents: lessons of the *Ondatra zibethicus* invasion. *Wildlife Biology* **2**: 213–220.

Darby, S. & Seer, D. (2008). *River Restoration: Managing the Uncertainty in Restoring Physical Habitat.* John Wiley & Sons Ltd, Chichester.

Darrigran, G. & Damborenea, C. (2005). A South American bioinvasion case history: *Limnoperna fortunei* (Dunker, 1857), the golden mussel. *American Malacological Bulletin* **20**: 105–112.

Darwall, W., Smith, K., Allen, D., Holland, R., Harrison, I. & Brooks, E. (2011a). *The Diversity of Life in African Freshwaters: Underwater, under Threat.* IUCN, Cambridge, UK & Gland, Switzerland.

Darwall, W.R.T., Holland, R.A., Smith, K.G., Allen, D.J., Brooks, E.G.E., Katarya, V., Pollock, C.M., Shi, Y., Clausnitzer, V., Cumberlidge, N., Cuttelod, A., Dijkstra, K.-D., Diop, M.D., García, N., Seddon, M.B., Skelton, P.H., Snoeks, J., Tweddle, D. & Vie, J.-C. (2011b). Implications of bias in conservation research and investment for freshwater species. *Conservation Letters* **4**: 474–482.

Darwall, W., Bremerich, V., De Wever, A., Dell, A.I., Freyhof, J., Gessner, M.O., Grossart, H.-P., Harrison, I., Irvine, K., Jähnig, S.C., Jeschke, J.M., Lee, J.J., Lu, C., Lewandowska, A.M., Monaghan, M.T., Nejstgaard, J.C., Patricio, H., Schmidt-Kloiber, A., Stuart, S.N., Thieme, M., Tockner, K., Turak, E. & Weyl, O. (2018). The Alliance for Freshwater Life: a global call to unite efforts for freshwater biodiversity science and conservation. *Aquatic Conservation: Marine and Freshwater Research* **28**: 1015–1022.

Datry, T., Fritz, K. & Leigh, C. (2016). Challenges, developments and perspectives in intermittent river ecology. *Freshwater Biology* **61**: 1171–1180.

Daufresne, M., Lengfellner, K. & Sommer, U. (2009). Global warming benefits the small in aquatic ecosystems. *Proceedings of the National Academy of Sciences of the United States of America* **106**: 12788–12793.

Dawson, J., Patel, F., Griffiths, R.A. & Young, R.P. (2016). Assessing the global zoo response to the amphibian crisis through 20-year trends in captive collections. *Conservation Biology* **30**: 82–91.

Dawson, W., Moser, D., van Kleunen, M., Kreft, H., Pergl, J., Pyšek, P., Weigelt, P., Winter, M., Lenzner, B., Blackburn, T.M., Dyer, E.E., Cassey, P., Scrivens, S.L., Economo, E.P., Guénard, B., Capinha, C., Seebens, H., García-Díaz, P., Nentwig, W., García-Berthou, E., Casal, C., Mandrak, N.E., Fuller, P., Meyer, C. & Essl, F. (2017). Global hotspots and correlates of alien species richness across taxonomic groups. *Nature Ecology and Evolution* **1**: 0186.

Day, J.J., Bills, R. & Friel, J.P. (2009). Lacustrine radiations in African *Synodontis* catfish. *Journal of Evolutionary Biology* **22**: 805–817.

Deacon, A.E., Ramnarine, I.W. & Magurran, A.E. (2011). How reproductive ecology contributes to the spread of a globally invasive fish. *PLoS ONE* **6**: e24416. https://doi.org/10.1371/journal.pone.0024416

Deines, A.M., Bunnell, D.B., Rogers, M.W., Beard, T.D. & Taylor, W.W. (2015). A review of the global relationship among freshwater fish, autotrophic activity, and regional climate. *Reviews in Fish Biology and Fisheries* **25**: 323–336.

Denic, M. & Geist, J. (2015). Linking stream sediment deposition and aquatic habitat quality in pearl mussel streams: implications for conservation. *River Research and Applications* **31**: 943–952.

de Senerpont Domis, L.N., Elser, J.J., Gsell, A.S., Huszar, V.L.M, Ibelings, B.W., Jeppesen, E., Kosten, S., Mooij, W.M., Roland, F., Sommer, U., van Donk, E., Winder, M. & Lürling, M. (2013). Plankton dynamics under different climatic conditions in space and time. *Freshwater Biology* **58**: 463–482.

Deutsch, C.A., Tewksbury, J.J., Huey, R.B., Sheldon, K.S., Ghalambor, C.K., Haak, D.C. & Martin, P.R. (2008). Impacts of climate warming on terrestrial ectotherms across latitude. *Proceedings of the National Academy of Sciences of the United States of America* **105**: 6668–6672.

Dias, M.S., Cornu, J.-F., Oberdorff, T., Lasso, C.A. & Tedesco, P.A. (2013). Natural fragmentation in river networks as a driver of speciation for freshwater fishes. *Ecography* **36**: 683–689.

Diaz, S., Demissew, S., Carabias, J., Joly, C., Lonsdale, M., Ash, N., Larigauderie, A., Adhikari, J.R., Arico, S., Baldi, A., Bartuska, A., Baste, I.A., Bilgin, A., Brondizio, E., Chan, K.M.A., Figueroa, V.E., Duraiappah, A., Fischer, M., Hill, R., Koetz, T., Leadley, P., Lyver, P., Mace, G.M., Martin-Lopez, B., Okumura, M., Pacheco, D., Pascual, U., Perez, E.S., Reyers, B., Roth, E., Saito, O., Scholes, R.J., Sharma, N., Tallis, H., Thaman, R., Watson, R., Yahara, T., Hamid, Z.A., Akosim, C., Al-Hafedh, Y., Allahverdiyev, R., Amankwah, E., Asah, S.T., Asfaw, Z., Bartus, G., Brooks, L.A., Caillaux, J., Dalle, G., Darnaedi, D., Driver, A., Erpul, G., Escobar-Eyzaguirre, P., Failler, P., Fouda, A.M.M., Fu, B., Gundimeda, H., Hashimoto, S., Homer, F., Lavorel, S., Lichtenstein, G., Mala, W.A., Mandivenyi, W., Matczak, P., Mbizvo, C., Mehrdadi, M., Metzger, J.P., Mikissa, J.B., Moller, H., Mooney, H.A., Mumby, P., Nagendra, H., Nesshover, C., Oteng-Yeboah, A.A., Pataki, G., Roue, M., Rubis, J., Schultz, M., Smith, P., Sumaila, R., Takeuchi, K., Thomas, S., Verma, M., Yeo-Chang, Y. & Zlatanova, D. (2015). The IPBES conceptual framework – connecting nature and people. *Current Opinion in Environmental Sustainability* **14**: 1–16.

Didžiulis, V. (2013). NOBANIS – Invasive Alien Species Fact Sheet – *Anguillicola crassus*. Online Database of the European Network on Invasive Alien Species – NOBANIS www.nobanis.org (www.nobanis.org/species-info/?taxaId=18268).

Deiner, K., Fronhofer, E., Mächle, E., Walser, C. & Altermatt, F. (2016). Environmental DNA reveals that rivers are conveyer belts of biodiversity information. *Nature Communications* **7**: 12544. https://doi.org/10.1038/ncomms12544

Ding, C., Jiang, X., Xie, Z. & Brosse, S. (2017). Seventy-five years of biodiversity decline of fish assemblages in Chinese isolated plateau lakes: widespread introductions and extirpations of narrow endemics lead to regional loss of dissimilarity. *Diversity and Distributions* **23**: 171–184.

Dinh Van, K., Janssens, L., Debecker, S., De Jonge, M., Lambret, P., Nilsson-Örtman, V., Bervoets, L. & Stoks, R. (2013). Susceptibility to a metal under global warming is shaped by thermal adaptation along a latitudinal gradient. *Global Change Biology* **19**: 2625–2633.

Domisch, S., Jähnig, S.C. & Haase, P. (2011). Climate-change winners and losers: stream macroinvertebrates of a submontane region in Central Europe. *Freshwater Biology* **56**: 2009–2020.

Dorts, J., Grenouillet, G., Douxfils, J., Mandiki, S.N.M., Milla, S., Silvestre, F. & Kestemont, P. (2012). Evidence that elevated water temperature affects the reproductive physiology of the European bullhead *Cottus gobio*. *Fish Physiology and Biochemistry* **38**: 389–399.

Douda, K., Vrtílek, M., Slavík, O. & Reichard, M. (2012). The role of host specificity in explaining the invasion success of the freshwater mussel *Anodonta woodiana* in Europe. *Biological Invasions* **14**: 127–137.

Douda, K., Liu, H.-L., Yu, D., Rouchet, R., Liu, F., Tang, Q.-Y., Methling, C., Smith, C. & Reichard, M. (2017). The role of local adaptation in shaping fish-mussel coevolution. *Freshwater Biology* **62**: 1858–1868.

Doughty, C.E., Roman, J., Faurby, S., Wolf, A., Haque, A., Bakker, E.S., Malhi, Y., Dunning, J.B. & Svenning, J. (2016). Global nutrient transport in a world of giants. *Proceedings of the National Academy of Sciences of the United States of America* **113**: 868–873.

Downing, A.S., Van Nes, E.H., Janse, J.H., Witte, F., Cornelissen, I.J., Scheffer, M. & Mooij, W.M. (2012). Collapse and reorganization of a food web of Mwanza Gulf, Lake Victoria. *Ecological Applications* **22**: 229–239.

Du, Y., Xue, H., Wu, S., Ling, F., Xiao, F. & Wei, X. (2011). Lake area changes in the middle Yangtze region of China over the 20th century. *Journal of Environmental Management* **92**: 1248–1255.

Ducatelle, R., Nishikawa, K., Nguyen, T.T., Kolby, J.E., Van Bocxlaer, I., Bossuyt, F. & Pasmans, F. (2014). Recent introduction of a chytrid fungus endangers Western Palearctic salamanders. *Science* **346**: 630–631.

Dudgeon, D. (1999). *Tropical Asian Streams: Zoobenthos, Ecology and Conservation*. Hong Kong University Press, Hong Kong.

Dudgeon, D. (2010). Requiem for a river: extinctions, climate change and the last of the Yangtze. *Aquatic Conservation: Marine and Freshwater Ecosystems* **20**: 127–131.

Dudgeon, D. (2011). Asian river fishes in the Anthropocene: threats and conservation challenges in an era of rapid environmental change. *Journal of Fish Biology* **79**: 1487–1524.

Dudgeon, D. (2013). Anthropocene extinctions: global threats to riverine biodiversity and the tragedy of the freshwater commons. In *River Conservation: Challenges and Opportunities* (S. Sabatier & A. Elosegi, eds), Fundación BBVA, Bilbao: pp. 129–167.

Dudgeon, D. (2014). Accept no substitute: biodiversity matters. *Aquatic Conservation: Marine and Freshwater Ecosystems* **24**: 435–440.

Dudgeon, D. & Morton, B. (1984). Site selection and attachment duration of *Anodonta woodiana* (Bivalvia: Unionacea) glochidia on fish hosts. *Journal of Zoology, London* **204**: 355–362.

Dudgeon, D. & Smith, R.E.W. (2006). Exotic species, fisheries, and conservation of freshwater biodiversity in tropical Asia: the case of the Sepik River, Papua New Guinea. *Aquatic Conservation: Marine and Freshwater Ecosystems* **16**: 203–215.

Dudgeon, D., Arthington, A.H., Gessner, M.O., Kawabata, Z., Knowler, D., Lévêque, C., Naiman, R.J., Prieur-Richard, A.-H., Soto, D., Stiassny, M.L.J. & Sullivan, C.A. (2006). Freshwater biodiversity: importance, threats, status and conservation challenges. *Biological Reviews* **81**: 163–182.

Dugan, P. (2008). Mainstream dams as barriers to fish migration: international learning and implications for the Mekong. *Catch and Culture* **14**: 9–15.

Dugan, H.A., Bartlett, S.L., Burke, S.M., Doubek, J.P., Krivak-Tetley, F.E., Skaff, N.K., Summers, J.C., Farrell, K.J., McCullough, I.M., Morales-Williams, A.M., Roberts, D.C., Ouyang, Z., Scordo, F., Hanson, P.C. & Weathers, K.C. (2017). Salting our freshwater lakes. *Proceedings of the National Academy of Sciences of the United States of America* **114**: 4453–4458.

Dugan, P.J., Barlow, C., Agostinho, A.A., Baran, A., Cada, G.F., Chen, D., Cowx, I.G.,Ferguson, J.W., Jutagate, T., Mallen-Cooper, M., Marmulla, G., Nestler, J.,Petrere, M., Welcomme, R.L. & Winemiller, K.O. (2010). Fish migration, dams, and loss of ecosystem services in the Mekong Basin. *Ambio* **39:** 344–348.

Duggan, I.C. (2010). The freshwater aquarium trade as a vector for incidental invertebrate fauna. *Biological Invasions* **12**: 3757–3770.

Dumont, H.J. (1994). The distribution and ecology of the fres – and brackish – water medusae of the world. *Hydrobiologia* **272**: 1–12.

Dunn, H. (2003). Can conservation assessment criteria developed for terrestrial systems be applied to riverine systems? *Aquatic Ecosystem Health and Management* **6**: 81–95.

Dunn, J.C. (2012). *Pacifastacus leniusculus* Dana (North American signal crayfish). In *A Handbook of Global Freshwater Invasive Species* (R.A. Francis, ed.), Earthscan, Oxford: pp. 195–205.

Durance, I. & Ormerod, S.J. (2010). Evidence for the role of climate in the local extinction of a cool-water triclad. *Journal of the North American Benthological Society* **29**: 1367–1378.

Durance, I., Vaughn, I.P. & Ormerod, S.J. (2009). *Evaluating Climatic Effects on Aquatic Invertebrates, Phase II: Review, Comparisons between Regions and Methodological Considerations.* Report: SC070047/R1, Environment Agency, Bristol. https://assets.publishing.service.gov.uk/government/uploads/system/uploads/attachment_data/file/291642/scho1209brjt-e-e.pdf

Eerkes-Medrano, D., Thompson, R.C. & Aldridge, D.C. (2015). Microplastics in freshwater systems: a review of emerging threats, identification of knowledge gaps and prioritisation of research needs. *Water Research* **75**: 63–82.

Ehrlich, P.R. & Pringle, R.M. (2008). Where does biodiversity go from here? A grim business-as-usual forecast and a hopeful portfolio of partical solutions. *Proceedings of the National Academy of Sciences of the United States of America* **105** (Suppl. 1): 11579–11586.

Elbrecht, V. & Leese, F. (2017). Validation and development of COI metabarcoding primers for freshwater macroinvertebrate bioassessment. *Frontiers in Environmental Science* **5**: 11. https://doi.org/10.3389/fenvs.2017.00011

El-Sabaawi, R.W., Frauendorf, T.C., Marques, P.S., Mackenzie, R.A., Manna, L.R., Mazzoni, R., Phillip, D.A.T., Warbanski, M.L. & Zandonà, E. (2016). Biodiversity and ecosystem risks arising from using guppies to control mosquitoes. *Biology Letters* **12**: 20160590; https://doi.org/10.1098/rsbl.2016.0590

Elliott, P., Aldridge, D.C. & Moggridge, G.D. (2008). Zebra mussel filtration and its potential uses in industrial water treatment. *Water Resources* **42**: 1664–1674.

Ellis, B.K, Stanford, J.A, Goodman, D., Stafford, C.P., Gustafson, D.L, Beauchamp, D.A., Chess, D.W., Craft, J.A., Delerray, M.A. & Hansen, B.S. (2011). Long-term effects of a trophic cascade in a large lake ecosystem. *Proceedings of the National Academy of Sciences of the United States* **108**: 1070–1075.

Elston, E., Anderson-Lederer, R., Death, R.G. & Joy, M.K. (2015). *The Plight of New Zealand's Freshwater Species*. Conservation Science Statement No. 1. Society for Conservation Biology (Oceania), Sydney. https://conbio.org/images/content_groups/Oceania/Scientific_Statement_1_.pdf

Elton, C.S. (1958). *The Ecology of Invasions by Animals and Plants*. Methuen, London.

Emde, S., Rueckert, S., Palm, H.W. & Klimpel, S. (2012). Invasive Ponto-Caspian amphipods and fish increase the distribution range of the acanthocephalan *Pomphorhynchus tereticollis* in the River Rhine. *PLoS ONE* **7**: e53218. https://doi.org/10.1371/journal.pone.0053218

Englund, J. & Wilkes, M.A. (2018). Does river restoration work? Taxonomic and functional trajectories at two restoration schemes. *Science of the Total Environment* **618**: 961–970.

Eva, B., Harmony, P., Gray, T., Guegan, F., Valentin, A., Miaud, C. & Dejean, T. (2016). Trails of river monsters: detecting critically endangered Mekong giant catfish *Pangasianodon gigas* using environmental DNA. *Global Ecology and Conservation* **7**: 148–156.

Everard, M. & Kataria, G. (2011). Recreational angling markets to advance the conservation of a reach of the Western Ramganga River, India. *Aquatic Conservation* **21**: 101–108.

Falkenmark, M. & Rockström, J. (2006). The new blue and green water paradigm: breaking new ground for water resources planning and management. *Journal of Water Resources Planning and Management* **132**: 129–132.

Fang, J., Wang, X., Zhao, S., Li, Y., Tang, Z., Yu, D., Ni, L., Liu, H., Xie, P., Da, L., Li, Z. & Zheng, C. (2006). Biodiversity changes in the lakes of the central Yangtze. *Frontiers in Ecology and the Environment* **4**: 369–377.

Farmer, T.M., Marschall, E.A., Dabrowski, K. & Ludsin, S.A. (2015). Short winters threaten temperate fish populations. *Nature Communications* **6**: 7724. https://doi.org/10.1038/ncomms8724

FAO (2010). *The State of World Fisheries and Aquaculture, 2010*. Food and Agriculture Organization of the United Nations, Rome.

FAO (2012). *The State of World Fisheries and Aquaculture, 2012*. Food and Agriculture Organization of the United Nations, Rome.

FAO (2014). *The State of World Fisheries and Aquaculture, 2014*. Food and Agriculture Organization of the United Nations, Rome.

FAO (2016). *The State of World Fisheries and Aquaculture 2016. Contributing to Food Security and Nutrition for All*. Food and Agriculture Organization of the United Nations, Rome.

Faria, V.V., McDavitt, M.T., Charvet, P., Wiley, T.R., Simpfendorfer, C.A. & Naylor, G.J.P. (2013). Species delineation and global population structure of critically endangered sawfishes (Pristidae). *Zoological Journal of the Linnean Society* **167**: 136–164.

Feldmeier, S., Schefczyk, L., Wagner, N., Heinemann, G., Veith, M. & Lötters, S. (2016). Exploring the distribution of the spreading lethal salamander chytrid fungus in its invasive range in Europe – a macroecological approach. *PLoS ONE* **11**: e0165682. https://doi.org/10.1371/journal.pone.0165682

Feng, L., Hu, C., Chen, X. & Zhao, X. (2013). Dramatic inundation changes of China's two largest freshwater lakes linked to the Three Gorges Dam. *Environmental Science & Technology* **47**: 9628–9634

Fernando, C.H. (2000). A view of the inland fisheries of Sri Lanka: past, present and future. *Sri Lanka Journal of Aquatic Science* **5**: 1–26.

Ficetola, G.F. (2013). Is interest toward the environment really declining? The complexity of analysing trends using internet search data. *Biodiversity and Conservation* **22**: 2983–2988.

Filipe, A.F., Markovic, D., Pletterbauer, F., Tisseuil, C., De Wever, A., Schmutz, S., Bonada, N. & Freyhof, J. (2015). Forecasting fish distribution along stream networks: brown trout (*Salmo trutta*) in Europe. *Diversity & Distributions* **19**: 1059–1071.

Finn, D.S., Räsänen, K. & Robinson, C.T. (2010). Physical and biological changes to a lengthening stream gradient following a decade of rapid glacial recession. *Global Change Biology* **16**: 3314–3326.

Fischer, J.R., Lewis-Weis, L.A., Tate, C.M., Gaydos, J.K., Gerhold, R.W. & Poppenga, R.H. (2006). Avian vacuolar myelinopathy outbreaks at a south-eastern reservoir. *Journal of Wildlife Diseases* **42**: 501–510.

Flecker, A.S., McIntyre, P.B., Moore, J.W., Anderson, J.T., Taylor, B.W. & Hall, R.O. (2010). Migratory fishes as material and process subsidies in riverine ecosystems. *American Fisheries Society Symposium* **73**: 559–592.

Fluet-Chouinard, E., Funge-Smith, S. & McIntyre, P.B. (2018). Global hidden harvest of freshwater fish revealed by household surveys. *Proceedings of the National Academy of Science of the United States of America* **115**: 7623–7628.

Foley, J.A, Ramankutty, N., Brauman, K.A, Cassidy, E.S, Gerber, J.S., Johnston, M., Mueller, N.D., O'Connell, C., Ray, D.K., West, P.C., Balzer, C., Bennett, E.M., Carpenter, S.R., Hill, J., Monfreda, C., Polasky, S., Rockström, J., Sheehan, J., Siebert, S., Tilman, D. & Zaks, D.P.M. (2011). Solutions for a cultivated planet. *Nature* **478**: 337–342.

Forero-Medina, G., Joppa, L. & Pimm, S.L. (2011). Constraints to species' elevational range shifts as climate changes. *Conservation Biology* **25**: 163–171.

Fox, A.D., Cao, L., Zhang, Y., Barter, M., Zhao, M.J., Meng, F.J. & Wang, S.L. (2011). Declines in the tuber-feeding waterbird guild at Shengjin Lake National Nature Reserve, China – a barometer of submerged macrophyte collapse. *Aquatic Conservation: Marine and Freshwater Ecosystems* **21**: 82–91.

Fox, R., Conrad, K.F., Parsons, M.S., Warren, M.S. & Woiwod, I.P. (2006). *The State of Britain's Larger Moths*. Butterfly Conservation and Rothamsted Research, Wareham, Dorset.

Francis, R.A. (2012). *A Handbook of Global Freshwater Invasive Species*. Earthscan, Oxford.

Franco, D., Sobrane Filho, S., Martins, A., Marmontel, M. & Botero-Arias, R. (2016). The piracatinga, *Calophysus macropterus*, production chain in the Middle Solimões River, Amazonas, Brazil. *Fisheries Management and Ecology* **23**: 109–118.

Freeman, M.C., Pringle, C.M. & Jackson, C.R. (2007). Hydrologic connectivity and the contribution of stream headwaters to ecological integrity at regional scales. *Journal of the American Water Resources Association* **43**: 5–14.

French, M., Alem, N., Edwards, S.J., Blanco Coariti, E., Cauthin, H., Hudson-Edwards, K.A., Luyckx, K., Quintanilla, J. & Sánchez Miranda, O. (2017). Community exposure and vulnerability to water quality and availability: a case study in the mining-affected Pazña Municipality, Lake Poopó Basin, Bolivian Altiplano. *Environmental Management* **60**: 555–573.

Freyhof, J. & Kottelat, M. (2008). *Hucho hucho. The IUCN Red List of Threatened Species 2008:* e.T10264A3186143. http://dx.doi.org/10.2305/IUCN.UK .2008.RLTS.T10264A3186143.en

Froese, R. & Pauly, D. (2018). *FishBase*. World Wide Web electronic publication, www.fishbase.org version (06/2018).

Fryxell, D.C., Arnett, H.A., Apgar, T.M., Kinnison, M.T. & Palkovacs, E.P. (2015). Sex ratio variation shapes the ecological effects of a globally introduced freshwater fish. *Proceedings of the Royal Society B* **282**: 20151970. http://doi.org/10 .1098/rspb.2015.1970

Fu, C., Wu, J., Chen, J., Wu, Q. & Lei, G. (2003). Freshwater fish biodiversity in the Yangtze River basin of China: patterns, threats and conservation. *Biodiversity and Conservation* **12**: 1649–1685.

Fukushima, M., Shimazaki, H., Rand, P.S. & Kaeriyama, M. (2011). Reconstructing Sakhalin taimen *Parahucho perryi* historical distribution and identifying causes for local extinctions. *Transactions of the American Fisheries Society* **140**: 1–13.

Funge-Smith, S. & Bennett, A. (2019). A fresh look at inland fisheries and their role in food security and livelihoods. *Fish and Fisheries*: in press. https://doi.org/10 .1111/faf.12403

Galbraith, H.S., Zanatta, D.T. & Wilson, C.C. (2015). Comparative analysis of riverscape genetic structure in rare, threatened and common freshwater mussels. *Conservation Genetics* **16**: 845–857.

Gallardo, B. & Aldridge, D.C. (2015). Is Great Britain heading for a Ponto–Caspian invasional meltdown? *Journal of Applied Ecology* **52**: 41–49.

Gallardo, B., Clavero, M., Sánchez, M.I. & Vilà, M. (2015). Global ecological impacts of invasive species in aquatic ecosystems. *Global Change Biology* **21**: 151–163.

Gao, H., Bohn, T.J., Podest, F., McDonald, K.C. & Lettenmaier, D.P. (2011). On the causes of the shrinking of Lake Chad. *Environmental Research Letters* **6**: 034021. http://doi.org/10.1088/1748-9326/6/3/034021

Gao, Z., Li, Y. & Wang, W. (2008). Threatened fishes of the world: *Myxocyprinus asiaticus* Bleeker 1864 (Catostomidae). *Environmental Biology of Fishes* **83**: 345–346.

Garcia, S.M., Kolding, J., Rice, J., Rochet, M.-J., Zhou, S., Arimoto, T., Beyer, J.E., Borges, L., Bundy, A., Dunn, D., Fulton, E.A., Hall, M., Heino, M., Law, R., Makino, M., Rijnsdorp, A.D., Simard, F. & Smith, A.D.M. (2012). Reconsidering the consequences of selective fisheries. *Science* **335**: 1045–1047.

García-Berthou, E., Alcaraz, C., Pou-Rovira, Q., Zamora, L., Coenders, G. & Feo, C. (2005). Introduction pathways and establishment rates of invasive aquatic species in Europe. *Canadian Journal of Fisheries and Aquatic Sciences* **62**: 453–463.

Garvey, J.E. (2012). Bigheaded carps of the genus *Hypophthalmichthys*. In *A Handbook of Global Freshwater Invasive Species* (R.A. Francis, ed.), Earthscan, Oxford: pp. 235–245.

Geerts, A.N., Vanoverbeke, J., Vanschoenwinkel, B., Van Doorslaer, W., Feuchtmayr, H., Atkinson, D., Moss, B., Davidson, T.A., Sayer, C.D. & De Meester, L. (2015). Rapid evolution of thermal tolerance in the water flea *Daphnia*. *Nature Climate Change* **5**: 665–668.

Gende, S.M., Edwards, R.T., Willson, M.F. & Wipfli, M.S. (2002). Pacific salmon in aquatic and terrestrial ecosystems. *BioScience* **52**: 917–928.

Gedney, N., Cox, P.M., Betts, R.A., Boucher, O., Huntingford, C. & Stott, P.A. (2006). Detection of a direct carbon dioxide effect in continental river runoff records. *Nature* **439**: 835–838.

Gerstner, C.L., Ortega, H., Sanchez, H. & Graham, D.L. (2006). Effects of the freshwater aquarium trade on wild fish populations in differentially-fished areas of the Peruvian Amazon. *Journal of Fish Biology* **68**: 862–875.

Gerten, D., Rost, S., von Bloh, W. & Lucht, W. (2008). Causes of change in 20th century global river discharge. *Geophysical Research Letters* **35**: L20405. http://doi.org/10.1029/2008GL035258

Gerten, D., Hoff, H., Rockström, J., Jägermeyr, J., Kummu, M. & Pastor, A.V. (2013). Towards a revised planetary boundary for consumptive freshwater use: role of environmental flow requirements. *Current Opinion in Environmental Sustainability* **5**: 551–558.

Geyer, R., Jambeck, J. & Law, K.L. (2017). Production, use, and fate of all plastics ever made. *Science Advances* **3**: e1700782. http://doi.org/10.1126/sciadv.1700782

Gherardi, F., Britton, J.R., Mavuti, K.M., Pacini, N., Grey, J., Tricarico, E. & Harper, D.M. (2011). A review of allodiversity in Lake Naivasha, Kenya: developing conservation actions to protect East African lakes from the negative impacts of alien species. *Biological Conservation* **144**: 2585–2596

Giam, X., Ng, H.T., Lok, A.F.S.L. & Ng, H.H. (2011). Local geographic range predicts freshwater fish extinctions in Singapore. *Journal of Applied Ecology* **48**: 356–363.

Giam, X., Koh, L.P., Tan, H.H., Miettinen, J., Tan, H.T.W. & Ng, P.K.L. (2012). Global extinctions of freshwater fishes follow peatland conversion in Sundaland. *Frontiers in Ecology and the Environment* **10**: 465–470.

Giam, X., Hadiaty, R.K., Tan, H.H., Parenti, L.R., Wowor, D., Sauri, S., Chong, K.Y., Yeo, D.C.J. & Wilcove, D.S. (2015). Mitigating the impact of oil-palm monoculture on freshwater fishes in Southeast Asia. *Conservation Biology* **29**: 1357–1367.

Giersch, J.J., Jordan, S., Luikart, G., Jones, L.A., Hauer, F.R. & Muhlfeld, C.C. (2015). Climate-induced range contraction of a rare alpine aquatic invertebrate. *Freshwater Science* **34**: 53–65.

Giesen, W. (1994). Indonesia's major freshwater lakes: a review of our current knowledge, development processes and threats. *Mitteilungen Internationale Vereinigung Limnologie* **24**: 115–128.

Geist, J. (2010). Strategies for the conservation of endangered freshwater pearl mussels (*Margaritifera margaritifera* L.): a synthesis of conservation genetics and ecology. *Hydrobiologia* **644**: 69–88.

Gilbert, M.A. & Granath, W.O. (2003). Whirling disease and salmonid fish: life cycle, biology, and disease. *Journal of Parasitology* **89**: 658–667.

Gilbert, N. (2015). Europe sounds alarm over freshwater pollution. *Nature News*: 2 March 2015. http://doi.org/doi:10.1038/nature.2015.17021

Glaubrecht, M. (2008). Adaptive radiation of thalassoid gastropods in Lake Tanganyika, East Africa: morphology and systematization of a paludomid species flock in an ancient lake. *Zoosystematics and Evolution* **84**: 71–122.

Gleick, P.H. (1996). Water resources. In *Encyclopedia of Climate and Weather* (S.H. Schneider, ed.), Oxford University Press, Oxford: pp. 817–823.

Goldberg, C.S., Sepulveda, A., Ray, A., Baumgardt, J. & Waits, L.P. (2013). Environmental DNA as a new method for early detection of New Zealand mudsnails (*Potamopyrgus antipodarum*). *Freshwater Science* **32**: 792–800.

Goldberg, C.S., Strickler, K.M. & Pilliod, D.S. (2015). Moving environmental DNA methods from concept to practice for monitoring aquatic macroorganisms. *Biological Conservation* **183**: 1–3.

Goldschmidt, T., Witte, F. & Wanink, J. (1993). Cascading effects of the introduced Nile perch on the detritivorous/planktivorous species in the sublittoral areas of Lake Victoria. *Conservation Biology* **7**: 686–700.

Goulsen, D. (2013). An overview of the environmental risks posed by neonicotinoid insecticides. *Journal of Applied Ecology* **50**: 977–987.

Gowdy, J. & Lang, H. (2016). *The Economic, Cultural and Ecosystem Values of the Sudd Wetland in South Sudan: An Evolutionary Approach to Environment and Development*. United Nations Environment Programme, Nairobi.

Gozlan, R.E. (2008). Introduction of non-native freshwater fish: is it all bad? *Fish and Fisheries* **9**: 106–115.

Gozlan, R.E., St-Hilaire, S., Feist, S.W., Martin, P. & Kent, M.L. (2005). Disease threat to European fish. *Nature* **435**: 1046.

Grabowski, Z.J., Chang, H., Granek, E.F. (2018). Fracturing dams, fractured data: empirical trends and characteristics of existing and removed dams in the United States. *River Research and Applications* **34**: 526–537.

Granek, E.F., Madin, E.M., Brown, M.A., Figueira, W., Cameron, D.S., Hogan, Z., Kristianson, G., de Villiers, P., Williams, J.E., Post, J., Zahn, S. & Arlinghaus, R. (2008). Engaging recreational fishers in management and conservation: global case studies. *Conservation Biology* **22**: 1125–1134.

Gray, M.J., Miller, D.L. & Hoverman, J.T. (2009). Ecology and pathology of amphibian ranaviruses. *Diseases of Aquatic Organisms* **87**: 243–266.

Greig, H.S., Kratina, P., Thompson, P.L., Palen, W.J., Richardson, J.S. & Shurin, J.B. (2012). Warming, eutrophication, and predator loss amplify subsidies between aquatic and terrestrial ecosystems. *Global Change Biology* **18**: 504–514.

Griffiths, A.M., Ellis, J.S., Clifton-Dey, D., Machado-Schiaffino, G., Bright, D., Garcia-Vazquez, E. & Stevens, J.R. (2011). Restoration versus recolonisation: the origin of Atlantic salmon (*Salmo salar* L.) currently in the River Thames. *Biological Conservation* **144**: 2733–2738.

Griffiths, R.A. & Pavajeau, L. (2008). Captive breeding, reintroduction, and the conservation of amphibians. *Conservation Biology* **22**: 852–861.

Griggs, D., Stafford-Smith., M., Gaffney, O., Rockström, J., Ohman, M.C., Shyamsunbdar, P., Steffen, W., Glaser, G., Kanie, N. & Noble, I. (2013). Sustainable development and goals for people and planet. *Nature* **495**: 305–307.

Grill, G., Lehner, B., Lumsdon, A.E., MacDonald, G.K., Zarfl, C. & Reidy Liermann, C. (2015). An index-based framework for assessing patterns and trends in river fragmentation and flow regulation by global dams at multiple scales. *Environmental Research Letters* **10**: 015001. http://dx.doi.org/10.1088/1748-9326/10/1/015001

Guimarães Frederico, R., Zuanon, J. & De Marco, P. (2018). Amazon protected areas and its ability to protect stream-dwelling fish fauna. *Biological Conservation* **219**: 12–19.

Gupta, N., Sivakumar, K., Mathur, V.B. & Chadwick, M.A. (2014). The 'tiger of Indian rivers': stakeholders' perspectives on the golden mahseer as a flagship fish species. *Area* **46**: 389–397.

Gustavsen, K., Hopkins, A. & Sauerbrey, M. (2011). Onchocerciasis in the Americas: from arrival to (near) elimination. *Parasites & Vectors* **4**: 205. http://doi.org/10.1186/1756-3305-4-205

Hadley, K.R., Patterson, A.M., Reid, R.A., Rusak, J.A., Somers, K.M., Ingram, R. & Smol, J.P. (2015). Altered pH and reduced calcium levels drive near extirpation of native crayfish, *Cambarus bartonii*, in Algonquin Park, Ontario, Canada. *Freshwater Science* **34**: 918–932.

Hall, S.J., Hilborn, R., Andrew, N.L. & Allison, E.H. (2013). Innovations in capture fisheries are an imperative for nutrition security in the developing world. *Proceedings of the National Academy of Sciences of the United States of America* **110**: 8393–8398.

Hallmann, C.A., Sorg, M., Jongejans, E., Siepel, H., Hofland, N., Schwan, H., Stenmans, W., Müller, A, Sumser, H., Hörren, T., Goulson, D. & de Kroonet, H. (2017). More than 75 percent decline over 27 years in total flying insect biomass in protected areas. *PLoS ONE* **12**: e0185809. https://doi.org/10.1371/journal.pone.0185809

Halls, A.S. & Kshatriya, M. (2009). *Modelling the Cumulative Barrier and Passage Effects of Mainstream Hydropower Dams on Migratory Fish Populations in the Lower Mekong Basin*. MRC Technical Paper No. 25, Mekong River Commission, Vientiane.

Hampton, S.E., Izmest'eva, L.R., Moore, M.V., Katz, S.L., Dennis, B. & Silow, E.A. (2008). Sixty years of environmental change in the world's largest freshwater lake – Lake Baikal, Siberia. *Global Change Biology* **14**: 1947–1958.

Hampton, S.E., Gray, D.K., Izmest'eva, L.R., Moore, M.V., Ozersky, T. & Ianora, A. (2014). The rise and fall of plankton: long-term changes in the vertical distribution of algae and grazers in Lake Baikal, Siberia. *PLoS ONE* **9**: e88920. https://doi.org/10.1371/journal.pone.0088920

Han, X., Feng, L., Hu, C. & Chen, X. (2018). Wetland changes of China's largest freshwater lake and their linkage with the Three Gorges Dam. *Remote Sensing of Environment* **204**: 799–811.

Hannah, L., Costello, C., Elliot, V., Owashi, B., Nam, S., Oyanedel, R., Chea, R., Vibrol, O., Phen, C. & McDonald, G. (2019). Designing freshwater protected areas (FPAs) for indiscriminate fisheries. *Ecological Modelling* **393**: 127–134.

Hansen, G.J.A., Hein, C.L., Roth, B.M., Vander Zanden, M.J., Gaeta, J.W., Latzka, A.W. & Carpenter, S.R. (2013). Food web consequences of long-term invasive crayfish control. *Canadian Journal of Fisheries and Aquatic Sciences* **70**: 1109–1122.

Hardiman, J.M. & Mesa, M.J. (2014). The effects of increased stream temperatures on juvenile steelhead growth in the Yakima River Basin based on projected climate change scenarios. *Climatic Change* **24**: 413–426.

Hardin, G. (1968). The tragedy of the commons. *Science* **162**: 1243–1248.

Harding, G., Griffiths, R.A. & Pavajeau, L. (2016). Developments in amphibian captive breeding and reintroduction programs. *Conservation Biology* **30**: 340–349.

Harrison, I.J. & Stiassny, M.L.J. (1999). The quiet crisis: a preliminary listing of the freshwater fishes of the world that are extinct or 'missing in action'. In *Extinctions in Near Time* (R.D.E. MacPhee, ed.), Kluwer Academic/Plenum Publishers, New York, USA, pp. 271–331.

Harrison, I.J., Green, P.A., Farrell, T.A., Juffe-Bignoli, D., Sáenz, L. & Vörösmarty, C.J. (2016). Protected areas and freshwater provisioning: a global assessment of freshwater provision, threats and management strategies to support human water security. *Aquatic Conservation: Marine and Freshwater Ecosystems* **26** (Suppl. 1): 103–120.

Harper, M.P. & Peckarsky, B.L. (2006). Emergence cues of a mayfly in a high altitude stream ecosystem: implications for consequences of climate change. *Ecological Applications* **16**: 612–621.

Hart, D.D. & Calhoun, A.J.K. (2010). Rethinking the role of ecological research in the sustainable management of freshwater ecosystems. *Freshwater Biology* **55** (Suppl. 1): 258–269.

Hassall, C. (2015). Odonata as candidate macroecological barometers for global climate change. *Freshwater Science* **34**: 1040–1049.

Hassall, C., Thompson, D.J., French, C.G. & Harvey, I.F. (2007). Historical changes in the phenology of British Odonata are related to climate. *Global Change Biology* **13**: 933–941.

Havel, J.E., Kovalenko, K.E., Thomaz, S.M., Amalfitano, S. & Kats, L.B. (2015). Aquatic invasive species: challenges for the future. *Hydrobiologia* **750**: 147–170.

Hayden, B., McLoone, P., Coyne, J. & Caffrey, J.M. (2014). Extensive hybridisation between roach, *Rutilus rutilus* L., and common bream, *Abramis brama* L., in Irish lakes and rivers. *Biology & Environment: Proceedings of the Royal Irish Academy* **114**: 1–5.

Hayes, K.A., Joshi, R.C., Thiengo, C.S. & Cowie, R.H. (2008). Out of South America: multiple origins of non-native apple snails in Asia. *Diversity and Distributions* **14**: 701–712.

Haynes, J.M., Tisch, N.A., Mayer, C.M. & Rhyne, R.S. (2005). Benthic macroinvertebrate communities in southwestern Lake Ontario following invasion of *Dreissena* and *Echinogammarus*. *Journal of the North American Benthological Society* **24**: 148–167.

Haynie, R.H., Bowerman, W.W., Williams, S.K., Morrison, J.R., Grizzle, J.R., Fischer, J.R. & Wilde, S.B. (2013). Are triploid grass carp suitable for aquatic vegetation management in systems affected by avian vacuolar myelinopathy? *Journal of Aquatic Animal Health* **25**: 252–259.

Hecky, R.E., Mugidde, R., Ramlal, P.S., Talbot, M.R. & Kling, G.W. (2010). Multiple stressors cause rapid ecosystem change in Lake Victoria. *Freshwater Biology* **55**: 19–42.

Heino, J., Virkkala, R. & Toivonen, H. (2009). Climate change and freshwater biodiversity: detected patterns, future trends and adaptations in northern regions. *Biological Reviews* **84**: 39–54.

Hekkala, E., Shirley, M.H., Amato, G., Austin, J.D., Charter, S., Thorbjarnarson, J., Vliet, K.A., Houck, M.L., Desalle, R. & Blum, M.J. (2011). An ancient icon reveals new mysteries: mummy DNA resurrects a cryptic species within the Nile crocodile. *Molecular Ecology* **20**: 4199–4215.

Helfield, J.M. & Naiman, R.J. (2006). Keystone interactions: salmon and bear in riparian forests of Alaska. *Ecosystems* **9**: 167–180.

Herbert, M.E., McIntyre, P.B., Doran, P.J., Allen, J.D. & Abell, R. (2010). Terrestrial reserve networks do not adequately represent aquatic ecosystems. *Conservation Biology* **24**: 1002–1011.

Hermoso, V., Abell, R., Linke, S. & Boon, P. (2016). The role of protected areas for freshwater biodiversity conservation: challenges and opportunities in a rapidly changing world. *Aquatic Conservation: Marine and Freshwater Ecosystems* **26**: 3–11.

Hermoso, V.L., Januchowski-Hartley, S., Linke, S., Dudgeon, D., Petry, P. & McIntyre, P.B. (2017). Optimal allocation of Red List assessments to guide conservation of biodiversity in a rapidly changing world. *Global Change Biology* **23**: 3525–3532.

Hermoso, V., Filipe, A.F., Segurado, P. & Beja, P. (2018). Freshwater conservation in a fragmented world: dealing with barriers in a systematic planning framework. *Aquatic Conservation: Marine and Freshwater Ecosystems* **28**: 17–25.

Herrmann, K.K. & Sorensen, R.E. (2009). Seasonal dynamics of two mortality-related trematodes using an introduced snail. *Journal of Parasitology* **95**: 823–828.

Hershner, C. & Havens, K.J. (2008). Managing invasive aquatic plants in a changing system: strategic consideration of ecosystem services. *Conservation Biology* **22**: 544–550.

Hes, D. & du Plessis, C. (2014). *Designing for Hope: Pathways to Regenerative Sustainability*. Routledge, Abingdon.

Hewitt, N., Klenk, N., Smith, A.L., Bazely, D.R., Yan, N., Wood, S., MacLellan, J.I., Lipsig-Mumme, C. & Henriques, I. (2011). Taking stock of the assisted migration debate. *Biological Conservation* **144**: 2560–2572.

Hicks, B.J., Ling, N. & Daniel, A.J. (2012). *Cyprinus carpio* L. (common carp). *A Handbook of Global Freshwater Invasive Species* (R.A. Francis, ed.), Earthscan, Oxford: pp. 247–260.

Hirschfeld, M., Blackburn, D.C., Doherty-Bone, T.M., Gonwouo, L.N., Ghose, S. & Rödel, M.-O. (2016). Dramatic declines of montane frogs in a Central African biodiversity hotspot. *PLoS ONE* **11**: e0155129. https://doi.org/10.1371/journal.pone.0155129

Hitt, N.P., Eyler, S. & Wofford, J.E. (2012). Dam removal increases American eel abundance in distant headwater streams. *Transactions of the American Fisheries Society* **141**: 1171–1179.

Hof, C., Araújo, M.B., Jetz, W. & Rahbek, C. (2011). Additive threats from pathogen, climate and land-use change for global amphibian diversity. *Nature* **480**: 516–519.

Hoffmann, R.C. (2001). Frontier foods for Late Medieval consumers: culture, economy, ecology. *Environment and History* **7**: 131–167.

Hoffmann, R.C. (2005). A brief history of aquatic resource use in medieval Europe. *Helgoland Marine Research* **59**: 22–30.

Hogan, Z. (2013a). *A Mekong Giant. Current Status, Threats and Preliminary Conservation Measures for the Critically Endangered Mekong Giant Catfish*. WWF, Gland. http://awsassets.panda.org/downloads/mgc_report_june2013.pdf

Hogan, Z. (2013b). *Catlocarpio siamensis. The IUCN Red List of Threatened Species 2013:* e.T180662A7649359. http://dx.doi.org/10.2305/IUCN.UK.2011-1.RLTS.T180662A7649359.en.

Hogan, Z. & Jensen, O. (2013). *Hucho taimen. The IUCN Red List of Threatened Species* 2013: e.T188631A22605180. http://dx.doi.org/10.2305/IUCN.UK.2013-1.RLTS.T188631A22605180.en

Holland, R.A., Darwall, W.R.T. & Smith, K.G. (2012). Conservation priorities for freshwater biodiversity: the key biodiversity area approach refined and tested for continental Africa. *Biological Conservation* **148**: 167–179.

Holmstrup, M., Bindesbøl, A.M., Oostingh, G.J., Duschl, A., Scheil, V., Köhler, H.R. & Spurgeon, D.J. (2010). Interactions between effects of environmental chemicals and natural stressors: a review. *Science of the Total Environment* **408**: 3746–3762.

Horton, A.A., Walton, A., Spurgeon, D.J., Lahive, E. & Svendsen, C. (2017). Microplastics in freshwater and terrestrial environments: evaluating the current understanding to identify the knowledge gaps and future research priorities. *Science of the Total Environment* **586**: 127–141.

Hortle, K.G. (2007). Consumption and yield of fish and other aquatic animals from the lower Mekong Basin. *MRC Technical Paper No. 16*. Mekong River Commission, Vientiane. http://archive.iwlearn.net/www.mrcmekong.org/down load/free_download/technical_paper16.pdf

Hortle, K.G. (2009). Fisheries of the Mekong River Basin. In *The Mekong: Biophysical Environment of a Transboundary River* (I.C. Campbell, ed.), Elsevier, New York: pp. 193–253.

Hossain, M.M., Islam, M.A., Ridgway, S. & Matsuishi, T. (2006). Management of inland open water fisheries resources of Bangladesh: issues and options. *Fisheries Research* **77**: 75–284.

Howard, S.D. & Bickford, D.P. (2014). Amphibians over the edge: silent extinction rate of data deficient species. *Diversity and Distributions* **20**: 837–846.

Howe, C., Suich, H., Vira, B. & Mace, G.M. (2014). Creating win-wins from trade-offs? Ecosystem services for human well-being: a meta-analysis of ecosystem service trade-offs and synergies in the real world. *Global Environmental Change* **28**: 263–275.

Hu, Z., Wang, S., Wu, H., Chen, Q., Ruan, R., Chen, L. & Liu, Q. (2014). Temporal and spatial variation of fish assemblages in Dianshan Lake, Shanghai, China. *Chinese Journal of Oceanology and Limnology* **32**: 799–809.

Huang, X.-C., Rong, J., Liu, Y., Zhang, M.-H., Wan, Y., Ouyang, S., Zhou, C.-H. & Wu, X.-P. (2013). The complete maternally and paternally inherited mitochondrial genomes of the endangered freshwater mussel *Solenaia carinatus* (Bivalvia: Unionidae) and implications for Unionidae taxonomy. *PLoS ONE* **8**: e84352. https://doi.org/10.1371/journal.pone.0084352

Hughes, R.M. (2015). Recreational fisheries in the USA: economics, management strategies, and ecological threats. *Fisheries Science* **81**: 1–9.

Hulme, P.E. (2003). Biological invasions: winning the science battles but losing the conservation war? *Oryx* **37**: 178–193.

Humphries, P. & Winemiller, K.O. (2009). Historical impacts on river fauna, shifting baselines and challenges for restoration. *BioScience* **59**: 673–684.

Hurlbert, A.H., Anderson, T.W., Sturm K.K. & Hurlbert, S.H. (2007). Fish and fish-eating birds at the Salton Sea: a century of boom and bust. *Lake and Reservoir Management* **23**: 469–499.

Hyatt, K.D., McQueen, D.J., Shortreed, K.S. & Rankin, D.P. (2004). Sockeye salmon (*Oncorhynchus nerka*) nursery lake fertilization: review and summary of results. *Environmental Reviews* **12**: 133–162.

ICEM (2010). *Strategic Environmental Assessment (SEA) of Hydropower of the Mekong Mainstream. Final Report.* International Centre for Environmental Management, Hanoi. www.mrcmekong.org/ISH/SEA/WEA-Main-Final.pdf/

Icochea, J., Reichle, S., De la Riva, I., Sinsch, U. & Köhler, J. (2004). *Telmatobius culeus*. The IUCN Red List of Threatened Species 2004: e.T57334A11623098. http://dx.doi.org/10.2305/IUCN.UK.2004.RLTS.T57334A11623098.en.

IPCC (2018). Summary for Policymakers. *Global warming of 1.5°C. An IPCC Special Report on the impacts of global warming of 1.5°C above pre-industrial levels and related global greenhouse gas emission pathways, in the context of strengthening the global response to the threat of climate change, sustainable development, and efforts to eradicate poverty* (V. Masson-Delmotte, P. Zhai, H.O. Pörtner, D. Roberts, J. Skea, P.R. Shukla, A. Pirani, W. Moufouma-Okia, C. Péan, R. Pidcock, S. Connors, J.B.R. Matthews, Y. Chen, X. Zhou, M.I. Gomis, E. Lonnoy, T. Maycock, M. Tignor & T. Waterfield, eds), World Meteorological Organization, Geneva. www.ipcc.ch/sr15/chapter/summary-for-policy-makers/

IPBES (2019). Summary for Policymakers of the Global Assessment Report on Biodiversity and Ecosystem Services of the Intergovernmental Science-Policy Platform on Biodiversity and Ecosystem Services. (S. Díaz, J. Settele, E.S. Brondizio, H.T. Ngo, M. Guèze, J. Agard, A. Arneth, P. Balvanera, K.A. Brauman, S.H.M. Butchart, K.M.A. Chan, L.A. Garibaldi, K. Ichii, J. Liu, S.M. Subramanian, G.F. Midgley, P. Miloslavich, Z. Molnár, D. Obura, A. Pfaff, S. Polasky, A. Purvis, J. Razzaque, B. Reyers, R.R. Chowdhury, Y.J. Shin, I. J. Visseren-Hamakers, K.J. Willis & C.N. Zayas, eds.), IPBES Secretariat, Bonn. www.ipbes.net/global-assessment-report-biodiversity-ecosystem-services

Ismail, G.B., Sampson, D.B. & Noakes, D.G. (2014). The status of Lake Lanao endemic cyprinids (*Puntius* species) and their conservation. *Environmental Biology of Fishes* **97**: 425–434.

Isaak, D.J., Wollrab, S., Horan, D.L. & Chandler, G. (2012). Climate change effects on stream and river temperatures across the northwest U.S. from 1980-2009 and implications for salmonid fishes. *Climatic Change* **113**: 499–524.

Isaak, D.J., Young, M.K., Nagel, D.E., Horan, D.L. & Groche, M.C. (2015). The cold-water climate shield: delineating refugia for preserving salmonid fishes through the 21st century. *Global Change Biology* **21**: 2540–2553.

ISSG (Invasive Species Specialist Group) (2015). The Global Invasive Species Database. Version 2015.1. www.iucngisd.org/gisd/

IUCN (2017). *The IUCN Red List of Threatened Species 2017.1.* International Union for Conservation of Nature and Natural Resources, Cambridge. www .iucnredlist.org/.

IUCN SSC Amphibian Specialist Group (2015). *Nectophrynoides asperginis. The IUCN Red List of Threatened Species* 2015: e.T54837A16935685. http://dx .doi.org/10.2305/IUCN.UK.2015-2.RLTS.T54837A16935685.en.

Izmest'eva, L.R., Silow, E.A. & Litchman, E. (2011). Long-term dynamics of Lake Baikal pelagic phytoplankton under climate change. *Inland Water Biology* **4**: 301. https://doi.org/10.1134/S1995082911030102

Izmest'eva, L.R., Moore, M.V., Hampton, S.E., Ferwerda, C.J., Gray, D.K., Woo, K.H., Pislegina, H.V., Krashchuk, L.S., Shimaraeva, S.V. & Silow, E.A. (2016). Lake-wide physical and biological trends associated with warming in Lake Baikal. *Journal of Great Lakes Research* **42**: 6–17.

Jackson, M.C. & Grey, J. (2012). Accelerating rates of freshwater invasions in the catchment of the River Thames. *Biological Invasions* **15**: 945–951.

Jackson, M., Loewen, C.J.G., Vinebrooke, R.D. & Chimimba, C.T. (2016). Net effects of multiple stressors in freshwater ecosystems: a meta-analysis. *Global Change Biology* **22**: 180–189.

Jackson, R.B., Carpenter, S.R., Dahm, C.N., McKnight, D.M., Naiman, R.J., Postel, S.L. & Running, S.W. (2001). Water in a changing world. *Ecological Applications* **11**:1027–1045.

Jacobsen, D., Milner, A.M., Brown, L.E. & Dangles, O. (2012). Biodiversity under threat in glacier-fed river systems. *Nature Climate Change* **2**: 361–364.

Jacobsen, D., Cauvy-Fraunie, S., Andino, P., Espinosa, R., Cueva, D. & Dangles, O. (2014). Runoff and the longitudinal distribution of macroinvertebrates in a glacier-fed stream: implications for the effects of global warming. *Freshwater Biology* **59**: 2038–2050.

Jacoby, D.M.P., Casselman, J.M., Crook, V., DeLucia, M., Ahn, H., Kaifu, K., Kurwie, T., Sasal, P., Silfvergrip, A.M.C., Smith, K.G., Uchida, K., Walker, M.M. & Gollock, M.J. (2015). Synergistic patterns of threat and the challenges facing global anguillid eel conservation. *Global Ecology and Conservation* **4**: 321–333.

Jacoby, S. (2008). *The Age of American Unreason.* Pantheon Books, New York.

Jakob, L., Axenov-Gribanov, D.V., Gurkov, A.N., Ginzburg, M., Bedulina, D.S., Timofeyev, M.A., Luckenbach, T., Lucassen, M., Sartoris, F.J., Pörtner, H.O. & Benstead, J. (2016). Lake Baikal amphipods under climate change: thermal constraints and ecological consequences. *Ecosphere* **7**: e01308.

Jansson, R., Backx, H., Boulton, A.J., Dixon, M., Dudgeon, D., Hughes, F., Nakamura, K., Stanley, E. & Tockner, K. (2005). Stating mechanisms and refining criteria for ecologically successful river restoration. *Journal of Applied Ecology* **42**: 218–222.

Jaramillo, F. & Destouni, G. (2015). Comment on 'planetary boundaries: guiding human development on a changing planet'. *Science* **348**: 1217–1218.

Jeppesen, E., Mehner, T., Winfield, I., Kangur, K., Sarvala, J., Gerdeaux, D., Rask, M., Malmquist, H., Holmgren, K., Volta, P., Romo, S., Eckmann, R., Sandstrom, A., Blanco, S., Kangur, A., Ragnarsson Stabo, H., Tarvainen, M., Ventela, A.M., Sondergaard, M., Lauridsen, T. & Meerhoff, M. (2012). Impacts of climate warming on the longterm dynamics of key fish species in 24 European lakes. *Hydrobiologia* **694**: 1–39.

Jeppesen, E., Meerhoff, M., Davidson, T.A., Søndergaard, M., Lauridsen, T.L., Beklioğlu, M., Brucet, S., Volta, P., González-Bergonzoni, I. & Nielsen, A. (2014). Climate change impacts on lakes: an integrated ecological perspective based on a multi-faceted approach, with special focus on shallow lakes. *Journal of Limnology* **73**: 84–107.

Jerde, C.L., Mahon, A.R., Chadderton, W.L. & Lodge, D.M. (2011). 'Sight-unseen' detection of rare aquatic species using environmental DNA. *Conservation Letters* **4**: 150–157.

Jerde, C.L., Chadderton, W.L., Mahon, A.R., Renshaw, M.A., Corush, J., Budny, M.L., Mysorekar, S. & Lodge, D.M. (2013). Detection of Asian carp DNA as part of a Great Lakes basin-wide surveillance program. *Canadian Journal of Fisheries and Aquatic Sciences* **70**: 522–526.

Jeschke, J.M., Gómez Aparicio L., Haider S., Heger, T., Lortie, C.J., Pyšek P. & Strayer, D.L. (2012). Support for major hypotheses in invasion biology is uneven and declining. *NeoBiota* **14**: 1–20.

Jeziorski, A., Yan, N.D., Paterson, A.M., Desellas, A.M., Turner, M.A., Jeffries, D.S., Keller, B., Weeber, R.C., McNicol, D.K., Palmer, M.E., McIver, K., Arseneau, K., Ginn, B.K., Cumming, B.F. & Smol, J.P. (2008). The widespread threat of calcium decline in fresh waters. *Science* **32**: 1374–1377.

Jeziorski, A., Tanentzap, A.J., Yan, N.D., Paterson A.M., Palmer, M.E., Korosi, J.B., Rusak, J.A., Arts, M.T., Keller, W., Ingram, R., Cairns, A. & Smol, J.P. (2015). The jellification of north temperate lakes. *Proceedings of the Royal Society B: Biological Sciences* **282**: 20142449. https://doi.org/10.1098/rspb.2014.2449

Jiang, Z. & Harris, R.B. (2016). *Elaphurus davidianus*. The IUCN Red List of Threatened Species 2016: e.T7121A22159785. http://dx.doi.org/10.2305/IUCN.UK.2016-2.RLTS.T7121A22159785.en.

Jiao, L. (2009). Scientists line up against dam that would alter protected wetlands. *Science* **326**: 508–509.

Jiguet, F., Godet, L. & Devictor, V. (2012). Hunting and the fate of French breeding waterbirds. *Bird Study* **59**: 474–482.

Johnson, P.T., Olden, J.D. & Vander Zanden, M.J. (2008). Dam invaders: impoundments facilitate biological invasions in freshwaters. *Frontiers in Ecology and the Environment* **6**: 357–363.

Jones, J.P.G., Rasamy, J.R., Harvey, A., Toon, A., Oidtmann, B., Randrianarison, M.H., Raminosoa, N. & Ravoahangimalala, O.R. (2008). The perfect invader: a parthenogenic crayfish poses a new threat to Madagascar's freshwater biodiversity. *Biological Invasions* **11**: 1475–1482.

Jones, L.A. & Ricciardi, A. (2014). The influence of pre-settlement and early post-settlement processes on the adult distribution and relative dominance of two invasive mussel species. *Freshwater Biology* **59**: 1086–1100.

Jones, R., Travers, C., Rodgers, C., Lazar, B., English, E., Lipton, J., Vogel, J., Strzepek, K. & Martinich, J. (2013). Climate change impacts on freshwater recreational fishing in the United States. *Mitigation & Adaptation Strategies for Global Change* **18**: 731–758.

Jonsson, M., Hedström, P., Stenroth, K., Hotchkiss, E.R., Vasconcelos, F.R., Karlsson, J. & Byström, P. (2015). Climate change modifies the size structure of assemblages of emerging aquatic insects. *Freshwater Biology* **60**: 78–88.

Jonsson, T. & Setzer, M. (2015). A freshwater predator hit twice by the effects of warming across trophic levels. *Nature Communications* **6**: 5992. https://doi.org/10.1038/ncomms6992

Juffe-Bignoli, D., Harrison, I., Butchart, S.H.M., Flitcroft, R., Hermoso, V., Jonas, H., Lukasiewicz, A., Thieme, M., Turak, E., Bingham, H., Dalton, J., Darwall, W., Deguignet, M., Dudley, N., Gardner, R., Higgins, J., Kumar, R., Linke, S., Milton, G.R., Pittock, J., Smith, K.G. & van Soesbergen, A. (2016). Achieving Aichi Biodiversity Target 11 to improve the performance of protected areas and conserve freshwater biodiversity. *Aquatic Conservation: Marine and Freshwater Ecosystems* **26**: 133–151.

Justus, J., Colyvan, M., Regan, H. & Maguire, L. (2009). Buying into conservation: intrinsic versus instrumental value. *Trends in Ecology & Evolution* **24**: 187–191.

Kano, Y., Musikasinthorn, P., Iwata, A., Tun, S., Yun, L., Win, S., Matsui, S., Tabata, R., Yamasaki, T. & Watanabe, K. (2016a). A dataset of fishes in and around Inle Lake, an ancient lake of Myanmar, with DNA barcoding, photo images and CT/3D models. *Biodiversity Data Journal* **4**: e10539. https://doi.org/10.3897/BDJ.4.e10539

Kano, Y., Dudgeon, D., Nam, S., Samejima, H., Watanabe, K., Grudpan, C., Magtoon, W., Musikasinthorn, P., Nguyen, P.T., Praxaysonbath, B., Sato, T., Shibukawa, K., Shimatani, Y., Suvarnaraksha, A., Tanaka, W., Thach, P., Tran, D.D., Yamashita, T. & Utsugi, K. (2016b). Impacts of dams and global warming on fish biodiversity in the Indo-Burma Hotspot. *PLoS ONE* **11**: e0160151. https://doi.org/10.1371/journal.pone.0160151

Karatayev, A.Y., Burlakova L.E., Padilla, D.K., Mastitsky, S.E & Olenin, S. (2009). Invaders are not a random selection of species. *Biological Invasions* **11**: 2009–2019.

Karraker, N.E., Gibbs, J.P. & Vonesh, J.R. (2008). Impacts of road deicing on the demography of vernal pool-breeding amphibians. *Ecological Applications* **18**: 724–734.

Karraker, N.E., Arrigoni, J. & Dudgeon, D. (2010). Effects of increased salinity and an introduced predator on lowland amphibians in Southern China: species identity matters. *Biological Conservation* **143**: 1079–1086.

Karraker, N.K. & Dudgeon, D. (2014). Invasive apple snails (*Pomacea canaliculata*) are predators of amphibians in South China. *Biological Invasions* **16**: 1785–1789.

Katunzi, E.F.B., Mbonde, A., Waya, R. & Mrosso, H.D.J. (2010). Minor water bodies around Lake Victoria – a replica of lost biodiversity. *Aquatic Ecosystem Health and Management* **13**: 277–283.

Kaufman, L. (1992). Catastrophic change in species-rich freshwater ecosystems. *BioScience* **42**: 846–858.

Kaushal, S.S., Likens, G.E., Jaworski, N.A., Pace, M.L., Sides, A.M., Seekell, D., Belt, K.T., Secor, D.H. & Wingate, R.L. (2010). Rising stream and river temperatures in the United States. *Frontiers in Ecology and the Environment* **8**: 461–466.

Kawarazuka, N. & Béné, C. (2011). The potential role of small fish in improving micronutrient deficiencies in developing countries: building the evidence. *Public Health Nutrition* **14**: 1927–1938.

Kazembe, J. & Makocho, P. (2004). *Oreochromis lidole. The IUCN Red List of Threatened Species 2004*: e.T61276A12456642. http://dx.doi.org/10.2305/IUCN.UK.2004.RLTS.T61276A12456642.en.

Kefford, B.J., Buchwalter, D., Cañedo-Argüelles, M., Davis, J.A., Duncan, R.P., Hoffman, A. & Thompson, R.M. (2016). Salinized rivers: degraded systems or new habitats for salt-tolerant faunas? *Biology Letters* **12**: http://doi: 10.1098/rsbl.2015.1072.

Keith Diagne, L. (2015). *Trichechus senegalensis. The IUCN Red List of Threatened Species 2015*: e.T22104A97168578. http://dx.doi.org/10.2305/IUCN.UK.2015-4.RLTS.T22104A81904980.en

Kellermann, V., Overgaard, J., Hoffmann, A.A., Fløjgaard, C., Svenning, J.C. & Loeschcke, V. (2012). Upper thermal limits of *Drosophila* are linked to species distributions and strongly constrained phylogenetically. *Proceedings of the National Academy of Sciences of the United States of America* **109**: 16228–16233.

Kemp, P.S. (2016). Meta-analyses, metrics and motivation: mixed messages in the fish passage debate. *River Research and Applications* **32**: 2116–2124.

Kemp, P.S., Worthington, T.A., Langford, T.E.L., Tree, A.R.J. & Gaywood, M.J. (2011). Qualitative and quantitative effects of reintroduced beavers on stream fish. *Fish and Fisheries* **13**: 158–181.

Kennedy, T.A., Muehlbauer, J.D., Yackulic, C.B., Lytle, D.A., Miller, S.W., Dibble, K.L., Kortenhoeven, E.W., Metcalfe, A.N. & Baxter, C.V. (2016). Flow Management for hydropower extirpates aquatic insects, undermining river food webs. *BioScience* **66**: 561–575.

Khoo, K.H., Leong, T.S., Soon, F.L., Tan, S.P. & Wong, S.Y. (1987). Riverine fishes in Malaysia. *Archiv für Hydrobiologie Beiheft, Ergebnisse Limnologie* **28**: 261–268.

Khosov, M. (1963; reprint of 1936 volume). *Lake Baikal and its Life*. Springer Science and Business Media, Dordrecht.

Kiernan, J.D., Moyle, P.B. & Crain, P.K. (2012). Restoring native fish assemblages to a regulated California stream using the natural flow regime concept. *Ecological Applications* **22**: 1472–1482.

King, A.J., Ward, K.A., O'Connor, P., Green, D., Tonkin, Z. & Mahoney, J. (2010). Adaptive management of an environmental watering event to enhance native fish spawning and recruitment. *Freshwater Biology* **55**: 17–31.

Kilpatrick, A.M., Salkeld, D.J., Titcomb, G. & Hahn, M.B. (2017). Conservation of biodiversity as a strategy for improving human health and well-being. *Philosophical Transactions of the Royal Society B: Biological Sciences* **372**: 2100131. http://doi.org/10.1098/rstb.2016.0131

King, J. & Pienaar, H. (2011). *Sustainable use of South Africa's inland waters: a situation assessment of resource directed measures 12 years after the 1998 National Water Act*. Water Research Commission Report TT 491/11, Pretoria.

King, J.M. & Brown, C. (2006). Environmental flows: striking the balance between development and resource protection. *Ecology and Society* **11**: 26. www .ecologyandsociety.org/vol11/iss2/art26/

King, J.M. & Brown, C. (2010). Integrated basin flow assessments: concepts and method development in Africa and South-east Asia. *Freshwater Biology* **55**: 127–146.

Kingsford, R.T. & Thomas, R.F. (2004). Destruction of wetlands and waterbird populations by dams and irrigation on the Murrumbidgee River in arid Australia. *Environmental Management* **34**: 383–396.

Kipp, R.M., Ricciardi, A., Larson, J., Fusaro, A. & Makled, T. (2013). *Hemimysis anomala*. USGS Nonindigenous Aquatic Species Database, Gainesville, FL. http://nas.er.usgs.gov/queries/factsheet.aspx?SpeciesID=2627

Kipp, R.M., Benson, A.J., Larson, J. & Fusaro, A. (2014). *Bithynia tentaculata*. USGS Nonindigenous Aquatic Species Database, Gainesville, FL. http://nas.er.usgs .gov/queries/FactSheet.aspx?speciesID=987

Klecka, G., Persoon, C. & Currie, R. (2010). Chemicals of emerging concern in the Great Lakes Basin: an analysis of environmental exposures. *Reviews in Environmental Contamination and Toxicology* **207**: 1–93.

Knapp, R.A. (2005). Effects of nonnative fish and habitat characteristics on lentic herpetofauna in Yosemite National Park, USA. *Biological Conservation* **121**: 265–279.

Knapp, R.A. & Sarnelle, O. (2008). Recovery after local extinction: factors affecting re-establishment of alpine lake zooplankton. *Ecological Applications* **18**: 1850–1859.

Knapp, R.A., Boiano, D.M. & Vredenburg, V.T. (2007). Removal of nonnative fish results in population expansion of a declining amphibian (mountain yellow-legged frog, *Rana muscosa*). *Biological Conservation* **135**: 11–20.

Knapp, R.A., Fellers, G.M., Kleeman, P.M., Miller, D.A.W., Vredenburg, V.T., Rosenblum, E.B. & Briggs, C.J. (2016). Large-scale recovery of an endangered amphibian despite ongoing exposure to multiple stressors. *Proceedings of the National Academy of Sciences of the United States of America* **113**: 11889–11894.

Knight, A.T. (2013). Reframing the theory of hope in conservation science. *Conservation Letters* **6**: 389–390.

Kobanova, G.I., Takhteev, V.V., Rusanovskaya, O.O. & Timofeyev, M.A. (2016). Lake Baikal ecosystem faces the threat of eutrophication. *International Journal of Ecology* **2016**: 6058082. http://dx.doi.org/10.1155/2016/6058082

Koblmüller, S., Duftner, N., Sefc, K.M., Aibara, M., Stipacek, M., Blanc, M., Egger, B. & Sturmbauer, C. (2007). Reticulate phylogeny of gastropod-shell-breeding cichlids from Lake Tanganyika – the result of repeated introgressive hybridization. *BMC Evolutionary Biology* **7**: 7. https://doi.org/10.1186/1471-2148-7-7

Koel, T.M., Mahony, D.L., Kinnan, K.L., Rasmussen, C., Hudson, C.J., Murcia, S. & Kerans, B.L. (2006). *Myxobolus cerebralis* in native cutthroat trout of the Yellowstone Lake ecosystem. *Journal of Aquatic Animal Health* **18**: 157–175.

Koel, T.M., Kerans, B.L., Barras, S.C., Hanson, K.C. & Wood, J.S. (2010). Avian piscivores as vectors for *Myxobolus cerebralis* in the Greater Yellowstone ecosystem. *Transactions of the American Fisheries Society* **139**: 976–988.

Kolby, J.E. (2014). Ecology: stop Madagascar's toad invasion now. *Nature* **509**: 563.

Koldewey, H., Cliffe, A. & Zimmerman, B. (2013). Breeding programme priorities and management techniques for native and exotic freshwater fishes in Europe. *International Zoo Yearbook* **47**: 93–101.

Kornis, M.S., Carlson, J., Lehrer-Brey, G. & Vander-Zanden, J. (2014). Experimental evidence that the ecological effects of an invasive fish are reduced at high densities. *Oecologia* **175**: 325–334.

Kornis, M.S., Weidel, B.C. & Vander-Zanden, M.J. (2017). Divergent life histories of invasive round gobies (*Neogobius melastomus*) in Lake Michigan and its tributaries. *Ecology of Freshwater Fish* **26**: 563–574.

Kosten S., Huszar, V.L.M., Bécares, E., Costa, L.S., van Donk, E., Hansson, L.-A., Jeppesen, E., Kruk, C., Lacerot, G., Mazzeo, N., De Meester, L., Moss, B., Lürling, M., Nõges, T., Romo, S. & Scheffer, M. (2012). Warmer climate boosts cyanobacterial dominance in shallow lakes. *Global Change Biology* **18**: 118–126.

Kostianoy, A.G. & Kosarev, A.N. (2009). *The Aral Sea Environment*. Springer-Verlag, Berlin.

Kostoski, G., Albrecht, C., Trajanovski, S. & Wilke, T. (2010). A freshwater biodiversity hotspot under pressure – assessing threats and identifying conservation needs for ancient Lake Ohrid. *Biogeosciences* **7**: 3999–4015. http://doi.org/10.5194/bg-7-3999-2010

Kottelat, M. & Chu, X. (1988). Revision of *Yunnanilus* with descriptions of a miniature species flock and six new species from China (Cypriniformes: Homalopteridae). *Environmental Biology of Fishes* **23**: 65–93.

Kovach, R.P., Gharrett, A.J. & Tallmon, D.A. (2012). Genetic change for earlier migration timing in a pink salmon population. *Proceedings of the Royal Society B: Biological Sciences* **279**: 3870–3878.

Kreps, T.A., Baldridge, A.K. & Lodge, D.M. (2012). The impact of an invasive predator (*Orconectes rusticus*) on freshwater snail communities: insights on habitat-specific effects from a multilake long-term study. *Canadian Journal of Fisheries and Aquatic Sciences* **69**: 1164–1173.

Kreuzberg-Mukhina, E.A. (2006). The Aral Sea basin: changes in migratory and breeding waterbird populations due to major human-induced changes to the region's hydrology. In *Waterbirds around the World* (G.C. Boere, C.A. Galbraith & D.A. Stroud, eds), The Stationery Office, Edinburgh: pp. 283–284.

Kristensen, T.K. & Stensgaard, A-S. (2010). *Gabbiella tchadiensis. The IUCN Red List of Threatened Species* 2010: e.T165387A6011471. http://dx.doi.org/10.2305/IUCN.UK.2010-3.RLTS.T165387A6011471.en

Kristofco, L.A. & Brooks, B.W. (2017). Global scanning of antihistamines in the environment: analysis of occurrence and hazards in aquatic systems. *Science of the Total Environment* **592**: 477–487.

Kuehne, L.M., Olden, J.D. & Duda, J.J. (2012). Costs of living for juvenile Chinook salmon (*Oncorhynchus tshawytscha*) in an increasingly warming and invaded world. *Canadian Journal of Fisheries and Aquatic Sciences* **69**: 1621–1630.

Kuemmerlen, M., Schmalz, B., Cai, Q., Haase, P., Fohrer, N. & Jähnig, S.C. (2015). An attack on two fronts: predicting how changes in land use and climate affect the distribution of stream macroinvertebrates. *Freshwater Biology* **60**: 1443–1458.

Kummu, M. & Sarkkula, J. (2008). Impact of the Mekong River flow alteration on the Tonle Sap flood pulse. *Ambio* **37**: 185–192.

Kummu, M., Penny, D., Sarkkula, J. & Koponen, J. (2008). Sediment: curse or blessing for Tonle Sap Lake? *Ambio* **37**: 158–163.

Kummu, M., Lu, X. X., Wang, J.J. & Varis, O. (2010). Basin-wide sediment trapping efficiency of emerging reservoirs along the Mekong. *Geomorphology* **119**: 181–197.

Kuparinen, A., Boit, A., Valdovinos, F.S., Lassaux, H. & Martinez, N.D. (2016). Fishing-induced life-history changes degrade and destabilize harvested ecosystems. *Scientific Reports* **6**: 22245. http://doi.org/10.1038/srep22245

Kwong, K.L., Chan, R.Y.K. & Qiu, J.-W. (2009). The potential of the invasive snail *Pomacea canaliculata* as a predator of various life-stages of five species of freshwater snails. *Malacologia* **51**: 343–356.

Kwong, K.L., Dudgeon, D., Wong, P.K. & Qiu, J.-W. (2010). Secondary production and diet of an invasive snail in freshwater wetlands: implications for resource utilization and competition. *Biological Invasions* **12**: 1153–1164.

Lai, G., Wang, P. & Li, L. (2016). Possible impacts of the Poyang Lake (China) hydraulic project on lake hydrology and hydrodynamics. *Hydrology Research* **47**: 187–205.

Lai, X., Huang, Q., Zhang, Y. & Jiang, J. (2014). Impact of lake inflow and the Yangtze River flow alterations on water levels in Poyang Lake, China. *Lake and Reservoir Management* **30**: 321–330.

Lake, P.S., Bond, N. & Reich, P. (2007). Linking ecological theory with stream restoration. *Freshwater Biology* **52**: 597–615.

Lake, P.S., Bond, N.R. & Reich, P. (2007). Linking ecological theory with stream restoration. *Freshwater Biology* **52**: 597–615.

Lakra, W.S., Sarkar, U.K., Dubey, V.K., Sani, R. & Pandey, A. (2011). River inter linking in India: status, issues, prospects and implications on aquatic ecosystems and freshwater fish diversity. *Reviews in Fish Biology and Fisheries* **21**: 463–479.

Lamer, J.T., Dolan, C.R., Petersen, J.L., Chick, J.H. & Epifanio, J.M. (2010). Introgressive hybridization between bighead carp and silver carp in the Mississippi and Illinois Rivers. *North American Journal of Fisheries Management* **30**: 1452–1461.

Lamouroux, N., Gore, J.A., Lepori, F. & Statzner, B. (2015). The ecological restoration of large rivers needs science-based, predictive tools meeting public expectations: an overview of the Rhône project. *Freshwater Biology* **60**: 1069–1084.

Landigran, P.J., Fuller, R., Acosta, N.J.R., Adeyi,O., Arnold, R., Basu, N., Baldé, A.B., Bertollini, R., Bose-O'Reilly, S., Boufford, J.I., Breysse, P.N., Chiles, T., Mahidol, C., Coll-Seck, A.M., Cropper, M.L., Fobil, J., Fuster, V., Greenstone, M., Haines, A., Hanrahan, D., Hunter, D., Khare, M., Krupnick, A., Lanphear, B., Lohani, B., Martin, K., Mathiasen, K.V., McTeer, M.A., Murray, C.J.L., Ndahimananjara, J.D., Perera, F., Potočnik, J., Preker, A.S., Ramesh, J., Rockström, J., Salinas, C., Samson, L.D., Sandilya, K., Sly, P.D., Smith, K.R., Steiner, A., Stewart, R.B., Suk, W.A., van Schayck, O.C.P., Yadama, G.N., Yumkella, K. & Zhong, M. (2017). *The Lancet* Commission on pollution and health. *The Lancet* **391**: 462–512.

Laramie, M.B., Pilliod, D.S. & Goldberg, C.S. (2015). Characterizing the distribution of an endangered salmonid using environmental DNA analysis. *Biological Conservation* **183**: 29–37.

Larigauderie, A. & Mooney, H.A. (2010). The Intergovernmental science-policy Platform on Biodiversity and Ecosystem Services: moving a step closer to an IPCC-like mechanism for biodiversity. *Current Opinion in Environmental Sustainability* **2**: 9–14.

Larsen, S., Muehlbauer, J.D. & Marti, E. (2016). Resource subsidies between stream and terrestrial ecosystems under global change. *Global Change Biology* **22**: 2489–2504.

Lassettre, N.S. & Kondolf, G.M. (2012). Large woody debris in urban stream channels: redefining the problem. *River Research and Applications* **28**: 1477–1487.

Latrubesse, E.M., Arima, E.Y., Dunne, T., Park, E., Baker, V.R., d'Horta, F.M., Wight, C., Wittmann, F., Zuanon, J., Baker, P.A., Ribas, C.C, Norgaard, R.B., Filizola, N., Ansar, A., Flyvbjerg, B. & Stevaux, J.C. (2017). Damming the rivers of the Amazon basin. *Nature* **546**: 363–369.

Law, A., Jones, K.C. & Willby, N.J. (2014). Medium vs. short-term effects of herbivory by Eurasian beaver on aquatic vegetation. *Aquatic Botany* **116**: 27–34.

Law, A., McLean, F. & Willby, N.J. (2016). Habitat engineering by beaver benefits aquatic biodiversity and ecosystem processes in agricultural streams. *Freshwater Biology* **61**: 486–499.

Lawler, J.J., Shafer, S.L., Bancroft, B.A. & Blaustein, A.R. (2010). Projected climate impacts for the amphibians of the Western Hemisphere. *Conservation Biology* **24**: 38–50.

Lawrence, D.J., Stewart-Koster, B., Olden, J.D., Ruesch, A.S., Torgersen, C.E., Lawler, J.J., Butcher, D.P. & Crown, J.K. (2014). The interactive effects of climate change, riparian management, and a nonnative predator on stream-rearing salmon. *Ecological Applications* **24**: 895–912.

Ledger, M.E. & Milner, A.M. (2015). Extreme events in fresh water. *Freshwater Biology* **60**: 1–6.

Leigh, C., Boulton, A.J., Courtwright, J.L., Fritz, K., May, C.L., Walker, R.H. & Datry, T. (2016). Ecological research and management of intermittent rivers: an historical review and future directions. *Freshwater Biology* **61**: 1181–1199.

Lele, S., Springate-Baginski, O., Lakerveld, R., Deb, D. & Dash, P. (2013). Ecosystem services: origins, contributions, pitfalls and alternatives. *Conservation and Society* **11**: 343–358.

Lehner, B. & Grill, G. (2013). Global river hydrography and network routing: baseline data and new approaches to study the world's large river systems. *Hydrological Processes* **27**: 2171–2186. (Data available at www.hydrosheds.org or www.hydrosheds.org/page/hydrobasins.)

Lehner, B., Liermann, C.R., Revenga, C., Vörösmarty, C., Fekete, B., Crouzet, P., Döll, P., Endejan, M., Frenken, K., Magome, J., Nilsson, C., Robertson, J.C., Rödel, R., Sindorf, N. & Wisser, D. (2011). High-resolution mapping of the world's reservoirs and dams for sustainable river-flow management. *Frontiers in Ecology and the Environment* **9**: 494–502.

Lenoir, J. & Svenning, J.C. (2015). Climate-related range shifts – a global multidimensional synthesis and new research directions. *Ecography* **38**: 15–28.

Leonard, P.B., Baldwin, R.F. & Hanks, R.D. (2017). Landscape-scale conservation design across biotic realms: sequential integration of aquatic and terrestrial landscapes. *Scientific Reports* **7**: 14556. https://doi.org/10.1038/s41598–017-15304-w.

Leprieur, F., Beauchard, O., Blanchet, S., Oberdorff, T. & Brosse, S. (2008). Fish invasion in the world's river systems: when natural processes are blurred by human activity. *PLoS Biol* **6**: e28. https://doi.org/10.1371/journal.pbio.0060028

Leprieur, F., Brosse, S., García-Berthou, E., Oberdorff, T., Olden, J.D. & Townsend, C.R. (2009). Scientific uncertainty and the assessment of risks posed by non-native freshwater fishes. *Fish and Fisheries* **10**: 88–97.

Leprieur, F., Tedesco, P.A., Hugueny, B., Beauchard, O., Dürr, H.H., Brosse, S. & Oberdorff, T. (2011). Partitioning global patterns of freshwater fish beta diversity reveals contrasting signatures of past climate changes. *Ecology Letters* **14**: 325–334.

Leung, B., Lodge, D.M., Finnoff, D., Shogren, J.F., Lewis, M.A. & Lamberti, G. (2002). An ounce of prevention or a pound of cure: bioeconomic risk analysis of invasive species. *Proceedings of the Royal Society of London, Series B: Biological Sciences* **269**: 2407–2413.

Leung, K.M.Y. & Dudgeon, D. (2008). Ecological risk assessment and management of exotic organisms associated with aquaculture activities. In *Understanding and Applying Risk Analysis in Aquaculture* (M.G. Bondad-Reantaso, J.R. Arthur & R.P. Subasinghe, eds), FAO Fisheries Technical Paper No. 519, FAO, Rome: pp. 67–100.

Leuven, R.S.E.W., van der Velde, G., Baijens, I., Snijders, J., van der Zwart, C., Lenders, H.J.R., bij de Vaate, A. (2009). The river Rhine: a global highway for dispersal of aquatic invasive species. *Biological Invasions* **11**: 1989–2008.

Li, C., Corrigan, S., Yang, L., Straube, N., Harris, M., Hofreiter, M., White, W.T. & Naylor, G.J.P. (2015). DNA capture reveals transoceanic gene flow in endangered river sharks. *Proceedings of the National Academy of Sciences of the United States of America* **112**: 13302–13307.

Li, Y., Cohen, J.M. & Rohr, J.R. (2013). Review and synthesis of the effects of climate change on amphibians. *Integrative Zoology* **8**: 145–161.

Limburg, K.E. & Waldman, J.B. (2009). Dramatic declines in North Atlantic diadromous fishes. *BioScience* **59**: 955–965.

Lindner, K., Cerwenka, A.F., Brandner, J., Gertzen, S., Borcherding, J. & Schliewen, U.K. (2013). First evidence for interspecific hybridization between invasive goby species *Neogobius fluviatilis* and *Neogobius melanostomus* (Teleostei: Gobiidae: Benthophilinae). *Journal of Fish Biology* **82**: 2128–2134.

Linke, S., Turak, E. & Nel, J. (2011). Freshwater conservation planning: the case for systematic approaches. *Freshwater Biology* **56**: 6–20.

Lips, K.R., Brem, F., Brenes, R., Reeve, J.D., Alford, R.A.,Voyles, J., Carey, C., Livo, L., Pessier, A.P. & Collins, J.P. (2006). Emerging infectious disease and the loss of biodiversity in a Neotropical amphibian community. *Proceedings of the National Academy of Sciences of the United States of America* **103**: 3165–3170.

Lira-Noriega, A., Aguilar, V., Alarcón, J., Kolb, M., Urquiza-Haas, T., González-Ramírez, L., Tobón, W. & Koleff, P. (2015). Conservation planning for freshwater ecosystems in Mexico. *Biological Conservation* **191**: 357–366.

Litchman, E. (2010). Invisible invaders: non-pathogenic invasive microbes in aquatic and terrestrial ecosystems. *Ecology Letters* **13**: 1560–1572.

Liu, L., Oza, S., Hogan, D., Perin, J., Rudan, I., Lawn, J.E., Cousens, S., Mathers, C. & Black, R.E. (2015). Global, regional, and national causes of child mortality in 2000–13, with projections to inform post-2015 priorities: an updated systematic analysis. *The Lancet* **385**: 430–440.

Liu, X., Cao, Y., Xue, T., Wu, R., Zhou, Y., Zhou, C., Zanatta, D.T., Ouyang, S. & Wu, X. (2017). Genetic structure and diversity of *Nodularia douglasiae* (Bivalvia: Unionida) from the middle and lower Yangtze River drainage. *PLoS ONE* **12**: e0189737. https://doi.org/10.1371/journal.pone.0189737

Lodge, D.M. (2010). It's the water, stupid. *BioScience* **60**: 6–7.

Lodge, D.M., Rosenthal, S.K., Mavuti, K.M., Muohi, W., Ochieng, P., Stevens, S.S., Mungai, B.N. & Mkoji, G.M. (2005). Louisiana crayfish (*Procambarus clarkii*) (Crustacea: Cambaridae) in Kenyan ponds: non-target effects of a potential biological control agent for schistosomiasis. *African Journal of Aquatic Science* **30**: 119–124.

Loehle, C. & Eschenbach, W. (2012). Historical bird and terrestrial mammal extinction rates and causes. *Diversity and Distributions* **18**: 84–91.

Lopes-Lima, M., Sousa, R., Geist, J., Aldridge, D.C., Araujo, R., Bergengren, J., Bespalaya, Y., Bódis, E., Burlakova, L., Van Damme, D., Douda, K., Froufe, E., Georgiev, D., Gumpinger, C., Karatayev, A., Kebapçi, Ü., Killeen, I., Lajtner, J., Larsen, B.M., Lauceri, R., Legakis, A., Lois, S., Lundberg, S., Moorkens, E., Motte, G., Nagel, K.-O., Ondina, P., Outeiro, A., Paunovic, M., Prié, V., von Proschwitz, T., Riccardi, N., Rudzīte, M., Rudzītis, M., Scheder, C., Seddon, M., Şerefliẞan, H., Simić, V., Sokolova, S., Stoeckl, K., Taskinen, J., Teixeira, A., Thielen, F., Trichkova, T., Varandas, S., Vicentini, H., Zajac, K., Zajac, T. & Zogaris, S. (2017). Conservation status of freshwater mussels in Europe: state of the art and future challenges. *Biological Reviews* **92**: 572–607.

López-Luna, M.A., Hidalgo-Mihart, M.G., Aguirre-León, G., González-Ramón, M.C. & Rangel-Mendoza, J.A. (2015). Effect of nesting environment on incubation temperature and hatching success of Morelet's crocodile (*Crocodylus moreletii*) in an urban lake of Southeastern Mexico. *Journal of Thermal Biology* **49**: 66–73.

Lorion, C.M. & Kennedy, B.P. (2009a). Relationships between deforestation, riparian forest buffers and benthic macroinvertebrates in Neotropical headwater streams. *Freshwater Biology* **54**: 165–180.

Lorion, C.M. & Kennedy, B.P. (2009b). Riparian forest buffers mitigate the effects of deforestation on fish assemblages in tropical headwater streams. *Ecological Applications* **19**: 468–479.

Lowe, W.H. (2012). Climate change is linked to long-term decline in a stream salamander. *Biological Conservation* **145**: 48–53.

Lowe-McConnell, R.H. (1993). Fish faunas of the African Great Lakes: origins, diversity, and vulnerability. *Conservation Biology* **7**: 634–643.

Lucentini, L., Puletti, M.E., Ricciolini, C., Gigliarelli, L., Fontaneto, D., Lanfaloni, L., Bilò, F., Natali, M. & Panara, F. (2011). Molecular and phenotypic evidence of a new species of genus *Esox* (Esocidae, Esociformes, Actinopterygii): the southern pike, *Esox flaviae*. *PLoS ONE* **6**: e25218. https://doi.org/10.1371/journal.pone.0025218

Lucifora, L.O., Balboni, L., Scarabotti, P.A., Alonso, F.A., Sabadin, D.E., Solari, A., Vargas, F., Barbini, S.A., Mabragaña, E. & Díaz de Astarloa, J.M. (2017). Decline or stability of obligate freshwater elasmobranchs following high fishing pressure. *Biological Conservation* **210**: 293–298.

Luiza-Andrade, A., Brasil, L.S., Benone, N.L., Shimano, Y., Justino Farias, A.P., Montag, L.F., Dolédec, S. & Juen, L. (2017). Influence of oil palm monoculture on the taxonomic and functional composition of aquatic insect communities in eastern Brazilian Amazonia. *Ecological Indicators* **82**: 478–483.

Lukács B.A., Vojtkó, A.E., Mesterházy, A., Molnár, V.A., Süveges, K., Végvári, Z., Brusa, G. & Cerabolini, B.E.L. (2017). Growth-form and spatiality driving the functional difference of native and alien aquatic plants in Europe. *Ecology and Evolution* **7**: 950–963.

Luke, S.H., Dow, R.A., Butler, S., Vun Khen, C., Aldridge, D.C., Foster, W.A. & Turner, E.C. (2017). The impacts of habitat disturbance on adult and larval dragonflies (Odonata) in rainforest streams in Sabah, Malaysian Borneo. *Freshwater Biology* **62**: 491–506.

Lukhaup, C. (2015). *Cherax* (*Astaconephrops*) *pulcher*, a new species of freshwater crayfish (Crustacea, Decapoda, Parastacidae) from the Kepala Burung (Vogelkop) Peninsula, Irian Jaya (West Papua), Indonesia. *Zookeys* **502**: 1–10.

Lundberg, G., Kottelat, M., Smith, G.R., Stiassny, M.L.J. & Gill, A.C. (2000). So many fishes, so little time: an overview of recent ichthyological discovery in continental waters. *Annals of the Missouri Botanical Gardens* **87**: 26–62.

Lürling, M., Eshetu, F., Faassen, E.J., Kosten, S. & Huszar, V.L.M. (2013). Comparison of cyanobacterial and green algal growth rates at different temperatures. *Freshwater Biology* **58**: 552–559.

Lydeard, C., Cowie, R.H., Ponder, W.F., Bogan, A.E., Bouchet, P., Clark, S.A., Cummings, K.S., Frest, T.J., Gargominy, O., Herbert, D.J., Hershler, R., Perez, K.E., Roth, B., Seddon, M., Strong, E.E. & Thompson, F.E. (2004). The global decline of nonmarine mollusks. *BioScience* **54**: 321–330.

Lymas, M. (2011). *The God Species*. Fourth Estate, London.

Lytle, D.A. & Poff, N.L. (2004). Adaptation to natural flow regimes. *Trends in Ecology & Evolution* **19**: 94–100.

Mccallum, M.L. & Bury, G.W. (2013). Google search patterns suggest declining interest in the environment. *Biodiversity and Conservation* **22**: 1355–1367.

MacDonald, R.J., Boon, S., Byrne, J.M., Robinson, M.D. & Rasmussen, J.B. (2014). Potential future climate effects on mountain hydrology, stream temperature, and native salmonid life history. *Canadian Journal of Fisheries and Aquatic Sciences* **71**: 189–202.

Mace, G.M., Cramer, W., Diaz, S., Faith, D.P., Larigauderie, A., Le Prestre, P., Palmer, M., Perrings, C., Scholes, R.J., Walpole, M., Walther, B.A., Watson, J.E.M. & Mooney, H.A. (2010). Biodiversity targets after 2010. *Current Opinion in Environmental Sustainability* **2**: 3–8.

Mace, G.M., Norris, K. & Fitter, A.H. (2012). Biodiversity and ecosystem services: a multilayered relationship. *Trends in Ecology & Evolution* **27**: 19–26.

MacKinnon, J., Verkuil, Y.I. & Murray, N. (2012). *IUCN Situation Analysis on East and Southeast Asian Intertidal Habitats, with Particular Reference to the Yellow Sea (including the Bohai Sea)*. Occasional Paper of the IUCN Species Survival Commission No. 47, IUCN, Gland and Cambridge.

MacNab, V. & Barber, I. (2012). Some (worms) like it hot: fish parasites grow faster in warmer water, and alter host thermal preferences. *Global Change Biology* **18**: 1540–1548.

Madsen, J., Christensen, T.K., Balsby, T.J.S., Tombre, I.M. (2015). Could have gone wrong: effects of abrupt changes in migratory behaviour on harvest in a waterbird population. *PLoS ONE* **10**: e0135100. https://doi.org/10.1371/journal.pone.0135100

Mandrak, N.E & Cudmore, B. (2010). The fall of native fishes and the rise of non-native fishes in the Great Lakes Basin. *Aquatic Ecosystem Health and Management* **13**: 255–268.

Mantua, N., Tohver, I. & Hamlet, A. (2010). Climate change impacts on streamflow extremes and summertime stream temperature and their possible consequences for freshwater salmon habitat in Washington State. *Climatic Change* **102**: 187–223.

Mantyka-Pringle, C.S., Martin, T.G., Moffatt, D.B., Linke, S. & Rhodes, J.R. (2014). Understanding and predicting the combined effects of climate change and land-use change on freshwater macroinvertebrates and fish. *Journal of Applied Ecology* **51**: 527–581.

Markovic, D., Carrizo, S., Freyhof, J., Cid, N., Lengyel, S., Scholz, M., Kasperdius, H., Darwall, W. (2014). Europe's freshwater biodiversity under climate change: distribution shifts and conservation needs. *Diversity and Distributions* **20**: 1097–1107.

Martel, A., Blooi, M., Adriaensen, C., Van Rooij,P., Beukema, W., Fisher, M.C., Farrer, R.A., Schmidt, B.R., Tobler, U., Goka, K., Lips, K.R., Muletz, C., Zamudio, K.R., Bosch, J., Lötters, S., Wombwell, E., Garner, T.W.J., Cunningham, A.A., Spitzen-van der Sluijs, A., Salvidio, S., Ducatelle, R., Nishikawa, K., Nguyen, T.T., Kolby, J.E., Van Bocxlaer, I., Bossuyt, F. & Pasmans, F. (2014). Recent introduction of a chytrid fungus endangers Western Palearctic salamanders. *Science* **346**: 630–631.

Martin, P., Dorn, N.J., Kawai, T., van der Heiden, C. & Scholtz, G. (2010). The enigmatic marmorkrebs (marbled crayfish) is the parthenogenetic form of *Procambarus fallax* (Hagen, 1870). *Contributions to Zoology* **79**: 107–111.

Martin, R.A. (2005). Conservation of freshwater and euryhaline elasmobranchs: a review. *Journal of the Marine Biological Association of the United Kingdom* **85**: 1049–1073.

Martin Österling, E. & Söderberg, H. (2015). Sea-trout habitat fragmentation affects threatened freshwater pearl mussel. *Biological Conservation* **186**: 197–203.

Matsusaki, S.S. & Kadoya, T. (2015). Trends and stability of inland fishery resources in Japanese lakes: introduction of exotic piscivores as a driver. *Ecological Applications* **25**: 1420–1432.

Matthews, J.H., Wickel, B.A.J. & Freeman, S. (2011). Converging currents in climate-relevant conservation: water, infrastructure, and institutions. *PLoS Biol* **9**: e1001159. https://doi.org/10.1371/journal.pbio.1001159

Matthews, T.G., Lester, R.E., Cummings, C.R. & Lautenschlager, A.D. (2015). Limitations to the feasibility of using hypolimnetic releases to create refuges for riverine species in response to stream warming. *Environmental Science & Policy* **54**: 331–339.

Matthews, W.J. & Marsh-Matthews, E. (2007). Extirpation of red shiners in direct tributaries of Lake Texoma (Oklahoma-Texas): a cautionary case history from a fragmented river-reservoir system. *Transactions of the American Fisheries Society* **136**: 1041–1062.

McCaffery, R.M. & Maxell, B. A. (2010). Decreased winter severity increases viability of a montane frog population. *Proceedings of the National Academy of Sciences of the United States of America* **107**: 8644–8649.

McCarthy, D.P., Donald, P.F., Scharlemann, J.P., Buchanan, G.M., Balmford, A., Green, J.M., Bennun, L.A., Burgess, N.D., Fishpool, L.D., Garnett, S.T., Leonard, D.L., Maloney, R.F., Morling, P., Schaefer, H.M., Symes, A., Wiedenfeld, D.A. & Butchart, S.H. (2012). Financial costs of meeting global biodiversity conservation targets: current spending and unmet needs. *Science* **338**: 946–949.

McCarthy, M. (2015). *The Moth Snowstorm*. John Murray, London.

McCully, P. (2001). *Silenced Rivers: The Ecology and Politics of Large Dams* (Enlarged and Updated Edition). Zed Books, London and York.

McDonald, D.B., Parchman, T.L., Bower, M.R., Hubert, W.A. & Rahel, F.J. (2008). An introduced and a native vertebrate hybridize to form a genetic bridge to a second native species. *Proceedings of the National Academy of Sciences* **105**: 10837–10842.

McDonnell, T.C., Sloat, M.R., Sullivan, T.J., Dolloff, C.A., Hessburg, P.F., Povak, N.A., Jackson, W.A. & Sams, C. (2015). Downstream warming and headwater acidity may diminish coldwater habitat in southern Appalachian mountain streams. *PLoS ONE* **10**: e0134757. https://doi.org/10.1371/journal.pone.0134757

McDowall, R.M. (2006). Crying wolf, crying foul, or crying shame: alien salmonids and a biodiversity crisis in the southern cool-temperate galaxioid fishes? *Reviews in Fish Biology and Fisheries* **16**: 233–422.

McIntosh, A., McHugh, P. & Budy, P. (2012). Salmo trutta L. A Handbook of Global Freshwater Invasive Species (R.A. Francis, ed.), Earthscan, Oxford: pp. 285–296.

McIntyre, P.B., Jones, L.E., Flecker, A.S. & Vanni, M.J. (2007). Fish extinctions alter nutrient recycling in tropical freshwaters. *Proceedings of the National Academy of Sciences of the United States of America* **104**: 4461–4466.

McIntyre, P.B., Reidy Liermann, C.A. & Revenga, C. (2016). Linking freshwater fishery management to global food security and biodiversity conservation. *Proceedings of the National Academy of Sciences of the United States of America* **113**: 12880–12885.

McKinstry, M.C., Caffrey, P. & Anderson, S.H. (2001). The importance of beaver to wetland habitats and waterfowl in Wyoming. *Journal of the American Water Resources Association* **37**: 1571–1577.

McKaye, K.R., Ryan, J.D., Stauffer, J.R., Lopez Perez, L.J., Vega G.L. & van den Berghe, E.P. (1995). African tilapia in Lake Nicaragua. *BioScience* **45**: 406–411.

McShane, T.O., Hirsch, P.D., Tran Chi, T., Songorwa, A.N., Kinzig, A., Monteferri, B., Mutekanga, D., Hoang Van, T., Dammert, J.L., Pulgar-Vidal, M., Welch-Devine, M., Brosius, J.P., Coppolillo, P., O'Connor, S. (2011). Hard choices: making tradeoffs between biodiversity conservation and human well-being. *Biological Conservation* **144**: 966–972.

Meerhoff, M., Teixeira-de Mello, F., Kruk, C., Alonso, C., González-Bergonzoni, I., Pacheco, J.P., Lacerot, G., Arim, M., Beklioğlu, M., Brucet, S., Goyenola, G., Iglesias, C., Mazzeo, N. & Kosten, S. (2012). Environmental warming in shallow lakes: a review of effects on community structure as evidenced from space-for-time substitution approaches. *Advances in Ecological Research* **46**: 259–350.

Mehner, T., Benndorf, J., Kasprzak, P. & Koschel, R. (2002). Biomanipulation of lake ecosystems: successful applications and expanding complexity in the underlying science. *Freshwater Biology* **47**: 2453–2465.

Mei, Z., Huang, S., Hao, Y., Turvey, S., Gong, W. & Wang, D. (2012). Accelerating population decline of Yangtze finless porpoise, *Neophocaena asiaeorientalis asiaeorientalis*. *Biological Conservation* **153**: 192–200.

Mei, Z., Zhang, X., Huang, S.L., Zhao, X., Hao, Y., Zhang, L., Qian, Z., Zheng, J., Wang, K. & Wang, D. (2014). The Yangtze finless porpoise: on an accelerating path to extinction? *Biological Conservation* **172**: 117–123.

Meis, S., Thackeray, S.J. & Jones, I.D. (2009). Effects of recent climate change on phytoplankton phenology in a temperate lake. *Freshwater Biology* **54**: 1888–1898.

Melero, Y., Palazón, S. & Lambin, X. (2014). Invasive crayfish reduce food limitation of alien American mink and increase their resilience to control. *Oecologia* **174**: 427–434.

Melis, T.S., Korman, J. & Kennedy, T.A. (2012). Abiotic and biotic responses of the Colorado River to controlled floods at Glen Canyon Dam, Arizona, USA. *River Research and Applications* **28**: 764–776.

Meybeck, M. (2003). Global analysis of river systems: from Earth system controls to Anthropocene syndromes. *Philosophical Transactions of the Royal Society B: Biological Sciences* **358**: 1935–1955.

Meyer, B.S., Matchiner, M. & Salzburger, W. (2013). A tribal level phylogeny of Lake Tanganyika cichlid fishes based on a genomic multi-marker approach. *Molecular Phylogenetics and Evolution* **83**: 56–71.

Micklin, P.P, Aladin, N.V. & Plotnikov, I. (2014). *The Aral Sea: The Devastation and Partial Rehabilitation of a Great Lake*. Springer-Verlag, Berlin.

Miettinen, J., Shi, C. & Liew, S.C. (2012). Two decades of destruction in Southeast Asia's peat swamp forests. *Frontiers in Ecology and the Environment* **10**: 124–128.

Miettinen, J., Shi, C. & Liew, S.C. (2016). Land cover distribution in the peatlands of Peninsular Malaysia, Sumatra and Borneo in 2015 with changes since 1990. *Global Ecology and Conservation* **6**: 67–78.

Miettinen, J., Hooijer, A., Vernimmen, R., Liew, S.C. & Page, S.E. (2017). From carbon sink to carbon source: extensive peat oxidation in insular Southeast Asia since 1990. *Environmental Research Letters* **12**: 024014.

Millennium Ecosystems Assessment (MEA) (2005). *Ecosystems and Human Well-Being: Biodiversity Synthesis*. World Resources Institute, Washington, DC.

Miller, G. (2010). In Central California, coho salmon are on the brink. *Science* **327**: 512–513.

Mills, E.L., Strayer, D.L., Scheuerell, M.D. & Carlton, J.T. (1996). Exotic species in the Hudson River Basin: a history of invasions and introductions. *Estuaries* **19**: 814–823.

Milly, P.C.D., Betancourt, J., Falkenmark, M., Hirsch, R.M., Kundzewicz, Z.W., Lettenmaier, D.P. & Stouffer, R. (2008). Stationarity is dead: whither water management? *Science* **319**: 573–574.

Minckley, W.L. (1995). Translocation as a tool for conserving imperiled fishes: experiences in the western US. *Biological Conservation* **72:** 297–309.

Mkumbo, O.C. & Marshall, B.E. (2015). The Nile perch fishery of Lake Victoria: current status and management challenges. *Fisheries Management and Ecology* **22**: 56–63.

Moe, S.J., De Schamphelaere, K., Clements, W.H., Sorensen, M.T., Van den Brink, P.J. & Liess, M. (2013). Combined and interactive effects of global climate change and toxicants on populations and communities. *Environmental Toxicology and Chemistry* **32**: 49–61.

Moilanen, A., Letherwick, J. & Edith, J. (2008). A method for spatial freshwater conservation prioritization. *Freshwater Biology* **53**: 577–592.

Montgomery, D.R. (2003). *King of Fish: The Thousand-Year Run of Salmon.* Westview Press, Cambridge, MA.

Moore, M.V., Hampton, S.E., Izmest'eva, L., Silow, E.A., Peshkova, E.V. & Pavlov, B.K. (2009). Climate change and the world's 'Sacred Sea' – Lake Baikal, Siberia. *BioScience* **59**: 405–417.

Moorkens, E.A. & Killeen, I.J. (2014). Assessing near-bed velocity in a recruiting population of the endangered freshwater pearl mussel (*Margaritifera margaritifera*) in Ireland. *Aquatic Conservation: Marine and Freshwater Ecosystems* **24**: 853–862.

Morgan I., McDonald, D.G. & Wood, C.M. (2001). The cost of living for freshwater fish in a warmer, more polluted world. *Global Change Biology* **7**: 345–355.

Morton, B. (1975). The colonization of Hong Kong's raw water supply system by *Limnoperna fortunei* (Dunker) (Bivalvia: Mytilacea) from China. *Malacological Review* **8**: 91–105.

Moss, B. (2010). Climate change, nutrient pollution and the bargain of Dr Faustus. *Freshwater Biology* **55** (Suppl. 1): 175–187.

Moss, B., Kosten, S., Meerhof, M., Battarbee, R., Jeppesen, E., Mazzeo, N., Havens, K., Lacerot, G., Liu, Z. & De Meester, L. (2011). Allied attack: climate change and eutrophication. *Inland Waters* **1**: 101–105.

Moyle, P.B., Li, H.W. & Barton, B.A. (1986). The Frankenstein effect: impact of introduced fishes on native fishes in North America. In *Fish Culture in Fisheries Management* (R.H. Stroud, ed.), American Fisheries Society, Bethesda, MD: pp. 415–426.

MRC (2002). *Annual Report 2001.* Meong River Commission, Phnom Penh.

MRC (2017). *The Council Study. Key Messages from the Study on Sustainable Management and Development of the Mekong River Basin, including Impact of Mainstream Hydropower Projects.* Mekong River Commission, Vientiane. mrcmekong.org/

assets/Publications/Council-Study/Council-study-Reports-discipline/CS-Key-Messages-long-v9.pdf

MRCS (2011a). *Proposed Xayaburi Dam Project – Mekong River. Prior Consultation Project Review Report.* Mekong River Commission Secretariat, Vientiane. mrcmekong.org/assets/Publications/Reports/PC-Proj-Review-Report-Xaiya buri-24-3-11.pdf

MRCS (2011b). *Mekong River Commission Strategic Plan 2011–2015.* Mekong River Commission Secretariat, Vientiane. www.mrcmekong.org/assets/Publications/strategies-workprog/Stratigic-Plan-2011-2015-council-approved25012011-final-.pdf

Mugue, N. (2010). *Pseudoscaphirhynchus kaufmanni. The IUCN Red List of Threatened Species 2010*: e.T18601A8498207. http://dx.doi.org/10.2305/IUCN.UK.2010-1.RLTS.T18601A8498207.en

Muhlfeld, C.C., Kalinowski, S.T., McMahon, T.E., Taper, M.L., Painter, S., Leary, R.F. & Allendorf, F.W. (2009). Hybridization rapidly reduces fitness of a native trout in the wild. *Biology Letters* **5**: 328–331.

Müller, W.E.G., Belikov, S.I., Kaluzhnaya, O.V., Perović-Ottstadt, S., Fattorusso, E., Ushijima, H., Krasko, A. & Schröder, H.C. (2007). Cold stress defense in the freshwater sponge *Lubomirskia baicalensis. FEBS Journal* **274**: 23–36.

Munsch, S.H., Cordell, J.R. & Toft, J.D. (2017). Effects of shoreline armouring and overwater structures on coastal and estuarine fish: opportunities for habitat improvement. *Journal of Applied Ecology* **54**: 1373–1384.

Myers, N., Mittermeier, R.A., Mittermeier, C.G., da Fonseca, G.A.B. & Kent, J. (2000). Biodiversity hotspots for conservation priorities. *Nature* **403**: 853–858.

Naeem, S. & Li, S. (1997). Biodiversity enhances ecosystem reliability. *Nature* **390**: 507–509.

Nahlik, A.M. & Fennessy, M.S. (2016). Carbon storage in US wetlands. *Nature Communications* **7**: 13835. https://doi.org/10.1038/ncomms13835

Naiman, R.J. & Dudgeon, D. (2011). Global alteration of freshwaters and influences on human and environmental well-being. *Ecological Research* **26**: 865–873.

Naiman, R.J., Johnston, C.A. & Lelley, J.C. (1988). Alteration of North American streams by beaver. *BioScience* **38**: 753–762.

Naithani, J., Plisnier, P. & Deleersnijder, E. (2011). Possible effects of global climate change on the ecosystem of Lake Tanganyika. *Hydrobiologia* **671**: 147–163.

Nakano, D. & Strayer, D.L. (2014). Biofouling animals in fresh water: biology, impacts, and ecosystem engineering. *Frontiers in Ecology and Environment* **12**: 167–175.

Nalepa, T.F., Fanslow D.L. & Lang, G.A. (2009). Transformation of the offshore benthic community in Lake Michigan: recent shift from the native amphipod *Diporeia* spp. to the invasive mussel *Dreissena rostriformis bugensis. Freshwater Biology* **54**: 466–479.

NatureServe (2013). *Thaleichthys pacificus. The IUCN Red List of Threatened Species 2013:* e.T202415A18236183. http://dx.doi.org/10.2305/IUCN.UK.2013-1.RLTS.T202415A18236183.en

Naylor, R.M. (1996). Invasions in agriculture: assessing the cost of the golden apple snail in Asia. *Ambio* **25**: 443–448.

Nel, J.L., Reyers, B., Roux, D.J. & Cowling, R.M. (2009). Expanding protected areas beyond their terrestrial comfort zone: identifying spatial options for river conservation. *Biological Conservation* **142**: 1605–1616.

Neuwald, J.L. & Valenzuela, N. (2011). The lesser known challenge of climate change: thermal variance and sex-reversal in vertebrates with temperature-dependent sex determination. *PLoS One* **6**: e18117. https://doi.org/10.1371/journal.pone.0018117

Nico, L. & Fuller, P. (2014a). *Hypophthalmichthys molitrix*. USGS Nonindigenous Aquatic Species Database, Gainesville, FL. http://nas.er.usgs.gov/queries/FactSheet.aspx?speciesID=549

Nico, L. & Fuller, P. (2014b). *Hypophthalmichthys nobilis*. USGS Nonindigenous Aquatic Species Database, Gainesville, FL. http://nas.er.usgs.gov/queries/FactSheet.aspx?speciesID=551

Nico, L.G. & Neilson, M.E. (2014). *Mylopharyngodon piceus*. USGS Nonindigenous Aquatic Species Database, Gainesville, FL. http://nas.er.usgs.gov/queries/factsheet.aspx?SpeciesID=573

Nico, L., Maynard, E., Schofield, P.J., Cannister, M., Larson, J., Fusaro, A. & Neilson, M. (2014a). *Cyprinus carpio*. USGS Nonindigenous Aquatic Species Database, Gainesville, FL. http://nas.er.usgs.gov/queries/FactSheet.aspx?SpeciesID=4

Nico, L.G., Fuller, P.L., Schofield, P.J. & Neilson, M.E. (2014b). *Ctenopharyngodon idella*. USGS Nonindigenous Aquatic Species Database, Gainesville, FL. http://nas.er.usgs.gov/queries/FactSheet.aspx?speciesID=514

Nico, L.G., Schofield, P.J. & Neilson, M. (2014c). *Oreochromis niloticus*. USGS Nonindigenous Aquatic Species Database, Gainesville, FL. https://nas.er.usgs.gov/queries/FactSheet.aspx?speciesID=468

Nilsson, C., Reidy, C.A., Dynesius, M. & Revenga, C. (2005). Fragmentation and flow regulation of the world's large river systems. *Science* **308**: 405–408.

Nilsson, C., Brown, R.L., Jansson, R. & Merritt, D.M. (2010). The role of hydrochory in structuring riparian and wetland vegetation. *Biological Reviews* **85**: 837–858.

Ng, P.K.L., Tay, J.B. & Lim, K.K.P. (1994). Diversity and conservation of black-water fishes in Peninsular Malaysia, particularly in the North Selangor peat swamp forest. *Hydrobiologia* **285**: 203–218.

Ng, T.H., Tan, S.K., Wong, W.H., Meier, R., Chan, S.-Y., Tan, H.H. & Yeo, D.C.G. (2016). Molluscs for sale: assessment of freshwater gastropods and bivalves in the ornamental pet trade. *PLoS ONE* **11**: e0161130. https://doi.org/10.1371/journal.pone.0161130

Ngor, P.B., McCann, K.S., Grenouillet, G., So, N., McMeans, B.C., Fraser, E. & Lek, S. (2018). Evidence of indiscriminate fishing effects in one of the world's largest inland fisheries. *Scientific Reports* **8**: 8947. https://doi.org/10.1038/s41598-018-27340-1

North, A.C., Hodgson, D.J., Price, S.J. & Griffiths, A.G.F. (2015). Anthropogenic and ecological drivers of amphibian disease (ranavirosis). *PLoS ONE* **10**: e0127037. https://doi.org/10.1371/journal.pone.0127037

Novak, P.A., Garcia, E.A., Pusey, B.J. & Douglas, M.M. (2017). Importance of the natural flow regime to an amphidromous shrimp: a case study. *Marine and Freshwater Research* **68**: 909–921.

Nyqvist, D., Nilsson, P.A., Alenäs, I., Elghagen, J., Hebrand, M., Karlsson, S., Kläpp, S. & Calles, O. (2017). Upstream and downstream passage of migrating adult Atlantic salmon: remedial measures improve passage performance at a hydropower dam. *Ecological Engineering* **102**: 331–343.

Obenour, D.R., Gronewold, A.D, Stow, C.A. & Scavia, D. (2014). Using a Bayesian hierarchical model to improve Lake Erie cyanobacteria bloom forecasts. *Water Resources Research* **50**: 7847–7860.

Oberdorff, T., Jézéquel, C., Campero, M., Carvajal-Vallejos, F., Cornu, J.F., Dias, M.S., Duponchelle, S., Maldonado-Ocampo, J.A., Ortega, H., Renno, J.F. & Tedesco, P.A. (2015). Opinion Paper: how vulnerable are Amazonian freshwater fishes to ongoing climate change? *Journal of Applied Ichthyology* **31**: 4–9.

O'Connor, J.E., Duda, J.J., Grant, G.E. (2015). 1000 dams down and counting. *Science* **448**: 496–498.

Olden, J.D. & Naiman, R.J. (2010). Incorporating thermal regimes into environmental flow assessments: modifying dam operations to restore freshwater ecosystem integrity. *Freshwater Biology* **55:** 86–107.

Olden, J.D., Hogan, Z.S. & Vander Zanden, J.V. (2007). Small fish, big fish, red fish, blue fish: size-biased extinction risk of the world's freshwater and marine fishes. *Global Ecology and Biogeography* **16**: 694–701.

Olden, J.D., Kennard, M., Lawler, J.J. & Poff, N.L. (2011). Challenges and opportunities in implementing managed relocation for conservation of freshwater species. *Conservation Biology* **25**: 40–47.

Olden, J.D., Konrad, C.P., Melis, T.S., Kennard, M.J., Freeman, M.C., Mims, M.C., Bray, E.N., Gido, K.B., Hemphill, N.P., Lytle, D.A., McMullen, L.E., Pyron, M., Robinson, C.T., Schmidt, J.C. & Williams, J.G. (2014). Are large-scale flow experiments informing the science and management of freshwater ecosystems? *Frontiers in Ecology and the Environment* **12**: 176–185.

O'Leary, J.K., Micheli, F., Airoldi, L., Boch, C., De Leo, G., Elahi, R., Ferretti, F., Graham, N.A.J., Litvin, S.Y., Low, N.H., Lummis, S., Nickols, K.J. & Wong, J. (2017). The resilience of marine ecosystems to climatic disturbances. *BioScience* **67**: 208–220.

Olivier, T.J., Handy, K.Q. & Bauer, R.T. (2013). Effects of river control structures on the juvenile migration of *Macrobrachium ohione. Freshwater Biology* **58**: 1603–1613.

Olrik, K., Cronberg, G. & Annadotter, H. (2013). Lake phytoplankton responses to global climate changes. In *Climatic Change and Global Warming of Inland Waters: Impacts and Mitigation for Ecosystems and Societies* (C. R. Goldman, M. Kumagai & R. D. Robarts, eds), John Wiley & Sons Ltd, Chichester: pp. 173–199.

Ondračková, M., Dávidová, M., Blažek, R., Gelnar, M., Jurajda, P. (2009). The interaction between an introduced fish host and local parasite fauna: *Neogobius kessleri* in the middle Danube River. *Parasitology Research* **105**: 201–208.

O'Reilly, C.M., Alin, S.R., Plisnier, P.-D., Cohen, A.S. & McKee, B.A. (2003). Climate change decreases aquatic ecosystem productivity of Lake Tanganyika, Africa. *Nature* **424**: 766–768.

O'Reilly, C.M., Sharma, S., Gray, D.K., Hampton, S.E., Read, J.S., Rowley, R.J., Schneider, P., Lenters, J.D., McIntyre, P.B., Kraemer, B.M., Weyhenmeyer, G.A., Straile, D., Dong, B., Adrian, R., Allan, M.G., Anneville, O., Arvola, L.,

Austin, J., Bailey, J.L., Baron, J.S., Brookes, J.D., Eyto, E., Dokulil, M.T., Hamilton, D.P., Havens, K., Hetherington, A.L., Higgins, S.N., Hook, S., Izmest'eva, L.R., Joehnk, K.D., Kangur, K., Kasprzak, P., Kumagai, M., Kuusisto, E., Leshkevich, G., Livingstone, D.M., MacIntyre, S., May, L., Melack, J.M., Mueller-Navarra, D.C., Naumenko, M., Noges, P., Noges, T., North, R.P., Plisnier, P., Rigosi, A., Rimmer, A., Rogora, M., Rudstam, L.G., Rusak, J.A., Salmaso, N., Samal, N.R., Schindler, D.E., Schladow, S.G., Schmid, M., Schmidt, S.R., Silow, E., Soylu, M.E., Teubner, K., Verburg, P., Voutilainen, A., Watkinson, A., Williamson, C.E. & Zhang, G. (2015). Rapid and highly variable warming of lake surface waters around the globe. *Geophysical Research Letters* **42**: 10773–10781. https://doi.org/10.1002/2015GL066235

Oreskes, N. & Conway, E.M. (2010). *Merchants of Doubt: How a Handful of Scientists Obscured the Truth on Issues from Tobacco Smoke to Global Warming*. Bloomsbury Press, New York.

Ormerod, S.J. (2014). Rebalancing the philosophy of river conservation. *Aquatic Conservation: Marine and Freshwater Ecosystems* **24**: 147–152.

Ormerod, S.J., Dobson, M., Hildrew, A.G. & Townsend, C.R. (2010). Multiple stressors in freshwater ecosystems. *Freshwater Biology* **55** (Suppl. 1): 1–4.

Orr, S., Pittock, J., Chapagain, A. & Dumaresq, D. (2012). Dams on the Mekong River: lost fish protein and the implications for land and water resources. *Global Environmental Change* **22**: 925–932.

Ospina-Álvarez, N. & Piferrer, F. (2008). Temperature-dependent sex determination in fish revisited: prevalence, a single sex ratio response pattern, and possible effects of climate change. *PLoS ONE* **3:** e2837. https://doi.org/10.1371/journal.pone.0002837

Owen, C.T., McGregor, M.A., Cobbs, G.A., Alexander, J.E. (2011). Muskrat predation on a diverse unionid mussel community: impacts of prey species composition size and shape. *Freshwater Biology* **56**: 554–564.

Pacini, N. & Harper, D.M. (2008). Aquatic, semi-aquatic and riparian vertebrates. *Tropical Stream Ecology* (D. Dudgeon, ed.), Academic Press, Amsterdam: pp. 147–197.

Page, L.M. & Hall, R.H. (2006). Identification of the sailfin catfishes (Teleostei: Loricariidae) in Southeast Asia. *Raffles' Bulletin of Zoology* **54**: 445–452.

Page, S.E., Rieley, J.O. & Banks, C.J. (2011). Global and regional importance of the tropical peatland carbon pool. *Global Change Biology* **17**: 798–818.

Pagnucco, K.S., Maynard, G.A., Fera, S.A., Yan, N.D., Nalepa, T.F. & Ricciardi, A. (2015). The future of species invasions in the Great Lakes-St. Lawrence River basin. *Journal of Great Lakes Research* **41** (Suppl. 1): 96–107.

Palmer, M.A., Filoso, S. & Fanelli, R.M. (2014). From ecosystems to ecosystem services: stream restoration as ecological engineering. *Ecological Engineering* **65**: 62–70.

Palmer, M.A., Menninger, H.L. & Bernhardt, E. (2010). River restoration, habitat heterogeneity and biodiversity: a failure of theory or practice? *Freshwater Biology*, **55** (Suppl. 1): 205–222.

Palmer, M.A., Reidy Liermann, C.A., Nilsson, C., Flörke, M., Alcamo, J., Lake, P.S. & Bond, N. (2008). Climate change and the world's river basins: anticipating management options. *Frontiers in Ecology and the Environment* **6**: 81–89.

Palmer, M.A., Bernhardt, E.S., Allan, J.D., Lake, P.S., Alexander, G., Brooks, S., Carr, J., Clayton, S., Dahm, C., Follstad Shah, J., Galat, D.L., Loss, S.G., Goodwin, P., Hart, D.D., Hassett, B., Jenkinson, R., Kondolf, G.M., Lave, R., Meyer, J.L., O'Donnell, T.K., Pagano, L. & Sudduth, E. (2005). Standards for ecologically successful river restoration. *Journal of Applied Ecology* **42**: 208–217.

Pandit, S.N., Maitland, B.M., Pandit, L.K., Poesch, M.S. & Enders, E.C. (2017). Climate change risks, extinction debt, and conservation implications for a threatened freshwater fish: carmine shiner (*Notropis percobromus*). *Science of the Total Environment* **598**: 1–11.

Paolucci, E.M., Thuesen, E., Cataldo, D. & Boltovskoy, D. (2010).Veligers of an introduced bivalve (*Limnoperna fortunei*) are a new food resource that enhances growth of larval fish in the Paraná River (South America). *Freshwater Biology* **55**: 1831–1844.

Parmesan, C. (2006). Ecological and evolutionary responses to recent climate change. *Annual Review of Ecology, Evolution, and Systematics* **37**: 637–669.

Parry, L., Barlow, J. & Pereira, H. (2014). Wildlife harvest and consumption in Amazonia's urbanized wilderness. *Conservation Letters* **7**: 565–574.

Pastor, A.V., Ludwig, F., Biemans, H., Hoff, H. & Kabat, P. (2014). Accounting for environmental flow requirements in global water assessments. *Hydrology and Earth System Sciences* **18**: 5041–5059.

Patoka, J., Kopecký, O., Vrabec, V. & Kalous, L. (2017). Aquarium molluscs as a case study in risk assessment of incidental freshwater fauna. *Biological Invasions* **19**: 2039–2046.

Patten, D.T., Harpman, D.A., Voita, M.I. & Randle, T.J. (2001). A managed flood on the Colorado River: background, objectives, design, and implementation. *Ecological Applications* **11**: 635–643.

Peart, C.R., Bills, R., Wilkinson, M. & Day, J.J. (2014). Nocturnal claroteine catfishes reveal dual colonisation but a single radiation in Lake Tanganyika. *Molecular Phylogenetics and Evolution* **73**: 119–128.

Pease, A.A. & Paukert, C.P. (2014). Potential impacts of climate change on growth and prey consumption of stream-dwelling smallmouth bass in the central United States. *Ecology of Freshwater Fish* **23**: 336–346.

Pelicice, F.M., Pompeu, P.S. & Agostinho, A.A. (2015). Large reservoirs as ecological barriers to downstream movements of Neotropical migratory fish. *Fish and Fisheries* **16**: 697–715.

Pennisi, E. (2014). The river masters: hippos are the nutrient kingpins of Africa's waterways. *Science* **346**: 802–805.

Pereira, H.M., Ferrier, S., Walters, M., Geller, G.N., Jongman, R.H.G., Scholes, R.J., Bruford, M.W., Brummitt, N., Butchart, S.H.M., Cardoso, A.C., Coops, N.C., Dulloo, E., Faith, D.P., Freyhof, J., Gregory, R.D., Heip, C., Hoft, R., Hurtt, G., Jetz, W., Karp, D.S., McGeoch, M.A., Obura, D., Onoda, Y., Pettorelli, N., Reyers, B., Sayre, R., Scharlemann, J.P.W., Stuart, S.N., Turak, E., Walpole, M. & Wegmann, M. (2013). Essential biodiversity variables. *Science* **339**: 277–278.

Petersen, T.A., Brum, S.M., Rossoni, F., Silveira, G.F.V. & Castello, L. (2016). Recovery of arapaima populations by community-based management in floodplains of the Purus River, Amazon. *Journal of Fish Biology* **89**: 241–248.

Petrovic, M., Ginebreda, A., Muñoz, I. & Barceló, D. (2013). The river drugstore: the threats of emerging pollutants to river conservation. In *River Conservation: Challenges and Opportunities* (S. Sabatier & A. Elosegi, eds), Fundación BBVA, Bilbao: pp. 105–126.

Peverell, S.C. (2005). Distribution of sawfishes (Pristidae) in the Queensland Gulf of Carpentaria, Australia, with notes on their ecology. *Environmental Biology of Fishes* **73**: 391–402.

Pezaro, N., Doody, J.S. & Thompson, M.B. (2016). The ecology and evolution of temperature-dependent reaction norms for sex determination in reptiles: a mechanistic conceptual model. *Biological Reviews* **92**: 1348–1364.

Phillimore, A.B., Hadfield, J.D., Jones, O.R. & Smithers, R.J. (2010). Differences in spawning date between populations of common frog reveal local adaptation. *Proceedings of the National Academy of Sciences of the United States of America* **107**: 8292–8297.

Pinder, A.C., Raghavan, R. & Britton, J.R. (2015). The legendary hump-backed mahseer *Tor* sp. of India's River Cauvery: an endemic fish swimming towards extinction. *Endangered Species Research* **28**: 11–15.

Poff, N.L. (2018). Beyond the natural flow regime? Broadening the hydro-ecological foundation to meet environmental flows challenges in a non-stationary world. *Freshwater Biology* **63**: 1011–1128.

Poff, N.L. & Zimmerman, J.K. (2010). Ecological responses to altered flow regimes: a literature review to inform the science and management of environmental flows. *Freshwater Biology* **55**: 194–205.

Poff, N.L., Richter, B.D., Arthington, A.H., Bunn, S.E., Naiman, R.J., Kendy, E., Acreman, M., Apse, C., Bledsoe, B.P., Freeman, M., Henriksen, J., Jacobson, R.B., Kennen, J.G., Merritt, D.M., O'Keeffe, J.H., Olden, J.D., Rogers, K., Tharme, R.E. & Warner, A. (2010). The ecological limits of hydrologic alteration (ELOHA): a new framework for developing regional environmental flow standards. *Freshwater Biology* **55**: 147–170.

Pollock, M.M., Heim, M. & Werner, D. (2003). Hydrologic and geomorphic effects of beaver dams and their influence on fishes. *American Fisheries Society Symposium* **31**: 1–21.

Polvi, L.E. & Wohl, E. (2012). The beaver meadow complex revisited – the role of beavers in post-glacial floodplain development. *Earth Surfaces Processes and Landforms* **37**: 332–346.

Poly, W.J. (2003). Design and evaluation of a translocation strategy for the fringed darter (*Etheostoma crossopterum*) in Illinois. *Biological Conservation* **113**: 13–22.

Pompeu, P.S., Agostinho, A.A. & Pelicice, F.M. (2012). Existing and future challenges: the concept of successful fish passage in South America. *River Research and Applications* **28**: 504–512

Pond, G.J. (2012). Biodiversity loss in Appalachian headwater streams (Kentucky, USA): Plecoptera and Trichoptera communities. *Hydrobiologia* **679**: 97–117.

Pool, T.K., Olden, J.D., Whittier, J.B. & Paukert, C.P. (2010). Environmental drivers of fish functional diversity and composition in the Lower Colorado River Basin. *Canadian Journal of Fisheries and Aquatic Sciences* **67**: 1791–1807.

Poole, G.C. & Berman, C.H. (2001). An ecological perspective on in-stream temperature: natural heat dynamics and mechanisms of human-caused thermal degradation. *Environmental Management* **27**: 787–802.

Poulsen, A.F., Ouch, P., Sintavong, V., Ubolratana, S. & Nguyen, T.T. (2002a). Fish migrations of the lower Mekong River Basin: implications for development, planning and environmental management. *MRC Technical Paper No. 8*. Mekong River Commission, Phnom Penh. www.mrcmekong.org/assets/Pub lications/technical/tech-No8-fish-migration-of-LMB.pdf

Poulsen, A., Poeu, O., Vivarong, S., Suntornratana, U. & Thanh Tung, N. (2002b). Deep pools as dry season fish habitats in the Mekong River Basin. *MRC Technical Paper No. 4*. Mekong River Commission, Phnom Penh. www .mrcmekong.org/assets/Publications/technical/tech-No4-Deep-pools-as-dry-season-fish-habitats.pdf

Price, S.J., Garner, T.W.J., Nichols, R.A., Balloux, F., Ayres, C., Mora-Cabello de Alba, A. & Bosch, J. (2014). Collapse of amphibian communities due to an introduced *Ranavirus*. *Current Biology* **24**: 2586–2591.

Price, S.J., Garner, T.W.J., Cunningham, A.A., Langton, T.E.S. & Nichols, R.A. (2016). Reconstructing the emergence of a lethal infectious disease of wildlife supports a key role for spread through translocations by humans. *Proceedings of the Royal Society B* **283**: 20160952. https://doi.org/10.1098/rspb.2016.0952

Pringle, R.M. (2005). The origins of the Nile perch in Lake Victoria. *BioScience* **55**: 780–787.

Pyke, G.H. (2008). Plague minnow or mosquitofish? A review of the biology and impacts of introduced *Gambusia* species. *Annual Review of Ecology, Evolution, and Systematics* **39**: 171–191.

Quinlan, E., Gibbins, C., Malcolm, I., Batalla, R., Vericat, D. & Hastie, L. (2014). A review of the physical habitat requirements and research priorities needed to underpin conservation of the endangered freshwater pearl mussel *Margaritifera margaritifera*. *Aquatic Conservation: Marine and Freshwater Ecosystems* **26**: 107–124.

Raghavan, R., Dahanukar, N., Tlusty, M.F., Rhyne, A.L., Krishna Kumar, K., Molur, S. & Rosser, A.M. (2013). Uncovering an obscure trade: threatened freshwater fishes and the aquarium pet markets. *Biological Conservation* **164**: 158–169.

Rahel, F.J. (2002). Homogenization of freshwater faunas. *Annual Review of Ecology and Systematics* **33**: 291–315.

Rahel, F. (2013). Intentional fragmentation as a management strategy in aquatic systems. *BioScience* **63**: 362–372.

Rahel, F.J. & Olden, J.D. (2008). Assessing the effects of climate change on aquatic invasive species. *Conservation Biology* **22**: 521–533.

Ramachandran, R., Kumar, A., Gopi Sundar, K.S. & Bhalla, R.S. (2017). Hunting or habitat? Drivers of waterbird abundance and community structure in agricultural wetlands of southern India. *Ambio* **46**: 613–620.

Rees, H.C., Maddison, B.C., Middleditch, D.J., Patmore, J.R. & Gough, K.C. (2014). The detection of aquatic animal species using environmental DNA – a review of eDNA as a survey tool in ecology. *Journal of Applied Ecology* **51**: 1450–1459.

Refsnider, J.M., Bodensteiner, B.L., Reneker, J.L. & Janzen, F.J. (2013). Nest depth may not compensate for sex ratio skews caused by climate change in turtles. *Animal Conservation* **16**: 481–490.

Refsnider, J.M., Milne-Zelman, C., Warner, D.A. & Janzen, F.J. (2014). Population sex ratios under differing local climates in a reptile with environmental sex determination. *Evolutionary Ecology* **28**: 977–989.

Reichard, M., Ondračková, M., Przybylski, M., Liu, H. & Smith, C. (2006). The costs and benefits in an unusual symbiosis: experimental evidence that bitterling fish (*Rhodeus sericeus*) are parasites of unionid mussels in Europe. *Journal of Evolutionary Biology* **19**: 788–796.

Reichard, M., Vrtílek, M., Douda, K. & Smith, C. (2012). An invasive species reverses the roles in a host–parasite relationship between bitterling fish and unionid mussels. *Biology Letters* **8**: 601–604.

Reid, A.J., Carlson, A.K., Creed, I.F., Eliason, E.J., Gell, P.A., Johnson, P.T., Kidd, K.A., MacCormack, T.J., Olden, J.D., Ormerod, S.J., Smol, J.P., Taylor, W.W., Tockner, K., Vermaire, J.C., Dudgeon, D. & Cooke, S.J. (2019). Emerging threats and persistent conservation challenges for freshwater biodiversity. *Biological Reviews* **94**: 849–873.

Reid, G.M. (1990). Captive breeding for the conservation of cichlid fishes. *Journal of Fish Biology* **37**: 157–166.

Reidy Liermann, C.R., Nilsson, C., Robertson, J. & Ng, R.Y. (2012). Implications of dam obstruction for global freshwater fish diversity. *Bioscience* **62**: 539–548.

Renöfält, B.M., Lejon, A.G., Jonsson, M. & Nilsson, C. (2013). Long-term taxon-specific responses of macroinvertebrates to dam removal in a mid-sized Swedish stream. *River Research and Applications* **29**: 1082–1089.

Revenga, C., Campbell, I., Abell, R., de Villiers, P. & Bryer, M. (2005). Prospects for monitoring freshwater ecosystems towards the 2010 targets. *Philosophical Transactions of the Royal Society B: Biological Sciences* **360**: 397–413.

Rhymer, J.M. & Simberloff, D. (1996). Extinction by hybridization and introgression. *Annual Review of Ecology and Systematics* **27**: 83–109.

Ricciardi, A. & Rasmussen, J.B. (1998). Predicting the identity and impact of future biological invaders: a priority for aquatic resource management. *Canadian Journal of Fisheries and Aquatic Sciences* **55**: 1759–1765.

Ricciardi, A. & Simberloff, D. (2009). Assisted colonization is not a viable conservation strategy. *Trends in Ecology & Evolution* **24**: 248–253.

Ricciardi, A., Neves, R.J. & Rasmussen, J.B. (1998). Impending extinctions of North American freshwater mussels (Unionoida) following the zebra mussel (*Dreissena polymorpha*) invasion. *Journal of Animal Ecology* **67**: 613–619.

Richardson, J.S., Taylor, E., Schluter, D., Pearson, M. & Hatfield, T. (2010). Do riparian zones qualify as critical habitat for endangered freshwater fishes? *Canadian Journal of Fisheries and Aquatic Sciences* **67**: 1197–1204.

Richter, B.D. & Thomas, G.A. (2007). Restoring environmental flows by modifying dam operations. *Ecology and Society* **12:** 2. www.ecologyandsociety.org/vol12/iss1/art12/

Richter, B.D., Warner, A.T., Meyer, J.L. & Lutz, K. (2006). A collaborative and adaptive process for developing environmental flow recommendations. *River Research & Applications* **22**: 297–318.

Richter, B.D., Postel, S., Revenga, C., Scudder, T., Lehner, B., Churchill, A. & Chow, M. (2010). Lost in development's shadow: the downstream human consequences of dams. *Water Alternatives* **3**: 14–42.

Ripple, W.J. & Beschta, R.L. (2012). Trophic cascades in Yellowstone: the first 15 years after wolf reintroduction. *Biological Conservation* **145**: 205–213.

Roberts, T.R. (2001a). Killing the Mekong: China's fluvicidal hydropower-cum-navigation development scheme. *Natural History Bulletin of the Siam Society* **49**: 143–159.

Roberts, T.R. (2001b). On the river of no returns: Thailand's Pak Mun Dam and its fish ladder. *Natural History Bulletin of the Siam Society* **49**: 189–230.

Rockström, J., Steffen, W., Noone, K., Persson, Å., Chapin III, F.S., Lambin, E., Lenton, T.M., Scheffer, M., Folke, C., Schellnhuber, H., Nykvist, B., De Wit, C.A., Hughes, T., van der Leeuw, S., Rodhe, H., Sörlin, S., Snyder, P.K., Costanza, R., Svedin, U., Falkenmark, M., Karlberg, L., Corell, R.W., Fabry, V.J., Hansen, J., Walker, B.H., Liverman, D., Richardson, K., Crutzen, C. & Foley, J. (2009). A safe operating space for humanity. *Nature* **461**: 472–475.

Rödder, D., Kielgast, J., Bielby, J., Schmidtlein, S., Bosch, J., Garner, T.W.J., Veith, M., Walker, S., Fisher, M.C. & Lötters, S. (2009). Global amphibian extinction risk assessment for the panzootic chytrid fungus. *Diversity* **1**: 52–66.

Rogers, K.H. (2008). Limnology and the post-normal imperative: an African perspective. *Verhandlungen Internationale Vereinigung für theoretische und angewandte Limnologie* **30**: 171–185.

Rohr, J.R. & Palmer, B.D. (2013). Climate change, multiple stressors, and the decline of ectotherms. *Conservation Biology* **27**: 741–751.

Rohr, J.R., Sesterhenn, T.M. & Stieha, C. (2011). Will climate change reduce the effects of a pesticide on amphibians? Partitioning effects on exposure and susceptibility to contaminants. *Global Change Biology* **17**: 657–666.

Roni, P., Beechie, T., Pess, G. & Hanson, K. (2015). Wood placement in river restoration: fact, fiction, and future direction. *Canadian Journal of Fisheries and Aquatic Sciences* **72**: 466–478.

Röpke, C.P., Amadio, S., Zuanon, J., Ferreira, E.J.G., de Deus, C.P., Pires, T.H.S. & Winemiller, K.O. (2017). Simultaneous abrupt shifts in hydrology and fish assemblage structure in a floodplain lake in the central Amazon. *Scientific Reports* **7**: 40170. https://doi.org/10.1038/srep40170

Rosenthal, S.K., Stevens, S.S. & Lodge, D.M. (2006). Whole-lake effects of invasive crayfish (*Orconectes* spp.) and the potential for restoration. *Canadian Journal of Fisheries and Aquatic Sciences* **63**: 1276–1285.

Rosell, F., Boszér, O., Collen, P. & Parker, H. (2005). Ecological impact of beavers *Castor fiber* and *Castor canadensis* and their ability to modify ecosystems. *Mammal Review* **35**: 248–276.

Roux, D.J., Nel, J.L., Ashton, P.J., Deacon, A.R., de Moor, F.C., Hardwick, D., Hill, L., Kleynhans, C.J., Maree, G.A., Moolman, J. & Scholes, R.J. (2008). Designing protected areas to conserve riverine biodiversity: lessons from a hypothetical redesign of the Kruger National Park. *Biological Conservation* **141**: 100–117.

Rowley, J.J.L., Emmett, D.A. & Voen, S. (2008). Harvest, trade and conservation of the Asian arowana *Scleropages formosus* in Cambodia. *Aquatic Conservation: Marine and Freshwater Ecosystems* **18**: 1255–1266.

Rowley, J., Brown, R., Bain, R., Kusrini, M., Inger, R., Stuart, B., Wogan, G., Thy, N., Chan-ard, T., Trung, C.T., Diesmos, A., Iskandar, D.T., Lau, M., Ming, L.T., Makchai, S., Truong, N.Q. & Phimmachak, S. (2010). Impending conservation crisis for Southeast Asian amphibians. *Biology Letters* **6**: 336–338.

Rühland, K.M., Paterson, A.M. & Smol, J.P. (2015). Lake diatom responses to warming: reviewing the evidence. *Journal of Paleolimnology* **54**: 1–35.

Russell, J.C., Sataruddin, N.S. & Heard, A.D. (2014). Over-invasion by functionally equivalent invasive species. *Ecology* **95**: 2268–2276.

Ryan, M.E., Johnson, J.R. & Fitzpatrick, B.M. (2009). Invasive hybrid tiger salamander genotypes impact native amphibians. *Proceedings of the National Academy of Sciences of the United States of America* **109**: 11166–11171.

Rybczynski, N. (2007). Castorid phylogenetics: implications for the evolution of swimming and tree-exploitation in beavers. *Journal of Mammalian Evolution* **14**: 1–35.

Rybczynski, N. (2008). Woodcutting behavior in beavers (Castoridae, Rodentia): estimating ecological performance in a modern and a fossil taxon. *Paleobiology* **34**: 389–402.

Saeed, F., Hagemann, S. & Jacob, D. (2009). Impact of irrigation on the South Asian summer monsoon. *Geophysical Research Letters* **36**: L20711. https://doi.org/10.1029/2009GL040625

Salles F.F., Gattolliat, J.-L., Angeli, K.B., De-Souza, M.R., Gonçalves, I.C., Nessimian, J.L., Sartori, M. (2014). Discovery of an alien species of mayfly in South America (Ephemeroptera). *Zookeys* **399**: 1–16.

Sandel, M.J. (2012). What isn't for sale? *The Atlantic April* **2012**: 1–20.

Sandin, L., Schmidt-Kloiber, A, Svenning, J., Jeppesen, E. & Friberg, N. (2014). A trait-based approach to assess climate change sensitivity of freshwater invertebrates across Swedish ecoregions. *Current Zoology* **60**: 221–232.

Santulli, G., Palazón, S., Melero, Y., Gosàlbez, J. & Lambin, X. (2014). Multi-season occupancy analysis reveals large scale competitive exclusion of the critically endangered European mink by the invasive non-native American mink in Spain. *Biological Conservation* **176**: 26–29.

Satgé, F., Espinoza, R., Zolá, R., Roig, H., Timouk, F., Molina, J., Garnier, J., Calmant, S., Seyler, F. & Bonnet, M.-P. (2017). Role of climate variability and human activity on Poopó Lake droughts between 1990 and 2015 assessed using remote sensing data. *Remote Sensing* **9**: 218. https://doi.org/10.3390/rs9030218

Schaller, G.B. (1993). *The Last Panda*. University of Chicago Press, Chicago.

Scheffers, B.R., De Meester, L., Bridge, T.C.L., Hoffman, A.A., Pandolfini, J.M., Corlett, R.T., Butchart, S.H.M., Pearce-Kelly, P.P., Kovacs, K.M., Dudgeon, D., Pacifici, M., Rondinini, C., Foden, W.B., Martin, T.G., Mora, C., Bickford, D. & Watson, J.E.M. (2016). The broad footprint of climate change from genes to biomes to people. *Science* **354** (6313): aaf7671–aaf7671. https://doi.org/10.1126/science.aaf7671

Schlaepfer, M.A., Sax, D.F. & Olden, J.D. (2011). The potential conservation value of non-native species. *Conservation Biology* **25**: 428–437.

Schlaepfer, P.M., Hoover, C. & Dodd, C.K. (2005). Challenges in evaluating the impact of the trade in amphibians and reptiles on wild populations. *BioScience* **55**: 256–264.

Schloesser, D.W., Nalepa, T.F. & Mackie, G.L. (1996). Zebra mussel infestation of unionid bivalves (Unionidae) in North America. *American Zoologist* **36**: 300–310.

Schmidt, J.C., Webb, R.H., Valdez, R.A., Marzolf, G.R. & Stevens, L.E. (1998). Science and values in river restoration in the Grand Canyon. *BioScience* **48**: 735–747.

Schmidt-Kloiber, A., Bremerich, V., De Wever, A., Jähnig, S.C., Martens, K., Strackbein, J., Tockner, K. & Hering, D. (2019). The Freshwater Information Platform: a global online network providing data, tools and resources for science and policy support. *Hydrobiologia* 838: 1–11.

Scholes, R.J., Mace, G.M., Turner, W., Geller, G.N., Jurgens, N., Larigauderie, A., Muchoney, D., Walther, B.A. & Mooney, H.A. (2008). Towards a global biodiversity observing system. *Science* **321**: 1044–1045.

Schwanz, L.E., Spencer, R.-J., Bowden, R.M. & Janzen, F.J. (2010). Climate and predation dominate juvenile and adult recruitment in a turtle with temperature-dependent sex determination. *Ecology* **91**: 3016–3026.

Schwartz, M.W., Hellman, J.J. & McLachlan, J.S. (2009). The precautionary principle in managed relocation is misguided advice. *Conservation Biology* **24**: 474.

Seehausen, O., van Alphen, J.J.M. & Witte, F. (1997). Cichlid fish diversity threatened by eutrophication that curbs sexual selection. *Science* **277**: 1808–1811.

Selz, O.M., Pierotti, M.E.R., Maan, M.E., Schmid, C. & Seehausen, O. (2014). Female preference for male color is necessary and sufficient for assortative mating in 2 cichlid sister species. *Behavioural Ecology* **25**: 612–626.

Serra, M.N., Albariño, R. & Villanueva, V.D. (2013). Invasive *Salix fragilis* alters benthic invertebrate communities and litter decomposition in northern Patagonian streams. *Hydrobiologia* **701**: 173–188.

Sheath, R.G. & Vis, M.L. (2013). Biogeography of Freshwater Algae, eLS. John Wiley & Sons, Chichester. https://doi.org/10.1002/9780470015902.a0003279 .pub3

Sheldon, A.L. (2012). Possible climate-induced shift of stoneflies in a southern Appalachian catchment. *Freshwater Science* **31**: 765–774.

Sheridan, J.A. & Bickford, D. (2011). Shrinking body size as an ecological response to climate change. *Nature Climate Change* **1**: 401–406.

Shiel, D., Ladd, B., Silva, L.C.R., Laffan, S.W. & Van Heist, M. (2016). How are soil carbon and tropical biodiversity related? *Environmental Conservation* **43**: 231–241.

Shiklomanov, I. (1993). World freshwater resources. In *Water in Crisis: A Guide to the World's Freshwater Resources* (P.H. Gleick, ed.), Oxford University Press, Oxford: pp. 13–24.

Shine, R. (2010). The ecological impact of invasive cane toads (*Bufo marinus*) in Australia. *Quarterly Review of Biology* **85**: 235–291.

Shirley, M.H. (2014). *Mecistops cataphractus*. *The IUCN Red List of Threatened Species 2014*: e.T5660A3044332. http://dx.doi.org/10.2305/IUCN.UK.2014-1.RLTS .T5660A3044332.en

Shu, F.-Y., Wang, H.-J., Pan, B.-Z., Liu, X.-Q. & Wang, H.-Z. (2009). Assessment of species status of Mollusca in the mid-lower Yangtze lakes. *Acta Hydrobiologica Sinica* **33**: 1051–1058.

Shumilova, O., Tockner, K., Thieme, M., Koska, A. & Zarfl, C. (2018). Global water transfer megaprojects: a solution for the water-food-energy nexus? *Frontiers in Environmental Science* **6**: 150. https://doi.org/10.3389/fenvs.2018.00150

Sierp, M., Qin, J. & Recknagel, F. (2008). Biomanipulation: a review of biological control measures in eutrophic waters and the potential for Murray cod *Maccullochella peelii peelii* to promote water quality in temperate Australia. *Reviews in Fish Biology and Fisheries* **19**: 143–165.

Sigsgaard, E.E., Carl, H., Møller, P.R. & Thomsen, P.F. (2015). Monitoring the near-extinct European weather loach in Denmark based on environmental DNA from water samples. *Biological Conservation* **183**: 46–52.

Silva, A.T., Lucas, M.C., Castro-Santos, T., Katopodis, C., Baumgartner, L.J., Thiem, J.D., Aarestrup, K., Pompeu, P.S., O'Brien, G.C., Braun, D.C., Burnett, N.J., Zhu, D.Z., Fjeldstad, H., Forseth, T., Rajaratnam, N.G., Williams, J.G. & Cooke, S.J. (2018). The future of fish passage science, engineering, and practice. *Fish and Fisheries* **19**: 340–362.

Silvertown, J. (2015). Have ecosystem services been oversold? *Trends in Ecology & Evolution* **30**: 641–648.

Simberloff, D. & Rejmánek, M. (2011). *Encyclopedia of Biological Invasions*. University of California Press, Berkeley.

Simberloff, D. & Von Holle, B. (1999). Positive interactions of nonindigenous species: invasional meltdown? *Biological Invasions* **1**: 21–32.

Simmons, M., Tucker, A., Chadderton, W.L., Jerde, C.L. & Mahon, A.R. (2016). Active and passive environmental DNA surveillance of aquatic invasive species. *Canadian Journal of Fisheries and Aquatic Sciences* **73**: 76–83.

Simoncini, M., Cruz, F.B., Larriera, A. & Piña, C.I. (2014). Effects of climatic conditions on sex ratios in nests of broad-snouted caiman. *Journal of Zoology* **293**: 243–251.

Skerratt, L.F., Berger, L., Speare, R., Cashins, S., McDonald, K.R., Phillott, A.D., Hines, H.B. & Kenyon, N. (2007). Spread of chytridiomycosis has caused the rapid global decline and extinction of frogs. *EcoHealth* **4**: 125–134.

Skyrienė, G. & Paulauskas, A. (2012). Distribution of invasive muskrats (*Ondatra zibethicus*) and impact on ecosystem. *Ekologija* **58**: 357–367.

Smart, A.C., Harper, D.M., Malaisse, F., Schmitz, S., Coley, S. & Gouder de Beauregard, A.-C. (2002). Feeding of the exotic Louisiana red swamp crayfish, *Procambarus clarkii* (Crustacea, Decapoda), in an African tropical lake: Lake Naivasha, Kenya. *Hydrobiologia* **488**: 129–142.

Smirnov, V.V., Smirnova-Zalumi, N.S. & Sukhanova, L.V. (2012). Fishery management of omul (*Coregonus autumnalis migratorius*) as part of the conservation of ichthyofaunal diversity in Lake Baikal. *Polish Journal of Natural Sciences* **27**: 203–214.

Smith, S.H. (1968). Species succession and fishery exploitation in the Great Lakes. *Journal of the Fisheries Research Board of Canada* **25**: 667–693.

Snyder, C.D., Hitt, N.P. & Young, J.A. (2015). Accounting for groundwater in stream fish thermal habitat responses to climate change. *Ecological Applications* **25**: 1397–1419.

So, N., Souvanny, P., Ly, V., Theerawat, S., Son, N.H., Malasri, K., Peng Bun, N., Sovanara, K., Degen, P. & Starr, P. (2015). Lower Mekong fisheries estimated to be worth around $17 billion a year. *Catch and Culture* **21** (3): 4–7.

Sokolow, S.H., Huttinger, E., Jouanard, N., Hsieh, M.H., Lafferty, K.D., Kuris, A.M., Riveau, G., Senghor, S., Thiam, C., N'Diaye, A., Sarr Faye, D. & De Leo, G.A. (2015). Reduced transmission of human schistosomiasis after restoration of a native river prawn that preys on the snail intermediate host. *Proceedings of the National Academy of the United States of America* **112**: 9650–9655.

Sokolow, S.H., Jones, I.J., Jocque, M., La, D., Cords, O., Knight, A., Lund, A., Wood, C.L., Lafferty, K.D., Hoover, C.M., Collender, P.A., Remais, J.V., Lopez-Carr, D., Fisk, J., Kuris, A.M. & De Leo, G.A. (2017). Nearly 400 million people are at higher risk of schistosomiasis because dams block the migration of snail-eating river prawns. *Philosophical Transactions of the Royal Society B: Biological Sciences* **372**: 2100127. https://doi.org/10.1098/rstb.2016.0127

Song, Z., Zhang, J., Jiang, X. Wang, C. & Xie, Z. (2013). Population structure of an endemic gastropod in Chinese plateau lakes: evidence for population decline. *Freshwater Science* **32**: 450–461.

Sopha, L., Pengby, N., Nam, S. & Hortle, K.G. (2010). With fewer fry from upstream, Tonle Sap *dai* fishery catch declines in latest season. *Catch and Culture* **16** (1): 8–9.

Soulé, M.E. (1985). What is conservation biology? *BioScience* **35**: 727–734.

Sousa, R., Antunes, C. & Guilhermino, L. (2008). Ecology of the invasive Asian clam *Corbicula fluminea* (Müller, 1774) in aquatic ecosystems: an overview. *Annales de Limnologie* **44**: 85–94.

Speed, R., Li, Y., Tickner, D., Huang, H., Naiman, R., Cao, J., Lei, G., Yu, L., Sayers, P., Zhao, Z. & Yu, W. (2016). *River Restoration: A Strategic Approach to Planning and Management*. UNESCO, Paris. http://unesdoc.unesco.org/images/0024/002456/245644e.pdf

Spencer, C.N., McClelland, B.R. & Stanford, J.A. (1991). Shrimp stocking, salmon collapse and eagle displacement. *BioScience* **41**: 14–21.

Spooner, D.E., Xenopoulos, M.A., Schneider, G. & Woolnough, D.A. (2011). Coextirpation of host–affiliate relationships in rivers: the role of climate change, water withdrawal, and host-specificity. *Global Change Biology* **17**: 1720–1732.

Stamm, C.K., Räsänen, K., Burdon, F.J., Altermatt, F., Jokela, J., Joss, A., Ackermann, M. & Eggen, R.I.L. (2016). Unravelling the impacts of micro-pollutants in aquatic ecosystems: cross-disciplinary studies at the interface of large-scale ecology. *Advances in Ecological Research* **24**: 183–223.

Stauffer, J.R., Jr., Madsen, H., McKaye, K., Konings, A., Bloch, P., Ferreri, C.P., Likongwe, J. & Macaula, P. (2006). Schistosomiasis in Lake Malawi: relationship of fish and intermediate host density to prevalence of human infection. *EcoHealth* **3**: 22–27.

Steffen, W., Richardson, K., Rockström, J., Cornell, S.E., Fetzer, I., Bennett, E.M., Biggs, R., Carpenter, S.R., de Vries, W., de Wit, C.A., Folke, C., Gerten, D., Heinke, J., Mace, G.M., Persson, L.M., Ramanathan, V., Reyers, B. & Sörlin, S. (2015). Planetary boundaries: guiding human development on a changing planet. *Science* **347**: 1259855. https://doi.org/10.1126/science.1259855

Steffensen, S.M., Thiem, J.D., Stamplecoskie, K.M., Binder, T.R., Hatry, C., Langlois-Anderson, N. & Cooke, S.J. (2013). Biological effectiveness of an inexpensive nature-like fishway for passage of warmwater fish in a small Ontario stream. *Ecology of Freshwater Fish* **22**: 374–383.

Stevens, L.E., Buck, K.A., Brown, B.T., Kline, N.C. (1997a). Dam and geomorphological influences on Colorado River waterbird distribution, Grand Canyon, Arizona, USA. *Regulated Rivers: Research & Management* **13**: 151–169.

Stevens, L.E., Shannon, J.P. & Blinn, D.W. (1997b). Colorado River benthic ecology in Grand Canyon, Arizona, USA: dam, tributary and geomorphological influences. *Regulated Rivers: Research and Management* **13**: 129–149.

Stevenson, R.J. & Esselman, P.C. (2013). Nutrient pollution: a problem with solutions. In *River Conservation: Challenges and Opportunities* (S. Sabatier & A. Elosegi, eds), Fundación BBVA, Bilbao: pp. 77–103.

Stewart-Koster, B., Bunn, S.E., MacKay, S.J., Poff, N.L., Naiman, R.J. & Lake, P.S. (2010). The use of Bayesian networks to guide investments in flow and catchment restoration for impaired river ecosystems. *Freshwater Biology* **55**: 243–260.

Stiassny, M.L.J. (1999). The medium is the message: freshwater biodiversity in peril. In *The Living Planet in Crisis: Biodiversity Science and Policy* (J. Cracraft & F.T. Grifo, eds), Columbia University Press, New York: pp. 53–71.

Still, D.A., Dickens, C., Breen C.M., Mamder, M. & Booth, A. (2010). Balancing resource protection and development in a highly regulated river: the role of conjunctive use. *Water SA* **36**: 371–378.

Stone, R. (2008). A new great lake – or dead sea? *Science* **320**: 1002–1005.

Strayer, D.L. (1999). Effects of alien species of freshwater mollusks in North America. *Journal of the North American Benthological Society* **18**: 74–98.

Strayer, D.L. (2006). Challenges for freshwater invertebrate conservation. *Journal of the North American Benthological Society* **25**: 271–287.

Strayer, D.L. (2010). Alien species in fresh waters: ecological effects, interactions with other stressors, and prospects for the future. *Freshwater Biology* **55** (Suppl. 1): 152–174.

Strayer, D.L. (2012). Eight questions about invasions and ecosystem functioning. *Ecology Letters* **15**: 1199–1210.

Strayer, D.L. (2017). What are freshwater mussels worth? *Freshwater Mollusk Biology and Conservation* **20**: 103–113.

Strayer, D.L. & Dudgeon, D. (2010). Freshwater biodiversity conservation: recent progress and future challenges. *Journal of the North American Benthological Society* **29**: 344–358.

Strayer, D.L., Downing, J.A., Haag, W.R., King, T.L., Layer, J.B., Newton, T.J. & Nichols, S.J. (2004). Changing perspectives on pearly mussels, North America's most imperiled animals. *BioScience* **54**: 429–439.

Strayer, D.L., Hattala, K.A. & Kahnle, A.W. (2004). Effects of an invasive bivalve (*Dreissena polymorpha*) on fish populations in the Hudson River estuary. *Canadian Journal of Fisheries and Aquatic Sciences* **61**: 924–941.

Strayer, D.L., Eviner, V.T., Jeschke, J.M. & Pace, M.L. (2006). Understanding the long-term effects of species invasions. *Trends in Ecology & Evolution* **21**: 645–651.

Sukenik, A., Hadas, O., Kaplan, A. & Quesada, A. (2012). Invasion of Nostocales (cyanobacteria) to subtropical and temperate freshwater lakes – physiological, regional, and global driving forces. *Frontiers in Microbiology* **3**: 86. https://doi .org/10.3389/fmicb.2012.00086

Sullivan, S.M.P. & Manning, D.W.P. (2017). Seasonally distinct taxonomic and functional shifts in macroinvertebrate communities following dam removal. *PeerJ* **5**: e3189. https://doi.org/10.7717/peerj.3189

Sullivan, S.M.P., Manning, D.W.P. & Davis, R.P. (2018). Do the ecological impacts of dam removal extend across the aquatic–terrestrial boundary? *Ecosphere* **9**: e02180. https://doi.org/10.1002/ecs2.2180

Sunday, J.M., Bates, A.E., Kearney, M.R., Colwell, R.K., Dulvy, N.K., Longino, J.K. & Huey, R.B. (2014). Thermal-safety margins and the necessity of thermoregulatory behavior across latitude and elevation. *Proceedings of the National Academy of Sciences of the United States of America* **111**: 5610–5615.

Sung, Y.H. & Fong, J.J. (2018). Assessing consumer trends and illegal activity by monitoring the online wildlife trade. *Biological Conservation* **227**: 219–225.

Swaisgood, R.R. & Sheppard, J.K. (2010). The culture of conservation biologists: show me the hope! *BioScience* **60**: 626–630.

Sweeney, B.W. & Blaine, J. (2016). River conservation, restoration, and preservation: rewarding private behavior to enhance the commons. *Freshwater Science* **35**: 755–763.

Sweeney, B.W. & Newbold, J.D. (2014). Streamside forest buffer width needed to protect stream water quality, habitat, and organisms: a literature review. *Journal of the American Water Resources Association* **50**: 560–584.

Sylvester, F., Boltovskoy, D. & Cataldo, D. (2007). The invasive bivalve *Limnoperna fortunei* enhances benthic invertebrate densities in South American floodplain rivers. *Hydrobiologia* **589**: 15–27.

Syvitski, J.P.M. & Kettner, A. (2011). Sediment flux and the Anthropocene. *Philosophical Transactions of the Royal Society A: Mathematical, Physical and Engineering Sciences* **369**: 957–975.

Syvitski, J.P.M. & Milliman, J.D. (2007). Geology, geography and humans battle for dominance over the delivery of sediment to the coastal ocean. *Journal of Geology* **115**: 1–19.

Syvitski, J.P.M., Kettner, A.J., Overeem, I., Hutton, E.W.H., Hannon, M.T., Brakenridge, G.R., Day, J., Vörösmarty, C., Saito, Y., Giosan, L. & Nicholls, R.J. (2009). Sinking deltas due to human activities. *Nature Geoscience* **2**: 681–686.

Tan, X., Li, X., Lek, S., Li, Y., Wang, C., Li, J. & Luo, J. (2010). Annual dynamics of the abundance of fish larvae and its relationship with hydrological variation in the Pearl River. *Environmental Biology of Fishes* **88**: 217–225.

Taylor, B.W., Flecker, A.S. & Hall, R.O. (2006). Loss of a harvested fish species disrupts carbon flow in a diverse tropical river. *Science* **313**: 833–836.

Taylor, C.A., Schuster, G.A., Cooper, J.E., DiStefano, R.J., Eversole, A.G., Hamr, H., Hobbs, H.H., Robinson, H.W., Skelton, C.E. & Thomas, R.F. (2007). A reassessment of the conservation status of crayfishes of the United States and Canada after 10+ years of increased awareness. *Fisheries* **32**: 372–389.

Tedesco, P.A., Oberdorff, T., Cornu, J.-F., Beauchard, O., Brosse, S., Dürr, H.H., Grenouillet, G., Leprieur, F., Tisseuil, C., Zaiss, R. & Hugueny, B. (2013).

A scenario for impacts of water availability loss due to climate change on riverine fish extinction rates. *Journal of Applied Ecology* **50**: 1105–1115.

Tedesco, P.A., Beauchard, O. Bigorne, R., Blanchet, S., Buisson, L., Conti, L., Cornu, J., Dias, M.S., Grenouillet, G., Hugueny, B., Jézéquel, C., Leprieur, F., Brosse, S. & Oberdorff, T. (2017). A global database on freshwater fish species occurrence in drainage basins. *Scientific Data* **4**: 170141. https://doi.org/10 .1038/sdata.2017.141

Tharme, R.E. (2003). A global perspective on environmental flow assessment: emerging trends in the development and application of environmental flow methodologies for rivers. *River Research and Applications* **19**: 397–441.

Thewlis, R.M., Timmins, R.J., Evans, T.D. & Duckworth, J.W. (1998). The conservation status of birds in Laos: a review of key species. *Bird Conservation International* **8** (Suppl. 1): 1–159.

Thieme, M.L., Abell, R., Stiassny, M.L.J., Lehner, B., Skelton, P., Teugels, G., Dinerstein, E., Kamden Toham, A., Burgess, B. & Olson, D. (2005*). Freshwater Ecoregions of Africa and Madagascar. A Conservation Assessment*. Island Press, Washington, DC.

Thieme, M.L., Lehner, B., Abell, R. & Matthews, R. (2010). Exposure of Africa's freshwater biodiversity to a changing climate. *Conservation Letters* **3**: 324–331.

Thomas, C.D. (2011). Translocation of species, climate change, and the end of trying to recreate past ecological communities. *Trends in Ecology & Evolution* **26**: 216–221.

Thompson, R.M. & Lake, P.S. (2010). Reconciling theory and practice: the role of stream ecology. *River Research and Applications* **26**: 5–14.

Thompson, R.M., King, A.J., Kingsford, R.M., Mac Nally, R. & Poff, N.L. (2018). Legacies, lags and long-term trends: effective flow restoration in a changed and changing world. *Freshwater Biology* **63**: 986–995.

Thorburn, D.C. & Morgan, D.L. (2005). Threatened fishes of the world: *Pristis microdon* Latham 1794 (Pristidae). *Environmental Biology of Fishes* **72**: 465–466.

Thorson, T.B. (1976a). Observations on the reproduction of the sawfish. *Pristis perotteti*, in Lake Nicaragua, with recommendations for its conservation. In *Investigations of the Ichthyofauna of Nicaraguan Lakes* (T.B. Thorson, ed.), School of Life Sciences, University of Nebraska-Lincoln, Lincoln, NE: pp. 641–650.

Thorson, T.B. (1976b). The status of the Lake Nicaragua shark: an updated appraisal. In *Investigations of the Ichthyofauna of Nicaraguan Lakes* (T.B. Thorson, ed.), School of Life Sciences, University of Nebraska-Lincoln, Lincoln, NE: pp. 561–574.

Thresher, R.E., Allman, J. & Stremick-Thompson, L. (2018). Impacts of an invasive virus (CyHV-3) on established invasive populations of common carp (*Cyprinus carpio*) in North America. *Biological Invasions* **20**: 1703–1718.

Tiegs, S.D., Levi, P.S., Rüegg, J., Chaloner, D.T., Tank, J.L. & Lamberti, G.A. (2011). Ecological effects of live salmon exceed those of carcasses during an annual spawning migration. *Ecosystems* **14**: 598–614.

Tierney, J.E., Mayes, M., Meyer, N., Johnson, C., Swarzenski, P.W., Cohen, A.S. & Russell, J.M. (2010). Late-twentieth-century warming in Lake Tanganyika unprecedented since AD 500. *Nature Geoscience* **3**: 422–425.

Tilman, D., Isbell, F. & Cowles, J.M. (2014). Biodiversity and ecosystem functioning. *Annual Review of Ecology, Evolution, and Systematics* **45**: 471–493.

Timoshkin, O.A., Bondarenko, N.A., Volkova, Y.A., Tomberg, I.V., Vishnyakov, V.S. & Malnk, V.V. (2015). Mass development of green filamentous algae of the genera *Spirogyra* and *Stigeoclonium* (Chlorophyta) in the littoral zone of the southern part of Lake Baikal. *Hydrobiological Journal* **51**: 13–23.

Todd, B.D., Scott, D.E., Pechmann, J.H. & Gibbons, J.W. (2010). Climate change correlates with rapid delays and advancements in reproductive timing in an amphibian community. *Proceedings of the Royal Society of London B: Biological Sciences* **278**: 2191–2197.

Tomazzoni, A.C., Pedó, E. & Hartz, S.M. (2005). Feeding associations between capybaras *Hydrochoerus hydrochaeris* (Linnaeus) (Mammalia, Hydrochaeridae) and birds in the Lami Biological Reserve, Porto Alegre, Rio Grande do Sul, Brazil. *Revista Brasileira de Zoologia* **22**: 712–716.

Tonkin, J.D., Merritt, D.M., Olden, J.D., Reynolds, L.V. & Lytle, D.A. (2018). Flow regime alteration degrades ecological networks in riparian ecosystems. *Nature Ecology & Evolution* **2**: 86–93.

Tonkin, J.D., Poff, N.L., Bond, N.R., Horne, A., Merritt, D.M., Reynolds, L.V., Olden, J.D., Ruhl, A., Lytle, D.A. (2019). Prepare river ecosystems for an uncertain future. *Nature* 570: 301–303.

Trolle, D., Nielsen, A., Rolighed, J., Thodsen, H., Andersen, H.E., Karlsson, I.B., Refsgaard, J.C., Olesen, J.E., Bolding, K., Kronvang, B., Søndergaard, M. & Jeppesen, S. (2015). Projecting the future ecological state of lakes in Denmark in a 6 degree warming scenario. *Climate Research* **64**: 55–72.

Tullos, D.D., Collins, M.J., Bellmore, J.R., Bountry, J.A., Connolly, P.J., Shafroth, P.B. & Wilcox, A.C. (2016). Synthesis of common management concerns associated with dam removal. *Journal of the American Water Resources Association* **52**: 1179–1206.

Turak, E., Dudgeon, D., Harrison, I.J., Freyhof, J., De Wever, A., Revenga, C., Garcia-Moreno, J., Abell, R., Culp, J.M., Lento, J., Mora, B., Hilarides, L. & Flink, S. (2017a). Observations of inland water biodiversity: progress, needs and priorities. In *The GEO Handbook on Biodiversity Observation Networks* (M. Walters & R.J. Scholes, eds), Springer, Cham: pp. 165–186. https://doi.org/ 10.1007/978-3-319-27288-7_7

Turak, E., Harrison, I., Dudgeon, D., Abell, R., Bush, A., Darwall, W., Finlayson, M., Ferrier, S., Freyhof, J., Hermoso, V., Juffe-Bignoli, D., Linke, S., Nel, J., Patricio, H., Pittock, J., Raghavan, R., Revenga, C. & Simaika, J. (2017b). Essential biodiversity variables for measuring change in global freshwater biodiversity. *Biological Conservation* **213**: 272–279.

Turner, G.F., Seehausen, O., Knight, M.E., Allender, C.J. & Robinson, R.L. (2001). How many species of cichlid fishes are there in African lakes? *Molecular Ecology* **10**: 793–806.

Turvey, S.T., Pitman, R.L., Taylor, B.L., Barlow, J., Akamatsu, T., Barrett, L.A., Zhao, X.J., Reeves, R.R., Stewart, B.S., Wang, K.X., Wei, Z., Zhang, X.F., Pesser, L.T., Richlen, M., Brandon, J.R. & Wang, D. (2007). First human-caused extinction of a cetacean species. *Biology Letters* **3**: 537–540.

Turvey, S.T., Barrett, L.A., Hao, Y., Zhang, L., Zhang, X., Wang, X., Hunag, Y., Zhou, K., Hart, T. & Wang, D. (2010). Rapidly shifting baselines in Yangtze fishing communities and local memory of extinct species. *Conservation Biology* **24**: 778–787.

Tweddle, D., Cowx, I.G., Peel, R.A. & Weyl, O.L.F. (2012). Challenges in fisheries management in the Zambezi, one of the great rivers of Africa. *Fisheries Management and Ecology* **22**: 99–111.

Tyus, H.M. & Saunders, J.F. (2000). Nonnative fish control and endangered fish recovery: lessons from the Colorado River. *Fisheries* **25**: 17–24.

UNEP (2016). *Transboundary River Basins: Status and Trends.* United Nations Environment Programme (UNEP), Nairobi.

United States Fish & Wildlife Service (2008). *Sonny Bono Salton Sea National Wildlife Refuge: Wildlife List.* www.fws.gov/saltonsea/pdf/SaltonSeaWildlifeList%2708 .6.pdf

United States Geological Survey (2014). *Marisa cornuarietis.* USGS Nonindigenous Aquatic Species Database, Gainesville, FL. http://nas.er.usgs.gov/queries/ FactSheet.aspx?speciesID=981

Utevsky, S., Zagmajster, M. & Trontelj, P. (2014). *Hirudo medicinalis. The IUCN Red List of Threatened Species 2014*: e.T10190A21415816. http://dx.doi.org/10 .2305/IUCN.UK.2014-1.RLTS.T10190A21415816.en

Vadadi-Fülöp, C., Sipkay, C., Mészáros, G. & Hufnagel, L. (2012). Climate change and freshwater zooplankton: what does it boil down to? *Aquatic Ecology* **46**: 501–519.

Vadeboncoeur, Y., McIntyre, P.B., Vander Zanden, M.J. (2011). Borders of bio-diversity: life at the edge of the world's large lakes. *Bioscience* **61**: 526–537.

Valdez, R.A., Hoffnagle, T.L., McIvor, C.C., McKinney, T. & Leibfried, W.C. (2001). Effects of the test flood on Fishes of the Colorado River in Grand Canyon, Arizona. *Ecological Applications* **11**: 686–700.

Valentini, A., Taberlet, P., Miaud, C., Civade, R., Herder, J., Thomsen, P.F., Bellemain, E., Besnard, A., Coissac, E., Boyer, F. & Gaboriaud, C. (2016). Next-generation monitoring of aquatic biodiversity using environmental DNA metabarcoding. *Molecular Ecology* **25**: 929–942.

Van der Velde, G., Paffen, B.G.P., Van den Brink, F.W.B., Bij de Vaate, A. & Jenner, H.A. (1994). Decline of zebra mussel populations in the Rhine. Competition between two mass invaders (*Dreissena polymorpha* and *Corophium curvispinum*). *Naturwissenschaften* **81**: 32–34.

Van Dijk, P.P. (2000). The status of turtles in Asia. In *Asian Turtle Trade: Proceedings of a Workshop on Conservation and Trade of Freshwater Turtles and Tortoises in Asia* (P.P. Van Dijk, B.I. Stuart & A.G.J. Rhodin, eds), Chelonian Research Monographs No. 2, Chelonian Research Foundation, Lunenberg: pp. 15–23.

van Lookeren Campagne, C. & Begum, S. (2017). *Red Gold and Fishing in the Lake Chad Basin.* Oxfam Briefing Note (February 2017), Oxfam GB, Oxford. www.oxfam.org/sites/www.oxfam.org/files/file_attachments/bn-red-gold-fishing-lake-chad-010217-en.pdf

van Rijssel, J.C. & Witte, F. (2013). Adaptive responses in resurgent Lake Victoria cichlids over the past 30 years. *Evolutionary Ecology* **27**: 253–267.

Van Vliet, M.T.H., Franssen, W.H.P., Yearsley, J.R., Ludwig, F., Haddeland, I., Lettenmaier, D.P. & Kabat, P. (2013). Global river discharge and water temperature under climate change. *Global Environmental Change* **23**: 450–464.

Vanderploeg, H.A., Nalepa, T.F., Jude, D.J., Mills, E.L., Holeck, K.T, Liebig, J.R., Grigorovich, I.A. & Ojaveer, H. (2002). Dispersal and emerging ecological impacts of Ponto-Caspian species in the Laurentian Great Lakes. *Canadian Journal of Fisheries and Aquatic Sciences* **59**: 1209–1228.

Vanham, D., Comero, S., Gawlik, B.M. & Bidoglio, G. (2018). The water footprint of different diets within European sub-national geographical entities. *Nature Sustainability* **1**: 518–525. http://doi.org/10.1038/s41893–018-0133-x

Varis, O. (2014). Curb vast water use in central Asia. *Nature* **514**: 27–29.

Vass, K.K., Tyagi, R.K., Singh, H.P. & Pathak, V. (2010). Ecology, changes in fisheries, and energy estimates in the middle stretch of the River Ganges. *Aquatic Ecosystem Health & Management* **13**: 374–384.

Vaughn, C.C. (2010). Biodiversity losses and ecosystem function in freshwaters: emerging conclusions and research directions. *BioScience* **60**: 25–35.

Vaughn, C.C. (2018). Ecosystem services provided by freshwater mussels. *Hydro-biologia* **810**: 15–27.

Verburg, P. & Hecky, R.E. (2009). The physics of the warming of Lake Tanganyika by climate change. *Limnology & Oceanography* **54**: 2418–2430.

Verburg, P., Hecky, R.E. & Kling, H. (2003). Ecological consequences of a century of warming in Lake Tanganyika. *Science* **301**: 505–507.

Verpoorter, C., Kutser, T., Seekell, D.A. & Tranvik, L.J. (2014). A global inventory of lakes based on high-resolution satellite imagery. *Geophysical Research Letters* **41**: 2014GL060641. https://doi.org/10.1002/2014GL060641

Verschuren, D., Johnson, T.C., Kling, H.J., Edgington, D.N., Leavitt, P.R., Brown, E.T., Talbot, M.R. & Hecky, R.E. (2002). History and timing of human impact on Lake Victoria, East Africa. *Proceedings of the Royal Society Biological Sciences Series B* **269**: 289–294.

Vidthayanon, C. (2013). *Tenualosa thibaudeaui*. The IUCN Red List of Threatened Species 2013: e.T21627A9303248. http://dx.doi.org/10.2305/IUCN.UK.2011-1.RLTS.T21627A9303248.en

Villéger, S., Blanchet, S., Beauchard, O., Oberdorff, T. & Brosse, S. (2011). Homogenization patterns of the world's freshwater fish faunas. *Proceedings of the National Academy of Sciences of the United States of America* **108**: 18003–18008.

Vishwanath, W. (2010). *Schizothorax richardsonii*. The IUCN Red List of Threatened Species 2010: e.T166525A135873256. http://dx.doi.org/10.2305/IUCN.UK.2010-4.RLTS.T166525A6228314.en

Vitule, J.R.S., Umbria, S.C. & Aranha, J.M.R. (2006). Introduction of the African Catfish *Clarias gariepinus* (Burchell, 1822) into Southern Brazil. *Biological Invasions* **8**: 677–681.

Vitule, J.R.S., Freire, C.A. & Simberloff, D. (2009). Introduction of non-native freshwater fish can certainly be bad. *Fish and Fisheries* **10**: 98–108.

Vitule, J.R.S., Freire, C.A., Vazquez, D.P., Nuñez, M.A. & Simberloff, D. (2012). Revisiting the potential conservation value of non-native species. *Conservation Biology* **26**: 1153–1155.

Vogel, G. (2017). Where have all the insects gone? *Science* **356**: 576–579.

Vörösmarty, C., Lettenmaier, D., Lévêque, C., Meybeck, M., Pahl-Wostl, C., Alcamo, J., Cosgrove, W., Grassl, H., Hoff, H., Kabat, P., Lansigan, F.,

Lawford, R. & Naiman, R.J. (2004). Humans transforming the global water system. *EOS, American Geophysical Union Transactions* **85**: 509–514.

Vörösmarty, C.J., McIntyre, P.B., Gessner, M.O., Dudgeon, D., Prusevich, A., Green, P., Glidden, S., Bunn, S.E., Sullivan, C.A., Reidy Liermann, C. & Davies, P.M. (2010). Global threats to human water security and river biodiversity. *Nature* **467**: 555–561.

Vouvoulis, N., Arpon, K.D. & Giakoumis, T. (2017). The EU Water Framework Directive: from great expectations to problems with implementation. *Science of the Total Environment* **575**: 358–366.

Vrtílek, M. & Reichard, M. (2012). An indirect effect of biological invasions: the effect of zebra mussel fouling on parasitisation of unionid mussels by bitterling fish. *Hydrobiologia* **696**: 205–214.

Wagner, M., Scherer, C., Alvarez-Muñoz, D., Brennholt, N., Bourrain, X., Buchinger, S., Fries, E., Grosbois, C., Klasmeier, J., Marti, T., Rodriguez-Mozaz, S., Urbatzka, R., Vethaak, A.D., Winther-Nielsen, M. & Reifferscheid, G. (2014). Microplastics in freshwater ecosystems: what we know and what we need to know. *Environmental Sciences Europe* **26**: 12. https://doi.org/10.1186/s12302-014-0012-7

Wake, D.B. & Vredenburg, V.T. (2008). Are we in the midst of the sixth mass extinction? A view from the world of amphibians. *Proceedings of the National Academy of Sciences of the United States of America* **105** (Suppl. 1): 11466–11473.

Wallace, J.S., Acreman, M.C. & Sullivan, C.A. (2003). The sharing of water between society and ecosystems: from conflict to catchment-based co-management. *Philosophical Transactions of the Royal Society of London Series B: Biological Sciences* **358**: 2011–2026.

Walpole, A.A., Bowman, J., Tozer, D.C. & Badzinski, D.S. (2012). Community-level response to climate change: shifts in anuran calling phenology. *Herpetological Conservation and Biology* **7**: 249–257.

Walsh, J.R., Carpenter, S.R. & Vander Zanden, M.J. (2016). Invasive species triggers a massive loss of ecosystem services through a trophic cascade. *Proceedings of the National Academy of Sciences of the United States of America* **113**: 4081–4085.

Walsh, M.R. & Reznick, D.N. (2008). Interactions between the direct and indirect effects of predators determine life history evolution in a killifish. *Proceedings of the National Academy of Sciences of the United States of America* **105**: 594–599.

Wang, D., Turvey, S.T., Zhao, X. & Mei, Z. (2013). *Neophocaena asiaeorientalis ssp. asiaeorientalis*. The IUCN Red List of Threatened Species 2013: e.T43205774A45893487. http://dx.doi.org/10.2305/IUCN.UK.2013-1.RLTS.T43205774A45893487.en

Wang, H. (2003). Biology, population dynamics, and culture of Reeves shad, *Tenualosa reevesii*. *American Fisheries Society Symposium* **35**: 77–83.

Wang, H., Wu, X., Bi, N., Li, S., Yuan, P., Wang, A., Syvitski, J.P., Saito, Y., Yang, Z. & Nittrouer, J. (2017). Impacts of the dam-orientated water-sediment regulation scheme on the lower reaches and delta of the Yellow River (Huanghe): a review. *Global and Planetary Change* **157**: 93–113.

Wang, S., Yue, P.Q. & Chen, Y.Y. (1998). *China Red Data Book of Endangered Animals. Pisces*. Science Press, Beijing.

Wang, X., Fox, A.D., Cong, P. & Cao, L. (2013). Food constraints explain the restricted distribution of wintering lesser white-fronted geese *Anser erythropus* in China. *Ibis* **155**: 576–592.

Wang, Z., Liu, G.C.S., Burton, G.A. & Leung, K.M.Y. (2019). Thermal extremes can intensify chemical toxicity to freshwater organisms and hence exacerbate their impact to the biological community. *Chemosphere* **224**: 256–264.

Ward, J.M. & Ricciardi, A. (2007). Impacts of *Dreissena* invasions on benthic macroinvertebrate communities: a meta-analysis. *Diversity and Distributions* **13**: 155–165.

Ward, J.M. & Ricciardi, A. (2010). Community-level effects of co-occurring native and exotic ecosystem engineers. *Freshwater Biology* **55**: 1803–1817.

Warkentin, I.G., Bickford, D., Sodhi, N.S. & Bradshaw, C.J.A. (2009). Eating frogs to extinction. *Conservation Biology* **23**: 1056–1059.

Waters, T.F. (1987). *The Superior North Shore. A Natural History of Lake Superior's Northern Lands and Waters*. University of Minnesota Press, Minneapolis.

Weber, C., Åberg, U. , Buijse, A.D., Hughes, F.M., McKie, B.G., Piégay, H., Roni, P., Vollenweider, S. & Haertel-Borer, S. (2018). Goals and principles for programmatic river restoration monitoring and evaluation: collaborative learning across multiple projects. *WIREs Water* **5**: e1257. https://doi.org/10.1002/wat2.1257

Weber, M.J. & Brown, M.L. (2009). Effects of common carp on aquatic ecosystems 80 years after 'carp as a dominant': ecological insights for fisheries management. *Reviews in Fisheries Science* **17**: 524–537.

Weber, M.J. & Brown, M.L. (2011). Relationships among invasive common carp, native fishes, and physicochemical characteristics in upper Midwest (USA) lakes. *Ecology of Freshwater Fish* **20**: 270–278.

Wei, Q. (2010). *Acipenser dabryanus*. The IUCN Red List of Threatened Species 2010: e.T231A13041556. http://dx.doi.org/10.2305/IUCN.UK.2010-1.RLTS.T231 A13041556.en

Welcomme, R.L., Cowx, I.G., Coates, D., Béné, C., Funge-Smith, S., Halls, A. & Lorenzen, K. (2010). Inland capture fisheries. *Philosophical Transactions of the Royal Society B: Biological Sciences* **365**: 2881–2896.

Welcomme, R.L., Baird, I.G., Dudgeon, D., Halls, A., Lamberts, D. & Mustafa, M.G. (2016). Fisheries of the rivers of Southeast Asia. In *Freshwater Fisheries Ecology* (J.F. Craig, ed.), John Wiley & Sons, Ltd., Chichester: pp. 363–376.

Wenger, S.J., Isaak, D.J., Luce, C.H., Neville, H.M., Fausch, K.D., Dunham, J.B., Dauwalter, D.C., Young, M.K., Elsner, M.M., Rieman, B.E., Hamlet, A.F. & Williams, J.E. (2011). Flow regime, temperature, and biotic interactions drive differential declines of trout species under climate change. *Proceedings of the National Academy of Sciences of the United States of America* **108**: 14175–14180.

West, D., David, B. & Ling, N. (2014). *Prototroctes oxyrhynchus*. The IUCN Red List of Threatened Species 2014: e.T18384A20887241. http://dx.doi.org/10 .2305/IUCN.UK.2014-3.RLTS.T18384A20887241.en

West, D.C., Walters, A.W., Gephard, S. & Post, D.M. (2010). Nutrient loading by anadromous alewife (*Alosa pseudoharengus*): contemporary patterns and predictions for restoration. *Canadian Journal of Fisheries and Aquatic Sciences* **67**: 1211–1220.

White, W.T., Appleyard, S.A., Sabub, B., Kyne, P.M., Harris, M., Lis, R., Baje, L., Usu, T., Smart, J.J., Corrigan, S., Yang, L. & Naylor, G.J.P. (2015). Rediscovery of the threatened river sharks, *Glyphis garricki* and *G. glyphis*, in Papua New Guinea. *PLoS ONE* 10: e0140075. https://doi.org/10.1371/journal.pone.0140075

WHO (2018a). *Drinking Water*. World Health Organization Fact Sheet, WHO, Geneva. www.who.int/en/news-room/fact-sheets/detail/drinking-water

WHO (2018b). *Sanitation*. World Health Organization Fact Sheet, WHO, Geneva. www.who.int/en/news-room/fact-sheets/detail/sanitation

WHO/UNICEF (2008). *Progress on Drinking-Water and Sanitation: Special Focus on Sanitation*. UNICEF, New York and WHO, Geneva.

Wiedner, C., Rücker, J., Brüggemann, R. & Nixdorf, B. (2007). Climate change affects timing and size of populations of an invasive cyanobacterium in temperate regions. *Oecologia* 152: 473–484.

Wiens, J.J. (2016). Climate-related local extinctions are already widespread among plant and animal species. *PLoS Biol* 14: e2001104. https://doi.org/10.1371/journal.pbio.2001104

Wikramanayake, E.D. (1990). Conservation of endemic rain forest fishes of Sri Lanka: results of a translocation experiment. *Conservation Biology* 4: 32–37.

Wilby, R.L., Orr, H., Watts, G., Battarbee, R.W., Berry, P.M., Chadd, R., Dugdale, S.J., Dunbar, M.J., Elliott, J.A., Extence, C., Knights, B., Milner, N.J., Ormerod, S.J., Solomon, D., Timlett, R., Whitehead, P.J. & Wood, P.J. (2010). Evidence needed to manage freshwater ecosystems in a changing climate: turning adaptation principles into practice. *Science of the Total Environment* 408: 4150–4164.

Wilcove, D.S., Giam, X., Edwards, D.P., Fisher, B. & Koh, L.P. (2013). Navjot's nightmare revisited: logging, agriculture, and biodiversity in Southeast Asia. *Trends in Ecology & Evolution* 28: 531–540.

Wilde, S.B., Johansen, J.R., Wilde, H.D., Jiang, P., Bartelme, B. & Haynie, R.S. (2014). *Aetokthonos hydrillicola gen. et sp. nov.*: epiphytic cyanobacteria on invasive aquatic plants implicated in avian vacuolar myelinopathy. *Phytotaxa* 181: 443–460.

Windsor, F.M., Tilley, R.M., Tyler, C.R. & Ormerod, S.J. (2019). Microplastic ingestion by riverine macroinvertebrates. *Science of the Total Environment* 646: 68–74.

Winemiller, K.O. & Jepsen, D.B. (2004). Migratory Neotropical fish subsidize food webs of oligotrophic blackwater rivers. In *Food Webs at the Landscape Level* (G.A. Polis, M.E. Power & G.R. Huxel, eds), University of Chicago Press, Chicago: pp. 115–132.

Winemiller, K.O., McIntyre, P.B., Castello, L., Fluet-Chouinard, E., Giarrizzo, T., Nam, S., Baird, I.G., Darwall, W., Lujan, N.K., Harrison, I., Stiassny, M.L.J., Silvano, R.A.M., Fitzgerald, D.B., Pelicice, F.M., Agostinho, A.A., Gomes, L.C., Albert, J.S., Baran, E., Petrere, M., Zarfl, C., Mulligan, M., Sullivan, J.P., Arantes, C.C., Sousa, L.M., Koning, A.A., Hoeinghaus, D.J., Sabaj, M., Lundberg, J.G., Armbruster, J., Thieme, M.L., Petry, P., Zuanon, J., Torrente Vilara, G., Snoeks, J., Ou, C., Rainboth, W., Pavanelli, C.S., Akama, A., van Soesbergen, A. & Sáenz, L. (2016). Balancing hydropower and biodiversity in the Amazon, Congo, and Mekong. *Science* 351: 128–129.

Winston, R.L., Schwarzländer, M., Hinz, H.L., Day, M.D., Cock, M.J.W. & Julien, M.J. (2014). *Biological Control of Weeds: A World Catalogue of Agents and Their Target Weeds*, 5th ed. USDA Forest Service, Morgantown, WV.

Witte, F., Welten, M., Heemskerk, M., van der Stap, I., Ham, L., Rutjes, H. & Wanink, J. (2008). Morphological changes in a Lake Victoria cichlid fish within two decades. *Biological Journal of the Linnean Society* **94**: 41–52.

Wittmann, M., Cooke, R.M., Rothlisberger, J.D., Rutherford, E.S., Zhang, H., Mason, D.M. & Lodge, D.M. (2015). Use of structured expert judgment to forecast invasions by bighead and silver carp in Lake Erie. *Conservation Biology* **29**: 187–197.

Wittmann, M.E., Ngai, K.L. & Chandra, S. (2013). Our new biological future? The influence of climate change on the vulnerability of lakes to invasion by non-native species. In *Climatic Change and Global Warming of Inland Waters: Impacts and Mitigation for Ecosystems and Societies* (C.R. Goldman, M. Kumagai & R.D. Robarts, eds), John Wiley & Sons Ltd, Chichester: pp. 255–270.

Woodward, G., Perkins, D.M. & Brown, L.E. (2010). Climate change and freshwater ecosystems: impacts across multiple levels of organization. *Philosophical Transactions of the Royal Society B* **365**: 2093–2106.

World Meteorological Organization (WMO) (2019). *United in Science*. High-level synthesis report of climate science information convened by the Science Advisory Group of the UN Climate Action Summit 2019, coordinated by the World Meteorological Organization, Geneva. https://public.wmo.int/en/resources/united_in_science

WWF (2018). *Living Planet Report – 2018: Aiming Higher* (M. Grooten & R.E.A. Almond, eds), WWF, Gland, Switzerland.

WWF/ZSL (2016). *The Living Planet Index database*. WWF and the Zoological Society of London, London. www.livingplanetindex.org

Wright, J. (2011). Conservative coevolution of Müllerian mimicry in a group of rift lake catfish. *Evolution* **65**: 395–407.

Wright, J.P., Jones, C.G. & Flecker, A.S. (2002). An ecosystem engineer, the beaver, increases species richness at the landscape scale. *Oecologia* **132**: 96–101.

Wu, J. (2015). Can changes in the distribution of lizard species over the past 50 years be attributed to climate change? *Theoretical and Applied Climatology* **125**: 785–798.

Xenopoulos, M.A., Lodge, D.M., Alcamo, J., Marker, M., Schulze, K. & Van Vuuren, D.P. (2005). Scenarios of freshwater fish extinctions from climate change and water withdrawal. *Global Change Biology* **11**: 1557–1564.

Xie, S., Li, Z., Liu, J., Xia, S., Wang, H. & Murphy, B.R. (2007). Fisheries of the Yangtze River show immediate impacts of the Three Gorges Dam. *Fisheries* **32**: 343–344.

Xiong, L., Ouyang, S. & Wu, X. (2012). Fauna and standing crop of freshwater mussels in Poyang Lake, China. *Chinese Journal of Oceanology and Limnology* **30**: 124–135.

Yamanishi, Y., Yoshida, K., Fujimori, N. & Yusa, Y. (2012). Predator-driven biotic resistance and propagule pressure regulate the invasive applesnail *Pomacea canaliculata* in Japan. *Biological Invasions* **14**: 1343–1352.

Ye, X., Li, Y., Li, X. & Zhang, Q. (2014). Factors influencing water level changes in China's largest freshwater lake, Poyang Lake, in the past 50 years. *Water International* **39**: 983–999.

Ye, X., Lin, M., Li, L., Liu, J., Song, L. & Li, Z. (2015). Abundance and spatial variability of invasive fishes related to environmental factors in a eutrophic Yunnan Plateau lake, Lake Dianchi, southwestern China. *Environmental Biology of Fishes* **98**: 209–224.

Yipp, M.W. (1990). Distribution of the schistosome vector snail, *Biomphalaria straminea* (Pulmonata: Planorbidae) in Hong Kong. *Journal of Molluscan Studies* **56**: 47–55.

Yoon, J., Kim, J., Yoon, J., Baek, S. & Jang, M. (2015). Efficiency of a modified Ice Harbor-type fishway for Korean freshwater fishes passing a weir in South Korea. *Aquatic Ecology* **49**: 417.

Youn, S., Taylor, W.W., Lynch, A.J., Cowx, I.G., Beard, T.D., Bartley, D. & Wu, F. (2014). Inland capture fishery contributions to global food security and threats to their future. *Global Food Security* **3**: 142–148.

Yousey, A.M., Chowdhury, P.R., Biddinger, N., Shaw, J.H., Jeyasingh, P.D. & Weider, L.J. (2018). Resurrected 'ancient' *Daphnia* genotypes show reduced thermal stress tolerance compared to modern descendants. *Royal Society Open Science* **5**: 172193. https://doi.org/10.1098/rsos.172193

Zalasiewicz, J., Williams, M., Haywood, A. & Ellis, M. (2011). The Anthropocene: a new era of geological time? *Philosophical Transactions of the Royal Society A* **369**: 835–841.

Zalasiewicz, J., Waters, C.N., do Sul, J.I., Corcoran, P.L., Barnosky, A.D., Cearreta, A., Edgeworth, M., Gałuszka, A., Jeandel, C., Leinfelder, R., McNeill, J.R., Steffen, W., Summerhayes, C., Wagreich, M., Williams, M., Wolfe, A.P. & Yonan, Y. (2016). The geological cycle of plastics and their use as a stratigraphic indicator of the Anthropocene. *Anthropocene* **13**: 4–17.

Zambrano, L., Reidl, P.M., McKay, J., Griffiths, R., Shaffer, B., Flores-Villela, O., Parra-Olea, G. & Wake, D. (2010). *Ambystoma mexicanum. The IUCN Red List of Threatened Species 2010: e.T1095A3229615.* http://dx.doi.org/10.2305/IUCN.UK.2010-2.RLTS.T1095A3229615.en

Zaret, T.M. & Paine, R.T. (1973). Species introduction in a tropical lake: a newly introduced piscivore can produce population changes in a wide range of trophic levels. *Science* **182**: 449–455.

Zarfl, C., Lumsdon, A.E., Berlekamp, J., Tydecks, L. & Tockner, K. (2015). A global boom in hydropower dam construction. *Aquatic Sciences* **77**: 161–170.

Zavaleta, E. (2000). The economic value of controlling an invasive shrub. *Ambio* **29**: 462–467.

Zehev, B.S., Vera, A., Asher, B. & Raimundo, R. (2015). Ornamental fishery in Rio Negro (Amazon region), Brazil: combining social, economic and fishery analyses. *Fisheries and Aquaculture Journal* **6**: 143–147.

Zhang, F., Tiyip, T., Johnson, V.C., Kung, H., Ding, J., Sun, Q., Zhou, M., Kelimu, A., Nurmuhammat, I. & Chan, N.W. (2015). The influence of natural and human factors in the shrinking of the Ebinur Lake, Xinjiang, China, during the 1972–2013 period. *Environmental Monitoring and Assessment* **187**: 4128. https://doi.org/10.1007/s10661-014-4128-4

Zhang, H., Wei, Q., Du, H., Shen, L., Li, Y. & Zhao, Y. (2009). Is there evidence that the Chinese paddlefish (*Psephurus gladius*) still survives in the upper Yangtze

River? Concerns inferred from hydroacoustic and capture surveys, 2006–2008. *Journal of Applied Ichthyology* **25** (Suppl. 2): 95–99.

Zhang, P., Zou, Y., Xie, Y., Zhang, H., Liu, X., Gao, D. & Feng, Y. (2018). Shifts in distribution of herbivorous geese relative to hydrological variation in East Dongting Lake wetland, China. *Science of the Total Environment* **636**: 30–238.

Zhang, Y., Jia, Q., Prins, H.H., Cao, L. & de Boer, W.F. (2015). Effect of conservation efforts and ecological variables on waterbird population sizes in wetlands of the Yangtze River. *Scientific Reports* **5**: 17136. https://doi.org/10.1038/srep17136

Zhao, X., Barlow, J., Taylor, B.L., Pitman, R.L., Wang, K., Wei, Z., Stewart, B.S., Turvey, S.T., Akamatsu, T., Reeves, R.R. & Wang, D. (2008). Abundance and conservation status of the Yangtze finless porpoise in the Yangtze River, China. *Biological Conservation* **141**: 3006–3018.

Zhou, Y., Michalak, A.M., Beletsky, D., Rao, Y.R. & Richards, R.P. (2015). Record-breaking Lake Erie hypoxia during 2012 drought. *Enviromental Science & Technology* **49**: 800–807.

Zieritz, A., Lopes-Lima, M., Bogan, A.E., Sousa, R., Walton, S., Rahim, K.A., Wilson, J.-J., Ng, P.-Y., Froufe, E. & McGowan, S. (2016). Factors driving changes in freshwater mussel (Bivalvia, Unionida) diversity and distribution in Peninsular Malaysia. *Science of the Total Environment* **571**: 1069–1078.

Ziv, G., Baran, E., Nam, S., Rodriguez-Iturbe, I. & Levin, S.A. (2012). Trading-off fish biodiversity, food security, and hydropower in the Mekong River Basin. *Proceedings of the National Academy of Sciences of the United States of America* **109**: 5609–5614.

Zonn, I.S., Glantz, M., Kosarev, A.N. & Kostianoy, A.G. (2009). *The Aral Sea Encyclopedia*. Springer-Verlag, Berlin.

Species Index

Note: Page numbers in *italics* refer to tables.

General Index

Note: Page numbers in *italics* refer to tables.